METHODS IN CELL BIOLOGY

VOLUME 35

Functional Organization of the Nucleus: A Laboratory Guide

Series Editor

LESLIE WILSON

Department of Biological Sciences
University of California, Santa Barbara
Santa Barbara, California

ASCB

METHODS IN CELL BIOLOGY

Prepared under the Auspices of the American Society for Cell Biology

VOLUME 35

Functional Organization of the Nucleus: A Laboratory Guide

Edited by

BARBARA A. HAMKALO
DEPARTMENT OF MOLECULAR BIOLOGY AND BIOCHEMISTRY
UNIVERSITY OF CALIFORNIA, IRVINE
IRVINE, CALIFORNIA

SARAH C. R. ELGIN
DEPARTMENT OF BIOLOGY
WASHINGTON UNIVERSITY
ST. LOUIS, MISSOURI

ACADEMIC PRESS, INC.
Harcourt Brace Jovanovich, Publishers

San Diego   New York   Boston
London   Sydney   Tokyo   Toronto

Academic Press, Inc.
San Diego, California 92101

United Kingdom Edition published by
ACADEMIC PRESS LIMITED
24-28 Oval Road, London NW1 7DX

Library of Congress Catalog Card Number: 64-14220

International Standard Book Number 0-12-564135-4 (Hardcover) (alk. paper)
International Standard Book Number 0-12-321920-5 (Paperback) (alk. paper)

PRINTED IN THE UNITED STATES OF AMERICA
91 92 93 94   9 8 7 6 5 4 3 2 1

CONTENTS

4. *Visualization of DNA Sequences in Meiotic Chromosomes*
Peter B. Moens and Ronald E. Pearlman

5. *Nucleic Acid Sequence Localization by Electron Microscopic in Situ Hybridization*
Sandya Narayanswami, Nadja Dvorkin, and Barbara A. Hamkalo

PART II. VISUALIZATION OF PROTEINS

6. *The Use of Autoantibodies in the Study of Nuclear and Chromosomal Organization*
W. C. Earnshaw and J. B. Rattner

PART IV. RECONSTITUTION OF FUNCTIONAL COMPLEXES

PART V. GENETIC APPROACHES

CONTRIBUTORS

Numbers in parentheses indicate the pages on which the authors' contributions begin.

STEPHEN A. ADAM, Department of Molecular Biology, Research Institute of Scripps Clinic, La Jolla, California 92037 (469)

VINCENT G. ALLFREY, Laboratory of Cell Biology, The Rockefeller University, New York, New York 10021 (315)

LYNNE M. ANGERER, Department of Biology, University of Rochester, Rochester, New York 14627 (37)

ROBERT C. ANGERER, Department of Biology, University of Rochester, Rochester, New York 14627 (37)

THELMA A. CHEN, Laboratory of Cell Biology, The Rockefeller University, New York, New York 10021 (315)

ROBERT F. CLARK, Department of Biology, Washington University, St. Louis, Missouri 63130 (203)

CAROLYN A. CRAIG, Department of Biology, Washington University, St. Louis, Missouri 63130 (203)

A. J. J. DIETRICH, Institute of Human Genetics, University of Amsterdam, NL-1105 AZ Amsterdam, The Netherlands (177)

NADJA DVORKIN, Department of Molecular Biology and Biochemistry, University of California, Irvine, Irvine, California 92717 (109)

W. C. EARNSHAW, Department of Cell Biology and Anatomy, The Johns Hopkins University School of Medicine, Baltimore, Maryland 21205 (135)

SARAH C. R. ELGIN, Department of Biology, Washington University, St. Louis, Missouri 63130 (203)

DAVID R. ENGELKE, Department of Biological Chemistry, University of Michigan, Ann Arbor, Michigan 48109 (383)

KAREN FINDLING, Biophysics Research Division, University of Michigan, Ann Arbor, Michigan 48109 (337)

BARBARA R. FISHEL, Department of Microbiology, University of Texas Southwestern Medical Center at Dallas, Dallas, Texas 75235 (525)

JAN FRONK, Biophysics Research Division, University of Michigan, Ann Arbor, Michigan 48109 (337)

W. T. GARRARD, Department of Biochemistry, University of Texas Southwestern Medical Center at Dallas, Dallas, Texas 75235 (525)

MAURIZIO GATTI, Dipartimento di Genetica e Biologia Molecolare, Università di Roma "La Sapienza," 00185 Rome, Italy (543)

LARRY GERACE, Department of Molecular Biology, Research Institute of Scripps Clinic, La Jolla, California 92037 (469)

DAVID S. GILMOUR, Department of Molecular and Cell Biology, Pennsylvania State University, University Park, Pennsylvania 16802 (369)

MICHAEL L. GOLDBERG, Section of Genetics and Development, Cornell University, Ithaca, New York 14853 (543)

T. GRIGLIATTI, Department of Zoology, University of British Columbia, Vancouver, British Columbia V6T 1Z4, Canada (587)

BARBARA A. HAMKALO, Department of Molecular Biology and Biochemistry, University of California, Irvine, Irvine, California 92717 (109)

C. HEYTING, Department of Genetics, Agricultural University, NL-6703 HA Wageningen, The Netherlands (177)

JON M. HUIBREGTSE, Laboratory of Tumor Virus Biology, National Cancer Institute, National Institutes of Health, Bethesda, Maryland 20892 (383)

MELISSA W. HULL, Department of Biological Chemistry, University of Michigan, Ann Arbor, Michigan 48109 (383)

CAROL VILLNAVE JOHNSON, Department of Cell Biology, University of Massachusetts Medical Center, Worcester, Massachusetts 01655 (73)

ROBERT E. KINGSTON, Department of Molecular Biology, Massachusetts General Hospital, Boston, Massachusetts 02114, and Department of Genetics, Harvard Medical School, Boston, Massachusetts 02115 (419)

SUSAN KLEIN, Biophysics Research Division, University of Michigan, Ann Arbor, Michigan 48109 (337)

JOHN P. LANGMORE, Biophysics Research Division, University of Michigan, Ann Arbor, Michigan 48109 (337)

JEANNE BENTLEY LAWRENCE, Department of Cell Biology, University of Massachusetts Medical Center, Worcester, Massachusetts 01655 (73)

JOHN T. LIS, Section of Biochemistry, Molecular and Cell Biology, Cornell University, Ithaca, New York 14853 (369)

PETER B. MOENS, Department of Biology, York University, Downsview, Ontario M3J 1P3, Canada (101)

SANDYA NARAYANSWAMI, Department of Molecular Biology and Biochemistry, University of California, Irvine, Irvine, California 92717 (109)

JOHN W. NEWPORT, Department of Biology, University of California, San Diego, La Jolla, California 92093 (449)

RONALD E. PEARLMAN, Department of Biology, York University, Downsview, Ontario M3J 1P3, Canada (101)

J. B. RATTNER, Department of Anatomy and Medical Biochemistry, University of Calgary, Calgary T2N 1N4, Canada (135)

ROBERT G. ROEDER, Laboratory of Biochemistry and Molecular Biology, The Rockefeller University, New York, New York 10021 (419)

SHARON Y. ROTH, Laboratory of Cellular and Developmental Biology, National Institute of Diabetes and Digestive and Kidney Diseases, National Institutes of Health, Bethesda, Maryland 20892 (289)

ANN E. ROUGVIE, Department of Cell and Development Biology, Harvard University, Cambridge, Massachusetts 02138 (369)

MARY C. RYKOWSKI, Department of Anatomy and The Arizona Cancer Center, College of Medicine, The University of Arizona, Tucson, Arizona 85724 (253)

H. SAUMWEBER, Institut für Entwicklungsphysiologie, Universität zu Köln, D-5000 Köln 41, Germany (229)

ROBERT T. SIMPSON, Laboratory of Cellular and Developmental Biology, National Institute of Diabetes and Digestive and Kidney Diseases, National Institutes of Health, Bethesda, Maryland 20892 (289)

ROBERT H. SINGER, Department of Cell Biology, University of Massachusetts Medical Center, Worcester, Massachusetts 01655 (73)

M. MITCHELL SMITH, Department of Microbiology, School of Medicine, University of Virginia, Charlottesville, Virginia 22908 (485)

CARL SMYTHE, Department of Biology, University of California, San Diego, La Jolla, California 92093 (449)

ANN O. SPERRY, Department of Cell Biology and Neuroscience, University of Texas Southwestern Medical Center at Dallas, Dallas, Texas 75235 (525)

RACHEL STERNE-MARR, Merck, Sharpe and Dohme, West Point, Pennsylvania 19486 (469)

GRAEME A. TANK, Biophysics Research Division, University of Michigan, Ann Arbor, Michigan 48109 (337)

IAN C. A. TAYLOR, Department of Molecular Biology, Massachusetts General Hospital, Boston, Massachusetts 02114, and Department of Genetics, Harvard Medical School, Boston, Massachusetts 02115 (419)

GRAHAM THOMAS, Department of Cellular and Developmental Biology, Harvard University, Cambridge, Massachusetts 02138 (383)

BARBARA J. TRASK, Biomedical Sciences Division, Lawrence Livermore National Laboratory, Livermore, California 94550 (3)

CLAUDIUS VINCENZ, Biophysics Research Division, University of Michigan, Ann Arbor, Michigan 48109 (337)

CYNTHIA R. WAGNER, Department of Biology, Washington University, St. Louis, Missouri 63130 (203)

JERRY L. WORKMAN, Department of Molecular Biology, Massachusetts General Hospital, Boston, Massachusetts 02114, and Department of Genetics, Harvard Medical School, Boston, Massachusetts 02115 (419)

PREFACE

During the last decade work in cell biology has flourished, in part because of the joining of molecular and genetic approaches with more classical techniques. In particular, studies on the complex nature of the relationship between nuclear structure and function benefits from a multidisciplinary, multifaceted attack. We think it is a propitious time to present in one place a selection of different methodological approaches that have been most useful in studies of the nucleus, as a convenience for our own generation, and as a spur to students interested in nuclear function.

Teachers for many years have compared the cell to a house we may not enter. We want to discern its internal structure, its three-dimensional floor plan, and the activities of its inhabitants. However, we have been restricted to peeking in the windows, to exploding the structure so that we can analyze or reassemble the remains, or to altering the blueprints at one site and examining the altered product. Although the basic problem of the precise relationship between nuclear structure and function remains, we are acquiring new tools and becoming more clever in how we use them to gain novel information. Molecular reagents such as nucleic acid probes and specific antibodies can be used to label both structural components and occupants, making our observations much more informative. In addition, we are developing new ways to maintain intermolecular complexes and interactions after cell lysis. It therefore becomes possible to use the same molecular reagents to select the parts we want away from the debris. We have had amazing success in putting back together dissociated parts. Clever utilization of genetic approaches allows us to identify the crucial players.

This volume focuses on the analysis of nuclear components to discern their function. It contains general discussions and detailed protocols for the visualization of specific nucleic acid sequences using hybridization probes; for the visualization of specific proteins using polyclonal antisera, monoclonal antibodies, and autoimmune sera; for isolation of defined chromatin fractions; for mapping protein–DNA interactions that exist *in vivo*; and for the reconstitution of functional templates and nuclear substructures. Finally, the application of genetic approaches to identify and characterize chromosomal components is described. A genetic test remains essential to confirm the biological significance of our hypotheses. Moreover, genetic screens are now available to identify potential components of the genetic apparatus.

The amount of detail presented varies among the chapters. We have assumed that the reader will have access to "Molecular Cloning: A Laboratory Manual," 2nd edition, by J. Sambrook, E. F. Fritsch, and T. Maniatis, and to "Antibodies: A Laboratory Manual" by E. Harlow and D. Lane, or to equivalent handbooks for standard molecular biological and immunological techniques. We have used our pages to present in detail specific modifications of these techniques which are unique to nuclear and chromatin structure and function studies.

Our reason for bringing together this material is the hope that the sum will be more than the parts, that investigators relying primarily on one approach will find in this volume both the motivation and the means to bring other approaches to bear on their particular study of the nucleus. A protein distribution pattern assessed visually on polytene chromosomes can be confirmed and analyzed at higher resolution using the UV-photocrosslinking technique. A functional association inferred from co-isolation can be confirmed and extended by microscopy. Cloning technologies are sufficiently routine so that identification of a gene in one system can rapidly lead to its identification in another. Thus the elegant genetic approaches available in yeast and *Drosophila* can be brought to bear on almost any question, regardless of the starting point. We hope that many investigators who pick up this book for details on one technique will find, with a little browsing, ideas for a second approach to the same problem, one that will provide complementary, but novel, information. If we are successful, this book will have served its purpose well, and indeed will have served the purpose of the ASCB, to be of service to its members.

BARBARA A. HAMKALO
SARAH C. R. ELGIN

Part I. Visualization of Nucleic Acids

COLOR PLATE 1. Applications of fluorescence *in situ* hybridization. (a) Hybridization of probe puC1.77 (from H. Cooke) to human peripheral blood lymphocytes. This probe binds intensely to the polymorphic heterochromatic region near the centromere of chromosome 1 and to a small region near the centromere of chromosome 9. The probe is labeled with biotin and detected with avidin-FITC; DNA is counterstained with propidium iodide. (b) A pool of sequences cloned from flow-sorted chromosome 4 material highlight the normal copy and translocation products of this chromosome in γ-irradiated human peripheral blood lymphocytes. (Reprinted with permission from Pinkel *et al.,* 1988.) (c) Human genomic DNA labels human chromosome 12 retained in a Chinese hamster × human somatic cell hybrid (HHW151, from J. Wasmuth). (d) The location of the original two copies of the DHFR gene as well as additional copies that have arisen by DHFR gene amplification are identified in methotrexate-resistant CHO cells using cosmids from the DHFR region (from J. Hamlin) as probe. (e) A cosmid representing the glucose-6-phosphate dehydrogenase gene (from J. Gitschier) is localized to the last band on the long arm of human chromosome X (q28). (f) Using two-color hybridization, the relative order of cosmids in G1 interphase chromatin of a male donor can be determined. Here, the color vision pigment genes are labeled with digoxigenin and FITC, and the factor VIII gene and a more proximal marker (33A31, DXS52) are labeled with biotin and TR. No DNA counterstain has been applied; the boundaries of the nucleus are evident from repetitive sequences in the probes whose hybridization has not been completely suppressed (see p. 30).

COLOR PLATE 2. Surface-spread mouse pachytene chromosomes were *in situ* hybridized to a biotinylated telomere DNA sequence. The biotin is detected by the fluorescence of FITC conjugated to avidin. The DNA is red fluorescent with the DNA-staining dye propidium iodide. The chromatin loops (ch) surround the bright axial structure, the synaptonemal complex (sc). The telomere DNA probe is detected by the bright orange (green FITC + red propidium iodide) dots (t) at the ends of the SCs (see p. 107).

COLOR PLATE 3. The pachytene chromosomes were hybridized with three probes simultaneously: A *Mus musculus* minor satellite DNA probe that hybridizes to DNA of the centromeres (c); probe 70–38, which hybridizes to low-frequency repeated sequences at the transition from the X chromosome centromeric heterochromatin (h) to the euchromatin; and probe 68–36, which hybridizes to a nearby region with a low-frequency repeated sequence. The minor satellite locates more precisely to the centromere when hybridization is done in the presence of major satellite DNA (not biotinylated) with which the minor satellite has several regions of homology (see p. 108).

2

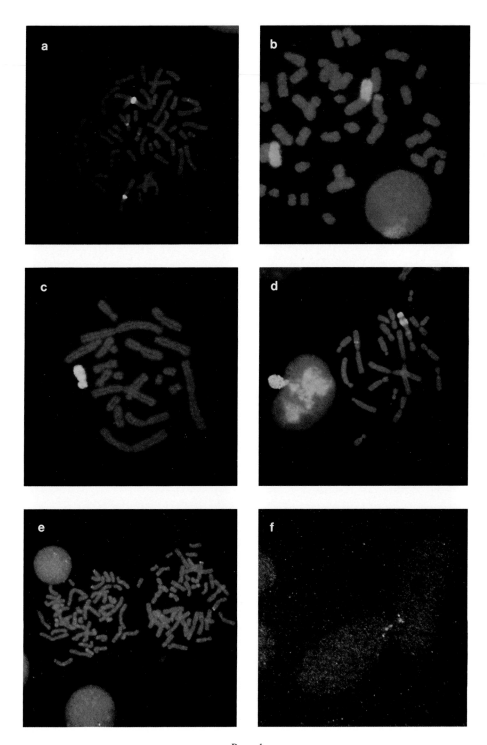

PLATE 1

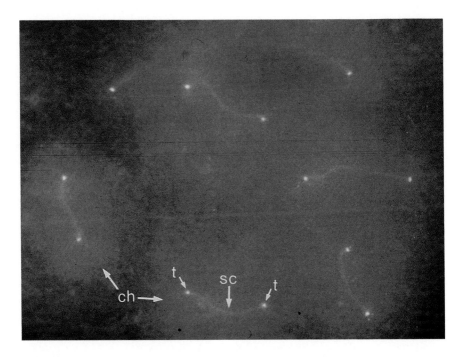

PLATE 2

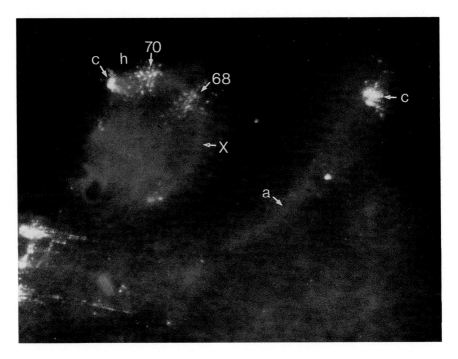

PLATE 3

Chapter 1

DNA Sequence Localization in Metaphase and Interphase Cells by Fluorescence in Situ Hybridization

BARBARA J. TRASK

Biomedical Sciences Division
Lawrence Livermore National Laboratory
Livermore, California 94550

METHODS IN CELL BIOLOGY, VOL. 35

I. Introduction

In situ hybridization provides a direct way to determine the relative location of specific DNA sequences in nuclei or chromosomes. Unique genes, specific chromosomal regions, entire chromosome types, or the genome of a particular species can be specifically highlighted with radioactive, absorptive, or fluorescent labels. Nonisotopic hybridization techniques are gradually supplanting those that require radioactively labeled probes due to rapid developments in methodology over the past 10 years. Nonisotopic techniques have advantages over isotopic techniques in safety, spatial resolution, and the speed with which results can be obtained (results can be obtained within 2 days). Target sequences of 40 kilobase pairs (kb) or more, for example, those cloned in cosmids or yeast artificial chromosomes, can be detected with $>90\%$ efficiency (e.g., Trask *et al.,* 1989a, 1991; Kievits *et al.,* 1990; Lichter *et al.,* 1990a; Wada *et al.,* 1990). Hybridization of repetitive sequences contained in these large fragments, which would override the signal from the unique portions of the probe, can be effectively suppressed by the addition of unlabeled genomic DNA (Landegent *et al.,* 1987). Most cosmid probes produce a single, bright, and highly localized fluorescent spot (<0.3 μm in diameter) at each site of hybridization (Trask *et al.,* 1989a). Targets as small as 2–3 kb can be mapped to chromosomal bands with nonradioactive plasmid

TABLE I

PROBE LABELING SCHEMES[a]

Probe label (partial list)	Detection scheme	Reference
Enzymatic incorporation		
Biotin	Avidin-F (biotinylated-anti-avidin, avidin-F)$_n$ or anti-biotin, anti-IgG-F	Langer-Safer *et al.* (1982)
Digoxigenin	Anti-digoxigenin, anti-IgG-F	Zischler *et al.* (1989)
Mercury	Sulfhydryl-ligand-DNP[b], anti-DNP-F	Hopman *et al.* (1986b)
Chemical modification		
AAF	Anti-AAF, anti-IgG-F	Landegent *et al.* (1984)
Sulfone[c]	Anti-sulfonate, anti-IgG-F	Morimoto *et al.* (1987)
DNP or TNP	Anti-D(T)NP, anti-IgG-F	Shroyer and Nakane (1983)

[a] F, Fluorochrome; AAF, aminoacetylfluorene; IgG, immunoglobulin; DNP, dinitrophenyl; TNP, trinitrophenyl.

[b] From Organon Technika.

[c] From FMC Bioproducts.

probes, although the efficiency of hybridization site detection obtained is often low (20–40%) (e.g., Albertson *et al.*, 1988; Lawrence *et al.*, 1988; Pinkel *et al.*, 1988; Viegas-Pequignot *et al.*, 1989). Also important are the variety of probe labeling schemes that allow analysis of multiple sequences in different colors in the same cell (Hopman *et al.*, 1986a, Trask *et al.*, 1988, 1991; Nederlof *et al.*, 1990) (Table I). To date, the most sensitive results are obtained with biotin- or digoxigenin-labeled probes. A variety of probe labels can be employed (Table II). Success has been achieved with fluorescein isothiocyanate (FITC), tetramethylrhodamine isothiocyanate (TRITC), Texas Red (TR), and 7-amino-4-methylcoumarin-3-acetic acid (AMCA) in terms of signal intensity and resistance to bleaching.

Because of these technical advancements, *in situ* hybridization has found increasing application in a number of research areas, including cytogenetics, prenatal diagnosis, tumor biology, nuclear organization, gene amplification,

TABLE II[a]

FLUOROCHROMES EMPLOYED IN HYBRIDIZATION SITE DETECTION

Fluorochromes[b]	Excitation max (nm)	Emission max (nm)
DNA counter stains		
(PI)	330 and 520	620
DAPI	350	460
Hoechst 33258	360	470
Quinacrine	455	495
Chromomycin	430	470
Hybridization site labels		
FITC	490	520
TRITC	554	573
XRITC	580	600
TR	596	620
AMCA	350	450
CY5[c]	646	663

Enzyme Precipitates[b] (Hybridization site labels)	
Peroxidase + DAB	Bright-field microscopy: brown
Alkaline phosphatase + NBT/BCIP	Bright-field microscopy: blue

[a] Partly after Waggoner (1990).
[b] PI, Propidium iodide; DAPI, 4',6'-diamidino-2-phenylindole; FITC, fluorescein isothiocyanate; TRITC, tetramethylrhodamine isothiocyanate; XRITC, rhodamine isothiocyanate analog X; TR, Texas Red; AMCA, 7-amino-4-methylcoumarin-3-acetic acid; CY5, pentamethine cyanine dye isothiocyanate CY5. 8-ITC; DAB, 3,3'-diaminobenzidine; NBT, nitro blue tetrazolium; BCIP, 5-bromo-4-chloro-3-indolyl phosphate.
[c] Mujumdar *et al.* (1989).

and gene mapping [recently reviewed by Lichter and Ward (1990), Lichter
et al. (1991), McNeil et al. (1991), Tkachuk et al. (1991), and Trask (1991a,b)].
Chromosomes can be identified, chromosomal abnormalities can be detected, and the chromosomal location of specific sequences can be determined in both metaphase and interphase chromatin. In situ hybridization
can aid detection of numerical and structural chromosome abnormalities
in tumors, aminocytes, and cells of individuals with genetic disorders.
Cloned sequences that are repeated up to 1000 times on specific chromosomes types, usually near their centromeres, produce intense hybridization
zones in metaphase and interphase. Aneuploidy of specific chromosome
types can be easily assayed with these chromosome-specific repeats (Cremer
et al., 1986; Pinkel et al., 1986; Hopman et al., 1988). Entire chromosomes
can be fluorescently tagged using collections of sequences derived from a
single chromosome type. Fluorescent domains produced by these collections can be counted to detect aneuploidy or transactions (Cremer et al.,
1988, 1990; Pinkel et al., 1988; Lichter et al., 1988a,b; Lucas et al., 1989).
Single-copy probes that flank breakpoints can be used to detect translocations, inversions, and deletions in metaphase and interphase (Tkachuk et al.,
1990; Dauwerse et al., 1990; Lux et al., 1990). Somatic-cell hybrid characterization is simplified by the fact that total nuclear DNA isolated from human
cells highlights human chromosomal material in rodent × human somatic-
cell hybrids (Schardin et al., 1985; Pinkel et al., 1986). Conversely, when DNA
from hybrids is used as a probe on human metaphases, the portions of the
human chromosomes contained in the hybrid are highlighted (Lichter et al.,
1990b). Cell-to-cell variation in the position of amplified genes relative to
parental gene copies can be easily studied using fluorescence in situ hybridization (Trask and Hamlin, 1989). Fluorescence in situ hybridization is perhaps
the most rapid means to place DNA sequences on the cytogenic and physical
map of a chromosome. Probes can be localized to within 3–10 megabase
pairs (Mb) by combining in situ hybridization and chromosomal banding
techniques (e.g., Bhatt et al., 1988; Lawrence et al., 1990; Cherif et al., 1990).
The relative order of sequences along the length of metaphase chromosomes
can be determined with a resolution of ∼1 Mb, if probes are labeled with
different fluorochromes (Trask et al., 1991). The proximity of hybridization
sites in the less condensed chromatin of interphase nuclei can be used to
determine probe order along the linear DNA molecule for sequences separated 20 kb to at least 1 Mb from each other (Lawrence et al., 1988, 1990;
Trask et al., 1989a, 1991). The organization of chromosomes in interphase
cells can be addressed by using optical sectioning, confocal microscopy, and
techniques for labeling sequences in cells suspended in fluid or in tissue sections (Manuelidis, 1985; Borden and Manuelidis, 1988; Manuelidis and Borden, 1988; Trask et al., 1988).

II. Technique Overview

The techniques for DNA sequence detection by nonisotopic *in situ* hybridization are straightforward and convenient. The probe is labeled with a reporter molecule, such as biotin, and broken into small fragments. Target chromatin is fixed to slides (for gene mapping or cytogenetic analyses) or is held in liquid suspension (for nuclear organization or flow cytometric analyses). The target chromatin is denatured and mixed with denatured, labeled probe DNA. If the probe contains repetitive sequences whose hybridization must be suppressed in order to make the hybridization sites of unique sequences visible [as is the case for cosmids, yeast artificial chromosomes (YACs), and chromosome-enriched sequence pools], unlabeled genomic DNA is added in excess to the hybridization. Hybridization takes place for a period of 2–16 hr in buffers and at temperatures designed to promote specific reannealing of probe to homologous sequences in the target. After washing to remove unbound probe, the sites of reporter molecules are made visible by incubating the target cells in immunoreagents that bind specifically to the reporter molecules. These reagents can carry a fluorescent tag, which can be viewed through suitable filter combinations using an epifluorescence microscope or quantified by a flow cytometer. Alternatively, the immunological reagents can carry an enzyme tag. Sites are later marked by an absorptive enzyme product that is made by incubating the slides in the enzyme substrate. This label can be viewed using bright-field or reflection contrast microscopy.

In this chapter, *in situ* hybridization techniques are detailed for labeling specific sequences in chromatin fixed to slides and in suspension. Procedures to label probes with biotin, digoxigenin, and aminoacetylfluorene (AAF) are described. Techniques for one-color fluorescent detection of these probe labels are given. Also described are techniques for the simultaneous detection of two probes (AAF and biotin, or digoxigenin and biotin) using two different fluorochromes. The AAF and biotin procedures have been adapted to label nuclei in suspension for quantitation of bound probe by flow cytometry or for analysis of nuclear organization by optical sectioning or confocal microscopy. A procedure for suspension labeling is given. The reader is referred to the literature (see reference list) for radioactive *in situ* hybridization techniques, for the procedures for enzyme deposition of absorptive label after nonisotopic hybridization, and for techniques by which tissue sections can be treated for hybridization.

The techniques described here have been drawn from procedures developed by a number of research groups (especially those of M. Harper, J. Lawrence, D. Pinkel, M. van der Ploeg, D. Ward, and J. Yunis). Protocols are evolving, and the reader is referred for alternative procedures to the methods sections of articles in the reference list. As success is affected by variables at all steps in the procedure, chromosome-specific repeat sequences should be used to test procedures and reagents before localization of cosmids or small probes is attempted.

III.   Probe Labeling

A.   Nick Translation (Biotin, Digoxigenin)

Overview.   The simplest and most reproducible means of labeling DNA sequence probes is by nick translation. Single-stranded nicks are randomly made in DNA by DNase. The frequency of nicks dictates the size of the fragments that ultimately make their way through chromatin and anneal at homologous sites in target. Experience has shown that double-stranded fragments of 200–400 bp, as measured on a conventional electrophoresis agarose gel, yield optimal results in terms of signal intensity and background (Lawrence and Singer, 1985). DNA polymerase I removes nucleotides starting at the site of nick through its $5'-3'$ exonuclease activity and then replaces removed nucleotides, using the other strand as template, through its $5'-3'$ polymerase activity. Biotin-dUTP or digoxigenin-dUTP is incorporated in place of thymidine. Incorporation levels can be optionally monitored by spiking the reaction with [^{3}H]dATP, but this addition necessitates subsequent disposal of reagents, plasticware, and washing solutions as hazardous waste.

The procedure described below uses the kit from BRL (cat. no. 8160SB), but nick translation kits are available from a number of other sources (Clontech, Boehringer Mannheim, and Enzo; see Section IX for further information on suppliers). The fragment size resulting from these kits varies and may require addition of DNase to the reaction to produce fragments in the 200–400 bp range. [BRL also markets a "Bionick" kit (cat. no. 8247SA) with which supplied biotin-7-dATP can be incorporated. The fragment size resulting from this kit is generally smaller than lots of BRL's regular "Nick Translation" kit.] Biotinylated nucleotides with varying linker arm length (7–21 carbons) can be used with equal success, in our experience, and are available from a number of sources (Enzo, BRL, Clontech, BM). The procedure is given for use of the 0.3 mM biotin-11-dUTP solution supplied by Enzo (cat. no. NU-806) and the 1 mM solution of digoxigenin-11-dUTP supplied by Boehringer-Mannheim (cat. no. 1093-088). The procedure assumes purified DNA probe is available at concentrations of ≥ 100 ng/μl. Minipreps (Sambrook *et al.,* 1989) treated with RNase and purified through Quiagen columns (cat. no. 50204) are suitable. Alternatively, DNA can be purified by cesium gradient as detailed in Sambrook *et al.* (1989). MgCl$_2$ is added to compensate for excess EDTA if the volume required to obtain 1 μg of DNA (x, see below) exceeds 10 μl.

Procedure. Proceed according to kit instructions, with several minor modifications.

1. (Optional) Dry down 1 μl of [^{3}H]dATP (deoxy(8-^{3}H)adenosine 5'-triphosphate, ammonium salt (Amersham, cat. no. TRK 347) in a 1.5-ml Eppendorf tube using a Speedvac centrifuge or by blowing N$_2$ gas over tube warmed to 68°C.
2. Prepare a 15–16°C water bath by adding a handful of ice to an ice bucket filled with water.
3. Add ingredients to tube in the order shown in the table below, mixing gently with pipet tip after each addition. Enzyme mixture should be kept in freezer until it is ready to be used and should be kept on ice at all times. Note: enzyme is in glycerol, and reaction mixture must be stirred well with pipet tip after all ingredients are added.

Reagents	Volume (μl)
1 μg of DNA in 10 mM Tris-HCl, 1 mM EDTA, pH 7.6 (TE)	x
Sterile H$_2$O	50 − sum of the volume of all other reagents
5 × nucleotide and reaction buffer mix [A4; 0.2 mM each of dATP, dCTP, and dGTP 500 mM Tris-HCl (pH 7.8, 50 mM MgCl$_2$, 100 mM 2-mercaptoethanol)]	5
⎡ 0.3 mM biotin-11-dUTP or	5 ⎤
⎣ 1 mM digoxigenin-11-dUTP	2 ⎦
MgCl$_2$ (100 mM, autoclaved), if $x > 10$ μl	2.5
DNase (BRL cat. no. 8046SA) diluted 1500-fold in H$_2$O just before use (use as needed, see above)	1
DNase–polymerase mixture [solution C, 0.4 U/ml DNA polymerase I, 40 pg/μl DNase I, 50 mM Tris-HCl (pH 7.5), 5 mM magnesium acetate, 1 mM 2-mercaptoethanol, 0.1 mM PMSF, 50% (v/v) glycerol, 100 μg/ml BSA]	5

4. Spin tubes briefly in Eppendorf centrifuge to collect solution at bottom.
5. Incubate for 90 minutes at 16°C by placing tubes in floating foam racks in water bath.
6. Stop reaction by the addition of 5 μl of 300 mM EDTA (pH 8.0, solution D).
7. (Optional) To determine the degree if incorporation, remove 1 μl of the reaction mixture and place in vial containing scintillation fluid.
8. Prepare spin columns in 1-ml tuberculin syringes to remove free nucleotides from DNA. Place a small quantity of silicone-treated glass wool [prepared by drenching glass wool in Sigmacote (Sigma, cat. no.

SL-2), wringing out, and autoclaving] at the bottom of spin column. Break the surface tension in the glass wool by addition of ~ 1 ml of 95% ethanol to column, followed by liberal rinses with sterile distilled water. Apply a slurry of Sephadex G-50 (DNA grade, Pharmacia cat. no. 17-0045-01) swollen in $1 \times$ TE and 0.1% SDS, pH 7.5, to fill column (with no air bubbles). Centrifuge columns at 200 g for 1 minute at room temperature (type of centrifuge not critical). Add more G-50 as needed to fill. Spin columns for 4 min to pack.

9. Place column over sterile Eppendorf tube in which labeled probe sample is to be collected and stored. Screw-cap 1.5-ml Eppendorf tubes are recommended. Apply sample to G-50 column. Spin for 4 minutes at 200 g. Determine volume collected. DNA concentration at this point is approximately 20 ng/μl.

10. (Optional) Transfer 1 μl of DNA solution to second scintillation vial.

11. (Optional) Calculate label incorporation:

$$\text{Fraction substituted} = \frac{\text{cpm recovered with DNA} \times 0.67 \times 100\%}{\text{cpm in reaction} \times [\text{AT}]}$$

where [AT] is the proportion of adenine–thymine base pairs in the DNA being labeled (0.6 for human DNA). Scintillation counts should be adjusted for the difference in volumes before and after the column separation. The formula assumes labeling of 1 μg of DNA with 10^{-9} mol of dATP (essentially all of which is cold) in the reaction mixture. These numbers are appropriate for the BRL kit. For successful hybridization, the modification should be $> 10\%$.

12. Check DNA size by agarose gel electrophoresis. Samples of 5–9 μl should be applied to minigel of 1% agarose run in $1 \times$ TAE ($1 \times$ TAE = 0.04 M Tris-acetate, 1 mM EDTA) against the molecular weight standards, Φ X digested with HaeIII and λ digested with $Hind$III (Sambrook et $al.$, 1989).

13. Store labeled DNA preparations in freezer. Thaw and refreeze as needed.

B. Aminoacetylfluorene Modification

Overview. DNA can be chemically modified with AAF through a chemical reaction at the C-8 carbon of guanine by the carcinogen N-acetoxy-2-aminoacetylfluorene (N-A-AAF) as described by Landegent and co-workers (1984). Approximately 20% of the guanine residues are labeled by this procedure. The DNA must be sonicated before use. The procedure assumes purified DNA probe is available at concentrations of ≥ 100 ng/μl. DNA can be purified by cesium gradient (Sambrook *et al.*, 1989). Also suitable are minipreps (Sambrook *et al.*, 1989) treated with RNase and purified through Quiagen columns (cat. no. 50204).

Warning. N-A-AAF is a carcinogen. Stock solution should be made up in carcinogen protection facility (glovebox). Conduct reaction using 10 mg/ml stock solution in fume hood while wearing protective clothing. Dispose of all waste solutions and equipment in proper carcinogen receptacle.

Procedure

1. This protocol is for 20 μg of DNA, but it can be scaled down to 5μg if necessary. Bring DNA to a total volume of 80μl with 1 × TE, pH 7.5.
2. Make up stock solution of N-A-AAF at 10 mg/ml in water-free dimethylsulfoxide. Stock solution should be stored at −20°C in dark. Keep N-A-AAF stock solution closed, as much as possible, dry, and in the dark. Any moisture that enters the tube will react with N-A-AAF and reduce its DNA-labeling capacity.
3. Warm N-A-AAF stock solution to 37°C in incubator.
4. Bring N-A-AAF and DNA solutions to room temperature. Spin briefly in Eppendorf centrifuge.
5. In fume hood under subdued light, add 5 μl of N-A-AAF to 20 μg of DNA. Mix briefly. Spin briefly.
6. Incubate DNA and N-A-AAF together at 37°C in the dark for 1 hour.
7. Add TE to 300 μl.
8. Extract DNA solution three times with phenol:chloroform: isoamylalcohol (25:24:1) that has been saturated with TE. The phenol removes free, unreacted, and still carcinogenic N-A-AAF from the aqueous, DNA-containing phase. See Sambrook *et al.* (1989) for details on phenol extraction. DNA is in the top, clear, aqueous phase. Transfer the aqueous phase after extraction to a new tube. Leave any white residue at interface with the phenol phase.
9. Extract twice with TE-saturated ether to remove traces of phenol. DNA is in the bottom, aqueous phase (see Sambrook *et al.*, 1989).
10. Let tube stand open at 37°C to allow ether to evaporate after last extraction. At this point, the unreacted carcinogen has been removed. The labeled DNA can be handled on the bench and in normal lighting.
11. Dilute DNA to 1.5 ml with TE and transfer to 15-ml polycarbonate tubes. (Large tubes prevent sample loss due to splattering and aerosol formation during sonication with some microtips. If your sonicator does not cause severe sample loss, bring AAF-labeled DNA up to 400 μl and sonicate in 1.5-ml Eppendorf tube.)
12. Sonicate sample to produce 200- to 400-bp fragments using a microtip three times for 40 seconds each at maximum speed. (Clean sonicator tip before use with 100% methanol.) Between the 40-second sonications,

put the tube on ice and carefully bubble N_2 gas through a pipet (plugged with sterile cotton) into the solution. Check DNA size after steps 13–16 (see step 17).

13. Spin to collect all liquid at the bottom of the 15-ml tube.
14. If necessary, divide sample among several Eppendorf tubes for ethanol precipitation. Add 0.1 vol of 2.5 M sodium acetate, pH 5.2. Mix well. Add 2.5 vol of cold 100% ethanol. Incubate at $-20°C$ for at least 30 minutes. Spin in an Eppendorf centrifuge at 4°C for 30 minutes. Remove supernate carefully. A very small pellet may be visible. Wash pellet in 70% ethanol. Repeat spin and supernate removal. Use a Speedvac centrifuge to remove remaining ethanol from pellet.
15. Resuspend pellet in TE to achieve a DNA concentration of ~ 100 ng/μl.
16. Measure optical density (OD) of DNA at 260 nm. A solution of 50 μg/ml gives an OD of 1.0 at 260 nm. Adjust concentration as necessary by dilution or ethanol precipitation (step 14).
17. Check DNA size by agarose gel electrophoresis to determine if soni-cated fragments are 200–400 bp. Samples of 1–2 μl should be ap-plied to a minigel of 1% agarose run in 1 × TAE against the molecular weight standards, Φ X digested with *Hae*III and λ digested with *Hind*III (Sambrook *et al.*, 1989). If fragments are still large, repeat steps 11–16.

IV.   Blocking DNA Preparation

Overview.   An excess of unlabeled DNA is added to the hybridization reaction for two purposes: (1) to reduce nonspecific binding of labeled probe DNA to chromatin, cell residue, or glass surfaces. For this purpose, DNA from an unrelated species that can be obtained cheaply, such as herring testes DNA (Sigma, #D6898), is used; (2) to anneal with and thereby suppress the hybridization of repetitive sequences in probe to those in the target and allow unique sequences to be more intensely labeled than the repetitive sequences. Suppression is necessary for cosmids, large-insert λ clones, YACs, or chromosome-enriched sequence pools. For suppression, genomic DNA from the species from which the probe DNA was derived is added to the hybridization. Alternatively, *Cot*I DNA, that fraction of genomic DNA that reanneals most rapidly and thus is enriched for repetitive sequences, can be used. Human genomic DNA (derived from placenta) is commercially avail-able (Sigma, cat. no. D7011). Genomic DNA from a variety of species is available from Clontech. Genomic DNA can also be prepared from tissue culture cells following procedures described in Sambrook *et al.* (1989). In all cases, the DNA is sonicated to 200- to 400-bp fragments for most effective blocking.

Procedure

1. Dissolve DNA in 1 × TE at ∼10 mg/ml in a polycarbonate tube.
2. Clean sonicator tip with 100% methanol. Position microtip just above the bottom of the tube. Sonicate solution four times for 40 sec at maximum speed. Between bouts of sonication, place tube in ice to cool.
3. Check DNA size by agarose gel electrophoresis to determine if sonicated fragments are in the 200–400 bp range as described above and repeat sonication if necessary.
4. Determine DNA concentration by measuring its optical density at 260 nm. Adjust DNA concentration by dilution or ethanol precipitation (see Sambrook *et al.*, 1989) to 10 mg/ml. Use 1 μl per 10 μl hybridization reaction.

V.   Slide Preparation

A.   Prometaphase Spreads from Blood Lymphocytes

Overview.   Phytohemagglutinin (PHA) stimulates cell division of peripheral blood lymphocytes in culture. When growth is maximal in terms of both the fraction of cycling cells and the total number of cells, the cells are synchronized by addition of methotrexate, which arrests cells in S phase. The cells are then released from arrest into thymidine-containing medium. Colcemid is added for ∼10 min when the cohort of cells enters mitosis, ∼5 hr after release from the methotrexate block. The cells are then harvested and swollen in hypotonic buffer. This step spreads chromosomes from each other within the cells and thins the cytoplasmic mass. The cells are fixed in acetic acid and methanol and dropped onto cleaned slides. Best spreading is obtained with fresh fixative. Baking of slides before storage serves to harden, or artificially age, them, which reduces chromosome fluffing and improves banding after hybridization. Slides stored at −20°C in a N_2-rich atmosphere remain suitable for hybridization for several years. Additional treatments to alter the accessibility of probe and detection reagents or to retain chromosome morphology (proteinase K, RNase, paraformadehyde) are used in some laboratories, but are not necessary for most applications.

Procedure (adapted from Yunis, 1976, and Harper *et al.,* 1981)

1. For 20 T-25 flasks, draw 10 ml of sterile peripheral blood into a vacutainer tube containing heparin (green top). *Warning*: dispose of

blood-contaminated plasticware and solutions as biohazardous waste.

2. Mix 100 ml of complete medium and distribute 5 ml into each flask. The complete medium can be made up in advance and stored at $-20°C$. Growth medium contains RPMI 1640 medium + HEPES (Irvine Sci. cat. no. 9159) supplemented with

    20% fetal bovine serum

    1% penicillin/streptomycin (Irvine Sci. cat. no. 9366)

    1% L-glutamine (Gibco, cat. no. 320-5030AG)

    13 units/ml heparin (Sigma cat. no. H8514)

    4% PHA (from a fresh hydrated stock made up by adding 10 ml of sterile distilled water to PHA from Gibco, cat. no. 670-0576AD).

3. Mix blood by inversion several times. Add 300 μl of blood to each flask, dropping the blood into the medium, not down the side of the flask. Mix gently.

4. Incubate at 37°C for 72 hours.

5. Swirl flasks to resuspend cells. Add 50 μl of 10^{-5} M methotrexate (Sigma cat. no. 6770) in Hanks balanced salt solution (HBSS; cat. no. 310-4020AJ) to each flask (final concentration = 10^{-7} M). Swirl solution to mix. Incubate at 37°C for an additional 17 hours. (Methotrexate stock can be stored at $-20°C$ in 1-ml aliquots.) *Warning*: methotrexate should be handled and disposed of as a carcinogen (see AAF, above).

6. Swirl flasks to mix. Pour the contents of two flasks into a 15-ml disposable centrifuge tube (capped). Repeat for all flasks. Centrifuge at 200 g for 8 minutes (the type of centrifuge is not critical). Remove all but ~ 0.5 ml of the supernatant without disturbing the pellet.

7. Add 10 ml of growth medium at room temperature to each tube, swirling tubes to slightly mix immediately before adding medium. Cap tubes and invert several times to mix.

8. Spin at 200 g for 8 minutes as above and repeat rinse.

9. After the second rinse, resuspend the contents of each tube in 10 ml of medium without heparin or PHA, but containing 10^{-5} M thymidine (Sigma, cat. no. T-9250). Invert to mix. (Thymidine stock is made up at 10^{-3} M in HBSS and can be frozen at $-20°C$ in 1-ml aliquots.)

10. Pour contents of each tube into two new T-25 flasks. Cap loosely and incubate for 5 to 6.5 hours as before. A number of different harvest times, separated by 30 minutes, can be tested to obtain optimal chromosome morphology.

11. To begin harvest, swirl flasks to resuspend cells and add 60 μl of colcemid (10 μg/ml stock in HBSS, frozen in aliquots at $-20°C$) to each flask (final concentration = 0.12 μg/ml). Swirl to mix and incubate as above for ~ 10–12 minutes.

12. Pour contents of two flasks into a 15-ml centrifuge tube as above. Repeat for all flasks. Spin at 200 g for 8 minutes as above.

13. Remove all but ~0.5 ml of the supernatant. Tap each tube several times to resuspend cells. Do this thoroughly to avoid cell clumping. Add 5 ml of 75 mM KCl, which has been freshly made and warmed to 37°C. Mix. Incubate in a 37°C water bath for 20 minutes.

14. Add 1 ml of freshly prepared 3:1 methanol:acetic acid to each tube while vortexing gently. Incubation at room temperature for 10–30 minutes of the suspension in this and other fixative steps may improve chromosome morphology. Spin down the cells at 200 g for 8 minutes. (Use fresh methanol that has not absorbed H_2O.)

15. Remove all but ~0.5 ml of the supernatant. Flick pellet to loosen. Slowly add a total of 5 ml of fixative while vortexing the tube gently.

16. Spin down cells at 200 g for 8 minutes.

17. Repeat steps 15 and 16 twice, making fresh fixative each time.

18. Clean glass microscope slides by soaking them for ~5 min in 95% ethanol containing a dash (2–3 ml/300 ml of ethanol) of ether. Remove slides from the bath and wipe dry with lint-free towels to avoid streaking.

19. After the last centrifugation, resuspend the cells in a small amount of fresh fixative until the suspension looks turbid. Drop one drop on a slide and observe under 16× magnification and phase to determine if the suspension needs to be diluted with fixative or concentrated by centrifugation. (Note: If the spreading of chromosomes from one another is not optimal, it may be improved if the slides are dipped in distilled H_2O or covered with a thin film of fixative solution just before the suspension is dropped on them.)

20. Let a single drop of suspension fall from Pasteur pipet held ~10 cm above a glass slide. It is possible to place two drops, which can be hybridized separately, on each slide. Continue quickly, to avoid introducing H_2O into the suspension, until the suspension is used up or until a sufficient number of slides has been made. Several hundred slides can be prepared from 20 5-ml blood cultures. The remaining suspension can be stored at −20°C. For reuse, centrifuge and resuspend cells in fresh fixative.

21. Put slides in plastic storage boxes. Bake slides in storage boxes for 4 hours at 65°C in a dry oven to harden chromatin.

22. Store plastic boxes at −20°C in sealed plastic bags that have been flushed with N_2.

23. Before use, bring bags containing slide boxes to room temperature. Remove the number of slides to be used and reseal the remainder in new bags flushed with N_2 after all traces of condensation have disappeared.

B. Metaphase Spreads from Fibroblast or Lymphoblast Cell Cultures

Overview.   Metaphase spreads can be obtained from cultured fibroblast, amniocyte, or lymphoblast cultures by adding colcemid to log-phase cultures. Mitotic cells can be selectively removed from monolayer fibroblast or amniocyte cultures by shake-off, leaving the majority of nonmitotic cells behind. Since lymphoblast cultures are in suspension, all cells are harvested. Although most chromosomes are short and condensed using the procedure described, a small fraction of cells contain long, prometaphase chromosomes like those obtained from synchronized peripheral blood cultures. Once cells are collected, they are swollen, fixed, and dropped onto slides as described for prometaphase spreads from peripheral blood lymphocyte cultures.

Procedure

1. Split or dilute cell cultures to obtain actively growing cultures. For fibroblasts, split 1 : 3 into new flasks the day before harvest. Cells should be 50–60% confluent at time of colcemid addition. For lymphoblasts, dilute to 2×10^5 ml into fresh medium 2 days before harvest.

2. Add colcemid. For fibroblasts, remove the growth medium and replace with fresh medium containing 0.1 μg/ml colcemid. Colcemid (Sigma, D6279) is added from a 10 μg/ml stock in HBSS (frozen in aliquots at $-20°C$). For lymphoblasts, add colcemid to 0.1 μg/ml final concentration.

3. Incubate at 37°C. For Chinese hamster or hamster × human somatic cell hybrid fibroblasts, incubate for 1–1.5 hours. For human fibroblasts or lymphoblasts, incubate for 4 hours. For mouse fibroblasts, incubate for 1–4 hours.

4. Collect mitotic cells. For fibroblasts, whack the side of the flask 1—3 times and pour out medium containing mitotic cells. For lymphoblasts, collect the entire culture.

5. Swell, fix, and drop cells. Centrifuge at 200 g for 10 minutes and follow steps 13–23 in the procedure for peripheral blood lymphocytes. Adjust volumes of KCl and fixative to the size of pellet.

C. Interphase Cells

Overview.   Cells arrested in G_1 are used for interphase mapping studies to avoid complex hybridization site patterns resulting after DNA replication. Fibroblast cultures can be arrested in G_1 by growing them to confluency and holding them without medium change for 4–7 days. The fraction of cells in S, G_2, and M can be determined by flow cytometric analysis of nuclear DNA content. Cells are trypsinized from the flask, hypotonically swollen, fixed, and dropped on slides as described for prometaphase preparations. Alternatively, cells can be synchronized following published protocols or fractionated by flow sorting to enrich for G_1 cells (e.g., D'Andrea *et al.,* 1983) (not described here).

Procedure

1. Grow fibroblast cultures to confluency and hold without a medium change at 37°C for 4–7 days. No rounded mitotic cells should have been visible for several days before harvest.

2. Trypsinize cells from flask following conventional tissue culture procedures. [Remove growth medium by aspiration. Add ∼ 10 ml of Puck's/ EDTA (5.4 mM KCl, 137 mM NaCl, 4.2 mM NaHCO$_3$, 5.6 mM glucose, 0.5 mM disodium EDTA, pH 7.2–7.6) and swirl over cells. Remove Puck's/EDTA by aspiration. Add 3 or 7 ml of trypsin solution (0.025% in Puck's/EDTA) per T-75 or T-150 flask, respectively. Incubate the flask for a few minutes at 37°C until the cells are released from the bottom of the flask. Add 7 or 13 ml of growth medium per T-75 or T-150 flask, respectively, to stop trypsin activity. Mix cells by pipetting.]

3. Centrifuge at 200 g for 10 minutes. Resuspend in 25–50 ml of PBS (0.9 mM CaCl$_2$, 2.7 mM KCl, 1.5 mM KH$_2$PO$_4$, 0.5 mM MgSO$_4$, 137 mM NaCl, 4.3 mM Na$_2$HPO$_4$, pH 7.6) containing 0.1% (w/v) bovine serum abumin (BSA). Repeat centrifugation and resuspension.

4. If possible, confirm arrest of cells in G_1 by flow cytometric analysis of nuclear DNA content.

    a. Remove several million cells. Hold the remainder on ice.

    b. Centrifuge and resuspend at 10^6 cells/ml in phosphate-buffered saline (PBS) containing 0.1% (w/v) BSA, 0.1% (v/v) Triton X-100, 40 μg/ml propidium iodide (Calbiochem or Sigma, from a 1 mg/ml stock in H$_2$O, stored at −20°C), and 0.15 mg/ml DNase-free RNase [prepared by boiling RNase stock (6 mg/ml) for 10 min].

    c. Incubate at 37°C for 30 minutes.

    d. Obtain the DNA content distribution by flow cytometry: measure nuclear fluorescence > 600 nm with an excitation laser tuned to 488 nm.

    b'. Alternatively, resuspend cells at 10^6 cells/ml in PBS containing 0.1% (w/v) BSA, 0.1% (v/v) Triton X-100, and 2 μg/ml Hoechst 33258 (from 200 μg/ml stock in H$_2$O, stored at −20°C).

    c'. Incubate on ice for 5 minutes.

    d'. Obtain the DNA content distribution by flow cytometry: measure nuclear fluorescence > 425 nm with an excitation laser tuned to the ultraviolet (multiline: 334, 351, and 363 nm).

5. If the DNA distribution indicates few or no cells in S or G_2/M, centrifuge the remaining cells.

6. For swelling, fixing, and dropping, follow steps 13–23 of the procedure for preparing prometaphase spreads from peripheral blood lymphocyte cultures.

VI.   Slide Hybridization

A.   Hybridization Mixture Preparation

Overview.   The hybridization buffer typically used contains formamide (each added percent decreases the T_m of DNA by 0.7°C), salt (increasing salt increases T_m), and dextran sulfate [increases the effective concentration of reagents (Wetmur, 1975)]. Thus, the specificity of accepted base pairing can be altered by increasing the formamide concentration and decreasing the salt concentration in the hybridization buffer. Two hybridization buffers are described below. The melting temperature of DNA in hybridization buffer (HB) 2 is approximately 8°C lower than it is in HB1. HB2 yields good specificity with a wide range of chromosome-specific repetitive probes. These same probes, when hybridized in HB1, can bind to many chromosomes.

Procedure

1. *HB1 preparation:* Mix 5 ml of formamide (BRL , cat. no. 5515UA/UB or other reagent-grade brand; if yellow color is present, the formamide should be deionized as described in Sambrook *et al.,* 1989), 1 ml of 20 × SSC (20 × SSC = 3.0 *M* NaCl, 0.3 *M* sodium citrate, pH 7.0), and 2 ml of 50% dextran sulfate (dissolve 25 g of dextran sulfate in water and bring the final volume to 50 ml. This takes some time and can be speeded by incubation at 70°C. Store 2-ml aliquots at −20°C. Warm to 37°C before use to decrease its viscosity). Bring the hybridization solution to pH 7.0 with 1.0 *N* HCl. This solution represents 80% of the final volume of the hybridization mixture. The remaining 20% of the volume is made up of probe and blocking DNA. Thus, HBI contains final concentrations of 50% formamide, 10% dextran sulfate, and 2 × SSC.

   HB2 preparation: The stringency of hybridization can be increased by using an alternative mixture that contains 55% formamide, 10% dextran sulfate, and 1 × SSC. This buffer is prepared by mixing 5.5 ml of formamide, 0.5 ml of 20 × SSC, and 2 ml of 50% dextran sulfate. Again, the remaining 20% of the final hybridization mixture volume is filled with probe and blocking DNA.

2. Mix 8 parts of HB1 or HB2 with one part of the appropriate type of sonicated blocking DNA (herring testes or genomic DNA stock at 10 mg/ml; see Section IV) in a volume sufficient for the number of slide

areas to be hybridized. The use of the same mixture reduces variation among slides, which can occur if HB and blocking DNA are mixed separately for each side.

3. For each slide area to be hybridized, put 9 μl of the HB + blocking DNA mix in 1.5-ml Eppendorf tube. Add a total of ≥ 20 ng of labeled probe(s) in 1 μl. We have found that this 20 ng can be divided among up to five different biotinylated or digoxigenin-labeled probes without significant signal reduction. If more probes are to be hybridized simultaneously, the probes should be mixed together, ethanol precipitated (Sambrook *et al.*, 1989), and resuspended in a smaller volume before addition to the hybridization mixture. To compensate for the lower signal intensity of the AAF labeling procedure compared to the biotin or digoxigenin labeling procedures, 100–200 ng (1 μl) of an AAF probe should be used per slide.

B.   Denaturation, Hybridization, and Washing

Overview.   In this step, double-stranded DNA in both target chromatin and probe is melted or denatured into single strands, which are then available for hybridization. The timing of probe and slide denaturation and the necessity for prehybridization in the presence of genomic DNA depend on the nature of the probes being used. Two procedures are described: one for repetitive sequence probes and one for probes for which repetitive sequence hybridization is to be suppressed. Hybridization takes place at 37°C for 2–16 hours, depending on the probe type. Mismatched and unhybridized probe molecules are then removed by rinsing the slides in wash fluids that contain the same concentrations of salt and formamide as the hybridization buffer. Washing at elevated temperature serves to remove probe bound with poor homology.

Procedure

1. Prepare denaturation solution by mixing 70 ml of formamide (BRL cat. no. 5515UA/UB), 10 ml of 20 × SSC, and 20 ml of distilled H$_2$O, and bringing to pH 7.0 with 1.0 *N* HCl. The final concentrations are 70% formamide, 2 × SSC, pH 7.0. Place two 50-ml Coplin jars filled with this solution in 72°C water bath. Fresh denaturation solution should be made every 2 weeks and stored at 4°C between use. Note: warm the Coplin jars and water bath simultaneously in order to avoid breaking cold Coplin jars. Use when the solution reaches 70°C.

2. Place two 50-ml Coplin jars containing 70% ethanol in an ice bucket. Allow the solution to cool to 0–4°C before use.

3. Mark the back of each slide with a diamond scribe to indicate the boundaries of the area where the cells have been dropped and where the

coverslips should be placed during hybridization. The cell drops can be seen by breathing lightly over the slide.

Repetitive sequence probes (chromosome-specific repeats, genomic DNA) or small probes containing only unique sequence (4–10 kb):

4. Immerse slides in denaturing solution for 2 minutes at 70°C. Each slide at room temperature that is put into a 50-ml Coplin jar will cause the temperature of the denaturing solution to drop by ~1°C. Therefore, immerse a maximum of three slides at any given time.

5. Transfer the slides quickly to a Coplin jar containing ice-cold 70% ethanol. Incubate for 1 minute. Agitate to rinse off the denaturing solution.

6. Dehydrate the slides by incubating for 3 minutes in 85, 90, and 100% ethanol. Dry slides with an air jet.

7. Denature probe mixture by placing tubes in a floating rack in 70°C waterbath for 5 minutes. Do not denature more probe tubes than can be applied to slides in ~2 minutes.

8. Quench probes for <2 minutes on ice.

9. Apply 10 µl of probe solution to each slide area. Slide can be at room temperature or on a 37°C slide warmer. Slides should not be left on the slide warmer for more than 2 minutes before probe is added.

Probes that contain unique sequence in addition to repetitive sequences, which must be suppressed with added genomic DNA:

4′. Denature probe mixture by placing tubes in a floating rack in 70°C water bath for 5 minutes.

5′. Transfer denatured probe tubes immediately to a 37°C incubator.

6′. Incubate probe for 15–60 minutes.

7′. During probe incubation, denature and dehydrate slides as described in steps 4–6 in the previous section.

8′. Place slides on a 37°C slide warmer.

9′. Apply 10 µl of probe solution to each slide area. Work with only a few probe tubes at any given time to avoid letting them cool below 37°C before being applied to slides.

10. Cover the hybridization solution immediately with a 22×22 mm² coverslip. Gently tap out air bubbles.

11. Liberally seal edges of coverslip with rubber cement. Apply using a 5-ml syringe.

12. Place slides immediately in a moist environment in a 37°C incubator. One possible arrangement is to place slides horizontally on racks above water in a sealed plastic box.

13. Incubate for 2–16 hours, depending on the probe. Repetitive sequence probes can produce a strong signal after as little as 2 hours incubation.

Unique sequence probes should be allowed to hybridize overnight or longer.

14. Prewarm wash solution 1 (WS1 = 50% formamide, $2 \times$ SSC, pH 7.0) and WS2 ($2 \times$ SSC = 1 part $20 \times$ SSC, 9 parts distilled H_2O, pH 7.0) to 42°C.

15. Remove slides from incubator and peel off rubber cement with forceps. *Do not allow slides to dry from this step on.* Immerse slides in Coplin jar containing WS1.

16. Wash slides in three 5-minute changes of WS1 at 42°C. Agitate slides periodically. The first and second wash solutions can be reused 1–2 times before being discarded. *Warning:* formamide is classed as a possible human teratogen and should be disposed as hazardous waste. Used wash solution can be stored at 4°C between uses.

17. Wash slides in three 5-minute changes of WS2 at 42°C.

18. Place slides in $4 \times$ SSC (1 part $20 \times$ SSC, 4 parts distilled H_2O, pH 7.0) at room temperature.

C.   Fluorescent Labeling

Overview.   Protocols are given in this section for labeling biotin, digoxigenin, and AAF probes with FITC. Procedures are also described for labeling, with FITC and TR, two or more probes that carry different reporter molecules and have been hybridized to the same slide. The buffers and procedures described are designed to minimize nonspecific binding of reagents. Amplification of biotin probe fluorescence through incubation in biotin–goat antiavidin antibody and a second avidin-FITC or -TR treatment is optional and should be performed only if probe signal is dim and background staining is negligible after first avidin-FITC or -TR treatment.

Procedures

General remarks: Slides should not be allowed to dry at any point in subsequent steps. Reagents can be rapidly applied to slides with minimal reagent waste by applying liquid to coverslips laid out on the bench top. A slide is then briefly drained and inverted over the drop of liquid on the coverslip. Surface tension causes the coverslip to adhere to the slide. Together, the slide and coverslip can be inverted and placed right-side-up on the bench where incubation takes place. To change buffer under the coverslip, apply new buffer to unused coverslips laid out on bench. Shake used coverslip off of slide, drain slide briefly, and briefly tap long edge of slide on blotting paper to remove excess liquid. Invert the slide over the drop of buffer on the new coverslip. Slides can be stored at 4°C in PN [0.1 M phosphate buffer, pH 8.0 (mix 0.1 M Na_2HPO_4 and 0.1 M NaH_2PO_4 to achieve pH 8.0), 0.5% NP-40] containing 0.02% sodium azide between steps.

FITC Staining of Biotinylated Probes

1. *Blocking:* Remove a slide from 4 × SSC buffer in which it has been stored, drain briefly, and blot excess liquid from the edge. Apply 100 μl of 4 × SSC/1% BSA (made up fresh daily) under a 22 × 60 mm² coverslip. Incubate at room temperature for ∼5 minutes.

2. *Avidin-FITC incubation:* Apply 100 μl of avidin-FITC solution (a good source is Vector cat. no. A-2011 used at 5 μg of avidin-FITC/ml 4 × SSC/1% BSA) under coverslip. Incubate at room temperature for 20–60 minutes.

3. *Washing:* Wash the slides for 5 minutes each in 50- to 100-ml volumes of 4 × SSC, 4 × SSC/0.1% Triton X-100, and then 4 × SSC at room temperature. Agitate occasionally. Rinse well in PN to remove salt. The slides can now be viewed, or can be further amplified.

4. *Amplification:* Amplification of probe fluorescence is accomplished by applying biotinylated goat anti-avidin antibody, followed by another layer of avidin-FITC. This can be done even after the slides have been viewed. In this case, the slides are washed in several changes of PN buffer. The amplification process (steps 4a–f) can be repeated as necessary.

   a. *Blocking:* Apply 100 μl of PNM buffer under coverslip. [PNM = PN buffer containing 5% (w/v) nonfat dry milk (e.g., Carnation) and 0.01% sodium azide; Mix in ≤500-ml quantities, let stand several days at 4°C, transfer clear upper solution to 15-ml centrifuge tubes, spin at 200 g for 15 minutes to remove milk solids, store at 4°C, use upper clear solution]. Incubate at room temperature for ∼5 minutes.

   b. *Biotinylated goat anti-avidin antibody incubation:* Apply 100 μl of antibody solution (e.g., Vector cat. no. BA-0300 used at 5 μg of antibody/ml PNM) under coverslip. Incubate at room temperature for 20–60 minutes.

   c. *Washing:* Wash in three 5-minute changes of PN buffer.

   d. *Blocking:* Repeat step 4a.

   e. *Avidin-FITC incubation:* Apply 100 μl of avidin-FITC solution (5 μg of avidin-FITC/ml PNM) under the coverslip. Incubate at room temperature for 20–60 minutes.

   f. *Washing:* Repeat step 4c.

FITC Staining of Digoxigenin-Labeled Probes

1. Rinse slides well in PN buffer.

2. *Blocking:* Remove slide from PN buffer in which it has been stored, drain briefly, and blot excess liquid from the edge. Apply 100 μl of PNM under the coverslip. Incubate at room temperature for ∼5 minutes.

3. *Sheep anti-digoxigenin antibody incubation:* Apply 100 μl of sheep anti-digoxigenin antibody solution (Fab fragments, Boehringer Mannheim,

cat. no. 1214-667, used at 15 μg/ml in PNM) under the coverslip. Incubate at room temperature for 1 hour.

4. *Washing:* Wash the slides in three 5-minute changes of PN.

5. *Blocking:* Repeat step 2.

6. *FITC conjugated rabbit anti-sheep IgG antibody incubation:* Apply 100 μl of FITC conjugated rabbit anti-sheep IgG (H + L) antibody solution (e.g., Vector, cat. no. FI-6000) used at 60–100 μg/ml PNM) under the coverslip. Incubate at room temperature for 1 hour.

7. *Washing:* Repeat step 4.

FITC Staining of AAF-Labeled Probes

1. Rinse slides in several 50-ml volumes of PN buffer.

2. *Blocking:* Apply 100 μl of PBS/0.1% Tween 20 containing 2% normal goat serum (e.g., Vector cat. no. S-1000) under the coverslip. Incubate at room temperature for ~5 minutes.

3. *Mouse anti-AAF antibody incubation:* Apply 100 μl of anti-AAF solution under coverslip (clone 4F, available from Dr. Robert Baan, is used at 750-fold dilution of antibody in PBS/0.1% Tween 20/2% normal goat serum). Incubate at 37°C for 1 hour.

4. *Washing:* Wash slides in three 5-minute changes PBS/0.1% Tween 20 at room temperature. Agitate occasionally.

5. *Blocking:* Repeat step 2.

6. *FITC conjugated goat anti-mouse IgG antibody incubation:* Apply 100 μl of antibody solution (e.g., Sigma used at 1000-fold dilution in PBS/0.1% Tween/2% normal goat serum) under the coverslip. Incubate at 37°C for 1 hour.

7. *Washing:* Repeat step 4.

Two-Color Labeling: FITC on Digoxigenin, TR on Biotin

1. *Blocking:* Remove the slide from 4× SSC buffer in which it has been stored, drain briefly, and blot excess liquid from the edge. Apply 100 μl of 4× SSC/1% BSA under the coverslip. Incubate at room temperature for ~5 minutes.

2. Avidin-TR incubation: Apply 100 μl of avidin-TR solution (e.g., Vector, cat. no. A-2006 used at 2.5 μg avidin-TR/ml 4× SSC/1% BSA) under the coverslip. Incubate at room temperature for 1 hour.

3. *Washing:* Wash the slides for 5 minutes each in 50- to 100-ml volumes of 4× SSC, 4× SSC/0.1% Triton X-100, 4× SSC at room temperature. Agitate occasionally. Rinse well in PN buffer to remove salt.

4. *Blocking:* Apply 100 μl of PNM buffer under the coverslip. Incubate at room temperature for ~5 minutes.

5. *Sheep anti-digoxigenin and (optional) biotinylated goat anti-avidin antibody incubation:* Apply 100μl of antibody solution (15 μg of sheep anti-digoxigenin antibody and 5 μg biotinylated anti-avidin antibody per

milliliter of PNM) under the coverslip. Incubate at room temperature for 1 hour.

6. *Washing:* Wash in three 5-minute changes of PN buffer as above.

7. *Blocking:* Repeat step 4.

8. *FITC conjugated rabbit anti-sheep IgG and avidin-TR incubation:* Apply 100 μl of antibody/avidin solution (60–100 μg of FITC-rabbit anti-sheep IgG and 2.5 μg of avidin-TR per milliliter of PNM) under the coverslip. Incubate at room temperature for 1 hour.

9. *Washing.* Repeat step 6.

Two-Color Labeling: FITC on AAF, TR on Biotin

1. Rinse in liberal volumes of PN.

2. *Blocking:* Apply 100 μl of PNM or PN/1% BSA under the coverslip. Incubate at room temperature for ∼5 minutes.

3. *Mouse anti-AAF antibody and avidin-TR incubation:* Apply 100 μl of antibody/avidin solution (750-fold dilution of mouse anti-AAF antibody and 2.5 μg of avidin-TR per milliliter PNM or PN/1% BSA) under the coverslip. Incubate at room temperature for 1 hour.

4. *Washing:* Wash the slides for 5 minutes each in 50- to 100-ml volumes of PN.

5. *Blocking:* Apply 100 μl of PNM or PN/1% BSA buffer under the coverslip. Incubate at room temperature for ∼5 minutes.

6. *FITC conjugated goat anti-mouse IgG and biotinylated goat anti-avidin antibody incubation:* Apply 100 μl of antibody solution (1000-fold dilution of FITC conjugated goat anti-mouse IgG (Sigma) and 5μg of biotinylated goat anti-avidin per milliliter of PNM or PN/1% BSA) under the coverslip. Incubate at room temperature for 1 hour.

7. *Washing:* Wash in three 5-minute changes of PN buffer (50- to 100-ml volumes).

8. *Blocking:* Repeat step 5.

9. *Avidin-TR incubation:* Apply 100 μl avidin solution (2.5 μg of avidin-TR/ml PNM) under the coverslip. Incubate at room temperature for 20–60 minutes.

10. *Washing.* Repeat step 7.

D.   DNA Counterstaining and Chromosome Banding

Overview.   DAPI (4′,6-diamidino-2-phenylindole) or propidium iodide (PI) can be used to counterstain DNA in nuclei or chromosomes. To view total DNA and probe fluorescence separately, DAPI is used as a counterstain as it can be excited independently from FITC and TR. To see DNA and FITC-labeled probe fluorescence simultaneously, use the red fluorescent DNA dye, PI. If probe fluorescence is dim, PI fluorescence can overwhelm the FITC, and

DAPI should be used. The counterstain is applied in a solution that reduces bleaching of fluorochromes (Johnson and Nogueira, 1981). Low-resolution bands can be produced in chromosomes by treating chromosomes with actinomycin after DAPI staining (Tucker *et al.*, 1988). Alternative banding procedures are described by Albertson *et al.* (1988), Bhatt *et al.* (1988), Viegas-Pequignot *et al.* (1989), Cherif *et al.* (1990), and Lawrence *et al.* (1990).

Warning. DAPI, PI, and actinomycin are possible carcinogens. They should be handled with protective clothing and disposed of as hazardous waste.

Procedures
1. Make up the antifade solution (Johnson and Nogueira, 1981). Dissolve 100 mg of p-phenylenediamine dihydrochloride (Sigma, cat. no. P1519) in 10 ml of PBS. Adjust to pH 8 with 0.5 M bicarbonate buffer (0.42 g of $NaHCO_3$ in 10 ml of water, pH to 9 with NaOH). Add to 90 ml of glycerol. Filter through a 0.22-μm filter to remove undissolved particulates. Store in the dark at $-20°C$. Although the solution darkens with time, it remains effective and can be stored up to 1 year.
2. Prepare stock solutions of DAPI (Sigma, cat no. D1388) and PI, both at 1 mg/ml in H_2O, and store in 1-ml aliquots at $-20°C$.
3. Prepare working antifade solutions. These can be stored at $-20°C$ for several weeks. They should be thawed to room temperature for use and kept closed to prevent bacterial growth.
   a. 0.25 μg DAPI/ml antifade, for use with dim FITC probe signals or with TR signals.
   b. 2.5 μg PI/ml antifade, for use with bright FITC probe signals produced by highly repetitive probes.
   c. 0.5 μg PI/ml antifade, for use with dim FITC probe signals or, optionally, in conjunction with DAPI banding.
4. Drain the slide well and blot excess moisture from back side and frosted area. Do not allow the slide to dry. Apply ~ 3 μl of antifade solution to each cell area and top with a 22 × 60 mm^2 coverslip. The slide can be viewed immediately.
5. (Optional) Chromosome banding procedure:
   a. Remove the slide from PN buffer, drain well.
   b. Apply 100 μl of DAPI solution prepared just before use from the DAPI stock solution [final concentration is 0.6 μg/ml in McIlvaine buffer (mix ~ 100 ml of 0.1 M citric acid and ~ 100 ml of 0.2 M Na_2HPO_4 to achieve pH 7.0)]. Incubate at room temperature for 20–30 minutes.
   c. Rinse the slide briefly in PN buffer.
   d. Apply 100 μl of actinomycin solution (Sigma cat. no. A-1410, use at 0.25 mg/ml in 10 mM NaH_2PO_4/1 mM EDTA) under a 22 × 60 mm^2

coverslip and incubate for 20 minutes at room temperature in the
dark.

   e. Apply dilute PI antifade solution and view immediately.

6. Slides can be stored at 4°C for several days with coverslips on, if drying of
antifade solution is prevented. Alternatively, coverslips can be removed
and the slides returned to PN buffer containing 0.02% sodium azide for
storage at 4°C.

E. Microscopy and Photography

Overview.   Slides can be viewed immediately after the application of antifade. Each field can be observed and photographed for several minutes before
substantial fading of the probe fluorescence occurs. Fluorescence microscopes
with high-quality optical components are required for viewing the small and
dim signals produced by single-copy cosmid, λ or plasmid probes. Table III
gives typical filter specifications for viewing common fluorochromes.

Photography.   Photographs can be taken using Ektachrome 400 ASA
color slide film (exposure times: DAPI <1 seconds, PI and/or FITC ~ 20
seconds, TR + FITC through dual-bandpass filter ~ 150 seconds). Scotch
640 T color slide film results in a slightly grainier image, some color distortion,
but significantly shorter exposure times (DAPI <0.5 seconds, PI and/or
FITC ~ 5 seconds, TR + FITC through double-bandpass filter ~ 15 seconds).

TABLE III

TYPICAL EXCITATION AND EMISSION FILTERS FOR MICROSCOPY[a]

Fluorochrome	Excitation filter	Reflector	Emission filter
DAPI or Hoechst	BP[b]360-371	FT395[c]	LP397[d]
PI[e] or TR	BP 540-552	FT580	LP590
FITC alone	BP450-490	FT510	BP515-565
PI + FITC	BP450-490	FT510	LP520
CA[f] or quinacrine	BP400-410	FT460	LP470
TR + FITC[g]	BP480-505 + BP560-595	BP505-555 + BP600-690	BP515-540 + BP610-660

  [a] The first five lines of the table give the specifications of Zeiss filter sets (filter sets with similar
specifications are available from other epifluorescence microscope manufacturers). The dual-band pass filter
for viewing TR and FITC simultaneously is manufactured by Omega Optical Inc.

  [b] Bandpass filter, i.e., passing wavelengths in indicated range (nm).

  [c] Dichroic reflector reflects wavelengths less than and passes wavelengths greater than indicated
wavelength (nm).

  [d] Long pass filter, i.e., passing wavelengths greater than indicated wavelength (nm).

  [e] PI also visible through DAPI filters.

  [f] Chromomycin A3.

  [g] Dual-bandpass filter, passing wavelengths in indicated ranges.

VII. Suspension Hybridization

Overview.   The positions and intensity of chromosome-specific repeat sequences can be determined in intact nuclei after hybridization in suspension. Similarly, the human chromosomal content of hybrid cells can be analyzed by microscopy and flow cytometry after hybridization of hybrid cell nuclei with human genomic DNA. Successful detection of single-copy sequences in intact nuclei has not yet been demonstrated. Cells are collected from suspension cultures or are removed from plates by trypsinization and fixed in ethanol so that they remain intact during denaturation and hybridization with labeled probe. Protein is removed by treatment with acid and detergent. The cells can be further fixed with paraformaldehyde for better preservation of the *in vivo* state of nuclei for microscopic analysis. Background probe and reagent binding is increased by this treatment. Therefore, it should be omitted if probe fluorescence is to be quantified by flow cytometry. Washing is performed by centrifugation and resuspension in fresh buffers. Procedures are given for FITC labeling of bound biotinylated and AAF-labeled probes. A technique for simultaneous detection of two probes, labeled with biotin and AAF, using TR and FITC is also described. Hybridization and detection of digoxigenin-labeled probes are theoretically possible, but have not yet been reported. The reader is referred to the literature for details on analyzing hybridized nuclei by optical sectioning and confocal laser scanning microscopy.

Procedure.   (Refer to slide hybridization sections for reagent ordering information.)

1. Prepare dimethylsuberimidate (DMS)-treated red blood cells to be used as a carrier particle to prevent loss of nuclei during the multiple wash and centrifugation steps of the procedure. Suspend washed erythrocytes, from which serum and white blood cells have been removed by centrifugation, in physiological salt solution at a concentration of 10^8 cells/ml. Treat three times with DMS to cross-link proteins (Pierce, cat. no. 20668). For each treatment, K_2CO_3 and DMS are added to the erythrocytes from a $5 \times$ concentrated stock solution, mixed immediately before use. Final K_2CO_3 concentration for each treatment is 20 mM. Final DMS concentrations for the three treatments are 3, 10, and 10 mM. (Addition of 100 mM K_2CO_3 may be required during the last two treatments to achieve a pH of 9–10.) Each DMS treatment takes place for 15 minutes at room temperature and is followed by pH adjustment from 10 to 8 by the addition of 50 μl of 100 mM citric acid per milliliter of cell suspension. The fixed erythrocytes are finally centrifuged and resuspended in $2 \times$ SSC at 10^8 ml. They can be stored indefinitely at 4°C after the addition of sodium azide to 0.1% final concentration.

2. Cultured cell isolation for nuclei:
   a. Collect several hundred million cells from suspension cultures or remove them by trypsinization from monolayer cultures as described above (Section V,C).
   b. Wash the cells twice with PBS by centrifugation at 200 g for 10 minutes. Pool and resuspend cells in a conical capped 50-ml tube after the final wash at 5×10^6 cells/ml. The volume of cell suspension is v.
3. Nuclei fixation and acid treatment:
   a. Add $(7/3 \times v)$ ml of cold 100% ethanol to PBS-washed cells while vortexing to achieve a final concentration of ethanol of 70%.
   b. Let the mixture stand on ice for 10 minutes.
   c. Centrifuge at 150 g for 10 minutes at 4°C.
   d. Remove the supernatant; flick the pellet to loosen.
   e. Add $(3 \times v)$ ml of 100% cold ethanol while vortexing.
   f. Let the mixture stand on ice for 10 minutes.
   g. Repeat steps c and d.
   h. Resuspend the pellet in $(1/2 \times v)$ ml of 0.1 N HCl/0.5% Triton X-100.
   i. Let stand at room temperature for exactly 10 minutes.
   j. Fill the 50-ml tube with IBT (IBT = 50 mM KCl, 10 mM MgSO$_4$, 5 mM HEPES, 0.25% Triton X-100, pH 8.0).
   k. Repeat steps c and d.
   l. Repeat steps j and k.
   m. (Optional) Resuspend in v ml of $2 \times$ SSC/0.1% Tween 20. Add v ml of a 2% paraformaldehyde solution [dissolve 2 g of paraformaldehyde powder in 25 ml of 100 mM MgCl$_2$; add several drops of 1 N NaOH and heat to 70°C to speed dissolution; when dissolved add 25 ml of $2 \times$ PBS (pH 7.5)]. Let stand for 10 minutes at room temperature. Fill the tube with IBT. Repeat steps c and d.
   n. Resuspend the pellet in IBT at 10^8 cells/ml. Count cells in hemacytometer after a 50-fold dilution of a sample of cells in IBT containing 2 μg/ml Hoechst 33258. The suspension should consist primarily of single, intact nuclei.
4. Denaturation and hybridization:
   a. Prepare the hybridization mixture by mixing 8 parts HB1 (see Section VI,A) with 1 part 10 mg/ml sonicated herring testes DNA.
   b. Mix nuclei suspension and hybridization mixture at a ratio of 1:18. Place 19 μl of this mixture, which contains 10^5 nuclei, in as many 1.5-ml Eppendorf tubes necessary for the experiment.
   c. Add probe to each tube: 1μl (100 ng) of AAF-labeled probe/tube or 1–2 μl (20–40 ng) of biotin-labeled probe/tube.
   d. Denature nuclei and probe together by placing tubes in floating rack at 70°C for 10 minutes.

e. Do not quench the suspension on ice, but bring tubes quickly to $\geq 37°C$, depending on probe, for optimal stringency. Incubate overnight.
5. Posthybridization wash procedure:
   a. Add 1.25 ml of WS1 (see Section VI,B) at 42°C to each tube.
   b. Let stand at 42°C for 10–15 minutes. Mix by vortexing occasionally.
   c. Bring the tubes to room temperature. Add 100 μl of DMS-treated red blood cells from a stock solution of 10^8 cells/ml. Mix by vortexing.
   d. Centrifuge at 150 g at room temperature for 10 minutes. Remove the supernatant by aspiration and flick the tube to loosen the pellet.
   e. Add 1.25 ml of 2 × SSC (pH 0.7) at 42°C, but let tubes stand at room temperature for 10–15 minutes.
   f. Repeat step 5d.
   g. Add 1.25 ml of IBT. Let the tubes stand at room temperature for 5 minutes.
   h. Repeat step 5d.
6. Fluorescent labeling of AAF probes:
   a. Add 200 μl of PTG (PTG = PBS containing 0.05% Tween 20 and 2% normal goat serum) to pellet. Vortex gently to mix. Let the tubes stand for 10 minutes at room temperature.
   b. Add 20 μl of a 1 : 100 dilution of mouse anti-AAF antibody. Mix. Incubate at 37°C for 45 minutes.
   c. Add 1.25 ml of PT (PT = PBS containing 0.05% Tween 20). Let stand with occasional mixing at room temperature for 10 minutes.
   d. Centrifuge as above, remove supernatant by aspiration, and flick pellet.
   e. Repeat step 6a.
   f. Add 20 μl of a 1 : 100 dilution of FITC conjugated goat anti-mouse IgG antibody (Sigma). Incubate at 37°C for 45 minutes.
   g. Repeat steps 6c and 6d.
7. Fluorescent labeling of biotin probes:
   a. Add 200 μl of STB (STB = 4 × SSC containing 0.1% Triton X-100 and 5% BSA) to the pellet. Vortex gently to mix. Let stand for 10 minutes at room temperature.
   b. Add 20 μl of a 15 μg/ml avidin-FITC solution. Incubate for 30 minutes at 37°C.
   c. Add 1.25 ml of ST (ST = 4 × SSC containing 0.1% Triton X-100). Mix. Let stand with occasional mixing at room temperature for 10 minutes.
   d. Centrifuge as above, remove the supernatant by aspiration, and flick the pellet.

     e. Add 1.5 ml of PT. Vortex. Let stand with occasional mixing at room
       temperature for 10 minutes.
     f. Repeat step 7d.
  8. Two-color labeling of biotin and AAF probes:
     a. Add 200 μl of PN/1% BSA. Vortex gently to mix. Let stand for 10
       minutes at room temperature.
     b. Add 20 μl of a solution containing 50 μg/ml streptavidin-TR
       (Molecular Probes) and 1:100 dilution of mouse anti-AAF (clone
       4F). Incubate for 30 minutes at 37°C.
     c. Add PN/1% BSA. Mix. Let stand with occasional mixing at room
       temperature for 10 minutes.
     d. Centrifuge as above, remove the supernatant by aspiration, and flick
       the pellet.
     e. Repeat step 8a.
     f. Add 20 μl of a 1:100 dilution of FITC conjugated goat anti-mouse
       IgG antibody (Sigma). Incubate at 37°C for 45 minutes.
     g. Repeat steps 8c and 8d.
     e. Add 1.5 ml of PT. Vortex. Let stand with occasional mixing at room
       temperature for 10 minutes.
     f. Repeat step 8d.
  9. Fluorescence microscopy:
     a. Resuspend nuclei in 250 μl of PT containing 2 μg/ml Hoechst 33258.
       Mix by vortexing.
     b. For photography, add an equal volume of anti-fade solution (see
       Section VI,D) to the nuclei suspension before covering with a
       coverslip.
     c. See Section VI,E for excitation and emission filter information.
     d. Photograph as described for slide hybridization (Section VI,E).

VIII.   Conclusions

    The procedures for DNA sequence localization in interphase and meta-
phase cells described in this chapter have a number of research applications. Several of these applications are illustrated in Color Plate 1.
Cytogenetic analysis is greatly simplified by the ability to identify chromosomes by hybridizing chromosome-specific probes to metaphase or interphase cells (e.g., Cremer *et al.*, 1986, 1988, 1990; Hopman *et al.*, 1988; Pinkel
et al., 1986, 1988; Lichter *et al.*, 1988a,b; Lucas *et al.*, 1989). Color Plate 1a
illustrates one such repetitive sequence that is specific for the heterochromatic
region near the centromere of chromosome 1. Normal heteromorphism in

the copy number of this sequence is evident after fluorescence *in situ* hybridization (Trask *et al.*, 1989b). Aneuploidy or translocations can be detected rapidly and simply in interphase or metaphase using these chromosome-specific repeats or collections of sequences derived from a single chromosome type. In Color Plate 1b, the normal and translocation products of chromosome 4 are highlighted in γ-irradiated peripheral blood cells using a pool of sequences cloned from flow-sorted chromosome 4 (Pinkel *et al.*, 1988). The nuclear organization of chromosomes can be studied after hybridization of chromosome-specific sequences to cells whose three-dimensional architecture has been preserved (e.g., Manuelidis, 1985; Borden and Manuelidis, 1988; Manuelidis and Borden, 1988; Trask *et al.*, 1988). Color Plate 1c shows that characterization of somatic cell hybrids, important for successful gene mapping or chromosome-specific library production, is simplified by using human genomic DNA to highlight human DNA retained in the hybrid (e.g., Schardin *et al.*, 1985; Pinkel *et al.*, 1986; Lichter *et al.*, 1990b). Color Plate 1d illustrates that the manifestations of gene amplification, in terms of the relative location of amplified and parental gene copies, can be readily studied in individual cells following fluorescence *in situ* hybridization of sequences in the amplified region (Trask and Hamlin, 1989). Fluorescence *in situ* hybridization is also making a contribution to the current effort to map the human and other genomes. Color Plate 1e shows that DNA sequences can be mapped readily along the chromosome by this technique (e.g., Albertson *et al.*, 1988; Bhatt *et al.*, 1988; Viegas-Pequignot *et al.*, 1989; Cherif *et al.*, 1990; Kievits *et al.*, 1990; Lichter *et al.*, 1990a; Lawrence *et al.*, 1990; Trask *et al.*, 1991). Color Plate 1f illustrates that the order of sequences that map to the same chromosomal region can be determined from the order of hybridization sites in interphase nuclei (Lawrence *et al.*, 1988, 1990; Trask *et al.*, 1989a, 1991).

IX. Supplier Information

Amersham, 2636 S. Clearbrook Dr., Arlington Heights, IL 60005-4692. Phone: (800) 323-9750 or (312) 364-7100.

Dr. Robert Baan, Molecular Biological Laboratories, Lange Kleiweg, Rijswijk, The Netherlands.

Boehringer Mannheim Biochemicals, P. O. Box 50414, Indianapolis, IN 46250. Phone: (800) 262-1640.

BRL/Gibco, P. O. Box 68, Grand Island, NY 14072-0068, or P. O. Box 6009, Gaithersburg, MD 20877. Phone: (800) 828-6686 or (716) 774-6700 or (301) 840-8000.

Calbiochem, P. O. Box 12087, San Diego, CA 92112-4180. Phone: (800) 854-3417 or (619) 450-9600.

Chemsyn Sciences Laboratories, 13605 W. 96th Terrace, Lenexa, KS 66215. Phone: (913) 542-0525.

Clontech, 4030 Fabian Way, Palo Alto, CA 94303. Phone: (800) 662-2566 or (415) 424-8188.

Enzo Diagnostics, Inc., 325 Hudson St., New York, NY 10013. Phone: (212) 741-3838 or (800) 221-7705.

FMC Bioproducts, 5 Maple St., Rockland, ME 04841. Phone: (207) 594-3353 or (800) 341-1574.

Irvine Scientific, Inc., 2511 Daimler St., Santa Ana, CA 92705-5588. Phone: (800) 437-5706 or (714) 261-7800.

Molecular Probes, Inc., P. O. Box 22010, Eugene, OR 97402. Phone: (503) 344-3007.

Omega Optical, Inc., P. O. Box 573, Brattleboro, BT 05301. Phone: (802) 254-2690.

Organon Technika, 100 Akzo Ave., Durham, NC 27704. Phone: (800) 682-2666 or (919) 620-2000.

Pharmacia, 800 Centennial Ave., Piscataway, NJ 03354. Phone: (800) 558-7110 or (201) 457-8000.

Pierce Chemical Co., P. O. Box 117, Rockford, IL 61105. Phone: (800) 874-3723 or (815) 968-0747.

Quiagen, Inc., P. O. Box 7401-737, Studio City, CA 92604. Phone: (800) 426-8157 or (818) 508-5258.

Sigma, P. O. Box 14508, St. Louis, MO 63178. Phone: (800) 325-3010 or (314) 771-5750.

Vector Laboratories, 30 Ingold Road, Burlingame, CA 94010. Phone: (415) 697-3600.

ACKNOWLEDGMENTS

Work was performed under the auspices of the U.S. Department of Energy by the Lawrence Livermore National Laboratory under contract number W-7405-ENG-48 with support from U.S.P.H.S. Grants HD-17665 and HG0025b. I thank Mari Christensen, Anne Fertitta, and Hillary Massa for their comments on an earlier version of the manuscript.

REFERENCES

Albertson, D. G., Fishpool, R., Sherrington, P., Nacheva, E., and Milstein, C. (1988). Sensitive and high resolution *in situ* hybridization to human chromosomes using biotin labeled probes: Assignment of the human thymocyte CD1 antigen to chromosome 1. *EMBO J.* **7,** 2801–2805.
Bhatt, B., Burns, J., Flannery, D., and McGee, J. O'D. (1988). Direct visualization of single copy genes on banded metaphase chromosomes by nonisotopic *in situ* hybridization. *Nucleic Acids Res.* **16,** 3951–3961.

Borden, J., and Manuelidis, L. (1988). Movement of the X chromosome in epilepsy. *Science* **242**, 1687–1691.

Cherif, D., Julier, C., Delattre, O., Derre, J., Lathrop, G. M., and Berger, R. (1990). Simultaneous localization of cosmids and chromosome R-banding by fluorescence microscopy: Application to regional mapping of human chromosome 11. *Proc. Natl. Acad. Sci. U.S.A.* **87**, 6639–6643.

Cremer, T., Landegent, J., Bruckner, A., Scholl, H. P., Schardin, M., Hager, H. D., Devilee, P., and Pearson, P. (1986). Detection of chromosome aberrations in the human interphase nucleus by visualization of specific target DNAs with radioactive and non-radioactive *in situ* hybridization techniques: Diagnosis of trisomy 18 with L1.84. *Hum. Genet.* **74**, 346–352.

Cremer, T., Lichter, P., Borden, J., Ward, D. C., and Manuelidis, L. (1988). Detection of chromosome aberrations in metaphase and interphase tumor cells by *in situ* hybridization using chromosome-specific library probes. *Hum. Genet.* **80**, 235–246.

Cremer, T., Popp, S., Emmerich, P., Lichter, P., and Cremer, C. (1990). Rapid metaphase and interphase detection of radiation-induced chromosome aberrations in human lymphocytes by chromosomal suppression *in situ* hybridization. *Cytometry* **11**, 110–118.

D'Andrea, A. D., Tantravahi, U., Lalande, M., Perle, M. A., and Latt, S. A. (1983). High resolution analysis of the timing of replication of specific DNA sequences during S phase of mammalian cells. *Nucleic Acids Res.* **11**, 4753–4774.

Dauwerse, J. G., Kievits, T., Beverstock, G. C., van der Keur, D., Smit, E., Wessels, H. W., Hagemeijer, A., Pearson, P. L., van Ommen, G.-J. B., and Breuning, M. H. (1990). Rapid detection of chromosome 16 inversion in acute nonlymphocytic leukemia, subtype M4: Regional localization of the breakpoint in 16p. *Cytogenet. Cell Genet.* **53**, 126–128.

Harper, M. E., Ullrich, A., and Saunders, G. F. (1981). Localization of the human insulin gene to the distal end of the short arm of chromosome 11. *Proc. Natl. Acad. Sci. U.S.A.* **78**, 4458–4460.

Hopman, A. H. N., Wiegant, J., Raap, A. K., Landegent, J. E., van der Ploeg, M., van Duijn, P. (1986a). Bi-color detection of two target DNAs by non-radioactive *in situ* hybridization. *Histochemistry* **85**, 1–4.

Hopman, A. H. N., Wiegant, J., Tesser, G. I., and van Duijn, P. (1986b). A non-radioactive *in situ* hybridization method based on mercurated nucleic acid probes and sulfhydryl-hapten ligands. *Nucleic Acids Res.* **4**, 6471–6488.

Hopman, A. H. N., Ramaekers, F. C. S., Raap, A. K., Beck, J. L. M., Devilee, P., van der Ploeg, M., and Vooijs, G. P. (1988). *In situ* hybridization as a tool to study numerical chromosome aberrations in solid bladder tumors. *Histochemistry* **89**, 307–316.

Johnson, G., and Nogueria, G. (1981). A simple method of reducing the fading of immuno-fluorescence during microscopy. *J. Immunol. Methods* **43**, 349–350.

Kievits, T., Devilee, P., Wiegant, J., Wapenaar, M. C., Cornelisse, C. J., van Ommen, G. J., and Pearson, P. L. (1990). Rapid subchromosomal localization of cosmids by nonradioactive *in situ* hybridization. *Cytometry* **11**, 105–109.

Landegent, J. E., Jansen in de Wal, N., Baan, R. A., Hoeijmakers, J. H. J. and van der Ploeg, M. (1984). 2-Acetylaminofluorene-modified probes for the indirect hybridocytochemical detection of specific nucleic acid sequences. *Exp. Cell Res.* **153**, 61–72.

Landegent, J. E., Jansen in de Wal, N., Dirks, R. W., Baas, F., and van der Ploeg, M. (1987). Use of whole cosmid cloned genomic sequences for chromosomal localization of by non-radioactive *in situ* hybridization. *Hum. Genet.* **77**, 366–370.

Langer-Safer, P. R., Levine, M., and Ward, D. C. (1982). Immunological method for mapping genes on *Drosophila* polytene chromosomes. *Proc. Natl. Acad. Sci. U.S.A.* **79**, 4281–4385.

Lawrence, J. B., and Singer, R. H. (1985). Quantitative analysis of *in situ* hybridization methods for the detection of actin gene expression. *Nucleic Acids Res.* **13**, 1777–1799.

Lawrence, J. B., Villnave, C. A., and Singer, R. H. (1988). Sensitive, high resolution chromatin and chromosome mapping *in situ*: Presence and orientation of two closely integrated copies of EBV in a lymphoma line. *Cell* **52**, 51–61.

Lawrence, J. B., Singer, R. H., and McNeil, I. A. (1990). Visual resolution of different distances within the human dystrophin gene in nuclei and chromosomes. *Science* **249**, 928–932.

Lichter, P., and Ward, D. C. (1990). Is non-isotopic *in situ* hybridization finally coming of age? *Nature (London)* **345**, 93–95.

Lichter, P., Cremer, T., Borden, J., Manuelidis, L., and Ward, D. C. (1988a). Delineation of individual human chromosomes in metaphase and interphase cells by *in situ* suppression hybridization using recombinant DNA libraries. *Hum. Genet.* **80**, 224–234.

Lichter, P., Cremer, T., Tang, C.-J. C., Watkins, P. C., Manuelidis, L., and Ward, D. C. (1988b). Rapid detection of human chromosome 21 aberrations by *in situ* hybridization. *Proc. Natl. Acad. Sci. U.S.A.* **86**, 9664–9668.

Lichter, P., Tang, C. C., Call, K., Hermanson, G., Evans, G. A., Housman, D., and Ward, D. C. (1990a). High-resolution mapping of human chromosome 11 by *in situ* hybridization with cosmid clones. *Science* **247**, 64–69.

Lichter, P., Ledbetter, S. A., Ledbetter, D. H., and Ward, D. C. (1990b). Fluorescence *in situ* hybridization with Alu and L1 polymerase chain reaction probes for rapid characterization of human chromosomes in hybrid cell lines. *Proc. Natl. Acad. Sci. U.S.A.* **87**, 6634–6638.

Lichter, P., Boyle, A. L., Cremer, T., and Ward, D. C. (1991). Analysis of genes and chromosomes by nonisotopic *in situ* hybridization. *GATA* **8**, 24–35.

Lucas, J. N., Tenjin, T., Straume, T., Pinkel, D., Moore II, D., Litt, M., and Gray, J. W. (1989). Rapid human chromosome aberration analysis using fluorescence *in situ* hybridization. *Intl. J. Radiat. Biol.* **56**, 35–44.

Lux, S. E., Tse, W. T., Menninger, J. C., John, K. M., Harris, P., Shalev, O., Chilcote, R. R., Marchesi, S. L., Watkins, P. C., Bennett, V., McIntosh, S., Collins, F. S., Francke, U., Ward, D. C., and Forget, B. G. (1990). Hereditary spherocytosis associated with deletion of human erythrocyte ankyrin gene on chromosome 8. *Nature (London)* **345**, 736–739.

Manuelidis, L. (1985). Individual interphase chromosome domains revealed by *in situ* hybridization. *Hum. Genet.* **71**, 288–293.

Manuelidis, L., and Borden, J. (1988). Reproducible compartmentalization of individual chromosome domains in human CNS cells revealed by *in situ* hybridization and three-dimensional reconstruction. *Chromosoma* **96**, 397–410.

McNeil, J. A., Johnson, C. V., Carter, K. C., Singer, R. H., and Lawrence, J. B. (1991). Localizing DNA and RNA within nuclei and chromosomes by fluorescence *in situ* hybridization. *GATA* **8**, 41–58.

Morimoto, H., Monden, T., Shimano, T., Higashiyama, M., Tomita, N., Murotani, M., Matsuura, N., Okuda, H., and Mori, T. (1987). Use of sulfonated probes for *in situ* detection of amylase mRNA in formalin-fixed paraffin sections of human pancreas and submaxillary gland. *Lab Invest.* **57**, 737–741.

Mujumdar, R. B., Ernst, L. A., Mujumdar, S. R., and Waggoner, A. S. (1989). Cyanine dye labeling reagents containing isothiocyanate groups. *Cytometry* **10**, 11–19.

Nederlof, P. M., van der Flier, S., Wiegant, J., Raap, A. K., Tanke, H. J., Ploem, J. S., and van der Ploeg, M. (1990). Multiple fluorescence *in situ* hybridization. *Cytometry* **11**, 126–131.

Pinkel, D., Straume, T., and Gray, J. W. (1986). Cytogenetic analysis using quantitative, high sensitivity, fluorescence hybridization. *Proc. Natl. Acad. Sci. U.S.A.* **83**, 2934–2938.

Pinkel, D., Landegent, J., Collins, C., Fuscoe, J., Segraves, R., Lucas, J., Gray, J. W. (1988). Fluorescence *in situ* hybridization with human chromosome-specific libraries: Detection of trisomy 21 and translocations of chromosome 4. *Proc. Natl. Acad. Sci. U.S.A.* **85**, 9138–9142.

Sambrook, J., Fritsch, E. F., and Maniatis, T. (1989). "Molecular Cloning: A Laboratory Manual," 2nd Ed., Cold Spring Harbor Laboratory, Cold Spring Harbor, New York.

Schardin, M., Cremer, T., Hager, H. D., and Lang, M. (1985). Specific staining of human chromosomes in Chinese hamster × man hybrid cell lines demonstrates interphase chromosome territories. *Hum. Genet.* **71**, 281–287.

Shroyer, K. R., and Nakane, P. K. (1983). Use of TNP-labeled cDNA for *in situ* hybridization. *J. Cell Biol.* **97**, 377a.

Tkachuk, D. C., Westbrook, C. A., Andreeff, M., Donlon, T. A., Cleary, M. L., Suryanarayan, K., Homge, M., Redner, A., Gray, J., and Pinkel, D. (1990). Detection of bcr-abl fusion in chronic myelogeneous leukemia by *in situ* hybridization. *Science* **250**, 559–562.

Tkachuk, D. C., Pinkel, D., Kuo, W.-L., Weier, H.-U., and Gray, J. W. (1991). Clinical applications of fluorescence *in situ* hybridization. *GATA* **8**, 67–74.

Trask, B. J. (1991a). Fluorescence *in situ* hybridization: applications in cytogenetics and gene mapping. *TIG* **7**, 149–154.

Trask, B. J. (1991b). Gene mapping by fluorescence *in situ* hybridization. *Curr. Opin. Gen. Dev.* **1**, 82–87.

Trask, B., and Hamlin, J. L. (1989). Early dihydrofolate reductase gene amplification events in CHO cells usually occur on the same chromosome arm as the original locus. *Genes Dev.* **3**, 1913–1925.

Trask, B., van den Engh, G., Pinkel, D., Mullikin, J., Waldman, F., van Dekken, H., and Gray, J. W. (1988). Fluorescence *in situ* hybridization to interphase cell nuclei in suspension allows flow cytometric analysis of chromosome content and microscopic analysis of nuclear organization. *Hum. Genet.* **78**, 251–259.

Trask, B., Pinkel, D., and van den Engh, G. (1989a). The proximity of DNA sequences in interphase cell nuclei is correlated to genomic distance and permits ordering of cosmids spanning 250 kilobase pairs. *Genomics* **5**, 710–717.

Trask, B., van den Engh, G., Mayall, B., and Gray, J. W. (1989b). Chromosome heteromorphism quantified by high-resolution bivariate flow karyotyping. *Am. J. Hum. Genet.* **45**, 739–752.

Trask, B., Massa, M., Kenwrick, S., and Gitschier, J. (1991). Mapping of human chromosome Xq28 by two-color fluorescence *in situ* hybridization of DNA sequences to interphase cell nuclei. *Am. J. Hum. Genet.* **48**, 1–15.

Tucker, J. D., Christensen, M. L., and Carrano, A. V. (1988). Simultaneous identification and banding of human chromosome material in somatic cell hybrids. *Cytogenet. Cell Genet.* **48**, 103–106.

Viegas-Pequignot, E., Dutrillaux, B., Magdelenat, H., and Coppey-Moisan, M. (1989). Mapping of single-copy DNA sequences on human chromosomes by *in situ* hybridization with biotinylated probes: Enhancement of detection sensitivity by intensified-fluorescence digital-imagining microscopy. *Proc. Natl. Acad. Sci. U.S.A.* **86**, 582–586.

Wada, M., Little, R. D., Abidi, F. Porta, G. Labella, T. Cooper, T. Della Valle, G., Urso, M., and Schlessinger, D. (1990). Human Xq24-Xq28: Approaches to mapping with yeast artificial chromosomes. *Am. J. Hum. Genet.* **46**, 95–106.

Waggoner, A. S. (1990). Fluorescent probes for cytometry. *In* "Flow Cytometry and Sorting" (M. R. Melamed, T. Lindmo, and M. L. Mendelsohn, eds.), pp. 209–226. Wiley-Liss, New York.

Wetmur, J. G. (1975). Acceleration of DNA renaturation rates. *Biopolymers* **14**, 2517–2524.

Yunis, J. J. (1976). High resolution of human chromosomes. *Science* **191**, 1268–1270.

Zischler, H., Nanda, I. Schafer, R., Schmid, M. , and Epplen, J. T. (1989). Digoxigenated oligonucleotide probes specific for simple repeats in DNA fingerprinting and hybridization *in situ. Hum. Genet.* **82**, 227–233.

Chapter 2

Localization of mRNAs by in Situ Hybridization

LYNNE M. ANGERER AND ROBERT C. ANGERER

Department of Biology
University of Rochester
Rochester, New York 14627

METHODS IN CELL BIOLOGY, VOL. 35

I. Introduction

In situ hybridization allows one to distinguish the cells in a complex tissue that express specific mRNAs and to make semiquantitative estimates of the relative concentration of these in different cell types. Other methods of analysis of RNA isolated from cells, tissues, or organisms, such as RNA blots or solution hybridization, average the content of a specific mRNA over all cell types in the sample. Thus, significant changes in gene expression within the component cell types are often obscured. For example, we have analyzed more than 75 different mRNAs expressed in sea urchin embryos and shown that the spatial patterns of expression of most change during development, and very few are distributed uniformly among different cell types. In other cases, expression of a gene shifts from one differentiating tissue to another at different developmental stages. In extreme cases, the average concentration of a mRNA does not change during development, but it is expressed in two different cell types at two different times (see, for example, actin CyIIa message in Cox *et al.,* 1986).

In situ hybridization offers several other potential advantages over solution or filter hybridization methods: (1) It is possible to analyze many tissues or cell types simultaneously, without purifying RNA from them. For example, we routinely assay, on a single microscope slide, the distribution of an individual mRNA in all regions of sea urchin embryos (which are about the size of mouse blastocysts) at 5–10 different development stages. On a larger scale, Rall *et al.* (1985) used saggital sections of an entire midgestional mouse embryo to analyze EGF receptor mRNA distributions. (2) *In situ* hybridization requires much less material since it is possible to analyze the distributions of a number of different mRNAs using only a few dozen embryos or small cell or tissue samples, such as those obtained from biopsy. Although double-labeling methods are not yet well developed, use of a series of adjacent thin sections allows comparison of distributions of several different mRNAs or of mRNA and protein in the same region and sometimes in the same cells, depending on cell size and section thickness (for example, see Angerer *et al.,* 1985, 1989). (3) Because *in situ* hybridization is done on fixed tissues, it often can be applied retrospectively as, for example, to archival samples of formalin-fixed, paraffin-embedded tissues with known clinical history (see, for example, Stoler *et al.,* 1986; Stoler, 1990). Furthermore, since fixed tissues can be stored for at least several years in paraffin, they can be reassayed with new probes for additional mRNAs as these become available. (4) In cases where a mRNA accumulates only in a small percentage of cells in a sample, *in situ* hybridization may actually be a more sensitive assay than blotting methods. For example, we easily detected, with ^{3}H-labeled probes, muscle-type actin mRNA in 10–15 of the 1500 cells of the sea urchin pluteus larva, whereas this message was barely detectable by blotting methods in RNA isolated from whole embryos (Cox *et al.*, 1986).

The aim and many of the advantages of *in situ* hybridization are shared with immunocytochemical detection of the encoded proteins. The basic difference is that they monitor gene expression at different levels. Measuring mRNA concentrations provides a more accurate estimate of gene activity because mRNA accumulates before the protein it encodes, and because relatively stable proteins may persist long after the gene is repressed and the mRNA has decayed. Probes for introns could provide even more direct assays of transcription because intron sequences have half-lives of only 10–20 minutes compared to several hours to several days for most mRNAs. To date, because the steady-state concentration of introns is low, this kind of analysis has been restricted to a few genes that are transcribed at very high rate (Fremeau *et al.*, 1986; Berman *et al.*, 1990), and transcripts appear to be less accessible to hybridization probes (L. M. Angerer and R. C. Angerer, unpublished observations). On the other hand, if knowing the actual concentration of a specific protein is more important than monitoring the activity of its gene, then an immunocytochemical approach should be used. It is worth noting that proteins are not always present at highest concentration in the cells that synthesize them as, for example, in the cases of secreted peptides, such as neurotransmitters, growth factors, and other signaling molecules.

There are also several practical differences between these two methods for localizing gene activity. First, protein antigenicity varies after fixation with different fixatives, and detection of different antigens may therefore require different tissue preparation methods. In contrast, the efficiency of hybridization to different mRNAs is similar for a given fixation condition so that it is possible to establish a single tissue preparation method for a given tissue. Second, cloned nucleic acid probes often are available before the corresponding antibodies, are essentially immortal, and are available in unlimited supply. Third, many interesting genes are members of small families that may encode almost identical proteins, for which it is difficult to raise specific antibodies. In contrast, it is often possible to obtain mRNA-specific probes from untranslated (non-protein-coding) regions of the mRNA at the 5' or 3' ends of the message because these usually diverge rapidly in sequence (for example, see separate detection of four different genes encoding cytoskeletal-type actins in Cox *et al.*, 1986). Fourth, although the sensitivity (molecules per cell) of immunochemical detection of proteins is not well characterized, as few as 5–10 copies of a mRNA per cell have been detected in well-optimized *in situ* hybridizations methods (see Hardin *et al.*, 1988). When a gene is activated, the mRNA can usually be detected before the protein it encodes.

The best *in situ* hybridization protocol would (1) preserve good histological detail; (2) produce quantifiable signals that accurately reflect target mRNA concentrations in different regions of the tissue; (3) produce reproducible signals; (4) be quick, inexpensive, and use a minimum of specialized equipment; (5) avoid hazardous substances; (6) have sufficient resolution so that, depending on the particular question, different tissues, different cell types, nucleus versus cytoplasm, or even subregions of cellular compartments can be distinguished (for examples, see Lawrence and Singer, 1986; Nash *et al.,* 1987; Yang *et al.,* 1989); (7) have high sensitivity.

Of these, sensitivity is most important. The abundance of different individual mRNAs in vertebrate tissues ranges from tens to tens of thousands of molecules per cell. Messenger RNAs encoding proteins that characterize terminally differentiated cellular phenotypes are generally relatively abundant, but those encoding many interesting proteins, such as regulatory molecules, are often much rarer. Clearly, the utility of the technique depends on the number of different kinds of mRNA that can be detected.

The sensitivity of *in situ* hybridization is a direct function of the signal-to-noise ratio that can be achieved. Signals are determined by the fraction of cellular mRNA retained in tissue, accessible to, and hybridized with probe, and the yield of signal (autoradiographic grains or colorimetric deposit) obtained for a given amount of hybridized probe. Noise results from binding of probe to unidentified cellular components and usually limits sensitivity.

Here we describe our current method, using it as a framework to discuss the role of each step of *in situ* hybridization with respect to the criteria outlined above, and considering options and potential pitfalls. This protocol was originally published by Cox *et al.* (1984) and has been updated by Angerer *et al.* (1987a,b; Angerer and Angerer, 1989). This technique employs hybridization of radiolabeled riboprobes to sections of aldehyde-fixed, paraffin-embedded tissue. It, or modifications, has been used successfully for localizing mRNAs in tissues of sea urchin embryos (Lynn *et al.,* 1983), *Drosophila* (Ingham *et al.,* 1985), plants (Cox and Goldberg, 1988), humans (Stoler *et al.,* 1986; Stoler, 1990), and mammalian embryos (Wilkinson *et al.,* 1987; Nomura *et al.,* 1988). Some major alternatives at different steps (fixation, protease pretreatment, probe type) are discussed below and represented in the methods of Lawrence and Singer (1985).

II. Tissue Preparation

A. Rationale

The first and most difficult aspect of developing an *in situ* hybridization procedure for a new system is identifying a good tissue preparation method.

Given appropriately prepared tissue, hybridization and posthybridization wash steps generally follow the same rules as for solution or blot hybridizations. Unlike blotting membranes, tissues are not chemically uniform and different ones respond differently to various fixation protocols. Procedures that preserve good morphology and retain RNA are usually not the same as those that afford high hybridization efficiency. Extensive fixation preserves histological detail and prevents loss of target RNA during hybridization and wash procedures. However, milder fixation allows increased accessibility of target RNA to hybridization probes (and to any other reporter molecules, such as enzymes and antibodies) and reduces the possibility of chemical modification of target RNA and concomitant loss of hybridizability. *Thus, for any new system, it is important to optimize fixation and prehybridization treatments of tissue to find the best compromise between these requirements.* It is crucial to understand that the steps in tissue preparation are interactive and interdependent. Some individual steps are beneficial in the context of one protocol, but lethal when used in a different protocol.

Two general classes of fixative are cross-linking agents, such as glutaraldehyde and paraformaldehyde, and precipitating fixatives, such as ethanol : acetic acid. Cross-linking fixatives are favored because they provide much better retention of cellular RNA (Angerer and Angerer, 1981; Lawrence and Singer, 1985); they probably also help to inactivate nucleases in tissues. Glutaraldehyde fixation leads to more extensive cross-linking and better preservation of morphology. Formaldehyde fixation is more frequently used, especially in clinical settings, is cheaper, and, in some cases, allows immunocytochemistry and *in situ* hybridization to be carried out on the same sections, although not necessarily under optimal conditions for both assays.

To improve probe penetration, tissue fixed with aldehydes is usually treated with low levels of protease (Brahic and Haase, 1978; Angerer and Angerer, 1981; Gee and Roberts, 1983; Brigati *et al.,* 1983). Depending on the tissue and fixation procedure used, treatment with proteinase K or pronase has been reported to increase, decrease, or have little effect on the magnitude of signals. Our early work showed that protease treatment of tissue fixed only with a precipitating fixative (ethanol : acetic acid) resulted in essentially complete loss from sections of RNA that had been labeled *in vivo* (Table I; Angerer and Angerer, 1981). Similar losses of signal and of histological detail were observed for tissue fixed for 30 minutes in 4% paraformaldehyde and treated with proteinase K (Fig. 1D), while omitting the protease digestion resulted in good signals and acceptable morphology (Fig. 1C). In contrast, proteinase K treatment of the same tissue fixed with glutaraldehyde increases signals at least several-fold (Fig. 1B) and as much as 10-fold to some target RNA (Table I; Angerer and Angerer, 1981). Such observations led to the generalization that the "tighter" fixation of glutaraldehyde requires opening up the tissue with a protease, whereas protease treatment of tissue fixed with

TABLE I

Effect of Fixative on RNA Retention and Signals[a]

	Ethanol/acetic acid	Glutaraldehyde
Relative RNA retention	1	5.5–9
Relative signals *in situ*		
– Proteinase K	1	1.4
+ Proteinase K	0	12.5

[a] Data from Angerer and Angerer (1981).

precipitating fixatives usually has a deleterious effect because it increases loss of RNA. Intermediate effects are observed with formaldehyde fixation, and vary with the exact fixation protocol and tissue (e.g., Lawrence and Singer, 1985).

B.   Fixation with Formaldehyde

Formaldehyde is the fixative used most commonly and has been applied in a wide variety of systems. It is especially convenient for human tissues obtained in a clinical setting. The following protocol has been used by Dr. Mark Stoler (Cleveland Clinic Foundation) for a variety of human tissues, in conjunction with an automated vacuum infiltration tissue processor (Stoler, 1990).

Protocol

Fix tissues for 1–8 hr in 10% neutral buffered formalin (Fisher Scientific; 4% formaldehyde in sodium phosphate buffer at pH 7). The processor then uses the following stages: formalin, 1.5 hours, 2 times; 60% ethanol, 30 minutes; 95% ethanol, 45 minutes, 2 times; 100% ethanol, 50 minutes; 100% ethanol, 30 minutes, 2 times; xylene, 60 minutes; xylene, 30 minutes; paraffin, 30 minutes, 4 times. All stages except the paraffin can be run at room temperature.

C.   Fixation with Glutaraldehyde

Direct comparison of glutaraldehyde and formaldehyde fixation in combination with various levels of proteinase K treatment, ranging from 0 to 3 μg/ml (37°C, 30 minutes) showed that glutaraldehyde provided comparable signals but significantly better morphology for sea urchin embryos (Fig. 1).

Fig. 1.   Sea urchin gastrulae were fixed either with 1% glutaraldehyde (A and B) or with 4% paraformaldehyde (C and D), embedded in paraffin, and sectioned at 5 μm thickness. Sections were incubated with proteinase K (1 μg/ml) (B and D) or with buffer only (A and C). Sections were then hybridized with a ^{3}H-labeled riboprobe (1 × 10^8 dpm/μg) complementary to an abundant mRNA (500 copies/expressing cell). Exposure time was 10 days. Bar in A, 10 μm.

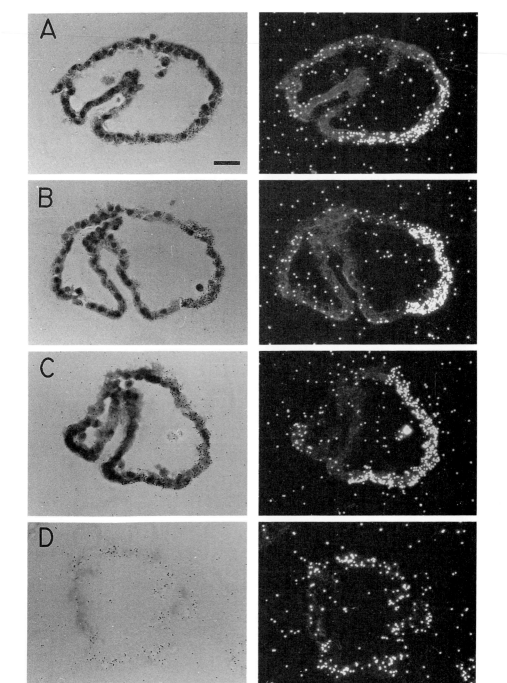

Therefore, we include this fixation protocol as a model, since it may also be superior in some other systems.

Protocol

Suspend cells or small tissue pieces in ≥ 5 volumes of 1% glutaraldehyde, buffered with 50 mM sodium phosphate, pH 7.5, at 4°C. Glutaraldehyde (Sigma; 25%) is purchased in 1-ml ampoules, used once, and discarded. The buffer should also include an appropriate concentration of NaCl to control the osmolality, in order to avoid excessive shrinking or swelling. The osmolality of 1% glutaraldehyde is equivalent to 0.375% NaCl (Millonig and Marinozzi, 1968). For example, for sea urchin embryos, addition of NaCl to 2.5% (w/v) and phosphate buffer (50 mM; about 80 mM Na$^+$) produces a fixative with the osmolality of sea water.

Change fixative once during a total fixation time of 1 hour. Optimal fixation time and concentration will depend on the tissue and size of the blocks. The times given are for suspensions of embryos of 100–200 μm in diameter. Larger pieces of tissue will require longer fixation, while samples such as single-cell layers of tissue culture cells require shorter fixation times and/or lower concentrations. For example, when we probed single cells spread on mircoscope slides, we reduced the glutaraldehyde concentration from 1 to 0.3% (Hurley *et al.*, 1989). More extensive fixation may increase the requirement for protease treatment (see below) but is preferable to underfixation and consequent loss of target RNA from sections.

Wash twice in NaCl-phosphate fixation buffer at 0°C for 30 minutes each. Wash once in isotonic NaCl to remove phosphate, which precipitates during ethanol dehydration steps.

D. Dehydration, Embedding, and Sectioning

1. Dehydration

Some dehydration protocols may damage osmotically sensitive tissue; for example, inclusion of a 30% ethanol step caused our embryos to explode. When in doubt, use higher concentrations for the first ethanol step even though they may cause some shrinkage.

Protocol

Dehydrate tissue through a graded series of ethanol concentrations, for 15 minutes each at room temperature: 50% twice, 70% twice, 85%, 95%, and 100% twice. (100% is 99% ethanol stored over molecular sieve.)

Dehydration may be interrupted at 70% ethanol and tissue can be stored in tightly capped tubes at 4°C for months.

2. EMBEDDING

We use paraffin embedding because it is easy, cheap, and gives good quality sections which are usually superior to frozen sections (for example, Ingham et al., 1985). Frozen sections that are postfixed have been used for embryos surrounded by impermeable membranes (Hafen et al., 1983) as well as for vertebrate tissues (Gee and Roberts, 1983; Rall et al., 1985; Davis et al., 1988). If it is necessary to work with large blocks of tissue, postfixation of frozen sections provides more uniform fixation. Other embedding media which have been used successfully, albeit infrequently, include methacrylate (Jamrich et al., 1984), polystyrene (Harkey et al., 1988), and diethylstearate (Howe and Steitz, 1986).

Protocol

Embed tissue in paraffin, using xylene as the intermediate solvent: Pass the tissue through two changes of xylene at room temperature for 20 minutes each; then two changes of 1 xylene : 1 paraffin [1 : 1 Tissue Tek : Paraplast II (mp 55–58°C)] for 20 minutes each at 60°C. Finally, infiltrate with three 15-minute changes of paraffin. Melt paraffin only a day or two before use because it deteriorates when stored melted for long periods. Place pieces of infiltrated tissue or droplets of concentrated samples in molds and, after the paraffin has begun to solify but is still soft, slowly fill the mold.

To locate small pieces of tissue in these blocks, it is helpful to stain them with a water-soluble stain, such as eosin, before embedding.

To avoid loss of very small numbers of small pieces of tissue, transfer them to boxes of 1% agarose before dehydration. This is conveniently done using wells of horizontal agarose gels sealed with agarose (Hough-Evans et al., 1987).

Paraffin blocks are easy to store, and tissue can be kept at 4°C for at least several years without noticeable deterioration in structure or in hybridization properties.

If possible, carry out the embedding steps, particularly the infiltration with paraffin, under reduced pressure, which is conveniently done in a vacuum oven; this reduces the times required by approximately a factor of two. If a vacuum oven is not available, at least degas the liquid paraffin under vacuum to prevent bubbles.

3. SLIDE PREPARATION

In our experience the simplest, cheapest, and best tissue adhesive is 3-aminopropyltriethoxysilane (TESPA; Aldrich). The hydrophobic silane moiety of TESPA binds to glass, while its amino group, activated by aldehyde, is crosslinked to tissue. In practice, we have found that the activation step usually is not required for tissue fixed with aldehyde. It is important to keep slides clean in order to avoid contamination by both RNase and by fine pieces of grit that interfere with spreading of the hybridization solution under coverslips. Handle slides by one end using forceps or gloved hands.

Protocol

To clean slides, place them in glass holders in staining dishes and immerse in Chromerge at least overnight. Rinse for at least 15 minutes each in running tap water and in running distilled (deionized) water. Let dry in a dust-free place.

To coat the cleaned sides, dip them for 5–10 seconds in 2% (v/v) TESPA in dry acetone. Rinse twice in acetone and once in distilled water. Dry in a 50°C oven. Slides can be used immediately or stored for at least 3 months.

An alternative adhesive that is better for some tissues is polylysine: Dip slides for 10 minutes in 50 μg/ml polylysine, 10 mM Tris, pH 8.0, and air-dry.

Coating with TESPA or polylysine leaves spots on the slides; this does not matter.

4. SECTIONING

Depending on the application, sections can be cut with steel knives on a standard rotary microtome at 5 μm nominal thickness or as thin as 1 μm using glass knives. For some applications as, for example, to compare the distributions of several different mRNAs within the same cells, adjacent 1-μm sections can be used (Angerer, *et al.*, 1985, 1989). These also provide better resolution of nucleus versus cytoplasm and minimize the quenching of ^{3}H decay by cytoplasm overlying nuclei. Sections cut with glass knives are of higher quality than those cut with steel knives.

Protocol

For most applications 5-μm sections are transferred on a moistened spatula tip to the surface of clean degassed 45°C water in a histological water bath. In a minute or less, the sections should expand to their original dimensions. Pick them up by inserting a TESPA-coated slide held under them at a 45°C angle, touching it to the edge of the ribbon, and then withdrawing it from the water. Alternatively, sections can be transferred directly to droplets of water on slides heated to 40°C on a slide warmer.

To cut sections with glass knives, prepare a reservoir or "boat" by wrapping a piece of tape around the knife, sealing it with paraffin, and filling it with water. Pick up sections floating on the water's surface with a silanized glass rod and transfer to droplets of water on slides, held on a 40°C warming plate.

After sections are mounted on slides, allow them to dry on a 40°C slide warmer for at least 2 hours or overnight.

E. Prehybridization Tissue Treatments

Both pronase and proteinase K have been used to partially remove proteins in order to increase target RNA accessibility to hybridization probes. We prefer proteinase K because it can be obtained as a pure enzyme with relatively constant specific activity, and, unlike pronase, is free of nucleases so that it does not have to be self-digested. For our tissue fixed in 1% glutaraldehyde, a proteinase K concentration of 1 μg/ml for 30 minutes at 37°C offers the best compromise between increasing signal and causing loss of histological detail. Similar conditions have been employed in combination with standard formalin fixation for vertebrate tissues, with optimal results usually at 1–10 μg/ml. Smears of cells usually require 0 to 1 μg/ml. In the special case of whole ciliated protoza surrounded by a thick pellicle (Yu and Gorovsky, 1986), signals continue to increase at proteinase K concentrations as high as 30 μg/ml.

Acetic anhydride was originally employed in hybridizations with ^{125}I-labeled probes to acetylate amino groups, reduce net positive charge on the sections, and thus decrease electrostatic binding of negatively charged probes (Hayashi et al., 1978). Although we found this treatment has a small beneficial effect on backgrounds, other workers using different fixation and protease protocols have reported a negligible effect (Lawrence and Singer, 1985). We continue to use acetic anhydride because it is easy, it cannot hurt, and it probably helps terminate proteinase K activity.

1. PROTEASE TREATMENT

Protocol
Remove the paraffin from sections by immersing the slides held in glass slide carriers in a staining dish containing fresh xylene (2 × 10 min) that is stirred with a magnetic stir bar. Paraffin must be completely removed, since residual fragments trap probe, leading to nonspecific signals.

Rehydrate the tissue by passing it through a graded series of ethanol concentrations (99, 99, 95, 85, 70, 50, 30%) and twice through distilled water, for 10–30 seconds each.

Place the slides in 100 mM Tris-HCl, 50 mM EDTA, pH 8.0, prewarmed at 37°C. Add proteinase K to 1 μg/ml. Mix thoroughly and incubate for 30 minutes. (Proteinase K is stored frozen in small aliquots as a 10 mg/ml stock; aliquots are discarded after one use.)

2. Acetylation

Protocol
Wash the slides briefly in distilled water and then in freshly prepared 0.1 M triethanolamine-HCl, pH 8.0.

To a clean, dry, staining dish, add sufficient acetic anhydride (supplied as a 100% liquid) to bring the final concentration to 0.25% (v/v). Add the slides in a glass carrier to the dish, and immediately add the triethanolamine buffer and mix by dipping the carrier up and down.

After 10 minutes, wash the slides briefly in 2 × SSC.

3. Dehydration

Protocol
Dehydrate the tissue through the same set of graded ethanols.

Place the slides, kept vertically in a test tube rack on a paper towel, to dry in a dust-free place.

Normally, we set up the hybridization on the same day, but slides may be stored for at least several days.

4. Other Pretreatments

Many protocols that use tissue fixed in formaldehyde, especially in combination with protease treatment, employ a second fixation step after the protease treatment (for example, see Brigati *et al.,* 1983). This is usually unnecessary with glutaraldehyde because of the greater cross-linking.

Some protocols for unsectioned material (whole mounts or cells) employ a wash in Triton X-100 to increase tissue permeability.

Some workers prehybridize sections in the hybridization cocktail in an effort to block nonspecific binding sites for probe. In our experience, this does not reduce backgrounds, and accurate determination of probe concentration requires that the slides subsequently be washed and dried.

With high-specific-activity ^{35}S-labeled probes, which potentially offer signals 50-fold higher than can be obtained with ^{3}H-labeled probes (see below), it is the background that usually limits sensitivity. Some methods suggest pretreatments with N-ethylmaleimide or iodoacetamide to block sulfhydryl groups suspected of reacting with thiolated polynucleotides (see "Current Protocols in Molecular Biology," Ausubel *et al.,* 1991), but we have not found these to have any reproducible effect. (See further discussion in Section IV,B).

Some studies have used pretreatments of tissue with DNase or RNase to establish that signals depend on the presence of DNA or RNA in the sections.

These are not as convincing or as convenient as other controls. Furthermore, depending on whether the probe is RNA or DNA, additional controls are required to establish that residual nuclease does not destroy the probe. A more direct way to determine whether the target is DNA or RNA is to compare sense probes which monitor hybridization to DNA to antisense probes which detect both DNA and RNA.

III.   Probes

A.   Comparison of Probe Types

The three major choices for probes are denatured, double-stranded DNAs (i.e., both complementary strands present); single-stranded, or "anti-sense" RNAs, which contain only the nucleic acid strand complementary to mRNA; and synthetic DNA oligonucleotides. A fourth choice, single-stranded DNA, is relatively difficult to produce and will not be discussed in detail here. The choice among these depends on (1) the abundance of the target mRNA, (2) whether qualitative or quantitative estimates of signals are desired, and (3) convenience.

The major difference between double-stranded and single-stranded probes is that the former self-reassociate during *in situ* hybridization, whereas antisense probes can only hybridize with the complementary mRNA in tissue. DNA probes are usually labeled by enzymatic reactions (nick-translation or random priming) which generate molecules with ends at random points in the sequence. When two such molecules hybridize, they will base-pair over only two-thirds of their length on average, leaving the remaining single-stranded portions to hybridize with an additional probe strand, or with the target RNA. Although, in theory, this "probe hyperpolymerization" offers the possibility of amplifying signals, in practice, this possibility is often not achieved because large hyperpolymers cannot easily penetrate the tissue to the target RNA. With tissue more highly cross-linked with glutaraldehyde, we found that double-stranded RNA probe self-reassociation in solution aborted the hybridization *in situ*, yielding 8-fold lower signals than the antisense RNA probes regardless of probe concentration and hybridization time (Cox *et al.*, 1984). Using tissue fixed with formaldehyde, which is less cross-linked, Lawrence and Singer (1985) demonstrated signal amplification by probe hyperpolymerization in some experiments. However, the degree of amplification was quite variable, presumably because the extent of hyperpolymer formation depends on fragment length, probe concentration, and hybridization time, and also is very sensitive to traces of DNase activity. Since antisense RNA probes make 1:1 duplexes with target RNA, they provide more consistent signals and are more reliable when quantitation is required.

Preparation of RNA probes has been described in detail by Melton *et al.,* (1984). To work with RNA probes, take standard precautions to avoid RNase acitivity. Stock solutions and distilled water are treated with diethylpyrocarbonate and autoclaved, where possible, and gloves should be used to handle slides and coverslips.

Methods for labeling DNA probes by nick-translation or random-primed DNA synthesis can be found in standard molecular techniques manuals [for example, "Molecular Cloning: A Laboratory Manual," Sambrook *et al.,* 1989, or "Current Protocols in Molecular Biology," Ausubel *et al.,* 1991].

DNA oligonucleotides are generally <50 nucleotides (nt) and penetrate tissue easily. Since olignucleotides are synthesized from unlabeled precursors, they are usually radioactively labeled by adding homopolymer tails of about 20 nt (usually dCTP) with terminal deoxynucleotidyltransferase (see standard cloning manuals). Note that the signal is proportional to the amount of hybridized probe, and hence to the sequence length (complexity) of the probe that is used. Thus, considerably more radioactivity can be hybridized to a target mRNA using a 1 to 2-kb RNA probe, compared to a single 50-nt oligonucleotide.

Recently, the polymerase chain reaction (PCR) has been used to generate DNA probes without cloning of sequences. Note that symmetric PCR generates equal quantities of the two complementary strands, which match perfectly on first collision and will be rapidly removed from participation in the *in situ* reaction. Therefore, probes should be generated by asymmetric PCR, using an excess of the oligonucleotide that primes synthesis of the antisense strand.

We have observed that probes for different mRNAs give signals approximately in proportion to their relative abundance as measured by solution hybridization methods. Signals were not detectably different for early histone mRNAs in early two-cell zygotes that are not loaded on polysomes and several hours later when these messages are almost quantitatively loaded (L. Angerer and R. Angerer, unpublished observations). On the other hand, using intron probes for nuclear targets, we observe signals that are significantly lower than predicted from their known concentration. Intron signals have only been reported for relatively abundant transcripts (e.g., Akam and Martinez-Arias, 1985; Fremeau *et al.,* 1986; Berman *et al.,* 1990). Furthermore, probes for different regions of an individual mRNA may exhibit considerably different hybridization efficiencies (Taneja and Singer, 1987). For several different mRNAs, we also have observed that probes complementary to protein coding sequences give higher signals that do those that hybridize to 3' untranslated regions (L. M. Angerer and R. C. Angerer, unpublished observations). This undoubtedly reflects decreased accessibility of target sequence due to secondary structure and/or RNA–protein interactions. An extreme case is U1 snRNP, which gives signals only about 1% of those for an equivalent amount of mRNA (Nash *et al.*, 1987).

B.   Adjusting Probe Fragment Length

Long probe fragments penetrate less effectively into fixed tissue. The magnitude of this effect depends on the tissue preparation and type of probe used. For antisense RNA probes, the effect of fragment length on probe penetration *per se* can be measured and we observed signals several-fold lower for a probe ~600 nt long than for one of ~200 nt (K. Cox, unpublished observations). From the above discussion, the effect of probe fragment length for tissue fixed with formaldehyde is probably less. For example, using double-stranded DNA probes, Lawrence and Singer (1985) reported relatively little effect of fragment length perhaps, in part, because poorer penetration by longer fragments is offset by their increased ability to form hyperpolymers. Brahic and Haase (1978) using the same probe type and fixation reported increased signals with shorter probes.

Cloned sequences transcribed *in vitro* by phage polymerases usually are a few hundred to a few thousand nucleotides long. The length of transcripts synthesized from a new template should be determined to be sure they are not prematurely terminated. We then routinely adjust the fragment length of RNA probes to an average of 150–200 nt, using controlled alkaline hydrolysis (see below). Fragments shorter than this do not improve sensitivity and the higher thermal stability of longer hybrids allows washes at higher temperatures which leads to improved signal-to-noise ratios. Standard methods for labeling DNA cited above routinely produce fragments several hundred nucleotides long which, in most cases, are suitable for *in situ* hybridization.

Protocol

After removal of template DNA by DNase digestion (50 μg/ml, 37°C, 30 minutes, 50 mM Tris-HCl, pH 7.4, 10 mM MgCl$_2$), extract RNA probes transcribed *in vitro* with phenol–chloroform and chloroform, and precipitate with ethanol.

Redissolve the probe in 50 μl of distilled water and adjust the pH to 10.2 by addition of 30 μl of 0.2 M Na$_2$CO$_3$ and 20 μl of 0.2 M NaHCO$_3$. (These stocks are frozen in small aliquots, tightly capped.)

Incubate at 60°C for the amount of time, in minutes, given by

$$t = \frac{\text{Starting length} - \text{Desired length}}{(0.11)(\text{Starting length})(\text{Desired length})}$$

where lengths are in kilobases.

Stop the hydrolysis by addition of 3 μl of $3M$ sodium acetate, pH 6.0 and 5 μl of 10% glacial acetic acid.

Add 10 μg of yeast tRNA carrier and precipitate the hydrolyzed probe with ethanol.

Resuspend it in a small volume of 10 mM Tris-HCl, 1 mM EDTA, pH 8.0. For probes labeled with ^{35}S, include 20 mM dithiothreitol.

Determine the final probe concentration by acid precipitation of a measured aliquot; at this stage, it should be $\geq 90\%$ acid precipitable.

C.   Hybridization Conditions

1.   HYBRIDIZATION COCKTAIL

The components of the cocktail used for *in situ* hybridization are the same as for RNA or DNA blot analysis. Reactions are carried out in high salt to increase the hybridization rate and reduce electrostatic binding of probe to sections. Formamide is added to reduce the T_m, and hence the temperature for optimum hybridization rate, which is $T_m - 25°C$; this reduces damage to sections and loss of tissue from slides. Dextran sulfate is added to increase the hybridization rate *in situ* and results in higher signals than when polyethylene glycol is used (Angerer *et al.,* 1987a). Different methods employ various combinations and concentrations of polymers designed to compete nonspecific binding of probe including Denhardt's solution (0.02% each polyvinylsulfate, polyvinylpyrollidone, and bovine serum albumin) and carrier RNA or DNA. For DNA probes, addition of 0.5 mg/ml yeast tRNA is sufficient; no further reduction in backgrounds occurs with additional carriers. Use RNA or DNA carriers for RNA or DNA probes, respectively. Add 100 mM DTT for ^{35}S-labeled probes.

2.   PROBE CONCENTRATION AND TIME

In situ hybridizations are done with a large sequence excess of probe over cellular target RNAs. In solution hybridization, such reactions are pseudo-first-order with respect to probe concentration, and the extent of reaction depends on the product of probe concentration and time (C_0t or R_0t). These two parameters are interchangeable since the same extent of reaction is achieved at all combinations of C_0 and t that yield the same product value. This reciprocity is not always observed for *in situ* hybridization, nor is it possible to predict rates of hybridization *in situ* from those observed in solution. Therefore, choices of probe concentration and hybridization time rest entirely on empirical data, which have been systematically derived in only a few cases (for example, Brahic and Haase, 1978; Cox et al., 1984; Lawrence and Singer, 1985; Yu and Gorovsky, 1986). It is likely that the combination of probe concentration and time required to achieve maximum signals will vary depending on tissue type and tissue preparation, and optimal conditions must be determined empirically to achieve maximum sensitivity.

The kinetics of in situ hybridization differ for double-stranded and single-stranded probes. For RNA probes we found that probe concentration and hybridization time were not interchangeable variables. For unknown reasons, the in situ hybridization terminates after 5–6 hours even though both unhybridized probe and hybridizable target remain; possible causes of this are discussed in Cox et al., 1984. Using a convenient overnight hybridization time, signals increase as a linear function of probe concentration until saturation of target RNAs is reached at approximately 0.3 μg/ml per kilobase of probe sequence complexity. Since nonspecific background depends linearly on probe concentration, signal/noise is approximately constant up to the saturating probe concentration and then decreases, because nonspecific binding continues to increase, while the signal does not. Therefore, for maximum sensitivity, use the concentration of probe which is just sufficient to saturate target RNAs.

Many workers have used concentrations of probe far below those required to saturate target RNAs when the targets are very abundant or when probes of very high specific activity are used (see discussion of ^{35}S-labeled probes below, in Section V,A). Since we do not understand what controls the rate of in situ hybridization, there is no theoretical basis for assuming that relative signals over different regions will accurately reflect relative target concentrations when saturation of targets has not been achieved. In fact, to our knowledge, in only one case has it been demonstrated that relative signals for two RNAs present at different concentration increase proportionally as a function of probe concentration and saturate at the same molar probe concentration (Cox et al., 1986). While this possible source of error is not of concern for general identification of predominant cell types containing a given mRNA, it could affect quantitative comparisons of signals that differ only several-fold. Therefore, hybridizations should be carried out at a concentration of probe and for a time sufficient to saturate all available target RNA, using higher concentrations of probe of lower specific activity. To ensure that there is no effect of differences in hybridization rate at subsaturating probe concentrations, measure relative signals over different tissue regions after hybridization at two different probe concentrations (Cox et al., 1984).

For double-stranded RNA or DNA probes, probe concentration and hybridization time appear to be interchangeable variables, i.e., the reaction appears to be "C_0t dependent." Using double-stranded RNAs, Cox et al. (1984)

demonstrated this phenomenon over a 10-fold range of concentration and time, but found that the reaction terminated when only 10–15% of the available target RNA was hybridized. This suggests that large hyperpolymers cannot penetrate to in situ targets, so that the "kinetics" of in situ hybridization with double-stranded probes actually measure the rate at which self-hybridization in solution removes probe from the in situ reaction. In cases where probe hyperpolymers can penetrate the tissue, their formation and hybridization to in situ targets will be favored by high probe concentrations and/or long hybridization times. Given the very different reactions involved, it is probably coincidental that the time course of hybridization observed with DNA probes by Lawrence and Singer (1985) was similar to that observed for single-stranded RNA probes by Cox et al. (1984).

Data on the time and concentration dependence of hybridization using single-stranded *DNA oligonucleotides* suggest that these probes behave like asymmetric RNAs (Taneja and Singer, 1987). For example, concentrations of probe required to saturate targets are in vast excess over cellular RNA, suggesting that much of the probe does not participate in the in situ reaction.

3.   TEMPERATURE

The hybridization temperature is selected to achieve maximum rate of hybridization and to suppress hybridization to sequences related, but not homologous to the probe. In all hybridization reactions the maximum rate of hybridization is observed at approximately 25°C below the melting temperature (T_m) of the duplexes formed. The approximate value of T_m is given by

$$T_m = 79.8 - D + 0.584(\% \ GC) + 18 \log[Na^+] - F(\% \ formamide) - L$$

D corrects for difference in the thermal stabilities of DNA–DNA, DNA–RNA, and RNA–RNA duplexes. Its value is approximately 10, 5, and 0°C, respectively (see discussion in Cox et al., 1984). % GC is the base composition of the probe sequence that hybridizes to target RNA (i.e., excluding any vector sequence), as percentage G + C base pairs. Na^+ is monovalent cation molarity. F is a constant describing the reduction in T_m provided by formamide, and has a value of 0.65, 0.5, and 0.35 for DNA–DNA, DNA–RNA, and RNA–RNA duplexes, respectively (discussed in Cox et al., 1984). L is a correction for the effect of duplex length on T_m. The thermal stability of duplexes decreases as an inverse function of their length. For probes ≥ 100 nt, L varies as a function of $[Na^+]$ and is given by B/l, where $B = 300 + 2000$ $[Na^+]$ (valid for $[Na^+]$ between 0.05 and 0.5M) and l is the length of duplexes formed, in nucleotides (Britten *et al.,* 1974). While the value of L is only a few degrees for longer probes, it rapidly becomes larger for probes shorter than 100 nt; therefore, check the probe length by denaturing gel electrophoresis before hybridization.

Cox *et al.* (1984) found that the T_m of duplexes formed *in situ* is 5°C lower than that of the same duplexes formed in solution and suggested that this difference results from a relatively short duplex length (about 50 nt) and/or steric constraints on duplex formation to target RNAs cross-linked in tissue by glutaraldehyde fixation.

Under the conditions of buffer, formamide, and salt concentration, probe type (RNA) and fragment length specified here, and probes of average (about 40%) G + C content, T_ms of duplexes formed *in situ* are approximately 70°C and optimum hybridization rate is observed at approximately 45°C. Note that hybridization rate varies only slightly over a 5 to 7°C range (Britten *et al.,* 1974), so that very precise estimates of T_m are not required to obtain good signals. When in doubt, choose higher temperatures, which helps to prevent nonspecific binding of probe during hybridization.

Appropriate temperatures for DNA oligonucleotide probes are considerably lower than for longer cloned DNA or RNA sequences, and the range of temperatures that provide good hybridization rate and good specificity is much narrower. DNA oligonucleotide probes may vary considerably in G + C content, and thus in T_m. Equations for predicting T_ms of different oligonucleotides are approximations in standard salts, such as SSC (SSC is 0.15 M NaCl, 0.015 M trisodium citrate). Because oligonucleotide hybridizations and/or posthybridization washes are often carried out at high stringency, i.e., within about 5°C of T_m to ensure specificity, it is essential to make some preliminary estimates of the T_m of hybrids formed with any individual oligonucleotide sequence. In contrast, in 3 M tetramethylammonium chloride, T_m is independent of base composition so that precise calculations can be made. For oligonucleotides between 16 and 32 nt, $T_m = 97 - 682(L^{-1})$ in this solution. However, to our knowledge, no *in situ* hybridizations have been reported in this solvent. For further discussion of hybridization using oligonucleotides, see "Current Protocols in Molecular Biology," Section 6.4 by Ausubel *et al.* (1991). Recently, a computer program, distributed by National Biosciences, has been developed that predicts oligonucleotide thermal stabilities quite accurately, taking into account their thermodynamic properties based not only on nucleotide composition but sequence and mismatch as well (Rychlik and Rhoads, 1989).

Protocol

Determine the total area in cm^2 to be covered by each different probe; multiply this by 4.5 to get the total volume, in microliters, of probe solution required.

Prepare the hybridization cocktail by combining the following components.

Component	Fraction final volume	Stock	Final concentration
Formamide	0.5	100%	50%
Buffer	0.1	$10\times$	$1\times$
Dextran sulfate	0.2	50%	10%
Denhardt's	0.01	$100\times$	$1\times$
($2\,M$ DTT)	0.05	$2M$	$0.1M$

The $10\times$ buffer is 3.0 M NaCl, 0.2 M Tris-HCl, 50 mM EDTA, pH 8.0, for probes labeled with ^{3}H. For probes labeled with ^{35}S, 200 mM sodium acetate, pH 5.0, is substituted for Tris-HCl (Bandtlow et al., 1987).

Since dextran sulfate is viscous, warm it at 50°C before pipetting. For accurate measuring, use silanized calibrated microcapillaries.

$1\times$ Denhardt's is 0.02% each of bovine serum albumin, Ficoll, and polyvinylpyrrolidone.

DTT is included only for probes labeled with ^{35}S.

Combine 0.05 final volume of 10 mg of yeast tRNA/ml, the required amount of probe, and sufficient distilled water to complete the final volume.

Heat this probe at 80°C for 2 minutes; cool on ice water.

Add a measured volume of a mix of the other components. Mix thoroughly on a vortex mixer and centrifuge briefly to collect the volume and eliminate air bubbles.

Count 1 μl of each probe mix in an aqueous scintillation cocktail to be sure the probe concentration is correct. (Dextran sulfate interferes with acid precipitation and quenches ^{3}H when precipitated or spotted on filters; therefore it is necessary to determine % precipitability of the probe stock and to use this value to convert total cpm to precipitable cpm.)

Apply probe solution to the center area of tissue on dry slides, using 4 μl per cm^2 of coverslip area, and gently lay a coverslip over the puddle with forceps. Coverslips are siliconized and baked at 150°C for at least 2 hours. It is important to use the indicated volume of probe because too thin a layer reduces signals; the amount indicated should just spread to the edges of the coverslip under its weight. Do not press on the coverslip to spread the probe. Slides and coverslips must be kept scrupulously clean, or bits of grit will interfere with formation of a uniform layer of probe free of bubbles.

Depending on the nature of the tissue and subsequent staining, it may be helpful to mark the position of strips of sections and/or of large air bubbles with a diamond pencil on the underside of the slide.

The hybridization reaction is sealed either by immersing the slides in a single layer under 1–2 cm of mineral oil in plastic boxes or by dribbling a thick layer of rubber cement at the joint of the coverslip with the slide and placing the slides in a humid chamber.

Hybridize for at least 5 hours; we typically use an overnight incubation because it is convenient. Slides can be stored after hybridization for at least several days at $-70°C$ before washing, if necessary.

IV. Posthybridization Washes

A. General Considerations

Posthybridization washes are designed to remove the large excess of unhybridized probe and, in some cases, to discriminate among closely related sequences that hybridize with the probe. The specificity of hybridization, i.e., whether the probe sequence hybridizes only to its homologous target RNA, is controlled by hybridization and wash temperatures through the relationship between the percentage base pair mismatch of a duplex and its T_m: 1% mismatched base pairs decreases the T_m of the duplex approximately $1°C$. Thus, when hybridization is carried out at maximum rates ($T_m - 25°C$), the probe may form duplexes, although at lower rate, with nonhomologous mRNAs as much as 20–25% different in sequence. Although these conditions do not appear to be very specific, in fact, most single-copy probe sequences only hybridize to their homologous targets at $T_m - 25°C$. The major source of background is not promiscuous hybridization, but probe sticking to unidentified tissue components. This also decreases when slides are washed at higher temperature, so we routinely include a wash at about $10°C$ below the T_m of homologous duplexes (see below).

For probes that hybridize to mRNAs encoded from members of small gene families of related sequences, extra precautions must be taken to accomplish, and to demonstrate, that the probe is specific for one gene. Duplexes formed with homologous or heterologous sequences each melt over a range of 4 to $5°C$ (for example, see Cox et al., 1984). Thus, if homologous and heterologous duplexes are mismatched by 8–10% or more, it is possible to find a temperature at which heterologous duplexes dissociate, but most of the homologous duplexes remain. To identify the best temperature, the simplest empirical approach is to hybridize multiple slides at $T_m - 25°C$ (for maximum rate) and then to wash individual slides at progressively higher temperatures in the high stringency wash step. For a more detailed description of this approach and some controls, see Angerer et al. (1985).

In general, all wash procedures are similar to those used for standard blotting assays. They include washes at high salt concentrations, which reduce electrostatic binding of probe, and at least one wash at high stringency (combination of low salt and high temperature, or inclusion of formamide). When RNA probes are used, treatment with RNase A markedly reduces backgrounds. Under conditions of high salt (0.5 M) and moderate temperature ($37°C$), RNA in duplex is resistant to RNase A, but single-stranded molecules are hydrolyzed. No enzymatic counterpart for removal of nonhybridized DNA probes has been used routinely with success. Although some protocols call for prolonged (more than 1 day) washes, we have not found that this decreases backgrounds with RNA probes and RNase washes. Our procedure takes about half a day; it is possible that even shorter washes (except for the initial high salt wash) can be used.

Finding effective washing procedures for probes labeled with ^{35}S has caused a number of laboratories considerable frustration. Over several years we lost about one-third of hybridizations with ^{35}S-labeled probes because of extremely high background binding. This phenomenon is clearly related to the chemistry of thiolated nucleotides, because the quantity of probe bound nonspecifically was much higher for those labeled with ^{35}S than with ^{3}H. To block this binding, we have tried preincubations with N-ethylmaleimide, iocloacetamide, S-UTP, S-polynucleotides, thiophosphate, posthybridization digestion with additional ribonucleases, and prolonged washes, all of which had no consistent effect. During the course of this work, we found that the single most important factor is to maintain the probe in a reducing environment provided by DTT or β-mercaptoethanol, but that this is not always sufficient. In addition, since we began about a year ago to lengthen the initial wash in $4\times$ SSC + 10 mM DTT to 1 hour, we have not had very high backgrounds. We have discussed elsewhere the use of ^{35}S-labeled probes in some detail (Angerer and Angerer, 1989).

B.   Wash Procedures

1.   Initial Washes

Protocol

For slides in mineral oil, remove as much oil as possible by draining them on paper towels, taking care to avoid sliding the coverslips, which causes a shear force that may damage the tissue.

Pass the slides through three changes of chloroform in Coplin jars to remove the remaining mineral oil. For each set of five slides, the last wash should be fresh.

Transfer the slides from chloroform to $4\times$ SSC in Coplin jars, covering the slides completely. For slides sealed with rubber cement, peel it off carefully and begin the washes at this step. Over a period of about 15 minutes, wash the slides in a total of 3×50 ml changes of $4\times$ SSC. The coverslips should loosen and come off in the first change. Peel the coverslips off with forceps only as a last resort. Virtually all of the probe is removed in the first few high-salt wash steps, and when radioactive probes are used only these need be treated as hazardous waste. Do not allow the slides to dry between these changes.

For probes labeled with ^{35}S, the $4\times$ SSC should contain 10 mM DTT and, as discussed above, the total time in this wash should be increased to at least 1 hour.

2. WASH FOR ^{35}S-LABELED PROBE

The following is an optional step for ^{3}H-labeled probes but always used for ^{35}S-labeled probes.

Protocol

Dehydrate the slides through ethanol (50, 70%,) containing 300 mM ammonium acetate, and then 95, 99, and 99% ethanol. Ammonium acetate is included in the ethanols as a precaution against denaturing duplexes.

When the ethanol has evaporated, place the slides in slide mailers or Coplin jars containing prewarmed (50–65°C) hybridization buffer [50% formamide, 0.3 M NaCl, 20 mM Tris-HCl, 5 mM EDTA, pH 8.0 (+ 10 mM DTT for ^{35}S-labeled probes]. It is safest to do these stringent washes in hybridization buffer, because then the exact relationship between hybridization and wash stringency is known. The higher temperature is only about 5°C below the *in situ* T_m.

Transfer the slides to a staining dish containing several hundred milliliters of $2 \times$ SSC at room temperature.

3. RNase AND FINAL WASHES

The most effective wash step when using riboprobes is digestion with ribonuclease in high-salt buffer as follows:

Protocol

Transfer the slides to slide mailers containing 20 μg/ml RNase A in 0.5 M NaCl, 10 mM Tris-HCl, 1 mM EDTA, pH 8.0. Incubate at 37°C for 30 minutes. Do not include DTT in the RNase digestion.

Complete the washes as follows, including 10 mM β-mercaptoethanol in the case of ^{35}S-labeled probes:

Wash three times with 200-ml changes of RNase buffer for a total of 30 minutes at 37°C in a staining dish.

Wash five times with 200 ml of $2 \times$ SSC for a total of 60 minutes at room temperature. It is important to wash with large volumes here to remove all of the ribonuclease before the final high-stringency, low-salt wash (see below). For experiments with more than 15 and up to 40 slides, this wash is done conveniently by inserting slides vertically in a plastic-coated test-tube rack, immersed in 4 liters of buffer stirred by a magnetic stirrer.

Wash once with 500 ml of $0.1 \times$ SSC/20 slides, 50°C for 15 minutes. This wash is approximately at $T_m - 25°C$; we do not use very high stringency washes here, because T_ms of the duplexes might be lower as a result of nicking during RNase digestion.

Dehydrate the sections through a graded series of ethanol concentrations containing 0.3 M ammonium acetate, as described above.

V. Detection Mechanisms

A. Considerations in Choosing Isotopes

We have used radioactively labeled probes and autoradiographic detection almost exclusively. Autoradiographic detection has three major advantages: (1) It currently offers the highest sensitivity: With ^{35}S-labeled probes we have been able to detect mRNAs present at only 5–10 copies per cell (Hardin, et al., 1988). (2) Detection is direct, i.e., it does not depend on any additional reagents such as antibodies or enzymes, and is therefore virtually foolproof. (3) Signals can be quantitated accurately and compared to backgrounds on separate slides. The major disadvantages of autoradiography are (1) the inconvenience, cost, and hazards of working with isotopes; (2) the time required for autoradiographic exposure; (3) lower resolution than is provided by methods based on fluorescent labels or histochemical reactions.

The important features of the four isotopes that have been used to label probes for *in situ* hybridization are compared in Table II. The specific activity

TABLE II

Comparison of Features of Four Isotopes Used to Label Probes

Isotope	$t_{1/2}$	Resolution	Specific activity (dpm/μg probe)	Relative autoradiographic efficiency	Grains/ day/kb
^{3}H	12.7 years	Excellent	2×10^8	1	0.003
^{35}S	87 days	Good	2×10^9	5	0.15
^{125}I	60 days	Very good	2×10^9	? (2–7)	0.06–0.21
^{32}P	14 days	Poor	$>5 \times 10^9$	1	0.075

of RNAs transcribed *in vitro* (dpm/μg probe) is determined by the specific activity of the precursor pool. Maximum specific activities for [^{3}H] nucleotides range from 20 to 60 Ci/mmol, while those for ^{35}S and ^{125}I are considerably higher, approximately 1000–1500 and 2000–2500 Ci/mmol, respectively. Although it is possible to use all four [^{3}H]nucleotides at maximum specific activity, i.e., without dilution by cold nucleotides, this is rather expensive for several reasons: (1) the nucleotide K_ms using phage RNA polymerase range from 10 to 60 μM and (2) in order to make a significant mass of probe needed to achieve saturation during hybridization, nucleotide concentrations of 50 μM are required. We use [^{3}H]UTP and [^{3}H]CTP because GTP is the least stable triphosphate and ATP has the highest K_m. Both UTP and CTP are available as ^{35}S-labeled nucleotides, whereas only CTP is iodinated and is available once per month from Dupont-NEN (Boston, MA).

Relative autoradiographic efficiency is the relative number of autoradiographic grains/disintegration, using Kodak NTB-2 emulsion diluted 1:1. The absolute efficiency for ^{3}H autoradiography is approximately 10% in the absence of quenching. In practice, for ^{3}H uniformly distributed in a 5-μm section, the large majority of grains are produced by decays in the upper 0.5–1μm of the section, so that the overall efficiency of detection is about 2% (Ada *et al.,* 1966; Pelc and Welton, 1967). Note that the higher energy of the ^{35}S β decay provides 5-fold more grains per disintegration (our empirical measurements), so one obtains a factor of 5 when ^{35}S precursors are diluted with unlabeled thiolated nucleotides to the same specific activity as obtained with undiluted ^{3}H-labeled precursors. Although ^{32}P is clearly superior to the other isotopes for exposing X-ray films (for an *in situ* hybridization application, see Rall *et al.,* 1985), the energy of ^{32}P β decay is so high that many disintegrations leave the NTB-2 emulsion without producing a latent grain; consequently, the overall efficiency is similar to that of ^{3}H.

^{3}H offers an advantage of very long half-life. We have used probes stored at $-70°$C for a year, if their length is still sufficient. Similarly, as long as ^{35}S-labeled probes are maintained in DTT, they are usable for at least a half-life (87 days). On the other hand, the instability of ^{125}I-labeled probes requires that they be used within a few weeks of the time when the nucleotides were iodinated. Thus, although iodinated probes offer the resolution of ^{3}H and much higher specific activity, they are inconvenient to use on a routine basis. Furthermore, use of ^{125}I requires additional precautions in handling, namely, use of lead shielding and thyroid monitoring.

The decay energy of these isotopes is inversely related to the resolution of signal that can be obtained using probes labeled with them. The resolution of ^{3}H and ^{125}I is sufficient, for example, to determine whether a signal is primarily nuclear or cytoplasmic, even in small cells (for example of nuclear signals, see Nash *et al.,* 1987) or to identify localized regions of mRNA concentration

TABLE III

COMPARISON OF SIGNALS OBTAINED USING ^{3}H- AND ^{35}S-LABELED PROBES

^{35}S probe input	Subsaturating		Saturating	
	^{3}H	^{35}S	^{3}H	^{35}S
Relative probe input	25	1	25	25
Relative probe specific activity	1	5	1	5
Relative autoradiographic efficiency	1	5	1	5
Relative grains/day	**25**	**25**	**25**	**625**
Use for	Abundant targets		Rare targets	

within a cell (Lawrence and Singer, 1986; Yang *et al.,* 1989). Resolution of ^{35}S-labeled probes is lower, but still sufficient to distinguish nucleus versus cytoplasm in most cases, and to determine which cells in a tissue are labeled (Angerer and Angerer, 1989). ^{32}P has poor resolution for liquid emulsion autoradiography and leaves a wide "spray" around the site of probe binding. Its use should be reserved for low-resolution cases where detection uses X-ray film.

The last column of Table II gives the approximate signal that can be expected as autoradiographic grains recorded/day/kilobase of probe of maximum specific activity hybridized in an optimized system. These numbers are derived from our work on sea urchin embryos, using probes whose target mRNA concentrations are known by independent solution hybridization measurements. For this tissue, a large fraction of target RNA is retained in sections and hybridized when probe concentrations are saturating (Angerer and Angerer, 1981; Cox *et al.,* 1984). Different mRNAs range in abundance from 0.3 to 75 molecules in a column 1 μm square by 5 μm deep, so that a probe for a 1-kb target yields 0.006 to 1.5 grains/μm^2 in a 1 week exposure for ^{3}H and 0.3 to 75 grains/μm^2 for ^{35}S. The labeled area in Fig. 1A has a grain density of 0.22 grains/μm^2. In practice, using ^{35}S-labeled probes of 5×10^8 dpm/μg and exposures of 2 weeks, we have been able to detect mRNAs present at about 5–10 copies per cell. Sensitivity using ^{35}S is currently limited by noise rather than by signal.

While ^{35}S potentially offers the highest sensitivity, it is not always the best choice for labeling probes. Because of the very high specific activity of ^{35}S-UTP, synthesis of probes of the maximum specific activity is very expensive. Consequently, some have used ^{35}S-labeled probes of very high specific activity (on the order of 1×10^9 dpm/μg) at concentrations far below (25- to 50-fold) those required to saturate target mRNAs. Table III compares signals that can be obtained using a ^{3}H-labeled probe at saturating concentration versus a ^{35}S-labeled probe of higher specific activity, but at subsaturating concentration. The magnitudes of the signals are equivalent, but the cost of probe preparation is lower and the resolution is greater for the ^{3}H-labeled probe. In practice, abundant or moderately abundant mRNAs can be detected in a few weeks with ^{3}H. We reserve ^{35}S for relatively rare messages (less than 0.1% of mRNA or 0.002% of total RNA), progressively increasing the specific activity for progressively rarer mRNAs.

B. Protocol for Autoradiography with ^{3}H- and ^{35}S-Labeled Probes

1. SETTING UP THE SLIDES FOR AUTORADIOGRAPHY

All steps are most easily carried out in a darkroom using the "Duplex Super Safelight" equipped with FDY filters in the top slots and FDW filters in the bottom slots (VWR TM72882-10). Alternatively, work in complete darkness, or use a safelight equipped with a Wratten red filter #1.

Prepare stock emulsion by melting Kodak NTB-2 in a 45°C water bath and diluting it with an equal volume of 600 mM ammonium acetate, prewarmed to 45°C. In handling emulsion, always swirl or pour it gently to avoid generating bubbles. Distribute 12-ml aliquots in plastic scintillation vials, wrap them in aluminium foil, and store in a light-tight container at 4°C.

Coat slides with emulsion by dipping them twice, about 1 second per cycle, without pausing. Although a slide mailer may be used to hold the emulsion, we recommend a specially shaped dipping chamber (Electron Microscopy Sciences, Ft. Washington, PA) which requires less emulsion and avoids scraping the slide edges.

First dip several test slides until all surface bubbles are removed and then dip those carrying tissue. After dipping, blot the bottom edge of the slides briefly on paper towels to remove excess emulsion, place the slides vertically in a plastic-coated test-tube rack, and allow them to dry for 30 minutes. Do not dry slides on paper towels or they will stick.

Transfer the rack with slides to a moist chamber and let them stand at room temperature for ≥3 hours. High humidity removes latent grains from the emulsion and thus reduces emulsion background. By the same token, it is important to expose the emulsion in completely dry air to avoid signals that are reduced and nonlinear as a function of exposure time.

Remove the slides from the moist chamber and allow them to dry slowly in air for 45 minutes. Photographic emulsions are sensitive to pressure and to stretching. A common artifact, if the emulsion is dried too quickly or if sections are too thick, is that lines of grains appear at the borders between tissue and empty space in the sections.

Place the slides in light-tight black plastic boxes, each carrying a small tube of dessicant, held in one end by a blank spacer slide. Seal the boxes with electrical tape and place them in plastic bags, either Zip-lock or heat-sealed, that also contain a vial of dessicant.

Expose the slides at 4°C. Approximate estimates of the exposure times required are as follows: ^{3}H (1 $\times$ 10^8 dpm/μg) allows detection of an abundant mRNA (0.5–1% of total mRNA) in 0.5 to 1 week; ^{35}S (5 $\times$ 10^8 dpm/μg) allows detection of rarer mRNAs (0.005–0.01% of total mRNA) in 2–4 weeks. It is always wise to prepare enough slides for at least two exposures for each probe and/or condition.

2. DEVELOPING

Allow the black boxes to come to about 15°C (about 15 minutes at room temperature). Keep them sealed to avoid moisture condensing on the cold slides.

Transfer the slides to glass slide carriers and immediately develop without agitation for 2.5 minutes in Kodak D-19 developer at 15°C, a temperature which provides better signal-to-noise ratios.

Stop for 10 seconds in 2% glacial acetic acid at 15°C.

Fix for 5 minutes in Kodak Fixer (not Rapid Fix) at 15°C.

Wash in distilled water for 15 minutes at 15°C.

Wash in cold running tap water for 30 minutes.

3. PRESERVATION

After the desired staining procedure, dehydrate slides through a graded series of ethanol concentrations, and then through two changes of fresh xylene.

Remove a slide from xylene, briefly drain off the excess, and place 3 drops of Permount over the tissue. Do not allow all xylene to dry from the slides, and be sure the xylene is clean and dry. Residual moisture results in a fine precipitate which interferes with visualizing grain patterns in dark-field illumination.

Carefully lay a 22 $\times$ 60 mm coverslip over the slide, allowing the Permount to spread. Apply even pressure over the coverslip to squeeze out bubbles and excess Permount which is removed by blotting with a paper towel.

Place the slide on a slide warmer (at 40°C) overnight to allow the Permount to dry and harden.

Remove excess Permount from the slides with a tissue moistened with xylene and the emulsion from the backs of slides with Formula 409 or Fantastic.

C. Nonisotopic Methods

Alternatives to autoradiographic detection use probes containing nucleotides that carry a ligand which can be coupled to a detector labeled with a fluorochrome or carrying a histochemical detection mechanism. The most frequently used ligand is biotin, which binds to streptavidin. Streptavidin can either be coupled directly to a fluorochrome, horseradish peroxidase, or alkaline phosphatase, or it can be bound by one of these reporter molecules also coupled to biotin. In theory, these many-layered sandwiches with enzymatic reporters offer the potential for considerable amplification of signals from bound probe, but in practice this has not been achieved. In unpublished experiments in which a single probe was labeled with both ^{3}H and biotin, we found that the substituted probes hybridize efficiently, and that the low nonisotopic signals were limited by the efficiency of the detection system, presumably as a consequence of poor penetration of large detector molecules into the sections. If this interpretation is correct, then efficiency of detection by non-isotopic methods will be higher under conditions of lower fixation and/or higher protease treatment, and using smaller detector molecules. For detecting mRNAs, current avidin–biotin-based detection systems appear to be about an order of magnitude less sensitive than ^{3}H-labeled probes, and several orders of magnitude less sensitive than ^{35}S-labeled probes. However, avidin–biotin-based methods are in routine use for relatively abundant mRNA targets in which they provide rapid detection and excellent resolution. It should be noted that nonisotopic methods coupled with the use of double-stranded DNA probes are extremely sensitive for detecting double-stranded DNA targets, as in viral or chromosomal *in situ* analyses (for examples, see Lawrence *et al.,* 1988; Pinkel *et al.,* 1988; Rykowski *et al.,* 1988).

Recently, digoxigenin has been employed as an alternative ligand, detected by anti-digoxigenin antibodies linked to reporter molecules (Tautz and Pfeifle, 1989). This system has lower background than the biotin–avidin system, perhaps because, unlike biotin, digoxigenin is not a normal component of tissues. Although no direct comparisons are available in the literature, empirical observation also suggests that signals are significantly higher with digoxigenin, for reasons that are unclear.

For further discussion of nonisotopic methods, see [3], this volume.

VI.   Comments on Microscopy

After the slides are developed, they can be stained with any method that is compatible with autoradiography. Staining procedures that use long incubations at low pH should be avoided, because this bleaches autoradiographic grains (see Rogers, 1979). Staining procedures that produce too much contrast may interfere with visualization of grains by bright-field or phase-contrast microscopy. We typically stain tissue lightly with eosin or with hematoxylin and eosin. In most cases we use phase-contrast illumination for examining morphology. Dark-field illumination (or epiluminescence) is very useful and almost essential for discerning patterns of autoradiographic grains. Furthermore, grains that cannot all be brought into focus simultaneously in bright field are easily visualized in dark field. It is much faster to scan for rare labeled cells in dark field and true dark field provides a "single blind" approach to analyzing data, since one can decide the pattern of grains first, and then examine the morphology under phase-contrast illumination.

VII.   Controls and Quantitative Measurements

It is important to learn the chracteristics of any new probe before trying *in situ* hybridization. RNA blots can be used to ensure that the probe is specific for a single mRNA and, more importantly, to give an estimate of the overall abundance of the mRNA, and identify differences in abundance among different tissues, stages of development, etc. This information provides useful quantitative checks on the specificity of *in situ* hybridization. RNA blots can also be used to identify approximately the correct wash temperatures when using heterologous probes or when trying to hybridize specifically to one member of a multigene family. If asymmetric RNA probes are used, sense and antisense probes can be hybridized separately to be sure the antisense strand has been identified correctly, and that no signal is obtained with the sense strand.

The most common controls are designed to demonstrate that the pattern of hybridization observed accurately reflects the relative concentration of target mRNA in different regions of the tissue. In most cases, strict quantitative proof that this goal has been achieved is impossible; furthermore, if one had data on concentrations of an mRNA in different cell types required to prove it, *in situ* hybridization would be unnecessary. It is also virtually impossible to prove that a given mRNA is absent from a given tissue, because different control probes give slightly different background binding levels so that "zero signal" cannot be precisely defined. However, usually differences in grain density over "expressing" and "nonexpressing" cell types are so large that whether "nonexpressing" is really "low-level expressing" is a separate question.

The most fundamental control is to use a "nonhomologous" probe that does not hybridize to any RNA in the cell. Because transcription of genes of eukaryotes is (almost always) asymmetric, a sense strand RNA probe provides a convenient control for estimating the level of nonspecific background binding. Corresponding control probes for double-stranded DNA often consist of prokaryotic plasmid molecules, although some of these sequences show levels of hybridization to tissue above the true background (for example, see Cox et al., 1984). As one examines the patterns for different probes in the same system, an inductive argument for specificity is gradually constructed by showing that some probes do not label all cell types and that different probes exhibit distinctive and reproducible patterns of hybridization.

In the case of complex patterns, a second question is whether the particular features observed on one section reflect the real distribution of target or are due to random variations in background over that particular section. An effective way to validate individual patterns is to compare hybridization to two adjacent sections sufficiently thin so that they pass through the same set of cells. This approach was used, for example, to demonstrate differences in abundance and distribution of metallothionein mRNA in embryos at the same stage of development from the same culture (Angerer et al., 1986).

Designing appropriate controls is most difficult for cases in which one is trying to make quantitative comparison of signals. These comparisons fall into two different categories. In the first, one is trying to authenticate differences in target RNA concentration *among different cell types* that vary only slightly (perhaps 2-fold or less) in apparent signal. The potential problem in such cases is that the observed difference may simply reflect general differences in hybridization efficiency to targets in different cell types and, for mRNAs present at low abundance, differences in nonspecific background binding. Unfortunately, no RNA exists that is known to be present in all cell types at uniform concentration (molecules/μm^3). However, the observation that one cell type always gives higher or lower signals for mRNAs that are widely distributed [for example, poly(A) or one encoding a "housekeeping" protein] might be an indication that this tissue responds differently to the particular tissue preparation method used. One useful control in such cases could be to show that relative signals over the two cell types are independent of the extent of prehybridization protease treatment.

The second type of quantitative comparison is to measure the levels of a particular mRNA in different samples *of the same cell type* that has been subjected to different experimental treatments. Because there is some variation in the magnitude of signal among different experiments for the same probe hybridized to the same kind of tissue, it is important to minimize differences in any step of the *in situ* protocol. Thus, tissue blocks of similar size should be fixed simultaneously and carried through the analysis together. Although signals can be very reproducible among different slides within an experiment, as shown, for example, by the fact that melting curves and kinetic data can be obtained with relatively little scatter (for example, see Brahic and Haase, 1978; Cox et al., 1984; Lawrence and Singer, 1985), the most accurate comparisons are made from adjacent tissue sections carried on the same slide, since slight variations in protease treatment, emulsion thickness, etc., lead to corresponding variations in signals.

VIII.   Concluding Remarks

Over the past 10 years, the fields of investigation that have become particularly dependent on *in situ* hybridization include cell biology, developmental biology, neurobiology, and medicine. Hundreds of laboratories have successfully adopted this approach to answer questions ranging from mechanisms of RNA compartmentalization in the cell, to mechanisms of regulation of differential gene expression in developing embryos as well as in complex tissues, to pathogenesis. Although there are variations in the details of various protocols, all are based essentially on the principles discussed here. For the greatest chance of success, the investigator who is beginning to set up this technique in his laboratory should follow a single procedure as precisely as possible. If a procedure is available for the tissue of interest, start with it. If not, then try a matrix of fixation conditions and prehybridization protease treatments. It is important to use (1) a *good positive control probe* that reacts with an abundant target RNA and affords the best chance of providing a *high signal*, and (2) *good negative controls*, such as heterologous nucleic acid probes and cells that lack the target RNA, to establish *low noise* levels. Once some signal is obtained, particular steps in the protocol can be modified.

Acknowledgments

This work has been supported by a grant (NIGMS) to RCA and LMA and by a Career Development Award (NICHD) to RCA. We wish to thank Dupont-NEN for their generous gifts of ^{35}S-UTP, ^{35}S-CTP, and ^{125}I-CTP.

References

Ada, G. L., Humphrey, J. H., Askonas, B. A., McDevitt, H. D., and Nossal, G. V. (1966). Correlation of grain counts with radioactivity (^{125}I and ^{3}H) in autoradiography. *Exp. Cell Res.* **41**, 557–572.

Akam, M. E., and Martinez-Arias, A. (1985). The distribution of *Ultrabiothorax* transcripts in *Drosophila* embryos. *EMBO J.* **4**, 1689–1700.

Angerer, L. M., and Angerer, R. C. (1981). Detection of polyA RNA in sea urchin eggs and embryos by quantitative *in situ* hybridization. *Nuclei Acids Res.* **9**, 2819–2840.

Angerer, L. M., and Angerer, R. C. (1989). *In situ* hybridization with ^{35}S-labeled RNA probes. *Biotech. Update* **4**, 1–6.

Angerer, L., DeLeon, D., Cox, K., Maxson, R., Kedes, L., Kaumeyer, J., Weinberg, E. and Angerer, R. (1985). Simultaneous expression of early and late histone messenger RNAs in individual cells during development of the sea urchin embryo. *Dev. Biol.* **112**, 157–166.

Angerer, L. M., Kawczynski, G., Wilkinson, D. G., Nemer, M., and Angerer, R. C. (1986). Spatial patterns of metallothionein mRNA expression in the sea urchin embryo. *Dev. Biol.* **116**, 543–547.

Angerer, L. M., Stoler, M. H., and Angerer, R. C. (1987a). *In situ* hybridization with RNA probes—an annotated recipe. In "In situ hybridization—Applications to Neurobiology" (K. L. Valentino, J. H. Eberwine, and J. D. Barchas, eds.), pp. 71–96. Oxford Univ. Press, Oxford.

Angerer, L. M., Cox, K. H., and Angerer, R. C. (1987b). Identification of tissue-specific gene expression by in situ hybridization. In "Methods in Enzymology, Guide to Molecular Cloning Techniques" (S. Berger, and A. Kimmel, eds.), Vol. 152, pp. 649–661. Academic Press, New York.

Angerer, L. M., Dolecki, G. J., Gagnon, M. L., Lum, R., Yang, Q., Humphreys, T., and Angerer, R. C. (1989). Progressively restricted expression of a homeobox gene within aboral ectoderm of developing sea urchin embryos. Genes Dev. 3, 370–383.

Ausubel, F. M., Brent, R., Kingston, R. E., Moore, D. D., Seidman, J. G., Smith, J. A., and Struhl, K., eds., (1991). "Current Protocols in Molecular Biology" Vols. 1 and 2. Greene Publishing Associates and Wiley-Interscience, New York.

Bandtlow, C. E., Heumann, R., Schwab, M. E., and Thoenen, H. (1987). Cellular localization of nerve growth factor synthesis by in situ hybridization. EMBO J. 6, 891–899.

Berman, S. A., Bursztajn, S., Bowen, B., and Gilbert, W. (1990). Localization of an acetylcholine receptor intron to the nuclear membrane. Science 247, 212–214.

Brahic, M., and Haase, A. T. (1978). Detection of viral sequences of low reiteration frequency by in situ hybridization. Proc. Natl. Acad. Sci. U.S.A. 75, 6125–6129.

Brigati, D. J., Myerson, D., Leary, J. J., Spalholz, B., Travis, S. Z., Fong, C. K. Y., Hsiung, G. D., and Ward, D. C. (1983). Detection of viral genomes in cultured cells and paraffin-embedded tissue sections using biotin-labeled hybridization probes. Virology 126, 32–50.

Britten, R. J., Graham, D. E., and Neufeld, B. (1974). Analysis of repeating DNA sequences by reassociation. In "Methods in Enzymology" (L. Grossman and K. Moldave, eds.), Vol. 29, pp. 363–418. Academic Press, New York.

Cox, K. H., and Goldberg, R. B. (1988). Analysis of plant gene expression. In "Plant Molecular Biology: A Practical Approach" (C. H. Shaw, ed.), pp. 1–36. IRL Press, New York.

Cox, K. H., DeLeon, D. V., Angerer, L. M., and Angerer, R. C. (1984). Detection of mRNAs in sea urchin embryos by in situ hybridization using asymmetric RNA probes. Dev. Biol. 101, 485–502.

Cox, K. H., Angerer, L. M., Lee, J. J., Davidson, E. H., and Angerer, R. C. (1986). Cell lineage-specific programs of expression of multiple actin genes during sea urchin embryogenesis. J. Mol. Biol. 188, 159–172.

Davis, C. A., Noble-Topham, S. E., Rossant, J., and Joyner, A. L. (1988). Expression of the homeobox-containing gene En-2 delineates a specific region of the developing mouse brain. Genes Dev. 2, 361–371.

Fremeau, R. T., Jr., Landblad, J. R., Pritchett, D. B., Wilcox, J. N., and Roberts, J. L. (1986). Regulation of pre-opiomelanocortin gene transcription in individual cell nuclei. Science 234, 1265–1269.

Gee, C. E., and Roberts, J. L. (1983). In situ hybridization histochemistry: A technique for the study of gene expression in single cells. DNA 2, 157–163.

Hafen, E., Levine, M., Garber, R. L., and Gehring, W. J. (1983). An improved in situ hybridization method for the detection of cellular RNAs in Drosophila tissue sections and its application for localizing transcripts of the homeotic Antennapedia gene complex. EMBO J. 1, 617–623.

Hardin, P. E., Angerer, L. M., Hardin, S. H., Angerer, R. C., and Klein, W. H. (1988). The Spec2 genes of Strongylocentrotus purpuratus: Structure and differential expression in embryonic aboral ectoderm cells. J. Mol. Biol. 202, 417–431.

Harkey, M. A., Whiteley, H. R., and Whiteley, A. H. (1988). Coordinate accumulation of primary mesenchyme-specific transcripts during skeletogenesis in the sea urchin embryo. Dev. Biol. 125, 381–395.

Hayashi, S., Gillam, I. C., Delaney, A. D., and Tener, G. M. (1978). Acetylation of chromosome squashes of *Drosophila melanogaster* decreases the background in autoradiographs from hybridization with [^{125}I]-labeled RNA. *J. Histochem. Cytochem.* **26,** 677–679.

Hough-Evans, B. R., Franks, R. R., Cameron, R. A., Britten, R. J., and Davidson, E. H. (1987). Correct cell type-specific expression of a fusion gene injected into sea urchin embryos. *Dev. Biol.* **121,** 576–579.

Howe, J. G., and Steitz, J. A. (1986). Localization of Epstein-Barr virus-encoded small RNAs by *in situ* hybridization. *Proc. Natl. Acad. Sci. U.S.A.* **83,** 9006–9010.

Hurley, D., Angerer, L. M., and Angerer, R. C. (1989). Altered expression of spatially regulated embryonic genes in the progeny of separated sea urchin blastomeres. *Development* **106,** 567–579.

Ingham, P. W., Howard, K. R., and Ish-Horowicz, D. (1985). Transcription patterns of the *Drosophila* segmentation gene *hairy. Nature (London)* **318,** 439–445.

Jamrich, M., Mahon, K. A., Gavis, E. R., and Gall, J. G. (1984). Histone RNA in amphibian oocytes visualized by *in situ* hybridization to methacrylate embedded tissue sections. *EMBO J.* **3,** 1939–1943.

Lawrence, J. B., and Singer, R. H. (1985). Quantitative analysis of *in situ* hybridization methods for the detection of actin gene expression. *Nucleic Acids Res.* **13,** 1777–1799.

Lawrence, J. B., and Singer, R. H. (1986). Intracellular localization of messenger RNA for cytoskeletal proteins. *Cell (Cambridge, Mass.)* **45,** 407–415.

Lawrence, J. B., Villnave, C. A., and Singer, R. H. (1988). Sensitive, high-resolution chromatin and chromosome mapping *in situ*: Presence and orientation of two closely integrated copies of EBV in a lymphoma cell line. *Cell (Cambridge, Mass.)* **52,** 51–61.

Lynn, D. A., Angerer, L. M., Bruskin, A. M., Klein, W. H., and Angerer, R. C. (1983). Localization of a family of mRNAs in a single cell type and its precursors in sea urchin embryos. *Proc. Natl. Acad. Sci. U.S.A.* **80,** 2656–2660.

Melton, D., Krieg, P., Rebagliati, M., Maniatis, T., Zinn, K., and Green, M. R. (1984). Efficient *in vitro* synthesis of biologically active RNA and RNA hybridization probes from plasmids containing a bacteriophage SP6 promoter. *Nucleic Acids Res.* **12,** 7035–7056.

Millonig, G., and Marinozzi, V. (1968). Fixation and embedding in electron microscopy. *In* "Advances in Optical and Electron Microscopy" (R. Barer and V. E. Cosslett, eds.), Vol. 2, pp. 251–336, Academic Press, London and New York.

Nash, M. A., Kozak, S. E., Angerer, L. M., Angerer, R. C., Schatten, H., Schatten, G. and Marzluff, W. F. (1987). Sea urchin maternal and embryonic U1 RNAs are spatially segregated in early embryos. *J. Cell Biol.* **104,** 1133–1142.

Nomura, S., Wills, A. J., Edwards, D. R., Heath, J. K., and Hogan, B. L. M. (1988). Developmental expression of 2ar (Osteopontin) and SPARC (Osteopectin) RNA as revealed by *in situ* hybridization. *J. Cell Biol.* **106,** 441–450.

Pelc, S. R., and Welton, M. G. E. (1967). Quantitative evaluation of tritium in autoradiography and biochemistry. *Nature (London)* **216,** 925–927.

Pinkel, D., Landigent, J., Collins, C., Fuscoe, J., Segraves, R., Lucas, J., and Gray, J. (1988). Fluorescence *in situ* hybridization with human chromosome-specific libraries: Detection of trisomy 21 and translocations of chromosome 4. *Proc. Natl. Acad. Sci. U.S.A.* **85,** 9138–9143.

Rall, L. B., Scott, J., Bell, G. I., Crawford, R. J., Penschow, J. D., Niall, H. D., and Coghlan, J. P. (1985). Mouse prepro-epidermal growth factor synthesis by the kidney and other tissues. *Nature (London)* **313,** 228–231.

Rogers, A. W. (1979). "Techniques in Autoradiography" Elsevier/North-Holland Biomedical Press, New York.

Rychlik, W., and Rhoads, R. E. (1989). A computer program for choosing optimal oligonucleotides for filter hybridization, sequencing and *in vitro* amplification of DNA. *Nucleic Acids Res.* **17,** 8543–8551.

Rykowski, M. C., Parmelee, S. J., Agard, D. A., and Sedat, J. W. (1988). Precise determination of the molecular limits of polytene chromosome band regulatory sequences for the *Notch* gene are in the interband. *Cell (Cambridge, Mass.)* **54,** 461–472.

Sambrook, J., Fritsch, E. F., and Maniatis, T. (1989). "Molecular Cloning: A Laboratory Manual" 2nd Ed., Vols. 1–3, Cold Spring Harbor Laboratory, Cold Spring Harbor, New York.

Stoler, M. H. (1990). *In situ* hybridization. *In* "Clinics in Laboratory Medicine: Tumor Markers in Diagnostic Pathology" (F. Gorstein and A. Thor, eds.), Vol. 10, pp. 215–236. Saunders, Philadelphia, Pennsylvania.

Stoler, M. H., Eskin, T. A., Benn, S., Angerer, R. C., and Angerer L. M. (1986). Human T lymphotrophic virus infection of the central nervous system: A preliminary *in situ* analysis. *J. Am. Med. Assoc.* **256,** 2360–2364.

Taneja, K., and Singer, R. H. (1987). Use of oligodeoxynucleotide probes for quantitative *in situ* hybridization to actin mRNA. *Anal. Biochem.* **166,** 389–398.

Tautz, D., and Pfeifle, C. (1989). A non-radioactive *in situ* hybridization method for the localization of specific RNAs in *Drosophila* embryos reveals translational control of the segmentation gene *hunchback. Chromosoma* **98,** 81–85.

Wilkinson, D. G., Bailes, A., and McMahon, A. P. (1987). Expression of the proto-oncogene *int-1* is restricted to specific neuronal cells in the developing mouse embryo. *Cell (Cambridge, Mass.)* **50,** 79–88.

Yang, Q., Angerer, L. M., and Angerer, R. C. (1989). Structure and tissue-specific developmental expression of a sea urchin arylsulfatase gene. *Dev. Biol.* **135,** 53–65.

Yu, S.-M., and Gorovsky, M. A. (1986). *In situ* dot blots: Quantitation of mRNA in intact cells. *Nucleic Acids Res.* **14,** 7597–7615.

Chapter 3

Fluorescent Detection of Nuclear RNA and DNA: Implications for Genome Organization

CAROL VILLNAVE JOHNSON, ROBERT H. SINGER, AND
JEANNE BENTLEY LAWRENCE

Department of Cell Biology
University of Massachusetts Medical Center
Worcester, Massachusetts 01655

In situ hybridization is a powerful and versatile methodological approach which makes it possible to visualize specific nucleic acid sequences directly within the morphological context of individual cells, nuclei, or chromosomes. Through the effort of several laboratories, technological advances have been made in the past 20 years which allow for sensitive, high-resolution detection of a single-copy DNA sequence in interphase or metaphase cells, as well as RNA sequences in the nucleus and cytoplasm. This technology has

METHODS IN CELL BIOLOGY, VOL. 35

numerous diverse applications in both biomedical research and clinical diagnostics. A major area in which it promises to have a large impact is in the investigation of genome organization, not only on chromosomes, but in the functional state within the interphase nucleus. This review focuses on methodology for fluorescence detection of individual genes and their nuclear transcripts, describing in detail the current technical protocols from our laboratory and illustrating recent and potential applications in the study of nuclear and chromosome organization.

I. Introduction to Technical Development

Gall and Pardue (1969) and John *et al.* (1969) first undertook hybridization directly to cellular material, using autoradiographic methods to detect abundant sequences. Much later, it was shown that it was possible to surmise the location of a single DNA sequence on chromosomes using statistical analysis of grain distributions from many cells (Harper *et al.*, 1981; Gerhard *et al.*, 1981). Autoradiographic *in situ* hybridization has been a highly valuable procedure which is still commonly used today. It has, however, several significant limitations, including poor resolution, lengthy development times, the necessity for statistical analysis for gene detection, and the hazard of handling radioactive material. Several laboratories contributed to the development of nonisotopic probe labeling methods to bypass the shortcomings inherent in autoradiographic methodologies (reviewed in Bauman *et al.*, 1989; Lawrence, 1990; Narayanswami and Hamkalo, 1991; Lichter *et al.*, 1991; McNeil *et al.*, 1991). These include direct fluorochrome labeling (Bauman *et al.*, 1981) sulfonation (Verdlov *et al.*, 1974), mercuration (Dale *et al.*, 1975; Hopman *et al.*, 1986), biotin (Manning *et al.*, 1975; Broker *et al.*, 1978; Langer *et al.*, 1981; Langer-Safer *et al.*, 1982), and digoxigenin (Boehringer Mannheim). Because of the convenience of biotin and the high resolution potentially obtainable with fluorescence, it became a major goal of our work over several years to develop sensitive, successful techniques for the localization of RNA and DNA, initially using biotinated probes and fluorescent detection.

Our work on *in situ* hybridization began with an interest in the expression of muscle-specific mRNAs within the cell cytoplasm (Lawrence and Singer, 1985, 1986). However, detection of even abundant cytoplasmic sequences met with limited and highly variable success. To develop improved *in situ* hybridization methods for this work, a rapid quantitative approach was devised to allow testing of a vast array of technical parameters, while controlling for the internal variations frequently encountered.

The process of hybridization and detection of specific nucleic acids within cytological material is conceptually complicated relative to the straightforward situation in which isolated nucleic acids stripped of protein are bound to a uniform substance, such as nitrocellulose. For analytical purposes, the steps of *in situ* hybridization can be divided into three main components, each of which is essential for success: (1) *preservation* of target sequences in a well-preserved but accessible state throughout hybridization, (2) *hybridization* of the probe to target molecules with high efficiency and without substantial nonspecific adherence, (3) *detection* using a reporter with sufficient specific activity or intensity to provide detectable signal with minimal background. Failure of any one parameter in any of the components results in a lower signal-to-noise ratio and a loss of sensitivity for a given detection system.

To quantitate each of these components individually, we used ^{32}P-labeled probes and scintillation counting to provide rapid evaluation of hybridization on dozens of samples per experiment. The initial model system used was detection of cytoplasmic β-actin mRNA in chicken muscle cultures. Preservation of cellular RNA in its native morphological context was a major objective; hence, we cultured cells with [^{3}H]uridine to monitor cellular RNA levels throughout the entire *in situ* process (Lawrence and Singer, 1985). Similarly, the effectiveness and causes of noise in the detection component were quantitated using ^{125}I-labeled avidin and probes labeled simultaneously with ^{32}P and biotin (Singer *et al.*, 1987).

Several principles and parameters emerged which proved key to improving aspects of the technique from cell preparation to hybridization and detection. A major technical advance was increased preservation of RNA throughout the procedure, which results showed had been poorly retained using several of the different commonly used protocols. For example, protease pretreatment of paraformaldehyde-fixed cultured cells did not improve signal-to-noise ratios and was actually highly destructive to mRNA, as was fixation in methanol: acetic acid (Lawrence and Singer, 1985). Glutaraldehyde, in contrast, can so extensively cross-link cellular proteins that RNA may become inaccessible to the probe. Cells fixed in glutaraldehyde are also extremely autofluorescent; hence most of our protocols use paraformaldehyde fixation. While cell preparation is of utmost importance, other key variables were identified in the hybridization and detection components. For instance, kinetic studies revealed that hybridization of even complex sequences was complete within 3–4 hours and that longer hybridization times can result in decreased signal, due to RNA degradation. The signal-to-noise ratio achieved was also found to be substantially dependent on the quality of reagents, in particular the brand and lot of formamide used. Detection with avidin was, surprisingly, found to be dramatically improved by simply changing the buffers used in staining (Singer *et al.*, 1987). Improvements in this first phase of technical development made it possible for us to demonstrate the nonhomogeneous distributions of cytoplasmic mRNAs in somatic cells (Lawrence and Singer, 1986) and to describe the precise temporal sequence of expression of muscle-specific mRNAs during myogenesis (Lawrence *et al.*, 1989b).

For several reasons, detailed in Section X, it was of interest to develop methodology for high-resolution detection of single genes or their primary transcripts within nuclei or chromosomes. Technically, the detection of low-abundance double-stranded DNA in the nucleus has quite different requirements than detection of more abundant but labile cytoplasmic mRNA. Initial failed attempts at detecting single genes established an important point, that the failure was not due to presumed limitations in the sensitivity of the biotin–avidin detection system, but in other components of the total process. An in-depth quantitative study was undertaken to identify ways of increasing hybridization efficiency while decreasing noise.

Key elements to the success of single-copy hybridization included probe fragment size after labeling and the use of high probe concentrations, well above theoretical saturation, which differed significantly from conditions used for detection of cytoplasmic mRNA. Attention to the various technical parameters described below allowed for increased hybridization efficiency and drastic reductions in background, making it possible to confidently visualize a single sequence in a single metaphase or interphase cell, without statistical analysis. (Lawrence *et al.*, 1988). The achievement of nonstatistical gene localization, as illustrated in Fig. 1, was pivotal in making possible many

---

FIG. 1. Fluorescence detection of single-copy DNA. These experiments utilized the DNA hybridization protocol using biotinylated probes detected with a one-step avidin stain. All photographs were taken by standard microscopy, without image processing. (A) Detection of a single human immunodeficiency virus (HIV) genome (∼ 8 kb probe) integrated into a D group chromosome of the 8E5 cell line (Lawrence *et al.*, 1990b). The chromosomes have been stained with propidium iodide which appears as a dimmer generalized fluorescence, above which the single-copy HIV signal is apparent in identical positions of each sister chromatid. (B) Hybridization of a neuoncogene cosmid sequence to interphase nuclei of a cytogenetic preparation, illustrating the high hybridization efficiency obtained. Each of the two homologs produces a signal in G_1 nuclei. We have found these human chromosome 17 sequences to be centrally localized in lymphocytes; hence, this photograph shows the maximal degree of pairing of any homologous sequences seen to date. As is evident, although the homologs are both in the central region, they are not closely paired (Lawrence *et al.*, 1990a). (C) Monolayer fibroblast cells hybridized with two phage probes for sequences separated by ∼ 1 Mb within the human dystrophin gene. In the interphase nucleus of this female G_1 cell, each of the two X chromosomes shows two closely paired signals. The intact cytoplasm is still present, but may not be clearly visible because of the very low background. As described previously (Lawrence *et al.*, 1990a), this type of analysis on paraformaldehyde-fixed cells demonstrates that the very high resolution that has been demonstrated in interphase nuclei is not a consequence of cytogenetic spreading methodology.

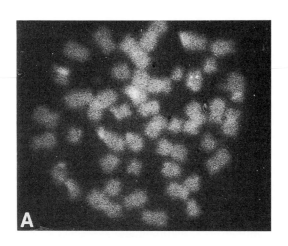

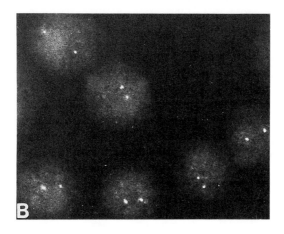

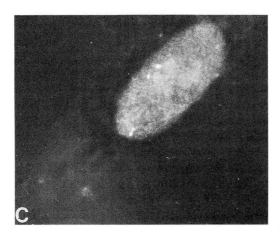

current and future applications of this technology. Without this, it would not be possible to detect single sequences in interphase cells confidently. When mapping genes on metaphase spreads, the high hybridization efficiency generates signal in identical positions on each sister chromatid, providing an internal control which allows unequivocal identification of signal on a single chromosome.

The methodology was initially shown using integrated Epstein-Barr Virus (EBV) sequences as a model system, demonstrating that it was possible to achieve nonstatistical detection of 5 kb of DNA with a one-step fluorescence detection without amplification or image processing. This was then extended for detection of normal human genomic sequences, with a wide range of implications for human gene mapping and cytogenetics. Moreover, nuclear RNA, representing primary transcripts emanating from an individual transcribed sequence, could also be visualized in cells carrying individual copies of latent EBV (Lawrence *et al.,* 1989a) (Fig. 2). This technology has been shown capable of detecting the focal concentration of nuclear RNA expressed from transfected sequences, as well as the primary nuclear transcripts of the HIV genome within just a few hours after infection (Lawrence *et al.,* 1990b). More recently, we have applied these techniques to visualization of intron and exon sequences from endogenous cellular genes (Xing *et al.,* submitted for publication). The power and diversity of these applications have far-reaching implications for a variety of genetic, cell biological, and developmental pursuits, which will be highlighted in Section X.

II. Methods

In this section, we described in detail methods that are currently used in our laboratory for nonisotopic *in situ* hybridization and fluorescent detection of nuclear DNA and RNA. The techniques for localizing DNA and RNA are very similar, but with some key differences. It is often necessary to distinguish rigorously between hybridization to DNA versus RNA; hence, it is useful to consider both protocols and their respective differences. "Skeleton" protocols for both RNA and DNA detection are compared in Table I.

III. Cell Preparations

The choice of cell preparation technique can be critical to the success of an experiment. However, the correct choice varies substantially depending on the

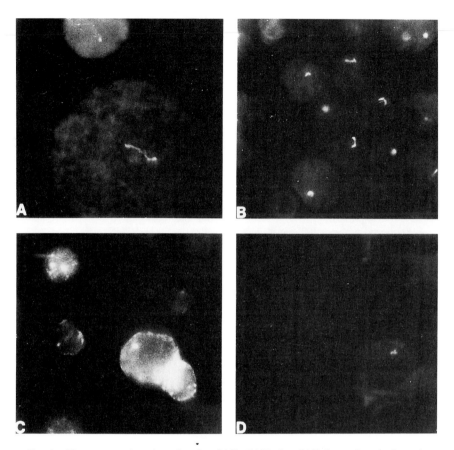

FIG. 2. Fluorescence detection of nuclear RNA. (A) Nuclear RNA for an Epstein-Barr virus (EBV) nuclear antigen expressed in latently infected Namalwa lymphoma cells. In a cytogenetic preparation, this concentrated nuclear RNA often appears in an elongated formation. (B) The same EBV RNA as above, but in paraformaldehyde-fixed intact cells. The sequence detected (BamHI W) is represented in primary transcripts in approximately 20 kb per molecule, but is of very low abundance in the cytoplasm and constitutes less than 1 kb of the mature mRNA (Lawrence et al., 1989a). (C) Detection of HIV RNA in lymphocytes infected in culture. In many cells a focus of newly synthesized transcripts is apparent, suggesting a single site of active transcription early in infection (Lawrence et al., 1990b). These cells were prepared as described in Section III,B and hybridized without denaturation. (D) Nuclear and cytoplasmic RNA transcribed from a neuoncogene transfected sequence. Cells are nonextracted intact 3T3 cells.

properties of the specific cells used and the objective of the experiment. For RNA detection, minimal manipulation of whole cells may be advantageous; however, some RNAs can be detected within isolated interphase nuclei of cytogenetic preparations, which contain both interphase and metaphase cells.

TABLE I

COMPARISON OF PROTOCOLS FOR RNA AND DNA DETECTION

RNA protocol summary (adapted from Lawrence *et al.*, 1989a)	DNA protocol summary (adapted from Lawrence *et al.*, 1988, 1990a)
1. Obtain cell preparations fixed for RNA retention. From 70% ethanol storage, cells may be either dehydrated through graded ethanol and air-dried or rehydrated in PBS	1. Obtain fixed cell preparations; if using cytogenetic preparation, bake at 65°C for at least 2 hours. From 70% ethanol storage, intact cells may be rehydrated in PBS
2. Nick-translate double-stranded DNA for incorporation of biotin-16-dUTP or digoxigenin-11-dUTP	2. If removal of RNA is necessary, samples can be treated with RNase A at this point, or with RNase H posthybridization (see Section VIII)
3. Aliquot and lyophilize probe and non-specific competitor (sheared salmon sperm DNA and *E. coli* tRNA)	3. Nick-translate double-stranded DNA for incorporation of biotin-16-dUTP or digoxigenin-11-dUTP
4. Resuspend probe in 100% formamide and heat for 10 minutes at 70°C	4. Aliquot and lyophilize probe and non-specific competitors (sheared salmon sperm DNA, *E. coli* tRNA). When using genomic DNA probes, add either total sheared species-specific DNA or Cot 1 DNA
5. Add an equal volume of hybridization buffer [containing vanadyl complex (from BRL) as a precaution against RNase] to heat-denatured probe and apply to cells	5. Denature cellular DNA: incubate for 2 minutes at 70°C in 70% formamide in 2 × SSC
6. Incubate in a humid chamber for 3 hours to overnight at 37°C	6. Dehydrate through cold ethanol and air-dry
7. Rinse	7. Resuspend lyophilized probe in 100% formamide and heat for 10 minutes at 70°C
8. Add fluorochrome-conjugated avidin, anti-digoxigenin, or both	8. Add an equal volume of hybridization buffer to heat-denatured probe and apply to dehydrated cells
9. Rinse	9. Incubate in a humid chamber overnight at 37°C
10. Counterstain with DAPI or propidium iodide if desired and mount in antifade solution before viewing	10. Rinse
	11. Add fluorochrome-conjugated avidin, anti-digoxigenin, or both
	12. Rinse
	13. Counterstain with DAPI or propidium iodide if desired and mount in antifade solution before viewing

A. Cytogenetic Preparations of Metaphase and Interphase Cells

It is advantageous to have actively dividing cell cultures for cytogenetic preparations to increase the frequency of metaphase spreads. Hence, splitting cells or feeding with fresh medium 1–2 days prior to fixation is recommended.

Longer incubation times in Colcemid result in greater numbers of cells arrested at metaphase but shorter chromosomes, making banding techniques more difficult. Addition of 3 mM bromodeoxyuridine (BrdU) to cultures 7 hours prior to harvesting alleviates this problem somewhat by elongating the chromosomes and enhancing banding (Vogel *et al.*, 1986; Lawrence *et al.*, 1990a). The presence of BrdU is also valuable for distinguishing G_1 from S/G_2 interphase cells using a fluorescently targeted anti-BrdU antibody (see Lawrence *et al.*, 1990a; McNeil *et al.*, 1991). The protocol described here is adapted from now standard cytogenetic technique.

Suspension and Attached Cells
1. Incubate log-phase cultures in 0.015 μg/ml Colcemid for 30 minutes to 2 hours.
2. *Attached*: Collect media containing floating mitotic cells, and set aside. For preparations greatly enriched in mitotic cells, add 1 × trypsin–EDTA (without Ca^{2+}, Mg^{2+}; Gibco) to flask and gently shake to collect loosely attached mitotic cells. Pool with previously collected cells. If mostly interphase cells are desired, trypsin should be kept in the flask at 37°C for 1–10 minutes to loosen remaining cells. Add medium to stop trypsin reaction.
   Suspension: Collect cells from actively dividing cultures containing both mitotic and interphase cells.
3. Pellet cells, remove most of the medium, leaving about 0.5 ml in which to resuspend the cell pellet completely.
4. Slowly and gently add 0.075 M KCl along the sides of the tube using a continual gentle agitation. Total volume should be 5–10 ml. Incubate at 37°C for 17 minutes.
5. Pellet cells at 3000 rpm, remove all but 0.5 ml of hypotonic solution in which to resuspend the cell pellet *completely*.
6. With continual tapping of the tube, add freshly prepared 3:1 methanol:acetic acid, drop by drop, to 1 ml. Add an additional 4–9 ml and incubate at 25°C for 5–10 minutes.
7. Pellet cells and repeat steps 5 and 6 at least twice. Repeated rinses in fixative may decrease cell debris and promote spreading. Cells can be stored at 4°C overnight and resuspended in fresh fixative before slide preparation.
8. Resuspend cells to working density in 3:1 methanol:acetic acid and drop from a distance onto dry, ethanol-washed glass slides.
9. Allow cells to air-dry overnight.
10. Store dessicated at −80°C.

Peripheral Blood Lymphocytes. For PBLs, 5—10 ml of blood is collected in a heparin-coated tube and 5—10 small drops are added to each 5 ml of prepared chromosome medium 1A (Gibco). After incubation for 72 hours (up to 96 hours has been successful) at 37°C, the procedure described above for suspension cell preparations is used.

Comments. Cells swollen in hypotonic solution prior to fixation are fragile, necessitating care in handling. It is important to resuspend cells well prior to fixation to avoid clumping. Fixative must be prepared fresh using high-quality methanol, preferably from a bottle that has not been opened for more than 2–3 weeks. Cell density of the slide can be fairly high without decreasing hybridization efficiency.

Preparation of well-spread metaphase chromosomes and nuclei free of residual cytoplasm is greatly dependent on the humidity of the room in which the cells are dropped. On warm, humid days, cells can often be dropped directly on the bench top. However, these conditions are not always present and it may be necessary to set up a chamber in which temperature and humidity can be roughly controlled. We are able to control these conditions by setting a covered slide warmer to 37–40°C, overlaying with premoistened paper towels, dropping the cells, and allowing to sit for approximately 60 seconds (not completely dry). Alternatively, a tissue culture incubator can be used. Slides should be checked using a phase microscope to assure that residual cytoplasm around chromosomes and nuclei is minimized.

B. Intact Cells in Suspension

Intact cells in suspension are ideal for flow sorting and provide good three-dimensional material for studies of nuclear organization, but may have somewhat increased background, making it more difficult to see small single-copy signals (< 10 kb). Hybridizing to intact cells on slides, however, is useful for detection of larger DNA sequences or for RNA of which multiple copies are present. For example, this type of preparation was optimal for detection of HIV in infected PBL cells in which a bright focus of active transcription, surrounded by punctate signal in the cytoplasm, was found to be characteristic of HIV-infected cells (Lawrence *et al.*, 1990b). In many types of suspension cells, we have found it unnecessary to extract the cells with detergent to obtain nuclear hybridization.

Protocol
1. Pellet cells.
2. Resuspend in isotonic phosphate-buffered saline (PBS) to a dense cell suspension (approximately $5 \times 500,000$ cells/ml).
3. Using a micropipettor, place a few drops of suspended cells onto multiwell serological slides (Cell-line), which are treated to promote cell adherence, and then draw up excess liquid.
4. Allow cells to air dry briefly to affix to the well surface (significant cell loss results if not allowed to dry).

5. Fix in 4% paraformaldehyde in 1 × PBS for 5–15 minutes (addition of 5 mM MgCl$_2$ may increase cell adhesivity). Paraformaldehyde can be made up in 1 × PBS without MgCl$_2$ and dissolved with NaOH. When the solution clears, add 5 mM MgCl$_2$ and adjust pH to 7.4 with HCl. We store paraformaldehyde at 4°C and use it within 2–3 weeks.

C. Monolayer Cells

Studying DNA and/or RNA in intact cells grown in a monolayer may be advantageous in that cells display dorsal–ventral polarity. The nucleus is generally more easily visualized than in suspension cells, and a minimum of manipulation in slide preparation is required so that morphology is extremely well maintained. The presence of extended cytoplasm also provides the opportunity to study RNA distribution within the cytoplasm, as well as RNA and DNA within the nucleus. Because of the extensive cytoplasm, in some cases it may be necessary to detergent-extract intact cells prior to fixation, allowing for nuclear probe penetration and lower background levels. While this extraction does not appear to impact nuclear RNA (Xing and Lawrence, 1991; Carter et al., in press), cytoplasmic mRNA is very labile; hence, the use of Triton X-100 extractions followed by paraformaldehyde fixation is not appropriate for cytoplasmic mRNA detection (Singer et al., 1989b).

Protocol

1. Plate cells onto 1.5-mm Goldseal glass coverslips that have been acid-washed and autoclaved in 0.5% gelatin. Cells should be subconfluent.
2. Wash cells on coverslips by immersing in Hanks' balanced salt solution (HBSS; Gibco). In cells in which nuclear penetration is limiting, cells may be extracted in 0.5% Triton X-100 in PBS or other isotonic salt solution (see Section III,E, below) for 60–90 seconds. It is convenient to process cells grown on coverslips in small coplin jars (Baxter Scientific).
3. Fix cells in 4% paraformaldehyde in 1 × PBS for 5–15 minutes. Store in 70% ethanol at 4°C for up to several months for RNA (or longer for DNA).

D. Cytospin Preparations

Cells in suspension can also be prepared for hybridization by cytospinning onto glass slides. Cytospinning flattens rounded cells and allows them to adhere well to the slide. The flatter morphology may make it easier to detect single-copy DNA or very weak RNA signals. However, the cytospinning procedure may be somewhat destructive to cell morphology and is likely to release cellular RNases. Despite this reservation, this protocol has been used successfully to visualize the distribution of intron and exon sequences in suspension cells (Xing et al., submitted for publication).

Protocol
1. Harvest cells and resuspend completely in HBSS or cell medium at working density.
2. Pipet 200 μl of cell suspension into cytospin funnel, spin for 5 minutes at 500 rpm.
3. Air-dry briefly to allow cells to adhere to the slide before immersion in fixative. Significant cell loss can be attributed to omitting the air-drying step.
4. Fix cells in 4% paraformaldehyde in 1 × PBS + 5 mM MgCl$_2$ for 5–15 minutes. Store cells in 70% ethanol at 4°C.

E. Nuclear Fractionation

Techniques which allow one to fractionate the nucleus in defined ways may be important in future studies directed at understanding the organization of the nucleus and chromosome. Below is a procedure for isolation of the nuclear "matrix" fraction, adapted from Fey *et al.* (1986). In recent work, we coupled this biochemical fractionation approach with *in situ* hybridization to provide direct visual evidence for the association of a specific nuclear RNA with the nuclear matrix (Xing and Lawrence, 1991).

Protocol
1. Cytospin cells as in Section III,D and air-dry briefly to allow for cell adhesion.
2. Wash cells in 1 × PBS at 4°C and incubate in cytoskeleton buffer [100 mM NaCl, 300 mM sucrose, 10 mM piperazine-N, N'-bis(2-ethanesulfonic acid) (PIPES), pH 6.8, 3 mM MgCl$_2$, 10 μM leupeptin, 2 mM vanadyl complex, and 0.5% Triton X-100] for 10 minuts at 4°C.
3. Extract cells in extraction buffer [250 mM ammonium sulfate, 300 mM sucrose, 10 mM PIPES, pH 6.8, 3 mM MgCl$_2$, 10 mM leupeptin (BMB), vanadyl complex, and 0.5% Triton X-100] for 10 minutes at 4°C.
4. To remove chromatin fraction, digest cells for 20–60 minutes at 20°C in cytoskeletal buffer containing 50 mM NaCl instead of 100 mM NaCl and 100 μg/ml bovine pancreas DNase I (Worthington Biochemical Corp).
5. Terminate reaction by immersion in cytoskeletal buffer containing 0.25 M ammonium sulfate.
6. Fix cells in 4% paraformaldehyde in 1 × PBS for 5 minutes and store in 70% ethanol at 4°C.

IV. Probe Preparation

Our protocols have been optimized for the use of double-stranded DNA probes labeled by nick-translation; however, end-labeled oligonucleotide probes can be used to detect sequences of sufficient abundance (Taneja and Singer, 1989; Lawrence *et al.*, 1989b; Carter *et al.*, in press). RNA probes, which are more labile, may also be used for *in situ* hybridization as described in [2], this volume. We currently prefer probes labeled by incorporation of either biotin-16-dUTP or digoxigenin-11-dUTP; however, many of the previously described detection techniques can provide high sensitivity if used under carefully controlled conditions promoting high hybridization efficiency. We expect that in the future these same hybridization conditions will be applicable to a new generation of probes directly conjugated to a fluorescent reporter.

Proper probe fragment size is imperative to a successful *in situ* hybridization experiment. We ultimately traced a major source of internal variation and high backgrounds with nonisotopic probes to the presence of large probe molecules after nick translation. Additionally, the probe fragment size must be small enough to penetrate the nucleus, which has been recognized as important in intact cells or tissues, but is generally not limiting in cytogenetic preparations. We have found that it is not necessary to cut out the insert from plasmid, phage, or cosmid clones. The presence of vector sequences can, in fact, amplify signal by network formations when large probes were used for cytoplasmic message hybridization. (Lawrence and Singer, 1985). While the ideal *average* probe size is approximately 200 base pairs, the *range* of probe fragment sizes is also extremely important, and should not extend above 700–800 nucleotides, since even small quantities of fragment in this range increase background significantly. Probe size is controlled by varying the concentration of DNase I in the nick-translation reaction, and the appropriate concentration of each new batch of DNase I should be determined, with 100–200 ng/ml providing a good starting point. It is convenient to aliquot DNase I in 0.15 M NaCl, (90%) glycerol and store at $-20°C$ once the working concentration has been determined. The probes are sized and checked for analog incorporation by running them on an alkaline 1.5% agarose gel, transferring to nitrocellulose, and staining with either alkaline phosphatase-conjugated avidin or anti-digoxigenin [Boehringer Mannheim Biochemicals (BMB)].

It is convenient to nick-translate large quantities of probe in one reaction because nonisotopically labeled probes are stable at 4°C for several months and because high concentrations are frequently used for each sample.

V.   Detailed Nick-Translation Procedure

The procedure given here is modified from Rigby *et al.*, 1977. To nick-translate 1 μg of double-stranded DNA for incorporation of biotin or digoxigenin nucleotide analogs:

1. Add the following ingredients in the indicated order:
   a. 10 μl of 10 × nick-translation buffer [0.5 M Tris-HCl, pH 7.5, 0.1 M MgSO$_4$, 500 μg/ml bovine serum albumin (molecular biology grade) 1 mM dithiothreitol (DTT); store at $-20°$C],
   b. 10 μl, 600 μM each of dATP, dCTP, dGTP Nucleotide stocks of 10 mM are supplied by P. L. Biochemicals or BMB and stored at $-80°$C. The working 600 μM diluted stocks should be stored at $-20°$C.
   c. 1 μg of DNA,
   d. 6 μl of either 1 mM biotin-16-dUTP or 1 mM digoxigenin-11-dUTP,
   e. H$_2$O to final reaction volume of 100 μl,
   f. 10 μl of 1–2 μg/ml DNase I (Worthington),
   g. 10 μl of DNA polymerase I, endonuclease free (5 U/μl) (BMB).
2. Mix and spin to pool solution.
3. Incubate for 2 hours at 15°C.
4. Stop reaction by addition of 1% SDS and by heating for 10 minutes at 65°C.
5. Remove unincorporated nucleotides by ethanol precipitation with addition of 10 μg of carrier salmon sperm DNA, 0.1 volume of 3 M sodium acetate and 2.5 volumes of 95% ethanol. Cool in the $-20°$C freezer for at least 1 hour.
6. Spin precipitated DNA at 10,000 rpm for 30 minutes.
7. Wash pellet with cold 70% ethanol.
8. Air-dry and resuspend probe in sterile double-distilled H$_2$O to desired concentration and store at 4°C.

VI.   Denaturation of Cellular DNA

A major difference between hybridization to DNA and RNA is that cellular DNA must be denatured before hybridization or essentially no hybridization is obtained. This denaturing step is critical to the success of an experiment; hence, the length of denaturation time, temperature, and pH of the denaturation solution should be carefully controlled.

Our Standard Protocol. (Modified from Harper *et al.,* 1981; Gerhard *et al.,* 1981).   Denature cells by placing in a 70°C preheated solution of 70% formamide 2 × SSC, pH 6.5–7.5, for exactly 2 minutes. It is helpful to preheat denaturing solution in a microwave and maintain 70°C temperature in a waterbath; however, prolonged incubations at 70°C cause the pH of the solution to drop, which can adversely affect results. Immediately immerse cells in cold 70% ethanol for 5 minutes, then cold 100% ethanol, and air-dry.

Sodium Hydroxide Denaturation (Raap *et al.,* 1986). Sodium hydroxide denaturation conveniently denatures cellular DNA while simultaneously removing RNA by alkaline hydrolysis.

Remove cells from storage (bake if using cytogenetic preparations and immerse in a solution of 0.07 N NaOH in 70% ethanol for 4–5 minutes at room temperature. (Treatment for 2 minutes may be sufficient for DNA denaturation but 4 minutes is often necessary to remove RNA fully). Wash cells in 70% and then 100% ethanol for 5 minutes each and air dry. (Recently, we have found NaOH concentrations of 0.2 N to be more reliable than 0.07 N.)

Denaturation of Probe and Cellular DNA Simultaneously. It is possible to denature cellular DNA and probe DNA simultaneously by applying probe in formamide–hybridization buffer directly to cells and heating at 65 to 100°C. We have tested denaturation using the higher temperatures and found that this is often destructive to chromosome morphology and increases the likelihood of high background. The lower temperature for simultaneous denaturation is used successfully by others (Lichter *et al.,* 1990).

VII. Hybridization

Genomic probes contain repetitive sequences which hybridize to small interspersed repeat sequences dispersed throughout the genome which obscure the "single-copy" signal. Fortunately, repetitive sequences can generally be sufficiently competed with unlabeled species-specific total DNA. Even more effective as a competitor, however, is Cot 1 DNA (Bethesda Research Labs), which essentially eliminates hybridization to repetitive sequences.

Many of the critical parameters fundamental to DNA hybridizations also apply to RNA hybridizations. Unlike DNA hybridizations, however, RNA hybridizations are preferably done for 3 hours to avoid potential RNA degradation problems. DNA hybridizations are routinely done overnight for convenience; however, high hybridization efficiency can be obtained within 3–4 hours. Hybridizations to highly repeated sequences or reactions using oligonucleotide probes at high concentration generally require only 30–60 minutes.

A problem which may be encountered in interphase DNA hybridization is the presence of transcripts generated from the gene of interest. Several approaches can be applied to overcome this, including RNase treatment and NaOH denaturing. However, we find this may not always be necessary, and prefer to omit it because RNase A treatment frequently results in degradation of DNA and morphology. This can be circumvented by the use of RNase H posthybridization, to digest RNA in an RNA–DNA hybrid (see Section VIII).

Long-term storage of cells (beyond a few months) in 70% ethanol may also pose RNA retention problems; however, this is not a concern for DNA hybridizations.

Detailed Hybridization Protocol

1. Aliquot and lyophilize 50 ng of each probe with 10 μg each of sonicated salmon sperm DNA and *E. coli* tRNA. If genomic clones are used, also aliquot either 10 μg of Cot 1 DNA or 10 μg of sheared total species-specific DNA. Multiple clones can be pooled and hybridized together without increasing background levels. This allows for double-label experiments in which multiple sequences labeled with biotin-16-dUTP or digoxigenin-11-dUTP can be detected simultaneously in two colors. Higher probe concentrations can be used for DNA detection, and lower amounts (10–20 ng) may be sufficient for RNA detection.

2. Resuspend lyophilized probe DNA in 10 μl of 100% formamide (nondeionized, Sigma). Just prior to step 3, heat denature probe for 10 minutes at 70°C.

3. To the heat-denatured probe mix, add an equal volume of hybridization buffer, consisting of 1 part 20 × SSC, 1 part nuclease-free BSA (Boehringer Mannheim), 1 part sterile double-distilled H_2O, and 2 parts autoclaved 50% dextran sulfate (500,000 MW, from Pharmacia). (For RNA hybridization, replace water with 1 part 2 mM vanadyl adenosine.) Immediately apply to cells. Lay a small piece of Parafilm over the hybridization solution to ensure all cells are covered with the solution and to eliminate air bubbles.

4. Cover all samples with a large sheet of Parafilm and completely seal around the edges.

5. Incubate in a 37°C humidified chamber for 3 hours to overnight for DNA hybridizations, and preferably for 3–4 hours for RNA hybridization.

6. Rinse samples for 30 minutes each in 50% formamide–2 × SSC at 37°C, 2 × SSC at 37°C, and 1 × SSC on a shaker at room temperature. Shorter rinse times have also been used successfully, but longer rinses generally have no advantage. Samples can be rinsed in small coplin jars, since large volumes of rinsing solutions are unnecessary.

VIII.   Controls Discriminating RNA versus DNA Hybridization

Hybridization to cellular DNA can result in detection of RNA sequences if the target DNA is expressed; therefore it may be desirable to remove RNA. To do so, cells can be pretreated with 100 μg/ml of RNase A for 1 hour in 2 × SSC prior to denaturation and hybridization, or, preferably, after hybridization with RNase H at 8 U/ml for 1–2 hours at 37°C in 100 mM KCl, 20 mM Tris-HCl (pH 7.5), 1.5 mM MgCl$_2$, 50 μg/ml of BSA, 1 mM DTT, 0.7 mM ethylenediaminetetraacetic acid (EDTA), and 13 mM N-2-hydroxy-ethylpiperazine-N'-2-ethanesulfonic acid (HEPES) (Minshull and Hunt; 1986; Lawrence et al., 1988). RNase A is readily effective in removing RNA; however, its use prior to denaturation sometimes can result in loss of DNA and nuclear morphology. Alternatively, NaOH denaturation, as mentioned, is effective in removal of RNA. RNase or NaOH treatment can also be used as a control for an RNA hybridization, the absence of signal in treated cells proving that hybridization is to RNA and not DNA. Other controls for RNA experiments include hybridization to cells that do not express the gene of interest, or the use of probes to nonexpressing genes in the same cell or to nonrelated sequences. Incubation of cell cultures with actinomycin D to inhibit transcription is another example of a negative control to confirm detection of primary transcripts (Lawrence et al., 1989a, 1990b).

As mentioned, the presence of signal on each sister chromatid provides indisputable evidence of bonafide signal in a DNA hybridization. However, such a "built-in" control is not provided in interphase cells. The easiest control for a DNA–DNA hybridization experiment would be the inclusion of a nondenatured cellular DNA sample (from which signal should be absent). Other controls could include the addition of a "nonrelevant" probe or, as a positive control, a probe of comparable target size that has been used successfully in previous experiments.

IX.   Detection

The procedures we use for biotinylated or digoxigenin-labeled probes are essentially identical, and are based on a rapid one-step reaction. Biotin is routinely detected with FITC–avidin (ENZO) and digoxigenin with rhodamine or fluorescein-conjugated sheep anti-digoxigenin (BMB). Texas-Red–avidin can be used successfully. Various modifications of avidin, such as streptavidin or avidin-DH, have also been tested, but were found to provide no advantage and in some cases resulted in less intense signal. Conveniently, probes labeled with these two different analogs can be hybridized and detected simultaneously, thus allowing for simple and efficient double-label (Johnson et al., 1991). Another method of simultaneous hybridization and detection can be done using probes covering different-size target sequences, such as a phage and a cosmid. The size and intensity of the signal are proportional to the size of the target sequence (Lawrence et al., 1988; Lawrence, 1990).

For single-label experiments, the choice of the probe label can depend on the type of cell preparation used. Because digoxigenin is detected with an antibody, the probe detection step generates more background than avidin in cells with thick cytoplasms, but not in cytogenetic preparations. However, digoxigenin is particularly valuable for the detection of small single-copy sequences (2 kb), because it provides stronger signal than does biotin–avidin without the use of multistep amplification procedures. While biotin-labeled probes for many intact cell types may have a lower background level, some cells (such as chick embryonic myoblasts) have endogenous biotin which can cause background. Alkaline phosphatase detection may also be used, but does not provide as high a resolution. For example, hybridization to HIV primary transcripts in PBL cells revealed the presence of a bright nuclear focus of RNA surrounded by more diffuse cytoplasmic RNA after fluorescent detection (Lawrence *et al.,* 1990b). These distinguishing characteristics were not apparent using the alkaline phosphatase system, in which the cells were evenly stained (Singer *et al.,* 1989). Alkaline phosphatase, however, does have an advantage over fluorescence in that there is no fading of the stained samples, which is useful when long-term storage of hybridized cells is needed (see for example, Garson *et al.,* 1987).

Our laboratory detects single-copy DNA sequences as small as 2 kb (Johnson *et al.,* submitted for publication) and multiple copies of RNA without amplification procedures or without image-processing technologies. However, if necessary, procedures for amplification of biotin–avidin signal have been described and are applied in many laboratories (Pinkel *et al.,* 1986). Cellular DNA is generally counterstained with either 4′,6-diamidino-2-phenylindole (DAPI) for total DNA staining or with propidium iodide for staining of double-stranded DNA and RNA. DAPI and propidium iodide can be added directly to antifade mounting medium [90% glycerol, 1 × PBS, 2.5% 1,4-Diazabicyclo[2.2.2]octane (DABCO) (Sigma)]. Slides are stored in the dark, and cold storage may help preserve fluorescence.

Detection Protocol

1. Equilibrate slides in 4 × SSC for 10 minutes.
2. Blot slide and add 100 μl of 2 μg/ml FITC–avidin and/or 2 μg/ml rhodamine or fluorescein sheep anti-dioxigenin in 4 × SSC + 1% BSA (see Note below).
3. Incubate for 30–60 minutes at 37°C.
4. Rinse slides at room temperature for 10 minutes each in 4 × SSC, 4 × SSC + 0.1% Triton X-100, 4 × SSC.
5. Mount cells in antifade medium containing counterstain.

Note.   If cells were incubated with BrdU prior to fixation, fluorescein or rhodamine anti-BrdU can be added at 1 : 1000 (Becton Dickinson, BMB) simultaneously with fluorescein–avidin or anti-digoxigenin. DNA, however, must be single-stranded for detection of BrdU; thus, cells must be denatured prior to staining if not previously denatured for hybridization (McNeil *et al.,* 1991).

Alkaline phosphatase detection is done with alkaline phosphatase–avidin (BMB) at 1:400 or with alkaline phosphatase anti-digoxigenin (BMB) at 1:500, according to vendor specifications described elsewhere (Singer *et al.,* 1989a). Staining with alkaline phosphatase should be carefully monitored because background levels increase with increasing incubation times.

Microscopy and Photography. All fluorescently stained samples are directly visualized at 630× or 1000× magnification on a Zeiss Axioplan or Axioscope microscope equipped with epifluorescence filters. Our laboratory generally uses standard fluorescence microscopy; however, a variety of more sophisticated instrumentation is currently available, but not necessary for the applications described here. The use of computerized image-processing techniques has been reviewed elsewhere (Lawrence, 1990; Lichter *et al.,* 1991). The laboratory routinely uses epifluorescence optics for DAPI, fluorescein, and rhodamine, and a recently developed dual-band filter for simultaneous visualization of fluorescein and Texas Red/rhodamine (Omega Corp.) This dual-band filter set allows for visualization of two fluorochromes simultaneously without the optical shift in relative image position that unavoidably occurs with the use of two separate filters. The precision of double-label hybridization and detection of two or more colors allows for ordering of multiple single-copy sequences along the length of a chromosome, or, for closer distances, in interphase cells. This methodology also allows for visualization of two different transcripts within the same cell.

Photography is done using either Tri-X (ASA 400 film) for black and white images or Ektachrome 400 film for color. Single-copy sequences are exposed for 60–120 seconds, multiple sequences for 30–120 seconds, and DAPI or propidium iodide for 5 seconds. A critical aspect of microscopy is fading of fluorescent signal. Samples must be mounted in an antibleach medium prior to visualization. Antibleach medium, however, may become ineffective over time and will provide the best results when freshly made. Cells can be stained for either total DNA by nuclear staining with DAPI or with propidium iodide for double-stranded RNA and DNA.

X. Applications and Discussion

It is clear that over the past few years nonisotopic *in situ* hybridization has emerged as a powerful and versatile approach for biomedical science and industry. However, this technology is not at an end point in its development, and new capabilities will continue to be innovated in the future. Now that it is established that this approach can reproducibly provide precise and unambiguous molecular information on individual cells, it is likely that new and improved ways of directly labeling probes will be successfully pursued in the near future. Eventually, it will be important to bring the technique to the point that single-copy oligonucleotide probes can be detected, and even small, low-abundance RNA molecules can be readily visualized. It is important to realize that there are both DNAs and RNAs which cannot be detected by fluorescence at this time. Further, any application of *in situ* hybridization should be extremely rigorous with regard to the inclusion of controls, since in our experience there are many sources of background which can mimic bona fide signal. For example, using some cDNA probes, we have found presumptive hybridization to highly abundant single-stranded nuclear sequences which appear to represent something other than the target mRNA. Additionally, we have seen several examples of probe sticking to specific cell types within a preparation. This "specific nonspecific" background can be very deceptive and must be carefully controlled.

Applications of nonisotopic hybridization to DNA are presented in [1] and [5], this volume and have been reviewed thoroughly from our laboratory in recent publications (Lawrence, 1990; McNeil *et al.*, 1991). Therefore, we will treat DNA applications briefly, considering its enormous importance, and include discussion of applications of RNA hybridization as well.

An obvious and important application of the ability to detect single-copy genes with fluorescence is to contribute to mapping of the human genome (and others). The resolution of fluorescence, as well as the speed and efficiency of this approach, is far superior to previous autoradiographic techniques which have a limit of resolution on the order of 10^4 or 10^5 kb. There has been a need for physical mapping techniques which provide resolution in the range of 1 or 2 Mb and below, and it would be particularly valuable if this approach could map genes across a broad range of distances, to help bridge the gap between lower and higher resolution techniques.

Recently, work from our laboratory as well as of others (Lawrence *et al.*, 1988, 1990a; Trask *et al.*, 1989, 1991; Lichter *et al.*, 1990) has shown that fluorescence hybridization can provide such an approach. Rigorous characterization of the limits and versatility of this technique showed that, in practice, resolution on chromosomes generally did not extend below 2 Mb. However, an important finding was that resolution could be greatly enhanced by analyzing the distance between two sequences in the interphase nucleus, where the chromatin is much less condensed (Lawrence *et al.*, 1988, 1990a; Trask *et al.*, 1989, 1991). This was initially based on the demonstration that one could detect two sequences at either end of an integrated EBV genome approximately 130 kb apart (Lawrence *et al.*, 1988). At the time, it was surprising that such molecular distances could be resolved with a light microscopic technique. Interphase distance was shown to exhibit a strong correlation with DNA distance over the ranges examined, for both dihydrofolate reductase coding sequences in Chinese hamster (Trask *et al.*, 1989) and human dystrophin sequences in normal human cells (Lawrence *et al.*, 1990a), providing a needed approach for determining physical order. At the same time, this approach provides new information and insights into the packaging of DNA, particularly within the interphase nucleus. Another major effort from several groups has involved the adaptation of fluorescence hybridization with chromosome banding techniques so that probes and chromosomes could be identified simultaneously, (Viegas-Pequignot *et al.*, 1989; Lawrence *et al.*, 1990a; Fan *et al.*, 1990; reviewed in McNeil *et al.*, 1991). As reviewed elsewhere (Lawrence, 1990), this approach allows direct measurement of the condensation of DNA as it folds along the chromatin fiber, and also allows analysis of somatic pairing for homologous genes, the position of active genes, and the presence and position of newly replicated DNA. It is particularly useful for determining the relative positions of multiple sequences that can be detected simultaneously when more than one probe is used. For the detection of the two ends of the EBV genome, we used two probes to targets of different sizes and distinguished each by relative fluorescence intensity. However, two fluorochromes of different colors are an advantage when evaluating the order of three sequences simultaneously, either on the chromosome or in the interphase nucleus. This is now made possible using the dual-band filter set described above, which allows the precise registration of red and green fluorescence by standard epifluorescence microscopy (see Tkachuk *et al.*, 1990; Lawrence *et al.*, 1990a; Johnson, 1991; Trask *et al.*, 1991).

The advent of nonstatistical single-copy gene detection in both individual metaphase and interphase cells makes possible a new era of "molecular cytogenetics." The speed, convenience, and resolution of nonisotopic probes makes it feasible to characterize cytogenetic aberrations cytologically with a much greater degree of precision. Cytogenetics had been limited to analysis, based on banding, of relatively gross chromosomal rearrangements involving 10 Mb or more of DNA. Several groups have shown the effectiveness of using chromosome libraries or repeat sequences to enumerate specific chromosomes involved in common aneuploidies or to detect translocations involving chromosome segments (Cremer *et al.*, 1986; Pinkel *et al.*, 1988; Lichter *et al.*, 1988). It is now possible to evaluate aberrations in specific genes (Tkachuk *et al.*, 1990; Ried *et al.*, 1990; Lawrence *et al.*, 1991). *In situ* probe technology can be used, advantageously, on a small number of cells, which need be neither dividing nor expressing, and each homolog can be evaluated rapidly for the presence, duplication, or deletion of a specific sequence. We are applying this approach to detect female carriers of Duchenne's muscular dystrophy in interphase cells (Lawrence *et al.*, 1991), which allows a direct qualitative evaluation of each homolog. This contrasts with Southern blots, in which the presence of the normal allele makes quantitative densitometry essential. Application to heterozygous deletions may prove particularly important in the detection and study of tumor suppressor genes, where it is believed that the progression of tumorigenesis may initially involve a heterozygous deletion at a specific site.

The ability to detect specific RNA sequences within the nucleus is relevant to cellular and developmental biology, and has potential for human genetics. Foremost, this approach is important for the detection of gene expression in general. Our results thus far suggest that the highest concentration of an individual RNA may reside at or near its site of transcription. The resolution inherent in fluorescence detection translates into greater sensitivity in that the ability to detect an RNA is a function not only of its abundance within the total cell, but of its focal concentration. Hence, the ability to detect an accumulation of primary transcripts enhances the ability to detect the expression of that gene and could prove important for any situation in which low-abundance sequences need to be detected or anonymous DNA clones need to be screened for expression. Furthermore, the expression of newly transcribed RNA within the nucleus represents the first indication of gene expression. Hence, studies directed at the regulation of gene expression can define precisely the timing of gene activation relative to cues in the cellular microenvironment by visually recording the initiation of transcript production. This type of analysis could reveal cellular parameters important for gene regulation, such as cell–cell interactions or cell cycle events.

Detection of nuclear RNA with high resolution is of fundamental importance for the study of nuclear structure, since this approach uniquely superimposes molecular and spatial information. The steady-state distribution of an RNA provides an indication of the pathway along which the RNA is transcribed, processed, and transported out of the nucleus. The viscosity of the nucleoplasm suggests that it could be compartmentalized in order to efficiently carry out its different biochemical functions. For example, it has been proposed that active genes are localized close to the nuclear envelope in order to facilitate transport of mRNA out of the nucleus (e.g., Hutchinson and Weintraub, 1985). The visualization of primary transcripts within the nucleus provides an approach to correlate spatial information with biochemical events and to begin to assess the extent to which this is true. For instance, as shown in Fig. 3, in the case of the integrated EBV genome we found that both the gene and its primary transcripts were in a position in the internal half of the nucleus and did not juxtapose the nuclear envelope (Lawrence *et al.*, 1988, 1989a). The highly localized distribution of this RNA, evidenced as a fluorescent "track" of

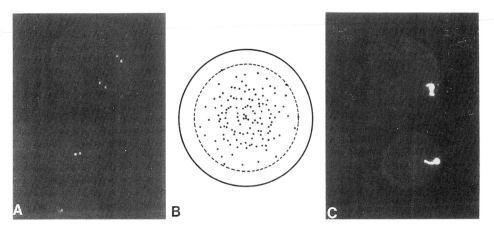

FIG. 3. Nuclear position of Epstein-Barr virus (EBV) DNA and RNA. Some three-dimensional information can be obtained by analyzing suspension cells that have random rotation before placing on slides. (A) Comparison of gene localization in diploid and tetraploid cells. The smaller (lower) nucleus has just one pair of signals representing two EBV genomes closely integrated on the chromosome. In larger, most likely tetraploid nuclei, two distinct pairs of spots are observed, with positions that appear to be nonrandom with respect to one another. (B) A composite showing the localization of the EBV genome with respect to the center in many randomly rotated nuclei. Analysis of freely rotated nuclei provides three-dimensional information indicating a clearly nonrandom placement of this sequence in the inner 50% of the nuclear volume (Lawrence et al., 1988). (C) EBV RNA in the nucleus of a lymphoma cell, illustrating that the track of nuclear RNA frequently extends into the peripheral region of the nucleus from which the corresponding gene, as shown in A and B, is excluded.

a few hundred transcript molecules, immediately gives insight into the basic structural organization of the nucleus. This observation does not support a free diffusion model of RNA metabolism, but is more consistent with a solid-state model wherein RNA and possibly even DNA are physically constrained by association with nuclear substructure. To investigate this possibility, we coupled nuclear fractionation with *in situ* hybridization (Xing and Lawrence, 1991) to demonstrate that the tracks of viral RNA were quantitatively and morphologically preserved throughout fractionation procedures which removed 95% of the nuclear protein, phospholipid, and DNA (Fey *et al.*, 1986; Berezney, 1980). The organization of the nucleus may be more fully understood by fractionating it or disrupting RNA metabolism in defined ways and then using *in situ* hybridization to determine these effects on the distribution of specific sequences.

In order to assess whether high concentrations of RNA are in the same locale as processing components, it is possible to couple *in situ* hybridization with immunofluorescent detection of processing components. We have found general compartmentalization of *Pol* II transcripts throughout the nucleus, which indicates that defined domains exist which are involved in RNA processing and, possibly, transcription (Carter *et al.,* in press). Because a variety of evidence indicates that processing and transcription are closely linked temporally, RNA concentrations in the nucleus may represent sites of both processing and transcription. While visualization of specific RNAs may suggest the location of the transcribing gene, ultimately the distribution of active genes will have to be rigorously demonstrated by direct hybridization to the genes themselves. Hence, in the future, a comprehensive understanding of nuclear structure as it relates to function will require an analytical approach which integrates DNA, RNA, and protein distributions within individual cells.

Finally, it is relevant to consider that the molecular basis of some genetic defects will occur at the RNA processing/transport level, given the large amount of processing a number of primary transcripts must undergo. It may in fact be possible that genetic disease may derive from defects that do not disrupt the amino acid sequence of a protein, but are a consequence of changes which affect chromatin organization or nuclear RNA metabolism. Hence, as we are better able to evaluate nuclear organization of nucleic acid sequences this may contribute to understanding the molecular basis for certain genetic defects. These defects will in turn provide an opportunity to study the effects of specific mutations on spatial organization of RNA and DNA within the nucleus.

REFERENCES

Bauman, J. G., Wiegant, J., and van Duijn, P. (1981). Cytochemical hybridization with fluorochrome-labeled RNA. *Histochemistry* **73,** 181–193.

Bauman, J., Pinkel, D., Trask, B., and Van der Ploeg, M. (1989). Flow cytometric measurement of specific DNA and RNA sequences. *In* "Flow Cytogenetics" (J. W. Gray, ed.), p. 275. Academic Press, New York.

Berezney, R. (1980). Fractionation of the nuclear matrix. *J. Cell Biol.* **85,** 641–650.

Broker, T. R., Angerer, L. M., Yen, P. H., Hershey, N. D., and Davidson, N. (1978). Electron microscopic visualization of tRNA genes with ferritin avidin: Biotin labels. *Nucleic Acids Res.* **5,** 363.

Carter, K. C., Taneja, K., and Lawrence, J. B. Discrete nuclear domains of poly (A) RNA and their relationship to the functional organization of the nucleus. *J. Cell Biol.* (in press).

Cremer, T., Landegent, J. E., Bruckner, H., Scholl, H. P., Schardin, M. D., Hager, H. D., Devilee, P., Pearson, P. L., van der Ploeg, M. (1986). Detection of chromosome aberrations in the human interphase nucleus by visualization of specific target DNAs using *in situ* hybridization diagnosis of trisomy 18 with probe L1.84. *Hum. Genet.* **74,** 346.

Dale, R. M. K., Martin, E., Livingstone, D. C., and Ward, D. C. (1975). Direct covalent mercuration of nucleotides and polynucleotides. *Biochemistry* **14,** 2447.

Fan, Y.-S., Davis, L. M., and Shows, T. B. (1990). Mapping small DNA sequences by fluorescence *in situ* hybridization directly on banded metaphse chromosomes. *Proc. Natl. Acad. Sci. U.S.A.* **87,** 6223.

Fey, E. G., Krochmalnil, G., and Penman, S. (1986). The non-chromatin substructure of the nucleus: The ribonucleoprotein-containing and RNP depleted matrices analyzed by sequential fractionation and resinless section EM. *J. Cell Biol.* **102**, 1654–1665.

Gall, J. G., and Pardue, M. L. (1969). Formation and detection of RNA–DNA hybrid molecules in cytological preparations. *Proc. Natl. Acad. Sci. U.S.A.* **63**, 378–383.

Garson, J. A., van den Berghe, J. A., and Kemshead, J. T. (1987). Novel nonisotopic *in situ* hybridization technique detects small (1 kb) unique sequences in routinely G-banded human chromosomes: Fine mapping of N-myc and β-NGF genes. *Nucleic Acids Res.* **15**, 4761.

Gerhard, D. S., Kawasaki, E. S., Carter Bancroft, F., and Szabo, P. (1981). Localization of a unique gene by direct hybridization *in situ*. *Proc. Natl. Acad. Sci. U.S.A.* **78**, 3755–3759.

Harper, M. E., Ullrich, A., and Saunders, G. R. (1981). Localization of the human insulin gene to the distal end of the short arm of chromosome 11. *Proc. Natl. Acad. Sci. U.S.A.* **78**, 4458–4460.

Hopman, A. H., Wiegant, J., Tesser, G. I., and van Duijn, P. (1986). A nonradioactive method based on mercurated nuclei acid probes and sulfhydryl-hapten ligands. *Nucleic Acids Res.* **14**, 6471.

Hutchinson, N., and Weintraub, H. (1985). Localization of DNase I-sensitive sequences to specific regions of interphase nuclei. *Cell (Cambridge, Mass.)* **43**, 471–482.

John, H., Birnstiel, M. L., and Jones, K. N. (1969). RNA–DNA hybrids at the cytological level. *Nature (London)* **223**, 582.

Johnson, C. V., McNeil, J. A., Carter, K. C., and Lawrence, J. B. (1991). A simple, rapid technique for precise mapping of multiple sequences in two colors using a single optical filter set. *Genet. Anal. Tech. Appl.* **8**(2), 75.

Johnson, C. V., Glaam, M., Green, N., Cool, D., Fisher, E., Hill, D., and Lawrence, J. B. (1992). Chromosomal localization of human PTPase sequences: specificity of fluorescence *in situ* hybridization using genomic vs. cDNA probes. (submitted for publication).

Langer, P. R., Waldrop, A. A., and Ward, D. C. (1981). Enzymatic synthesis of biotin-labeled polynucleotides: Novel nucleic acid affinity probes. *Proc. Natl. Acad. Sci. U.S.A.* **78**, 6633–6637.

Langer-Safer, P. R., Levine, M., and Ward, D. C. (1982). Immunological method for mapping genes on *Drosophila* polytene chromosomes. *Proc. Natl. Acad. Sci. U.S.A.* **79**, 4381–4385.

Lawrence, J. B. (1990). A fluorescence *in situ* hybridization approach for gene mapping and the study of nuclear organization. *In* "Genome Analysis Volume 1: Genetic and Physical Mapping" (K. Davies and S. Tilghman, eds.), pp. 1–38. Cold Spring Harbor Laboratory, Cold Spring Harbor, New York.

Lawrence, J. B., and Singer, R. H. (1985). Quantitative analysis of *in situ* hybridization methods for detection of actin gene expression. *Nucleic Acids Res.* **13**, 1777–1799.

Lawrence, J. B., and Singer, R. H. (1986). Intracellular localization of messenger RNAs for cytoskeletal proteins. *Cell (Cambridge, Mass.)* **45**, 407–415.

Lawrence, J. B., Villnave, C. A., and Singer, R. H. (1988). Sensitive high-resolution chromatin and chromosome mapping *in situ*: Presence and orientation of two closely integrated copies of EBV in a lymphoma line. *Cell (Cambridge, Mass.)* **52**, 51–61.

Lawrence, J. B., Singer, R. H., and Marselle, L. M. (1989a). Highly localized tracks of specific transcripts within interphase nuclei visualized by *in situ* hybridization. *Cell (Cambridge, Mass.)* **57**, 493–502.

Lawrence, J. B., Taneja, K., and Singer, R. H. (1989b). Temporal resolution and sequential expression of muscle-specific genes revealed by *in situ* hybridization. *Dev. Biol.* **133**, 235–246.

Lawrence, J. B., Singer, R. H., and McNeil, J. A. (1990a). Interphase and metaphase resolution of different distances within the human dystrophin gene. *Science* **249**, 928–932.

Lawrence, J. B., Marselle, L. M., Byron, K. S., Johnson, C. V., Sullivan, J. L., and Singer, R. (1990b). Subcellular localization of low-abundance human immunodeficiency virus nucleic acid sequences visualized by fluorescence *in situ* hybridization. *Proc. Natl. Acad. Sci. U.S.A.* **87**, 5420–5424.

Lawrence, J. B., Caskey, T., and McNeil, J. A. (1991). Carrier detection for Duchenne's muscular dystrophy by *in situ* hybridization. *In* "Miami Short Reports, Volume 1—Advances in Gene Technology: The Molecular Biology of Human Genetic Disease" (Fazal Ahmad *et al.,* eds.), p. 114. Boehringer Mannheim Biochemicals, Indianapolis, Indiana.

Lichter, P., Cremer, T., Tang, C.-J. C., Hatkins, P. C., Manuelidis, L., and Ward, D. C. (1988). Rapid detention of human chromosome 21 aberrations by *in situ* hybridization. *Proc. Natl. Acad. Sci. U.S.A.* **85,** 9664.

Lichter, P., Tang, C. C., Call, K., Hermanson, G., Evans, G., Housman, D., and Ward, D. C. (1990). High-resolution mapping of human chromosome II by *in situ* hybridization with cosmid clones. *Science* **247,** 64.

Lichter, P., Boyle, A. L., Cremer, T., and Ward, D. C. (1991). Analysis of genes of chromosomes by non-isotopic *in situ* hybridization. *Genet. Anal. Tech. Appl.* **8**(2), 24–35.

McNeil, J. A., Johnson, C. V., Carter, K. C., Singer, R. H., and Lawrence, J. B. (1991). Localizing DNA and RNA within nuclei and chromosomes. *Genet. Anal. Tech. Appl.* **8**(2), 41–58.

Manning, J. E., Hershey, N. D., Broker, T. R., Pellegrini, M., Mitchell, H. K., and Davidson, N. (1975). A new method of *in situ* hybridization. *Chromosoma* **53,** 107.

Minshull, J., and Hunt, T. (1986). The use of the single-stranded DNA and RNase H to promote hybrid arrest of translation of mRNA/DNA hybrids in reticulocyte lysate cell-free translations. *Nucleic Acids Res.* **14,** 6433–6451.

Narayanswami, S., and Hamkalo, B. (1991). DNA sequence mapping using electron microscopy. *Genet. Anal. Tech. Appl.* **8**(1), 14–23.

Pinkel, D., Straume, T., and Gray, J. W. (1986). Cytogenetic analysis using quantitative, high-sensitivity fluorescence hybridization. *Proc. Natl. Acad. Sci. U.S.A.* **83,** 2934.

Pinkel, D., Landegent, C., Collins, C., Fuscoe, J., Seagraves, R., Lucas, J., and Gray, J. (1988). Fluorescence *in situ* hybridization with human chromosome-specific libraries: Detection of trisomy 21 and translocations of chromosome 4. *Proc. Natl. Acad. Sci. U.S.A.* **85,** 9138.

Raap, A. K., Marijnen, J. G. J., Vrolijk, J., and van der Ploeg, M. (1986). Denaturation, renaturation, and loss of DNA during *in situ* hybridization procedures. *Cytometry* **7,** 235–242.

Ried, T., Mahler, V., Vogt, P., Blonden, L., Van Ommen, G. J. B., Cremer, T., and Cremer, M. (1990). Direct carrier detection by *in situ* suppression hybridization with cosmid clones of the Duchenne/Becker muscular dystrophy locus. *Hum. Genet.* **85,** 581–586.

Rigby, P. W., Dickman, M., Rhodes, C., and Berg, P. E. (1977). Labelling deoxyribonucleic acid to high specific activity *in vitro* by nick translation with DNA polymerase. *J. Mol. Biol.* **113,** 237–251.

Singer, R. H., Lawrence, J. B., and Rashtchian, R. N. (1987). Toward a rapid and sensitive *in situ* hybridization methodology using isotopic and nonisotopic probes. *In* "*In Situ* Hybridization: Application to the Central Nervous System" (K. Valentino, J. Eberwine, and J. Barchas, eds.), p. 71. Oxford Univ. Press, New York.

Singer, R. H., Byron, K. S., Lawrence, J. B., and Sullivan, J. L. (1989a). Detection of HIV-1 infected cells from patients using non-isotopic *in situ* hybridization. *Blood* **74,** 2295.

Singer, R. H., Langevin, G. L., and Lawrence, J. B. (1989b). Ultrastructual visualization of cytoskeletal mRNAs and their associated proteins using double-label *in situ* hybridization. *J. Cell Biol.* **108,** 2343.

Taneja, K., and Singer, R. H. (1989). Detection and localization of actin mRNA isoforms in chicken muscle cells by *in situ* hybridization using biotinated oligonucleotide probes. *J. Cell. Biochem.* **44,** 241–252.

Tkachuk, D. C., Westbrook, C. A., Andreeff, M., Donlon, T. A., Cleary, M. L., Suryanarayan, K., Homage, M., Redner, A., Gray, J., and Pinkel, D. (1990). *Science* **250,** 559.

Tkachuk, D., Pinkel, D., Kuo, W.-L., Weir, U., and Gray, J. (1990). Clinical application of fluorescein of *in situ* hybridization. *Genet. Anal. Tech. Appl.* **8**(2), 24–35.

Trask, B. J., Pinkel, D., and van den Engh, G. (1989). The proximity of DNA sequences in interphase cell nuclei is correlated to genomic distances and permits ordering of cosmids spanning 250 kilobase pairs. *Genomics* **5,** 710.

Trask, B. J., Mass, H., Kenwrick, S., and Gitschier, J. (1991). Mapping of human chromosome Xq 28 by two-color fluorescence *in situ* hybridization of DNA sequences to interphase cell nuclei. *Human Genet.* **48,** 1.

Verdlov, E., Monastyrskaya, G. S., Guskova, L. I., Levitan, T. L., Sheichenko, V. I., and Budowsky, E. (1974). Modification of cytidine residues with a bisulfite-o-methylhydroxylamine mixture. *Biochim. Biophys. Acta* **340,** 153.

Viegas-Pequignot, E., Li, Z., Dutrillaux, B., Apiou, F., and Paulin, D. (1989). Assignment of human desmin gene to band 2q35 by non-radioactive *in situ* hybridization. *Hum. Genet.* **83,** 33–36.

Vogel, W., Autenreith, M., and Speit, G. (1986). Detection of bromodeoxyuridine-incorporation in mammalian chromosomes by a bromodeoxyuridine-antibody. I. Demonstration of replication patterns. *Hum. Genet.* **72,** 129.

Xing, Y., and Lawrence, J. B. (1991). Preservation of specific RNA distribution within the chromatin-depleted nuclear substructure demonstrated by *in situ* hybridization coupled with biochemical fractionation. *J. Cell Biol.* **112,** 1055.

Xing, Y., Johnson, C. V., Dobner, P., and Lawrence, J. B. (1992). Nuclear distribution of the pre-mRNAs for neurotensin and fibronectin demonstrated by *in situ* hybridization with exon and intron probes (submitted for publication).

Chapter 4

Visualization of DNA Sequences in Meiotic Chromosomes

PETER B. MOENS AND RONALD E. PEARLMAN

Department of Biology
York University
Downsview, Ontario M3J 1P3, Canada

I. Introduction

Because of the formation of an axial core structure, meiotic chromosomes become individually distinct at early prophase while they are still in the extended state and while the chromatin is still not condensed. In this state, the chromosomes are in an unusually favorable conformation for DNA sequence localization. The length of the pachytene chromosome permits a high resolution of sequence positions along the chromosome and the decondensed chromatin loops permit discrimination between sequences localized at the core or in the chromatin. Good chromosome preparations can be obtained rapidly by the surface spreading of fresh spermatocytes and the hybridization sites of biotinylated probes are well visualized with avidin conjugated with fluorescein isothiocyanate (FITC). Limitations include the use of nonhuman spermatocytes, the lack of banding patterns, and demonstration of repeated sequences only to date. The surface spreading is adapted from Counce and

101

Meyer (1973) and Dresser and Moses (1980). The *in situ* hybridization follows Lawrence *et al.* (1988) and the *in situ* labeling is taken from Koch *et al.* (1989). The use of these techniques with pachytene chromosomes has previously been reported by Moens and Pearlman (1989, 1990b).

II. Surface Spreading of Meiotic Chromosomes

Materials. Paraformaldehyde; sodium dodecyl sulfate (SDS); Photo-Flo (Kodak); phenol red; NaCl; Eagle's minimal essential medium, with Hanks' salts, without L-glutamine (MEM) (Gibco); coverslips; filters (22-μm pore size); 1 N NaOH; stock solution of 0.05 M sodium borate (pH 9.2 with 1 N NaOH).

Prepare a 1% paraformaldehyde solution by dissolving 1 g in 100 ml of distilled water at 60°C (or less) in the presence of 100 μl of 1% phenol red solution (to allow monitoring of the pH) and one drop of 1 N NaOH. When dissolved, cool, filter through a 22-μm pore size membrane, and adjust pH to 8.2 with a 0.01 M sodium borate solution. Cool in ice or refrigerator. Use within days.

Prepare a stock solution of SDS by dissolving 60 mg of SDS per milliliter of distilled water. Add 100 μl of 1% phenol red to allow monitoring of the pH and bring the pH of the stock to 8.2 with borate buffer. Store at room temperature and use this stock to add the SDS to the paraformaldehyde when indicated.

Prepare a 4% (v/v) Photo-Flo solution in distilled water, pH 8.0 with 0.01 M sodium borate, and cool before use.

The spreading solution consists of 0.5% (w/v) NaCl in distilled water.

Methods

1. Sacrifice a rat or mouse by 1 minute of etherization and 1 minute of CO_2 asphyxiation. Make a ventral cut into the abdomen and extract the testes with a pair of forceps. Remove all adhering tissue, especially fat, which will interfere with the spreading of the cells on the water surface.

2. Cut the tunica and collect seminiferous tubules in a drop of MEM on a block of dental wax. Cut the tubules with a clean razor blade, and squeeze repeatedly with broad tweezers. Transfer the cell suspension to a 15-ml plastic centrifuge tube and add MEM to a volume of 15 ml. Gently draw the cells up and down 10 times using a plastic pipette to disperse cell clusters. Allow pieces of tubules to settle by leaving the tube undisturbed for 1–2 minutes. When the tubules have settled, collect the supernatant and spin at low speed (160 g, in a bench-top centrifuge) to sediment the spermatocytes. Discard the supernatant and gently shake cells loose.

3. Clean coverslips with a commercial window cleaner such as Windex.

Rinse well under running water then distilled water. Wipe dry and set aside for use in spreading.

4. Fill a 35-mm petri dish with the 0.5% NaCl spreading solution until a convex surface is formed. Using a Pipetman, draw up 5 μl of cell suspension, wipe the pipette tip clean, and touch the expelled droplet to the convex spreading solution surface. With fine tweezers hold a clean coverslip and touch it to the surface so that the cells adhere to the coverslip.

5. Place the coverslip in a coplin jar filled with 1% paraformaldehyde and 0.03% SDS in ice for 3 minutes. Transfer to paraformaldehyde without SDS for 3 minutes, then 3 × 1 minute rinses in Photo-Flo to spread the cells on the slide.

6. Air-dry at least overnight before use. Coverslips can be placed flat in a petri dish, sealed with Parafilm, and stored in a −80°C freezer for a few weeks.

7. For the preparation of additional coverslips, rinse the petri dish with distilled water and use fresh 0.5% salt solution each time.

Notes. The SDS concentration affects the amount of spreading of the spermatocyte nuclei and chromosomes. At 0.06%, the material becomes highly dispersed and may fail to adhere to the coverslip. At 0.03% SDS, mouse spermatocytes are more compact but rat spermatocytes are quite well spread. Although most spreading protocols use 4% paraformaldehyde for fixation, the 1% solution appears to be adequate and it is the preferred concentration for immunocytochemistry because it is less likely to distort the configuration of the antigen.

III. *In Situ* Hybridization

Materials
HEPES (*N*-2-hydroxyethylpiperazine-*N'*-2-ethanesulfonic acid) (Sigma)
Random oligonucleotide primer pd(N)$_6$ (Pharmacia)
dGTP, dCTP, dTTP (Boehringer Mannheim)
Biotin-7-dATP or biotin-14-dATP (Gibco/BRL)
Large (Klenow) fragment of DNA polymerase I of *Escherichia coli* (Amersham)
Sephadex G-25 or G-50
Phosphate-buffered saline (PBS): 0.01 M phosphate, 0.15 M NaCl, pH 7
20× SSC: 0.3 M sodium citrate, 3 M NaCl, pH 7
TE buffer: 10 mM Tris-HCl, pH 8.0, 1 mM EDTA
Salmon sperm DNA: 1 mg/ml in TE, sonicate 3 × 10 seconds until no apparent viscosity remains, freeze in 20-μl aliquots
E. coli tRNA: 1 mg/ml in TE, use 20 μl/probe

RNase A, 1 mg/ml in 2 × SSC: dilute 1:10 before use and boil for 30 minutes to inactivate contaminating DNase

Hybridization buffer: 4 × SSC, 2% bovine serum albumin (BSA) (DNase free), 20% dextran sulfate

Deionized formamide prepared with analytical 20–50 mesh Bio-Rad mixed bed ion-exchange resin (AG 501-X8) according to Sambrook *et al.* (1989)

Avidin conjugated to fluorescein (Sigma): 2 μg/ml in 4 × SSC with 1% BSA

Antibleach mounting solution: Dissolve 100 mg of *p*-phenylenediamine in 10 ml of PBS, add 90 ml of glycerol, adjust pH to 9.0 by using litmus paper, store at $-20°C$

FITC visualization: epifluorescence exciter filter 450–490 nm, reflector 510 nm, barrier short pass 515–560 nm

DAPI visualization of DNA: 4′,6-diamidino-2-phenylindole at 0.1 mg/ml in PBS, use 5–10 μl/1 ml of mounting solution. Exciter filter 330–380 nm, reflector 420 nm, barrier 420 nm

Propidium iodide visualization of DNA: 0.1 mg/ml in PBS, use 5–10 μl/ml of mounting solution. Exciter filter 450–495 nm, reflector 510 nm, barrier 515 nm long pass will show FITC as well as propidium iodide

Methods

Probe Formation

1. We recommend using isolated inserts rather than linearized plasmids as probes since in our hands, signal intensity and reproducibility seem to be better. Plasmids containing the DNA of interest are digested with the appropriate restriction endonucleases to release the inserts and the inserts are purified by agarose gel electrophoresis (Sambrook *et al.*, 1989). DNA from large-scale CsCl-purified preparations or mini plasmid preparations are suitable provided sufficient starting material is used to allow isolation of at least 100 ng of pure insert DNA.

2. The insert DNA is recovered from the gel by electrophoresis on to NA45 ion-exchange paper (Schleicher and Schuell, Keane, New Hamshire), followed by elution from the paper in high salt (0.1 M NaCl, 0.1 mM EDTA, 0.02 M Tris-HCl, pH 8) at 60°C for 1 hour according to supplier's directions. DNA is precipitated with 2 volumes of 95% ethanol, washed once with 70% ethanol, dried under vacuum at room temperature, and dissolved in an appropriate volume of sterile distilled water. Alternatively, the proper region of the gel can be excised with a razor blade and centrifuged through siliconized sterile glass wool (Heery *et al.*, 1990) to isolate insert DNA. Glass wool is soaked for 5 minutes in a small volume of 5% dimethyl dichlorosilane (Sigma) in carbon tetrachloride, rinsed extensively with distilled water, dried in air, and autoclaved.

3. The amount of purified DNA insert is estimated by visual comparison with a photograph of an ethidium bromide-stained agarose gel with a standard amount of DNA.

4. Approximately 20 to 50 ng of gel-isolated DNA is biotinylated using the random priming reaction of Feinberg and Vogelstein (1983), although one can label up to 300 ng.

5. The DNA is denatured by heating in a boiling water bath for 5 minutes and instantly cooling on ice. The 25 μl of reaction mixture in an Eppendorf tube contains, in addition to the DNA:

200 mM HEPES, pH 6.6
50 mM Tris-HCl, pH 7.4
5 mM MgCl$_2$
0.8 mM 2-mercaptoethanol
0.14 OD, 260 units of random oligonucleotide primer
10 μM each of dGTP, dCTP, and dTTP
25 μM biotin-7 or biotin-14-dATP
10 μg of BSA
2–3 units Klenow fragment

Incubation is generally at 20°C for 16 hours but 1 hour at 37°C is satisfactory. The reaction is stopped by the addition of 2.5 μl of 1 M EDTA and 72.5 μl of STE (TE containing 0.1 M NaCl).

6. The biotinylated DNA is separated from the unincorporated nucleotides by centrifugation through Sephadex G-25 or G-50 (Sambrook et al., 1989). The extent of biotinylation is assessed on dot blots using streptavidin-conjugated alkaline phosphatase visualized with nitroblue tetrazolium (NBT) and 5-bromo-4-chloro-3-indolyphosphate (BCIP). Colour development is compared with that produced by known amounts of biotinylated DNA (Gibco/BRL). The length of the single-stranded biotinylated DNA prepared by this protocol is about 50 to 100 nucleotides.

Hybridization

1. Add 5 μl of stock salmon sperm DNA and 20 μl of stock tRNA to approximately 10 ng of probe in 20 μl of STE. Mix and precipitate by the addition of 0.1 volume of 3 M sodium acetate, pH 5.5, and 2 volumes of 95% ethanol and freeze in dry ice or at -70°C. Collect the precipitate by centrifugation, wash once in 70% ethanol, and dry under vacuum at room temperature.

2. Just before use, add 5 μl of 100% formamide and denature at 75°C for 10 minutes. Cool instantly on ice.

3. Add 5 μl of hybridization buffer, stir well, and spin briefly in a microcentrifuge. Place the 10-μl drop on Parafilm which covers the bottom of a 75-mm petri dish. When using half coverslips, split into two 5-μl drops.

4. Identify coverslips by cutting one or more of the corners with a diamond glass cutter. Half coverslips usually suffice and aid in the recognition of the

side to which the cells are bound. This is determined by focusing with a phase-contrast microscope using a 40 × objective.

5. Coverslips may be incubated at 37°C for 30 minutes in boiled RNase, and then washed three times in 2 × SSC.

6. They are denatured in 70% formamide in 2 × SSC at 70°C for 2 minutes, dehydrated immediately through cold 70, 95, and 100% alcohol, and air dried.

7. The coverslip is placed, cells down, on the drop of probe. Some tissue paper soaked in 2 × SSC is placed in the petri dish, and the dish is closed and sealed with Parafilm, and placed in a 37°C oven for 3 hours or overnight.

8. Coverslips are washed in 50% formamide in 2 × SSC at 37°C for 30 minutes, in 2 × SSC at 37°C for 30 minutes, and 1 × SSC at room temperature for 30 minutes to remove unhybridized probe. Use a lower temperature (e.g., room temperature) for lower stringency.

Detection

1. Incubate in avidin–FITC for 30 minutes at room temperature.

2. Wash for 10 minutes each in 4 × SSC, 4 × SSC with 0.1% Triton-X, 4 × SSC. Rinse for 1 minute in 4% Photo-Flo in water.

3. On a clean glass slide place a drop of mounting medium with DAPI or propidium iodide and place coverslip on top. Absorb excess mounting agent with blotting paper and seal the edges with rubber cement. After 1 hour the DNA stains are fully developed. The FITC gradually fades after several days.

Photography. Use Kodak TMax 400 film for photographs or Fuji Colour 400 film for making slides directly. Expose for 15 to 30 seconds with 100 × objective (Colour Plates 2 and 3). Use commercial development. Pictures can be printed on Cibachrome paper with no filters used.

IV.   Primed *in Situ* Labeling

Materials. Oligonucleotide primers (17 to 21 nucleotides long) for the primed *in situ* labeling (PRINS) reaction (Koch *et al.,* 1989) are synthesized using standard procedures with either an Applied Biosystems or a Pharmacia gene synthesizer.

Reaction mixture contains
200 mM HEPES, pH 6.6
50 mM Tris-HCl, pH 7.4
5 mM MgCl$_2$
10 μg of BSA
67 μM each of dGTP, dCTP, and dTTP
75 μM biotin-14- dATP

~ 320 pmol of oligonucleotide

H_2O to 24 μl

Termination buffer: 50 mM EDTA, 50 mM NaCl

BN buffer (washing): 100 mM $NaHCO_3$, pH 8.0, 0.01% Nonidet P-40
(10 μl/100 ml)

Avidin and mounting agent as for *in situ* hybridization

Methods

1. Surface-spread and denature spermatocytes on coverslips as described for *in situ* hybridization.

2. Heat separately: coverslips, a Parafilm-covered surface, and reaction mix to proper temperature (37–53°C). The temperature used is 5 to 10°C below the calculated T_m for the oligonucleotide used. The approximate T_m for oligonucleotide hybrids is calculated using the formula of Sambrook *et al.* (1989):

$$T_m = 2 \times \text{the number of A and T residues}$$

$$+ 4 \times \text{the number of G and C residues}$$

3. Mix 24 μl of reaction mix with 5 U (1 μl) of Klenow polymerase and place on the Parafilm. Put the coverslip, cells down, on the drop. Incubate for 30 minutes.

4. Stop the reaction by placing the coverslip in termination buffer for 1 minute at 65°C. Wash in 1 × BN buffer 3–4 × 10 minutes at 4°C.

5. Stain with avidin–FITC as described for *in situ* hybridization.

V.   Results

Surface-spread pachytene chromosomes stained with the DNA binding dye propidium iodide (Color Plate 2) reveal the central core, the synaptonemal complex (SC) surrounded by chromatin loops. The loop structure is evident in EM images (e.g., Weith and Traut, 1980; Rattner *et al.*, 1980). The structure is clearly different from a somatic metaphase chromosome because of the distinct core, decondensed chromatin, and extended length. These characteristics make the surface-spread pachytene chromosome attractive for the analysis of higher order chromatin organization. For example, in Color Plate 2, the biotinylated telomeric probe (TTAGGG)$_n$ hybridizes to the end (t) of mouse SCs (sc) rather than to the surrounding chromatin (ch) (for full details, see Moens and Pearlman, 1990b).

Unlike the telomere and centromere DNA probes, the X-specific probes 70–38 and 68–36 (Disteche *et al.*, 1985) hybridize to low-frequency repeated sequences which are contained in the chromatin surrounding the X core rather than the core itself (Color Plate 3, details in Moens and Pearlman, 1990a). The location of the probes along the core of the X chromosome reveals another valuable aspect of the use of the surface-spread pachytene chromosomes. Whereas in mitotic metaphase chromosomes the centromere and regions 70–38 and 68–36 are close together, in the pachytene chromosome they are sufficiently separated that the three probes can be applied simultaneously and still be recognized individually.

REFERENCES

Counce, S. J., and Meyer, G. F. (1973). Differentiation of the synaptonemal complex and the kinetochore in *Locusta* spermatocytes studied by whole mount electron microscopy. *Chromosoma* **44**, 231–253.

Disteche, C. M., Trantravahi, U., Gandy, S., Eisenhard, M., Adler, D., and Kunkel, L. M. (1985). Isolation and characterization of two repetitive DNA fragments located near the centromere of the mouse X chromosome. *Cytogen. Cell Genet.* **39**, 262–268.

Dresser, M. E., and Moses, M. J. (1980). Synaptonemal complex karyotyping of the Chinese hamster (*Cricetulus griseus*). IX. Light and electron microscopy of synapsis and nucleolar development by silver staining. *Chromosoma* **76**, 1–22.

Feinberg, A. P., and Vogelstein, B. (1983). A technique for radio labelling DNA restriction endonuclease fragments to high specific activity. *Anal. Biochem.* **132**, 6–13.

Heery, D. M., Gannon, F., and Powell, R. (1990). A simple method for subcloning DNA fragments from gel slices. *Trends Genet.* **6**, 173.

Koch, J. E., Kolvraa, S., Petersen, K. B., Gregersen, N., and Bolund, L. (1989). Oligonucleotide-priming methods for the chromosome-specific labelling of alpha satellite DNA *in situ*. *Chromosoma* **98**, 259–265.

Lawrence, J. B., Villnave, C. A., and Singer, R. H. (1988). Sensitive, high-resolution chromatin and chromosome mapping *in situ*: Presence and orientation of two closely integrated copies of EBV in a lymphoma line. *Cell (Cambridge, Mass.)* **52**, 51–61.

Moens, P. B., and Pearlman, R. E. (1989). Satellite DNA I in chromatin loops of rat pachytene chromatin and in spermatids. *Chromosoma* **98**, 287–294.

Moens, P. B., and Pearlman, R. E. (1990a). *In situ* DNA sequence mapping with surface-spread mouse pachytene chromosomes. *Cytogenet. Cell Genet.* **53**, 219–220.

Moens, P. B., and Pearlman, R. E. (1990b). Telomere and centromere DNA are associated with the cores of meiotic prophase chromosomes. *Chromosoma* **100**, 8–14.

Rattner, J. B., Goldsmith, M., and Hamkalo, B. A. (1980). Chromatin organization during meiotic prophase of *Bombyx mori*. *Chromosoma* **79**, 215–224.

Sambrook, J., Fritsch, E. F., and Maniatis, T. (1989). "Molecular Cloning, A Laboratory Manual," 2nd Ed., Cold Spring Harbor Laboratory, Cold Spring Harbor, New York.

Weith, A., and Traut, W. (1980). Synaptonemal complexes with associated chromatin in a moth *Ephestiá kuehniella* Z. *Chromosoma* **78**, 275–291.

Chapter 5

Nucleic Acid Sequence Localization by Electron Microscopic in Situ Hybridization

SANDYA NARAYANSWAMI, NADJA DVORKIN, AND
BARBARA A. HAMKALO

Department of Molecular Biology and Biochemistry
University of California, Irvine
Irvine, California 92717

METHODS IN CELL BIOLOGY, VOL. 35

I. Introduction

In situ hybridization is a pivotal genome mapping technique that provides the cytological location of a cloned sequence. Its use, in conjunction with physical mapping, permits the ordering of linked probes relative to each other and relative to chromosomal structures. The original protocol employed radioactive DNA or RNA probes which, after hybridization to denatured cytological preparations, were detected by autoradiography. This classic procedure has generated a large body of information on the locations of sequences in the chromosomes of a variety of species, from protozoans to man, as well as defining sites of transcript accumulation (this volume [2] and [3]).

The development of nonisotopic labeling regimens, in conjunction with detection of specific ligand interactions by fluorescence or immunoenzymatic assays obviates problems inherent with radioactive detection. These labeling regimens are rapid, specific, and efficient. Using these detection systems, numerous single-copy sequences have been mapped at the light microscope (LM) level over the past few years to mitotic and meiotic chromosomes and interphase nuclei (this volume [1] and [4]).

Development of equivalent mapping techniques at the electron microscope (EM) level should present the opportunity to determine the relative map positions of sequences which cannot be discriminated readily in the LM. In addition, it permits high-resolution mapping within chromosomal landmarks such as centromeres, telomeres, and secondary constrictions. Finally, EM localization is particularly well suited to mapping sequences on small chromosomal structures (e.g., double minutes) and for subnuclear localization in small nuclei such as yeast.

The EM *in situ* hybridization technique was originally described by Hutchison *et al.* (1982). It used biotin-substituted probes and immunogold tagging of hybrid sites. Since that work appeared, various aspects of the original protocol have been modified to increase the signal strength and reduce the background (Narayanswami *et al.,* 1989). The modified technique described in detail below is capable of detecting moderate (~ 50–100 copies) and highly repetitive sequences both efficiently and reproducibly. Unique sequences cannot yet be efficiently located with this protocol, due in part to differences in specimen preparation for electron microscopy which result in more intact chromosomes compared to LM protocols (S. Narayanswami and B. A. Hamkalo, unpublished observations). However, current work is directed toward achieving efficient single-copy detection.

II.   Specimen Preparation

A.   Metaphase Chromosomes

Specimen preparation methods for EM *in situ* hybridization (EMISH) to whole-mount metaphase chromosomes involve the generation of large numbers of chromosomes from either established cell lines or diploid tissue, and the deposition of these chromosomes onto EM grids (Rattner and Hamkalo, 1978). It is essential to begin with preparations containing large numbers of metaphase cells, since the small size of an EM grid limits the amount of material that can be examined in a single specimen and the large size of contaminating interphase nuclei reduces one's ability to analyze large numbers of individual chromosomes. Established adherent cell lines are an excellent source of metaphase chromosomes because they can be synchronized at metaphase by the use of agents such as Colcemid or nocodazole and mitotic cells can be preferentially collected by shake-off. If cell lines are unavailable, one can use diploid cells derived from blood or spleen cultures. Although these cells grow in suspension and do not divide very rapidly, it is possible to obtain sufficient numbers of dividing cells from both phytohemagglutinin (PHA)-stimulated human peripheral blood blocked with methotrexate (Watt and Stephen, 1986, Narayanswami and Hamkalo, in press) and from interleukin-2-stimulated mouse splenic lymphocytes after mitotic arrest with Colcemid (Narayanswami *et al., in preparation*).

A detailed protocol for a typical mouse cell line (L929) follows. The density of a culture which gives optimal metaphase arrest and the amount of time required for detergent lysis of arrested cells vary among cell lines and must be determined empirically.

1. Logarithmic cells are blocked in metaphase by treatment with Colcemid (50–80 ng/ml; Gibco) or nocodazole (100 ng/ml; Sigma) for 6–18 hours.

2. Metaphase cells are selectively detached and collected by gently shaking the culture flask followed by pelleting in a 15-ml disposable conical centrifuge tube in a table-top centrifuge (740 g) and resuspension in 0.1–0.2 × the original volume of the same culture medium to a final concentration of about 1×10^5 cells/ml.

3. Cells are lysed by the addition of Nonidet P-40 (NP40) at 25°C. The stock solution typically used is 1% in double distilled (dd) H_2O with the pH adjusted to 8–9 with 0.1N NaOH, followed by filtering through a 0.22 μm

Millipore filter. One drop of the cell suspension is placed in one drop of 1% NP40 on a clean microscope slide and, after careful mixing, cell lysis is monitored by phase-contrast microscopy. One should assess the mitotic index and determine the amount of time for at least 70% cell lysis and release of chromosomes that are visible as individual structures. When lysis is carried out with a 1:1 dilution of NP40, nearly 100% of L929 cells lyse within 1 minute of detergent addition.

4. After optimal lysis conditions are determined, the chambers used for centrifuging the chromosomes onto EM grids are prepared (Fig. 1). Chambers are made out of Plexiglas and each can be used to prepare four specimens at once. The chambers fit the 50-ml swinging buckets of any centrifuge. Chambers are filled with 1 *M* sucrose (pH 8.5) until a convex meniscus is formed; gold EM grids (400 mesh, hexagonal; Polysciences) covered with a

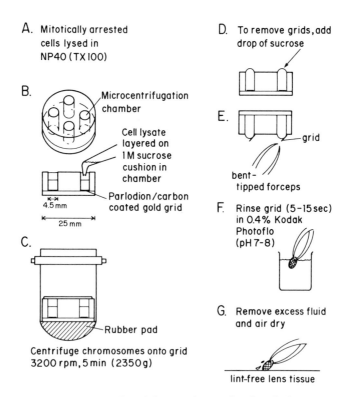

A. Mitotically arrested cells lysed in NP40 (TX 100)

B. Microcentrifugation chamber

Cell lysate layered on 1 M sucrose cushion in chamber

Parlodion/carbon coated gold grid

4.5 mm

25 mm

C. Centrifuge chromosomes onto grid 3200 rpm, 5 min (2350 g)

Rubber pad

D. To remove grids, add drop of sucrose

E. grid

bent-tipped forceps

F. Rinse grid (5–15 sec) in 0.4% Kodak Photoflo (pH 7–8)

G. Remove excess fluid and air dry

lint-free lens tissue

FIG. 1. A schematic representation of the procedure used to deposit chromosomes or nuclei onto EM grids.

Parlodion/carbon support film are rendered hydrophilic by placing in 95% ethanol for 1 minute; they are then dipped briefly in 1 M sucrose and dropped into one of the four wells of the microcentrifugation chamber carbon film side up; they will fall to the bottom and rest on the Plexiglas surface. About half of the sucrose solution is removed, leaving the rest to form a cushion in the chambers. Gold grids are used because copper grids release ions into solution during EMISH, resulting in a marked reduction in preparation quality.

5. When the chambers are ready, cells are lysed as determined in step 3 in a siliconized test tube. If necessary, the lysate is mixed carefully to separate released chromosomes. A 30- to 50-μl aliquot of the chromosome suspension is layered with a Pasteur pipette on top of the sucrose cushion in each well and preparations are centrifuged for 5 minutes at 2500 g at 25°C in 50-ml swinging buckets in a table-top centrifuge.

6. After centrifugation, 1 M sucrose is added to each chamber to form a convex meniscus; upon inverting the chamber, the grids float away from the bottom of the chamber and are held in the hanging droplet. Grids are picked from the droplets individually with forceps and rinsed in Photoflo solution (0.4% Kodak Photoflo 200, pH 8.5, Millipore-filtered) to reduce surface tension and remove residual sucrose. Preparations are then air-dried.

7. Prior to carrying out EMISH, each grid is checked by phase-contrast microscopy (40 ×) in order to evaluate chromosome morphology and density. Optimal preparations should consist of well-spread groups of chromosomes, with few interphase nuclei. Approximately one metaphase per grid square is an optimal chromosome density.

B.   Yeast Nuclei

Saccharomyces cerevisiae nuclei are prepared for EMISH by a protocol adapted from that published by Hurt *et al.,* (1988). Isolated nuclei are used instead of lysed protoplasts because the presence of cytoplasmic contaminants in lysates causes nonspecific binding of probe and detection reagents. If nascent RNA is to be localized, nuclear isolation is carried out in the presence of sodium azide, which is added to the culture immediately before harvesting at a concentration of 10 ml of 20% sodium azide (w/v) per liter of culture medium (Szent-Gyorgi *et al.,* 1987). Poisoning energy metabolism prevents nascent transcript runoff during spheroplast formation and nuclear isolation.

Centrifugation in steps 1–2 is carried out in a refrigerated centrifuge at 4200 g for 5–10 minutes at 25°C in a fixed-angle rotor.

1. Cells are grown to an A_{600} of 1.0, (1–4 × 10⁷ cells/ml). A 200-ml culture yields nuclei for more than 50 grids. Cells are harvested by centrifugation in a

preweighed tube, washed with 50 ml of cold (4°C) ddH$_2$O, and repelleted. The cell pellet is weighed by removing the supernatant, inverting the tube for 30 minutes to allow excess liquid to drain off, and weighing the tube and pellet. The weight of the empty tube is substracted from this figure to give the weight of the pellet.

2. Cells are incubated in 10 ml of 0.1 M Tris-HCl, pH 9.4, 10 mM dithiothreitol (DTT) per gram of cells at 25°C for 10 minutes, pelleted as in step 1, and the pellet is then washed in 20 ml/g cells spheroplast buffer (SB: 20 mM potassium phosphate, pH 7.4, 1.2 M sorbitol) and the centrifugation is repeated.

3. The pellet is resuspended in 10 ml SB/g cells containing 4–5 mg Zymolase-20T/g cells (ICN Immunobiologicals) and incubated at 25°C for 20–60 minutes or until spheroplasting is complete. This process is monitored by adding 20 μl of the cell suspension to 1 ml of 10% sodium dodecyl sulfate (SDS); clearing of the solution indicates that spheroplasting is complete. Spheroplasts are pelleted at 2100 g for 5 minutes at 25°C in a table-top centrifuge.

4. The pellet is resuspended in 10 ml/g cells 20 mM potassium phosphate, pH 6.45, 0.5 mM MgCl$_2$, 18% Ficoll DL-400 at 25°C. The pH of this solution is critical for obtaining intact spheroplasts: after dissolving the Ficoll, the pH is adjusted to 6.45 with HCl. The spheroplast suspension is transferred to a precooled Dounce homogenizer and lysis is effected with several strokes of a tight-fitting pestle until all clumped material is gone.

5. The lysate is diluted with an equal volume of ice-cold 20 mM potassium phosphate, pH 6.45, 0.5 mM MgCl$_2$, 2.4 M sorbitol. Spheroplast lysis and release of the nuclei should be checked in the LM.

The following steps are done on ice or at 0°C.

6. The lysate is centrifuged at 1500 g for 10 minutes in a refrigerated centrifuge in a fixed-angle rotor to remove unbroken cells and cell wall debris, which form a pellet. The supernatant, containing nuclei, is pelleted at 17,000 g for 25 minutes. This pellet is carefully resuspended in 4 ml of 20 mM potassium phosphate, pH 6.45, 0.5 mM MgCl$_2$, 16.6% Ficoll, 0.3 M sucrose, Dounce-homogenized with three strokes of a loose-fitting pestle, and centrifuged at 1100 g for 5 minutes in a fixed-angle rotor.

7. The supernatant is layered on a sucrose step gradient in two ultraclear centrifuge tubes (Beckman, cat. no. #344060). The gradient consists of 2 ml each of 2, 1.8, 1.5, 1.3, and 1.2 M sucrose in 20 mM potassium phosphate, pH 6.45, 0.5 mM MgCl$_2$, 9% Ficoll. Gradients are centrifuged in a refrigerated ultracentrifuge using a swinging bucket rotor (e.g., Beckman SW40 rotor) for 1 hour at 106,000 g. Nuclei are concentrated at the interface between 1.5 and 1.8 M sucrose and are visible as a distinct white turbid band. This band is removed with a Pasteur pipette and placed on ice.

III.   Fixation

Metaphase chromosomes deposited onto EM grids as described above are exposed to the cross-linking fixative glutaraldehyde. Glutaraldehyde fixes the chromosomes to the grid films, preserves their morphology better than paraformaldehyde, and results in essentially complete retention of chromosomes on grids during subsequent steps.

Yeast nuclei are fragile and lyse if centrifuged onto grids without prior fixation. Therefore, they are fixed in suspension as described below. Although several different fixatives can be used (e.g., paraformaldehyde, glutaraldehyde, or mixtures of formaldehyde and glutaraldehyde), glutaraldehyde results in the best morphological preservation.

A.   Metaphase Chromosomes

1. The fixative is prepared fresh each day by diluting 1.25 ml of a newly opened ampoule of 8% glutaraldehyde (EM grade, Polysciences) into 100 ml of freshly prepared 2 × SSC (20 × SSC is 3 M NaCl, 0.3 M sodium citrate, pH 7.0) in a beaker on a magnetic stirrer.

2. Grids are placed in single grid holders (Fullam, cat. no. 14890) and immersed in the fixative, avoiding the formation of air bubbles on the grid surfaces. Air bubbles can be removed by carefully aspirating them with a Pasteur pipette.

3. Grids are fixed for 20 minutes at 25°C with gentle stirring. These conditions appear to be optimal, since increasing the concentration of glutaraldehyde or the time of fixation does not improve morphology. It is possible to fix for as little as 5 minutes, although chromosome morphology is less well preserved.

4. After fixation, individual grids are rinsed in the Photoflo solution and air-dried. Fixed specimens can be stored at 25°C for at least 3 months without noticeable loss of signal or deterioration in chromosome morphology.

B.   Yeast Nuclei

1. Nuclear concentration is checked under phase contrast in order to determine the appropriate dilution. Since the nuclei are in 1.5–1.8 M sucrose and will subsequently be layered over 1 M sucrose for deposition onto grids, dilution is necessary. Nuclei are diluted with SB to a density of 2–3 × 10⁶ nuclei/ml.

2. Glutaraldehyde at 0.3% in SB is added to the dilute nuclei in a 1:3 ratio to give a final glutaraldehyde concentration of 0.1%. Nuclei are mixed gently and incubated at 25°C for 5 minutes.

3. The fixed nuclear suspension is layered onto a 1 *M* sucrose cushion in microcentrifugation chambers and nuclei are deposited on grids as described above. Each grid should have about 10–20 nuclei per grid square for optimal data collection. Specimens can be stored for at least 1 month at 25°C without obvious loss of signal or morphology.

IV.   Prehybridization Treatments

Prehybridization treatments for EMISH are the same as those employed for LM *in situ* hybridization. These include RNase A, proteinase K, and denaturation. RNase digestion is typically included when transcribed DNA sequences are localized, since nascent RNA can reduce the hybridization to template sequences. Proteinase K is employed to increase accessibility of target sequences to probe ([2] and [3], this volume). Protease treatment of yeast nuclei results in a random labeling pattern, at least with biotin-substituted probes, possibly as a result of unmasking endogeneous biotin-containing sites.

A.   RNase Digestion

1. After fixation, grids are incubated in 400 µg/ml of pancreatic RNase (Sigma) in 2 × SSC for 30 minutes at 25°C. The grids are placed in 50-µl droplets of enzyme on the surface of a clean petri dish.

2. RNase is removed by washing the grids, in single-grid holders, once in 2 × SSC for 10 minutes at 25°C, followed by a Photoflo rinse and air-drying.

B.   Proteinase K Digestion

Grids are placed in 50-µl droplets of 1 µg/ml autodigested proteinase K (Sigma) in 2 × SSC for 30 minutes at 37°C in a petri dish. Grids are washed, rinsed, and dried as above.

C.   Denaturation

Denaturation is routinely performed using alkaline 2 × SSC. This method gives the largest *in situ* signal in our protocol, despite the fact that there is some DNA loss. Denaturation at 70°C in 2 × SSC/70% formamide results in better morphological preservation but gives about a twofold lower signal. Denaturation with 70% ethanol/0.14 *M* NaOH results in appreciably lower signals (about fourfold) than the standard alkali treatment. Denaturation times used vary with the base composition of the target since G + C-rich DNAs ($>60\%$ G + C) require more extensive denaturation treatments (up to 30 minutes) to achieve a reproducible signal. Denaturation should be carried out immediately prior to hybridization since overnight storage of denatured preparations results in a reduced signal, presumably as a consequence of the reassociation of denatured DNA.

1. METAPHASE CHROMOSOMES

1. Denaturation is carried out at 25°C with the grids in single-grid holders. The denaturation solution is prepared just prior to use by adding 10 N NaOH to 100 ml of 2 × SSC to a final pH of 12.0 (this is approximately 24 drops using a Pasteur pipette, or 1 ml).

2. Grids in single-grid holders are immersed in this solution for 2 minutes, avoiding the formation of air bubbles as noted above.

3. Grids are then rinsed in Photoflo to remove alkali and air-dried. The air-drying step in the three preceding sections is not obligatory, but represents a convenient stopping point during the procedure.

2. YEAST NUCLEI

If DNA is to be detected, nuclei are denatured for 10 minutes as described above. However, the denaturation time is reduced to 2 minutes for RNA detection. Some denaturation appears to be required to expose the RNA targets optimally. If both DNA and RNA are to be detected simultaneously, denaturation is essential and, while brief denaturation is sufficient to detect both DNA and RNA (Dvorkin *et al.,* 1991), we do not know the relative efficiency of DNA detection compared to longer denaturation times.

V. Hybridization

A. Probe Types and Labeling

As with LM *in situ* hybridization, a variety of probes and labeling regimens can be employed in EMISH. The majority of localizations on metaphase chromosomes have used double-stranded (ds) DNA probes labeled with biodUTP by nick-translation (Narayanswami and Hamkalo, 1987). Other methods of labeling DNA and substituted RNA probes also give positive results with variations in signal intensity.

Nick-Translation with BiodUTP.   A commercial nick-translation kit (e.g., from Bethesda Research Laboratories) is used according to the supplier's instructions. Each reaction contains 2 μg of DNA in 100 μl, and, in addition to the other dNTPs, 5 μM biodUTP; 20 μCi [^{3}H]dATP is included so that incorporation of biotin can be monitored.

1. For each reaction, 20 μCi of [^{3}H]dATP (17 Ci/mmol) is lyophilized in an Eppendorf tube.

2. To each tube on ice, dNTPs and 10 μl of the *Pol*I-DNase I stock solution are added, and the reactions are mixed well and incubated at 15°C for 90 minutes. The reaction is stopped by addition of 10 μl of STOP buffer (300 mM EDTA, pH 8.0) and 2 μl of 5 M NaCl.

3. In order to determine the specific activity, a 5-μl aliquot is removed from each reaction and spotted on a GF/C filter (Whatman). The filter is dried, placed in cold 5% trichloroacetic acid (TCA) for at least 1 hour, washed with five changes of cold 5% TCA followed by two rinses in 95% ethanol, and then air-dried. The filter is counted in a liquid scintillation counter and the specific activity expressed as cpm/μg DNA.

4. While the filters are drying, unincorporated nucleotides are removed from the reaction by Sephadex G-50 chromatography using the spin column procedure of Maniatis *et al.* (1982). A 100-μl reaction can be loaded on a 1-ml G-50 column. The volume of the column eluate is measured and the amount of incorporated radioactivity determined by scintillation counting of a 5-μl aliquot of the eluate.

5. The amount (micrograms) of DNA recovered in the column eluate is determined by dividing the total radioactivity in the eluate by the specific activity of the probe preparation as determined in step 3. The probe is diluted to 4 μg/ml in either 100 mM NaCl, 10 mM Tris, pH 8.0, 1 mM EDTA, or distilled water and stored frozen.

Several biotinylated nucleotides containing linker arms of lengths ranging from C_4 to C_{21} are available commercially (ENZO, Sigma, Clontech). The shortest linker (C_4) provides suboptimal labeling in some circumstances but we have not seen major differences among the C_{11}, C_{16}, or C_{21} linker-containing nucleotides. Thus, probes labeled with any of these bionucleotides are suitable for EMISH.

In our experience, biodUTP is somewhat more efficiently incorporated than biodCTP in nick-translation reactions. However, this is not necessarily true in the case of other labeling regimes, and it has been reported that biodCTP is incorporated more efficiently than biodUTP in terminal deoxyribonu-cleotidyltransferase-directed tailing reactions (Moyzis *et al.*, 1988).

Probes can be biotinylated by methods other than nick-translation and tailing. Photoactivable forms of biodUTP are commercially available

(McInnes *et al.,* 1987) and can be directly coupled to DNA under a sun lamp. Although probes labeled in this manner can be used for EMISH, we find that they are less sensitive than probes labeled by nick-translation, presumably because the degree of substitution tends to be lower.

Finally, oligonucleotide probes can be biotinylated by the polymerase chain reaction (PCR) (Weier *et al.,* 1990). Briefly, oligonucleotides are labeled with 30 cycles of standard PCR, followed by 10 PCR cycles in the presence of bio-11-dUTP. This protocol results in high levels of incorporation of the modified nucleotide. PCR-generated probes can also be used for EM localizations, although in our hands different batches of probe give varying amounts of signal (Narayanswami *et al.,* 1989).

B.   Alternative Labeling Systems

Several alternative DNA labeling systems have been developed in recent years. These include labeling with dinitrophenyl (DNP) (Lichter *et al.,* 1990), digoxigenin (Boehringer Mannheim), and *N*-acetoxy-2-acetylaminofluorene (AAF, Landegent *et al.,* 1984; see [1], this volume). AAF is available from the National Cancer Institute Carcinogen Standard Reference Repository (Bethesda, MD). A combination of one or more of these with biotin permits the simultaneous localization of two or more probes. DNP and digoxigenin labeling is accomplished by nick-translation in the presence of DNPdUTP or digdUTP (Narayanswami *et al.,* 1989) as described above for biodUTP. In the case of AAF, the labeling reaction involves direct modification of probe DNA or RNA by the covalent addition of AAF. In all instances, labeled probes are stable for at least a year when stored frozen at $-20°C$.

Direct Probe Modification with AAF

1. DNA is digested with DNase I to a mean length of about 3 kb. It is denatured by boiling at 100°C for 5 minutes, and fast-cooled on ice.

2. DNA is diluted to 0.2 mg/ml in 1.6 mM sodium citrate, pH 7, 20% (v/v) ethanol. AAF is added from a stock solution in dimethyl sulfoxide (6 mg/ml) to a final concentration of 60 μg/ml (for approximately 5% modification). The final concentration of AAF determines the amount of substitution of the DNA; however, the final concentration of DMSO should not exceed 0.8% (v/v).

3. Reactions are run for 1 hour at 37°C in the dark, during which time AAF covalently attaches to G residues in the DNA.

4. Unreacted AAF is removed by extracting the reactions six times with water-saturated diethyl ether.

5. The ratio of A_{305}/A_{260} is used to calculate the percentage modification of the DNA, based on a standard curve (Fuchs and Daune, 1972).

Since AAF is carcinogenic, it must be handled with great care, although modified probes are innocuous and uncoupled AAF can be inactivated by bleach after the reaction is completed. Both DNA and RNA can be labeled with AAF. DNA can be labeled in both the single-stranded and double-stranded form. However, we find that modification of single-stranded DNA (denaturated before labeling, as above) is more reliable than modification of double-stranded DNA, perhaps because the formation of intrastrand cross-links during coupling (Fuchs and Daune, 1972) prevents subsequent denaturation. RNA can be modified with AAF using the conditions given above.

There are differences in the relative sensitivities of differently substituted DNA probes as assayed by EMISH. Although DNP- and digoxigenin-labeled nucelotides are incorporated into DNA to similar levels to biotin (S. Narayanswami, unpublished results) they give a several-fold lower hybridization signal compared to biotin-labeled probes under equivalent conditions (Narayanswami *et al.*, 1989). Nevertheless, these three labels can be used for the facile detection of highly and moderately repeated sequences by EMISH. AAF-labeled probes give substantially less signal under equivalent conditions, but the lower signal can be compensated for by increasing probe concentration 5- to 10-fold relative to biotinylated probes. There is some increase in background under these conditions so that optimal reaction/detection conditions will vary with the probe, label, and target material. In addition, because direct labeling of probes is possible with AAF (as opposed to nick-translation in the presence of DNase I and *Pol* I), AAF can be used to label relatively small pieces of DNA without degradation of the probe. Thus, AAF is useful when one of a pair or group of probes is smaller than the other(s).

Finally, although all modifications reduce the thermal stability of hybrids, this effect is most pronounced with AAF, where each percent substitution is accompanied by a reduction in the T_m of $1°C$ (Landegent *et al.*, 1984). Because of this, levels of substitution in the $5-10\%$ range are routinely used. This is in contrast to bio-dUTP, where even totally substituted probes give specific and intense signals under normal hybridization conditions. A benefit of a reduction in T_m in all cases is the reduction in signal at cross-hybridizing, nonidentical sites.

C. Hybridization Conditions

Hybridization of chromosomes or nuclei to dsDNA probes labeled with biotin, DNP, digoxigenin, or AAF is carried out in a standard hybridization buffer containing high salt, formamide, and dextran sulfate. Final probe concentrations typically are about 800 ng/ml for biotin, DNP, and digoxigenin, and 5- to 10-fold higher for AAF. Hybridizations are carried out at 30°C except in the case of AAF-substituted probes, where hybridization is performed at 27°C (Landegent *et al.*, 1984), due to the significantly reduced stability of hybrids containing AAF residues. In general, overnight hybridization is most convenient, although a good signal is obtained after 6–7 hours with repetitive sequence targets (Lundgren, 1989).

The conditions detailed are used with double-stranded probes, regardless of modification. Hybridization buffer (HB) contains 50% formamide, 10% dextran sulfate, 1 mM trisodium EDTA, 10 mM Tris (pH 7.6), 0.2% Ficoll, 1 mg/ml BSA (bovine serum albumin, nuclease-free, BRL), 40 μg/ml *Escherichia coli* DNA, and 0.6 M NaCl. Stock solutions of 0.5 M EDTA, 1 M Tris, 2% Ficoll, and 5 M NaCl are Millipore-filtered before use.

1. For 100 μl of hybridization solution, mix 79 μl of HB, 1 μl of 7 mg/ml yeast tRNA (phenol extracted, reprecipitated, and dissolved in ddH$_2$O), and 20 μl of probe DNA (4 μg/ml) in a siliconized Eppendorf tube.

2. This mixture is placed in a boiling water bath for 5 minutes to denature the DNA, followed by rapid cooling on ice for 2 minutes. Tubes are spun briefly in a microfuge (up to maximum speed and then down) in order to bring droplets of condensation to the bottom of the tube, and tubes are kept on ice prior to use.

3. A 50-μl aliquot of the hybridization solution is pipetted into a siliconized Reactivial (Kontes, cat. no. 749001-0000).

4. Two grids, back-to-back (i.e., specimen side facing out), are placed in each vial using forceps. The vials are capped and incubated overnight in a waterbath at the temperature selected for hybridization.

5. After hybridization, grids in individual grid holders are rinsed in a beaker with stirring three times for 20 minutes each in 100 ml of 2 × SSC at 25°C to remove unhybridized probe. Grids are kept moist between this and subsequent steps to retain antibody reactivity.

D. Multiple Hybridization

Incorporation of minor modifications in the protocol described above permits localization of multiple probes. If probes are cloned in the same vector, it is essential to purify the cloned insert from all but one probe so that vector sequences do not cross-hybridize. In this case, inserts are prepared for probes which give the largest signals because vector sequences allow network formation and hence signal enhancement. Alternatively, differently tagged oligonucleotides can be used to locate all but one probe. Various considerations also dictate the choice of differently substituted probes for a particular experiment. Because of the differences in sensitivity among the various modified probes (see Section V,B), we recommend biotinylation of the probe for the sequence that is expected to give the smaller signal, since it is the most sensitive in EMISH, in order to maximize the chance of detecting more than one probe with different target sizes.

Probes are added to the hybridization mixture simultaneously and denatured together. As a result, probes used for multiple localizations should be at higher stock concentrations ($\sim 20 \, \mu g/ml$) than the $4 \, \mu g/ml$ recommended for a nick-translated probe used alone. They can be stored without dilution after purification over Sephadex G-50. Several probes can then be added to hybridization buffer without drastically changing the concentrations of buffer components. When AAF is used to label one probe, hybridization is performed at 27°C, which is optimal for AAF hybrid formation.

VI.   Signal Detection

Hybridized preparations are first reacted with specific antibodies to the ligands used to modify DNA probes, essentially as in fluorescence *in situ* hybridization (FISH), followed by incubation with an appropriate secondary antibody immobilized on colloidal gold. These reactions are analogous to those used for FISH detection. In addition, streptavidin or avidin can be used to locate biotin-containing hybrids. Colloidal gold and colloidal gold complexed with antibodies or streptavidin/avidin in sizes from 1 to 40 nm are available from Janssen Pharmaceutica, E-Y Labs, Sigma, and Polysciences. Colloidal gold can be prepared in the laboratory with minimal equipment. The protocol involves reduction of gold chloride by appropriate reducing agents selected according to the size of colloid desired (Frens, 1973). A protocol for the conjugation of streptavidin to gold is given below, followed by detailed instructions for antibody and gold labeling of hybridized preparations.

A.   Coupling of Streptavidin to Colloidal Gold

This protocol is derived from De Waele *et al.* (1983). It gives details for coupling streptavidin to 20-nm gold, but it can be adapted to the adsorption to colloidal gold of any other protein.

1. All glassware and test tubes used in this protocol are soaked and washed in a hot 1% solution of Micro cleaning solution (Fullam, cat. no. 1547), rinsed 10 times with tap water, 10 times with distilled water, and either autoclaved or baked at 200°C for 12 hours. Tubes used for the centrifugation and storage of colloidal gold should be polycarbonate (Sorvall) to prevent adsorption of the gold to the tubes and resultant flocculation.

2. Prior to protein adsorption, the minimum amount of protein required to protect the colloid from salt-induced flocculation is determined. The pH of 10–15 ml of a colloidal gold sol is adjusted with 0.2 M K_2CO_3 to 0.4 pH units above the isoelectric point of the protein (IEP of streptavidin is 7.0; colloidal gold is adjusted to 7.4). Dilute buffer is used for pH adjustment because high salt concentrations cause unprotected colloids to flocculate. Colloid pH can be read directly with a gel-filled electrode. If one is unavailable, 10 drops (250 μl) of 1% polyethylene glycol (PEG)-6000 is added to 0.5-ml aliquots of gold before reading, in order to prevent flocculation on the electrode.

3. The appropriate protein concentration for coupling is determined by adding increasing amounts of protein (e.g., 0.1, 1, 2, 5 μg) to a series of 10 tubes. The source of streptavidin routinely used is Bethesda Research Laboratories (cat. no. 5532LB, 1 mg/ml).

4. Water is added to each tube to bring the volume to 100 μl followed by the addition of 1 ml of colloid. After 10 minutes incubation at 25°C in order to allow the protein to adsorb to the gold, 200 μl of 10% NaCl is added to each tube, followed by incubation at 25°C for at least 1 hour.

5. The A_{520} of each sample is read and plotted against the amount of protein. The plot will show a linear increase up to a certain protein concentration where the A_{520} plateaus. The amount of protein at the plateau plus 10% is the quantity of protein that should be added to 1 ml of gold in order to stabilize it against salt-induced flocculation.

6. A solution of 8% BSA (Sigma, fraction V) is prepared prior to making the bulk preparation; BSA is used to quench the conjugation. The BSA is dissolved in 2 mM sodium borate, pH 9 (Millipore-filtered), dialyzed overnight at 4°C against the same buffer, and centrifuged in a refrigerated centrifuge at 100,000 g for 1 hour at 4°C in a swinging bucket rotor to clarify before use.

7. For the bulk preparation, the pH of 20–30 ml of gold sol is adjusted to 7.4 as above; an appropriate amount of streptavidin is added, followed by incubation for 2 minutes at 25°C with stirring.

8. The 8% BSA solution is then added to a final concentration of 1% and the gold is washed twice in 1% BSA buffer [1% BSA (fraction V, Sigma), 0.9% NaCl, 0.02 M sodium azide, 20 mM Tris, pH 8.2] in order to remove free streptavidin. Next, 5-ml aliquots of the gold are placed in polycarbonate centrifuge tubes and the gold pelleted by centrifugation at 23,400 g in a refrigerated centrifuge in a fixed-angle rotor for 50 minutes at 4°C. Most of the supernatant is removed and 5 ml of fresh 1% BSA buffer is added. The pellet is resuspended and the wash is repeated. In each case, the pellet should be soft and loose in appearance. A hard compact pellet should be discarded, since it signifies that the gold preparation is massively aggregated and will fail to resuspend properly.

9. The second resuspended pellet is centrifuged as above and the soft pellet resuspended in one-tenth the original volume of 1% BSA buffer. The gold preparation is stored at 4°C and is stable for several months. Certain commercial sources (BioCell) are available in glycerol and can be stored frozen at −20°C.

B. Primary Antibody Reactions

Antibodies to biotin, digoxigenin, and AAF-G are commercially available from ENZO, Vector, Boehringer Mannheim, and Robert Baan (Landegent *et al.,* 1984). The anti-DNP antibody used in our laboratory (Narayanswami *et al.,* 1989) was the gift of Dr. D. Segal. Primary antibodies are diluted to an appropriate concentration in PBS supplemented with BSA and NaCl. The presence of BSA and high salt reduces nonspecific binding of antibodies to the grid film to virtually zero. Such nonspecific adsorption represented the major source of background observed during the development of this technique. Each antibody batch is titrated using a test system (in our case, mouse satellite DNA and mouse metaphase chromosomes) in order to determine optimal antibody dilution and reaction time. The conditions given are suitable for multiple-step antibody sandwich reactions and multiple labeling.

1. Grids are placed in 50-μl droplets of affinity-purified primary antibody (e.g., for anti-biotin antibodies, 0.8 μg/ml in PBS, 2 mg/ml BSA, 0.5 M NaCl) in petri dishes with the specimen side up and incubated in a moist atmosphere for 4 hours at 37°C. A plastic slide box lined with dampened paper towels forms a convenient moist chamber.

2. Unbound antibody is removed by rinsing grids at 25°C three times with 100 ml of PBS/NaCl for 10 minutes each in a single grid holders in a beaker with stirring. Grids should not be allowed to dry between this step and the next step, below.

C. Secondary Antibody/Gold Reactions

1. Five hundred microliters of the 15- to 20-nm gold stock is pelleted at 12,000 g for 15 minutes at 4°C in a fixed-angle rotor in a refrigerated centrifuge. The 5-nm gold stocks are pelleted at 23,400 g for 45–60 minutes in order to separate unadsorbed protein which will be in the supernatant. In both cases, the pellet should be soft and loose in appearance.

2. Most of the supernatant, containing released protein, is discarded, except for a small quantity directly over the gold. The remaining pellet is resuspended in the same volume of Millipore-filtered (8 μm followed by

0.22 μm filters) 1% BSA buffer. The BSA can be substituted with 1% fish gelatin (Sigma, cat. no. G-7765) with comparable results (Birrell *et al.*, 1987).

3. Gold is diluted in 1% BSA to a concentration determined empirically for each batch, followed by low-speed centrifugation of the diluted sample in a table-top centrifuge with a swinging bucket rotor (400-700 g for 10 minutes at 25°C) in order to remove aggregates. Aliquots of 50 μl of the supernatant are carefully removed with a P200 Pipetman and placed in clean petri dishes. Grids are individually placed in the 50-μl droplets and incubated in a moist chamber at 25°C. Although overnight incubation is convenient, the time can be shortened, based on that required to give a visible signal. Highly repeated sequences (e.g., mouse satellite) exhibit a good signal after as little as 2 hours incubation in gold.

4. Grids are rinsed free of unbound gold at 25°C in single grid holders three times for 20 minutes each in 100 ml of 1% BSA buffer in a beaker with stirring. They are then rinsed in Photoflo and air-dried. Specimens can be viewed in the EM without further contrasting since chromosomes and gold are sufficiently electron dense. However, greater contrast can be obtained by staining or heavy metal shadowing.

In general, 15- to 20-nm gold is used because the particles are readily identified even at low magnifications (2600–3300 $\times$). However, smaller or larger particles can also be used. The low electron density of 5-nm particles makes them most suitable for use when a large signal is expected, since large clusters of these particles are readily located even at low magnifications. One-nanometer particles are most useful when accessibility is a problem. Although they are invisible at the magnifications used to examine chromosomes, they can be identified by silver intensification (see below).

D. Multiple Labeling

Signal detection schemes for multiple probe labeling are basically the same as for single probe labeling except that precautions are necessary to avoid cross-reaction among immunological reagents. Avidin or streptavidin can be used to label biotin-probes in conjunction with antibodies that recognize other modifications. Primary antibodies generated in enough different species are available so that differently substituted probes can be combined, at least pairwise. In this case, grids are incubated with a mixture of two or more primary antibodies, followed by incubation with a mixture of appropriately labeled secondary antibodies on different-sized gold particles.

The choice of gold particle sizes for multiple probe labeling depends on parameters such as signal size and the distance between sequences of interest.

At the magnifications used to examine metaphase chromosomes, 5-nm gold is difficult to detect unless the signal is large. However, 10-, 15-, 20-, and 30-nm particles are all visible and distinguishable from each other in the EM. Although we have only carried out double labeling to date, the availability of four modified probes and different sizes of gold should permit simultaneous detection of four or more loci.

E.   Signal Amplification

It is possible to enhance a small signal by one of two amplification approaches.

1.   ANTIBODY SANDWICH

Amplification by this approach is analogous to that used in FISH. It is applicable as described for biotin-labeled probes, but can be generalized by applying layers of primary and secondary antibody when other ligands are used.

1. After hybridization as described, grids are incubated in anti-biotin followed by incubation with biotin-coupled secondary antibody (Jackson ImmunoResearch) under the same conditions. These steps can be repeated one or more times if necessary.

2. Amplified signals ultimately are detected with streptavidin gold.

Although moderately repeated sequences ($\sim$ 100 copies) are detectable without amplification, two cycles of amplifications are sufficient to produce a signal which is visible in the LM (Narayanswami *et al.,* 1989). One drawback of several rounds of signal amplification in EMISH is a loss of spatial resolution.

2.   SILVER INTENSIFICATION

The second approach to signal amplification, silver intensification, involves increasing the size, and hence the visibility, of individual particles (Holgate *et al.,* 1983; Narayanswami and Hamkalo, 1987). Layers of metallic silver are deposited directly on colloidal gold particles. Amplification by silver intensification is particularly useful when reduced accessibility dictates the use of very small particles and is imperative when 1-nm gold is used (Cooke *et al.,* 1990). The procedure is rapid and results in a several-fold increase in particle size within about 10 minutes. Several different kits are available (Janssen Pharmaceutica, BioCell), and any one can be used according to the supplier's instructions.

VII. Typical Results

Figures 2, 3, and 4 show typical examples of EMISH, and illustrate most of the modifications described above. In Figure 2, yeast 25S ribosomal RNA was localized to a dense, crescent-shaped region of a yeast nucleus thought to be the nucleolus (Dvorkin *et al.,* 1991). Figure 3 shows the localization of the mouse major centrometric satellite DNA, a typical highly repeated sequence, on mouse metaphase chromosomes. Chromosomes label heavily at the primary constriction with this probe. However, the spotty labeling pattern suggests that other sequences may be interspersed with the major satellite (Radic *et al.,* 1987).

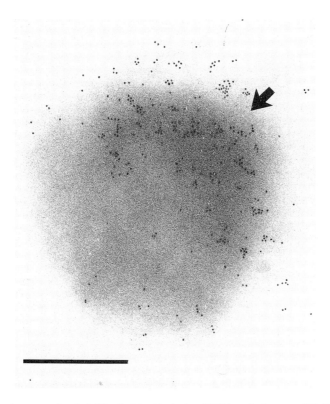

Fɪɢ. 2. A yeast nucleus hybridized with a biotinylated DNA probe for yeast 25S rRNA under DNA–DNA hybridization conditions. Signal detection was with rabbit anti-biotin antibody and goat anti-rabbit-conjugated 15-nm colloidal gold. The nucleolus is heavily labeled (arrow), showing the location of the rRNA genes. Bar, 1 μm.

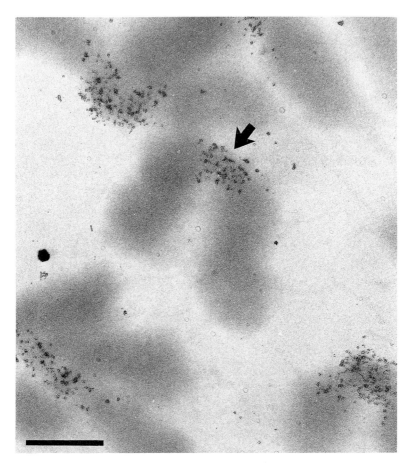

FIG. 3. An example of metaphase chromosomes from a concanavalin A/phenylmercuric acetate-stimulated mouse lymphocyte culture after EMISH with a biotinylated probe for the mouse major centromeric satellite DNA. Hybrids were detected with rabbit anti-biotin antibody followed by 15-nm goat anti-rabbit-conjugated colloidal gold. The arrow denotes a labeled centromere. Bar, 2 μm.

Finally, Fig. 4 shows a typical example of double-label EMISH. Chromosomes from a mouse–human hybrid cell line selected to retain human chromosome 1 were hybridized with probes for human alphoid sequence pE25A (Carine *et al.*, 1989), which was biotinylated and detected with 15-nm gold, and human satellite III (Cooke and Hindley, 1979), which was AAF-labeled and detected with 5-nm gold. Both probes are highly concentrated on chromosome 1 and do not cross-react with mouse DNA. Differences in signal

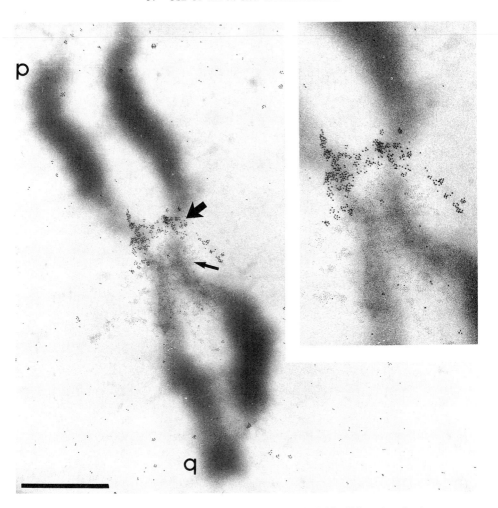

FIG. 4. A human chromosome 1 from a mouse–human hybrid cell line after simultaneous EMISH with biotinylated pE25A (a human alphoid sequence, detected with 15-nm gold, large arrow) and AAF-labeled human satellite III purified insert (detected with 5-nm gold, small arrow). p and q arms are marked. The inset shows the labeled regions at higher magnification. Bar, 2 μm.

size dictated both the choice of ligand used to modify the respective probes and the size of gold particle used to detect them. A large amount of satellite III is located on the q arm of human chromosome 1, whereas the pE25A signal is smaller and more easily detectable after labeling with 15-nm gold. In all cases, both sister chromatids are labeled, often with similar patterns of gold particle distribution.

VIII.   Summary and Future Applications

EMISH has been applied successfully to the localization of DNA and RNA sequences in both whole-mount metaphase chromosomes and nuclei in organisms from yeast to man. The methodology is analogous to that used at the LM level with a few modifications: EM grids are used in place of glass slides, and material is centrifuged through a sucrose cushion onto the grid rather than dropping a whole-cell lysate onto a slide. The centrifugation step is useful because it virtually eliminates cytoplasmic components which can give rise to nonspecific labeling. Chromosomes are not pretreated with methanol/acetic acid and appear to retain greater integrity than those treated in this way (S. Narayanswami, unpublished results). Finally, detection is via colloidal gold rather than fluorochromes, although enzyme-linked detection also has been used in the EM (Manuelidis and Ward, 1984).

The main source of background in EMISH appears to be nonspecific binding of antibodies to grid films. Even when this background is high, chromosomes typically are surrounded by gold-free halos, probably due to charge repulsion between negatively charged gold and chromosomes. As noted above, background can be dramatically reduced with appropriate blocking solutions.

As with any other experimental protocol, reproducible positive and negative controls are essential when attempting EMISH on new material. Highly repetitive, previously localized sequences are excellent positive controls and prokaryotic DNAs represent good negative controls. Once a positive signal is obtained, any and all steps can be modified to improve it.

At the present time, EMISH is less sensitive than FISH, in which single-copy sequences are routinely localized by chromosomal *in situ* suppression hybridization (CISS, Pinkel *et al.,* 1988, Lichter *et al.,* 1990); see [1] and [3], this volume). This difference is due in part to differences in how chromosomes are prepared. For example, chromosomes prepared for EMISH and hybridized and detected by FISH do not give reproducible positive fluorescent signals with single-copy probes. Thus, current efforts are directed toward enhancing detection by modifications in preparative procedures. In addition, detection using relatively large (>5 nm) gold particles undoubtedly reduces sensitivity, so that the use of 1-nm particles and silver intensification may be required for efficient single-copy detection (Cooke *et al.,* 1990).

In addition to modifying EMISH, the use of primer extension in the presence of biodUTP after hybridization of an oligonucleotide to chromosome preparations is being investigated. This technique, named PRINS (Koch *et al.,* 1989), has been successfully used with FISH to label single-copy sequences on meiotic chromosomes (see [4], this volume). Furthermore, it may be possible to develop PCR-like reactions on chromosomes to effect signal amplification. The future applications of EMISH overlap those described for FISH (see [1] and [3], this volume). However, it is particularly suited for precise localizations below the level of resolution of the LM and for studies on the organization of small nuclei (e.g., yeast).

ACKNOWLEDGMENTS

Research was supported by NIH grant GM 23241 to Barbara A. Hamkalo; Nadja Dvorkin is supported by training grant NIH GM 07311. We would like to thank the relevant individuals for probes, cell lines, chemicals, and antibodies, and Mrs. Agnes Demetrescu for technical assistance.

REFERENCES

Birrell, B. C., Hedberg, K. K., and Griffith, O. H. (1987). Pitfalls of immunogold labeling: Analysis of light microscopy, transmission electron microscopy, and photoelectron microscopy. *J. Histochem. Cytochem.* **35**, 843–853.

Carine, K., Jacquemin-Sablon, A., Walter, E., Mascarello, J., and Scheffler, I. E. (1989). Molecular characterization of human minichromosomes with centromere from chromosome 1 in human-hamster hybrid cells. *Somatic Cell Mol. Genet.* **15**(5), 445–460.

Cooke, H. J., and Hindley, J. (1979). Cloning of human satellite III DNA: Different components are on different chromosomes. *Nucleic Acids Res.* **6**, 3177–3197.

Cooke, C. A., R. L. Bernat and W. C. Earnshaw (1990). CENP-B; A major centromere protein located beneath the kinetochore *J. Cell Biol.* **110**, 1475–1488.

De Waele, M., de Mey, J., Moeremans, M., de Brabander, M., and Van Camp, B. (1983). Immuno-gold staining method for the light microscopic detection of leukocyte cell surface antigens with monoclonal antibodies. *J. Histochem. Cytochem.* **31**, 376–381.

Dvorkin, N., Clark, M. W., and Hamkalo, B. A. (1991). Ultrastructural localization of nucleic acid sequences in *Saccharomyces cerevisiae* nucleoli. *Chromosoma* (in press).

Fostel, J., Narayanswami, S., Hamkalo, B. A., Clarkson, S. G., and Pardue, M. L. (1984). Chromosomal location of a major tRNA gene cluster of *Xenopus laevis*. *Chromosoma* **90**, 254–260.

Frens, G. (1973). Controlled nucleation for the regulation of the particle size in monodisperse gold suspension. *Nature Phys. Sci.* **241**, 20–21.

Fuchs, R., and Daune, M. (1972). Physical studies on deoxyribonucleic acid after covalent binding of a carcinogen. *Biochemistry* **11**, 2659–2666.

Harper, M. E., and Saunders, G. F. (1981). Localization of single copy DNA sequences on G-banded human chromosomes by *in situ* hybridization. *Chromosoma* **83**, 431–439.

Holgate, C. S., Jackson, P., Cowen, P. N., and Bird, C. C. (1983). Immunogold-silver staining: A new method of immunostaining with enhanced sensitivity. *J. Histochem. Cytochem.* **31**, 938–944.

Hurt, E. C., McDowell, A. M., and Schimmang, T. (1988). Nucleoli and nuclear envelope proteins of the yeast *Saccharomyces cerevisiae*. *Eur. J. Cell. Biol.* **46**, 554–563.

Hutchinson, N. J., Langer-Safer, P. R., Ward, D. C., and Hamkalo, B. A. (1982). *In situ* hybridization at the electron microscope level: Hybrid detection by autoradiography and colloidal gold. *J. Cell Biol.* **95**, 609–618.

Koch, J. E., Kolvraa, S., Petersen, K. B., Gregersen, N., and Bolund, L. (1989). Oligonucleotide priming methods for the chromosome specific labeling of alpha-satellite DNA *in situ. Chromosoma* **98,** 259–265.

Landegent, J. E., Jansen in de Wal, N., Baan, R. A., Hoeijmakers, J. H. J., van der Ploeg, M. (1984). The use of 2-acetylaminofluorene modified probes for the indirect hybridocytochemical detection of specific nucleic acid sequences in microscopic preparations. *Exp. Cell. Res.* **153,** 61–72.

Lichter, P., Tang, C.-J. C., Call, K., Hermanson, G., Evans, G. A. Housman, D., and Ward, D. C. (1990). High resolution mapping of human chromosome 11 by *in situ* hybridization with cosmid clones. *Science* **247,** 64–69.

Lundgren, K. (1989). Studies in eukaryotic chromosome organization using electron microscopic *in situ* hybridization. Ph. D. Thesis, University of California, Irvine, California.

McInnes, J. L., Vise, P. D., Habili, N., and Symons, R. H. (1987). Chemical biotinylation of nucleic acids with photobiotin and their use as hybridization probes. *Focus* **9,** 1–4.

Maniatis, T., Fritsch, E. F., and Sambrook, J. (1982). "Molecular Cloning. A Laboratory Manual," pp. 466, 467. Cold Spring Harbor Laboratory, Cold Spring Harbor, New York.

Manuelidis, L., and Ward, D. C. (1984). Chromosomal distribution and nuclear distribution of the HindIII 1.9 kb human DNA repeat. *Chromosoma* **91,** 28–38.

Moyzis, R. K., Buckingham, J. M., Cram, L. S., Dani, M., Deaven, L. L., Jones, M. D., Meyne, J., Ratliff, R. C., and Wu, J. R. (1988). A highly conserved repetitive DNA sequence (TTAGGG)n, present at the telomeres of human chromosomes. *Proc. Natl. Acad. Sci. U.S.A.* **85,** 6622–6626.

Narayanswami, S., and Hamkalo, B. A. (1987). Hybridization to chromatin and whole chromosome mounts. *In* "Electron Microscopy in Molecular Biology. A Practical Approach" (J. Sommerville and U. Scheer, eds.), pp. 215–232. IRL Press, Oxford.

Narayanswami, S., and Hamkalo, B. A. (1990). High resolution mapping of *Xenopus laevis* 5S and ribosomal RNA genes by EM *in situ* hybridization. *Cytometry* **11,** 144–152.

Narayanswami, S., and Hamkalo, B. A. (1991). DNA sequence mapping using electron microscopy. *Gen. Anal. Tech. App.* **8,** 14–23.

Pinkel, D., Landgent, J., Collins, C., Fuscoe, J., Segraves, R., Lucas, J., and Gray, J. (1988). Fluorescence *in situ* hybridization with human chromosome-specific libraries: Detection of trisomy 21 and translocations of chromosome 4. *Proc. Natl. Acad. Sci. U.S.A.* **85,** 9138–9142.

Radic, M. Z., Lundgren, K., and Hamkalo, B. A. (1987). Curvature of mouse satellite DNA and condensation of heterochromatin. *Cell (Cambridge, Mass.)* **50,** 1101–1108.

Rattner, J. B., and Hamkalo, B. A. (1978). Higher-order structure in metaphase chromosomes I. The 250A fiber. *Chromosoma* **59,** 363–372.

Rattner, J. B., Saunders, C., Davie, J. R., and Hamkalo, B. A. (1982). Ultrastructural organization of yeast chromatin. *J. Cell Biol.* **92,** 217–222.

Szent-Gyorgi, C., Finkelstein, D. B., and Garrard, W. T. (1987). Sharp boundaries demarcate the chromatin structure of a yeast heat-shock gene. *J. Mol. Biol.* **193,** 71–80.

Watt, J. L., and Stephen, G. S. (1986). Lymphocyte culture for chromosome analysis. *In* "Human Cytogenetics, A Practical Approach" (D. E. Rooney, and B. H. Czepulkowski, eds.), pp. 39–55. IRL Press, Oxford.

Weier, H.-U., Segraves, R., Pinkel, K., and Gray, J. W. (1990). Synthesis of Y-chromosome specific, labeled DNA probes by *in vitro* DNA amplification. *J. Histochem. Cytochem.* **38,** 421–426.

Part II. Visualization of Proteins

Chapter 6

The Use of Autoantibodies in the Study of Nuclear and Chromosomal Organization

W. C. EARNSHAW

Department of Cell Biology and Anatomy
Johns Hopkins University School of Medicine
Baltimore, Maryland 21205

J. B. RATTNER

Department of Anatomy and Medical Biochemistry
University of Calgary
Calgary T2N 1N4, Canada

METHODS IN CELL BIOLOGY, VOL. 35

I.   Introduction

Autoantibodies and Autoimmune Disease

The first clear indication that antigen–antibody reactions were a feature of systemic rheumatic diseases was established by the work of Hargraves and Kunkel more than 30 years ago (Hargraves *et al.,* 1948; Holman and Kunkel, 1957). Since then, it has become apparent that circulating serum antibodies to nuclear antigens are indeed a hallmark of systemic rheumatic disease. While the clinician is interested in these antibodies as diagnostic markers, the biologist sees them as useful probes for the study of nuclear and chromosomal organization and function. In general, antinuclear antibodies (ANAs) are found in a series of related diseases that include systemic lupus erythematosus (SLE), discoid lupus erythematosus (DLE), rheumatoid arthritis (RA), Sjøgren's syndrome, systemic sclerosis (SS), polymyositis and dermatomyositis [PM/DM], mixed connective tissue disease [MCTD], connective tissue syndromes, undifferentiated systemic rheumatic diseases, as well as diseases of inflammatory, neoplastic, or drug origin (Fritzler, 1986; Tan, 1982). In addition, it is not uncommon to find ANAs in individuals with no apparent disease (Fritzler, 1986).

In this chapter, we provide guidelines for the identification and utilization of autoantibodies in the study of nuclear and chromosomal architecture, composition, and function.

II.   Identification and Initial Characterization of Autoimmune Sera

A.   Identifying the Appropriate Serum

Autoantisera are typically selected for further study because they recognize an interesting structural pattern by indirect immunofluorescence or an interesting polypeptide by immunoblotting. This section describes the initial stages of obtaining, storing, and characterizing autoantisera.

Since ANA testing is commonly carried out as part of the diagnosis and management of rheumatic disease, the intial identification of sera with antibodies to specific nuclear components is often carried out in a clinical setting. Establishment of a suitable collaboration with a clinician is therefore often a first step for cell biologists in beginning to work with autoantibodies. When beginning this process, it is important to learn whether it will be possible to obtain further samples of sera that turn out to be of interest. A number of very interesting antigens have been characterized initially, only for the investigator to learn that further serum samples will not be forthcoming.

Our rule of thumb is to study only those sera available in at least several milliliters.

It may also be important to note, at least in the United States, that for a worker to have access to human sera, it is necessary to have filed the requisite human subjects reassurances with the National Institutes of Health (NIH). The assurances are not necessary if the cell biologist is provided with coded human sera by the collaborating clinician and does not have access to the patient names. These regulations differ in other countries.

For the purposes of this discussion, we assume that sera have been obtained from a clinician. These sera should be handled with care. Human sera may harbor significant pathogens such as human immunodeficiency virus (HIV) or hepatitis viruses, and diagnosis of such agents is not yet a routine procedure for most large clinical serum banks. Therefore, it is necessary to take care to always wear gloves and minimize the production of aerosols during opening of tubes. Detergents such as are used during the immunoblotting procedure are generally believed to inactivate HIV. Care should be taken during immunofluorescence, however, where the antibody solution may never come in contact with detergent.

Antibodies to the more common nuclear components are also available from a variety of commercial sources both individually or as part of diagnostic kits.

Table I summarizes the types of antibodies commonly detected, the disease with which they are associated, and the corresponding indirect immunofluorescent (IIF) pattern. In addition, Fig. 1 illustrates some of the more common IIF patterns as seen on Indian muntjac chromosomes and HeLa nuclei. If one is interested in antibody probes for the centromere, for example, one can start by requesting sera from a patient with the calcinosis, Raynauds phenomenon, esophageal dysmotility, telangiectasiae (CREST) variant of scleroderma that shows a discrete speckled pattern in interphase nuclei and at the metaphase plate (Fig. 1D). Often, however, autoantibodies to interesting or unique nuclear structures are not placed in standard laboratory categories. An example is chromatid linking proteins (CLiP) antigens that reside at the interface between sister kinetochores (Rattner et al., 1988). Antibodies to such novel nuclear components are often found in patients who are designated as having unusual or unclassified patterns.

Once the relevant patient group has been identified, it is necessary to aliquot the serum and to establish the titer of the antibody of interest.

B. Serum Storage

Typically, the sera are provided in aliquots of one to several milliliters. It is very important to aliquot these samples into "single use" aliquots of 50–100μl. Aliquoting is important for two reasons. First, repeated freezing and

TABLE I

NUCLEAR AND CHROMOSOMAL ANTIGENS RECOGNIZED BY ANTINUCLEAR AUTOANTIBODIES

Antigen	M_r (kDa)	Comments	Ref.[a]
Nucleosome			
DNA		Both double-stranded and single-stranded	1–5
Histones		Most commonly H1 and H2B, but H3 and H4 are also targets	6–11
Histone octamer		Specific for histones assembled into the octamer	12
HMG17	9.3		13
Ubiquitin	7	Conjugated to H2A and H2B	14
Various RNP Antigens			
Ro	60		15, 16
RNP proteins	11	A, A', B, B', B'', C, D, E, and F proteins are all targeted	11, 17, 18
Sm	11		19, 20
68-kDa U1 RNP protein	68		21
Nucleolus			
Topoisomerase I	100	Largely, but not exclusively, nucleolar	22–25
Fibrillarin	34		26
RNA polymerase I	90	Proteins of 65, 120, 42, and 25 kDa are recognized	27
NOR-90	90		28
B23	37		29
Centromere			
CENP-A	~17	Centromere-specific histone variant	30–33a
CENP-B	80	Centromere, located beneath kinetochore	30, 32, 33b
CENP-C	140	Centromere	30, 32
CENP-D	50	Centromere. May be human homolog of RCC1 protein	30, 34, 35
CLiP		Antigen unknown	36

Centromere-associated proteins			
Tubulin	52	Kinetochore in Colcemid-blocked cells	36a
Calmodulin	17	Kinetochore in Colcemid-blocked cells	37, 37a
Other [see (Tan, 1989) for a recent review]			
RNA			38
Topoisomerase II	170		39, 40
PCNA/cyclin	36	Auxiliary cofactor of DNA polymerase δ	41–43
Poly(ADP-ribose) polymerase	116	Reactivity with nucleoli and metaphase chromosomes	44, 45
Nuclear lamins	60, 70		46
Perichromin	33		47
NuMa protein	250–300	Nuclear at interphase, spindle poles during mitosis	48
RANA nuclear antigen	80		49
Nuclear matrix	Numerous	Staining throughout the nucleus	50, 51
NSP-1, NSP-2	Unknown	Variable numbers of nuclear speckles at interphase	52

[a] Key to references: 1, Ceppelini et al. (1957); 2, Robbins et al. (1957); 3, Stollar et al. (1962); 4, Rothfield and Stollar (1967); 5, Arana and Seligmann (1967); 6, Tan et al. (1976); 7, Hannestad and Stollar (1978); 8, Hardin and Thomas (1983); 9, Thomas et al. (1984); 10, Costa et al. (1986); 11, Tan (1989); 12, Rekvig and Hannestad (1980); 13, Bustin et al. (1982); 14, Muller et al. (1988); 15, Alspaugh and Maddison (1979); 16, Deutscher et al. (1988); 17, Pettersson et al. (1984); 18, Query and Keene (1988); 19, Wieben et al. (1985); 20, Stanford et al. (1987); 21, Guldner et al. (1988); 22, Douvas et al. (1979); 23, Shero et al. (1986); 24, Guldner et al. (1986); 25, Maul et al. (1986); 26, Lischwe et al. (1985); 27, Stetler et al. (1982); 28, Rodriguez-Sanchez et al. (1987); 29, Li et al. (1989); 30, Moroi et al. (1980); 31, Guldner et al. (1984); 32, Earnshaw and Rothfield (1985); 33, Palmer and Margolis (1987); 33a, Palmer et al. (1991); 33b, Cooke et al. (1990); 34, Kingwell and Rattner (1987); 35, Bischoff et al. (1990); 36, Rattner et al. (1988); 36a, Mitchison and Kirschner (1985); 37, J. B. Rattner (unpublished); 37a, Dedman et al. (1980); 38, Eilat et al. (1978); 39, Hoffmann et al. (1989); 40, Meliconi et al. (1989); 41, Takasaki et al. (1981); 42, Matsumoto et al. (1987); 43, Almendral et al. (1987); 44, Yamanaka et al. (1987); 45, Cherney et al. (1987); 46, McKeon et al. (1983); 47, McKeon et al. (1984); 48, Price et al. (1984); 49, Billings et al. (1983); 50, Fritzler et al. (1984a); 51, Solden et al. (1982); 52, Fritzler et al. (1984b).

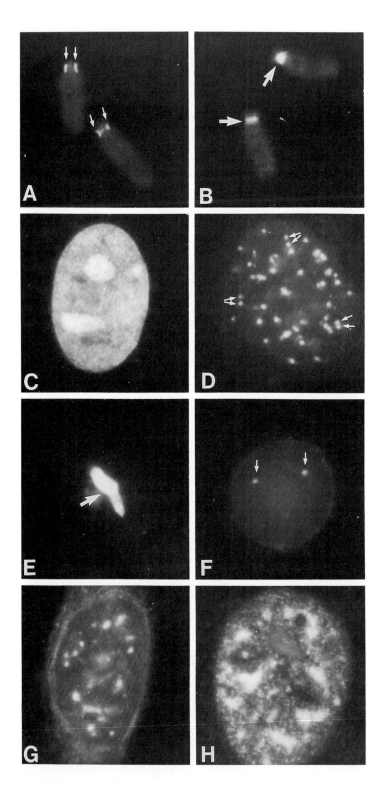

thawing results in a decrease in antibody reactivity. Second, this eliminates the possibility of accidental bacterial contamination of the entire sample. It is preferable to store aliquotted sera frozen at −80°C. Storage at −20°C in frost-free freezers should be avoided, since these freezers eliminate frost by automatic activation of periodic defrost cycles. Such cycles are sufficient to thaw small aliquots, and will reduce reactivity of the auto-antibodies after extended periods.

C.   Serum Titer and Isotype

Antibody titer is a crucial factor in determining the serum to be chosen for further study. The titer of some ANAs is higher in certain disease groups than others. For example, in sera from patients with the CREST variant of scleroderma it is not uncommon to find titers of anti-centromere antibodies in the 1:500 to 1:10,000 range, while in patients with ankylosing spondulitis titers of these antibodies reach only the 1:100–1:250 range (Martin *et al.*, 1991). In addition, the titer of antibodies within a given disease group varies both between individuals and between samples taken from the same individual at different times (Martin *et al.*, 1991).

The titer of the autoantibody against a particular nuclear antigen is established by making serial dilutions of the serum and then assaying for reactivity using both IIF and immunoblotting protocols. The end-point titer is the dilution at which reacitivity is no long detected. It must be remembered that patient sera are polyclonal and often contain antibodies to multiple nuclear antigens. Each of these autoantibodies may have different titers. Thus, it is not uncommon to see different IIF and immunoblot patterns at different dilutions of the autoimmune serum. For example, in immunoblots against nuclear proteins with anti-centromere antibodies, the titer of antibodies against CENP-A (centromere protein A) was 1:1,638,400, the titer against CENP-B was 1:3,276,800, and the titer against CENP-C was 1:12,800 (Earnshaw *et al.*, 1987a).

Fig. 1.   Indirect immunofluorescence pattern of human autoantibodies on chromosomes and nuclei. (A, B) Isolated Indian muntjac chromosomes were treated with sera containing auto-antibodies that react with either the kinetochore domain (A, arrows) or the entire centromere (B, arrows). (C–H) A variety of nuclear patterns produced by autoimmune sera on HeLa cells. (C) Complete nuclear staining typically produced by anti-DNA and anti-histone autoantibodies. (D) Anti-centromere (kinetochore) staining pattern. (E) Reactivity with the nucleolus (arrow). (F) NSP-1 (nuclear speckle pattern), typically 1–3 small dots in the nucleus. The antigen(s) responsible for this pattern is currently unknown. (G) NSP-2, characterized by numerous spots of various sizes. The antigen(s) responsible for this pattern is currently unknown. (H) Extensive nuclear staining pattern commonly associated with nuclear matrix antigens.

It is also important to be aware that different autoantisera may have differing isotype distributions. Autoantibodies are not always IgG. In fact, significant levels of both IgM and IgA anti-centromere antibodies have been detected in addition to IgG (Hildebrandt *et al.,* 1990), and anti-DNA is often IgM. Isotype characterization may be important for two reasons. First, not all secondary reagents will detect all isotypes equally. (For example, commercial secondary antibodies are often anti-IgG. Also protein A does not detect human IgG_3.) In addition, the isotype may influence the immunofluorescence patterns observed with certain antibodies. For example, IgMs may have difficulty penetrating into regions of very compact structure.

III.   Characterization of Autoantigens by Light Microscopy

Autoantibodies have proved to be powerful reagents for characterization of the intracellular distribution and molecular properties of nuclear and chromosomal antigens. This section is concerned with the determination of the intracellular distribution of antigens recognized by autoantibodies.

The goal of light microscopic techniques employing autoantibodies is to (1) maximize the ability to detect antibody binding, (2) achieve maximal spatial resolution in the distribution of antibody-reactive sites, (3) maintain *in vivo* cellular architecture, (4) minimize cross-reactivity and the production of artifactual antibody-binding patterns.

To achieve these goals, it is necessary to consider (A) the selection and preparation of substrates, (B) fixation protocols, (C) antibody detection systems, and (D) mounting and photographic protocols. We begin by discussing substrate selection and preparation. We then present a sample protocol for immunofluorescence analysis. Finally, we discuss the fine points (B–D) in detail.

A.   Substrate Selection

The determination of the appropriate cell substrate is determined by (1) the goal of the study, (2) the characteristics and unique features of cells from different origins, and (3) the degree to which the epitope recognized by the antibody is conserved in different tissues and species. The staining will typically be carried out on tissue culture cells grown in monolayer culture, cell suspensions, or isolated component suspensions deposited on coverslips or slides. Cryopreserved sections derived from various organs and smear preparations may also be used.

1. CULTURED CELL SUBSTRATES

Cells growing in monolayer culture are particularly useful in studying the distribution of autoantigens at high resolution and as a function of cell cycle phase.

Since autoantibodies are typically thought to have arisen in response to human antigens, these studies often begin by using human cell lines, such as HeLa. A number of companies supply autoantibody screening kits to clinical laboratories that include prepared slides with cells of human origin. These slides are often used in the clinic for the classification of autoantibody patterns. A human liver cell line, HEp-2 (Immunoconcepts, Sacramento, CA), is most often employed in these studies. The advantage of this line is that its rapid growth produces abundant mitotic figures within a single microscope field. However, the small size of human chromosomes and the tendency for human cells to round up during mitosis often makes them less than ideal substrates for antibody studies. For this reason, other cell lines with special characteristics have frequently been employed.

Human autoantibodies have been applied to cells as diverse as mammals and plants. Favorites among these alternative cell types include cells of the Indian muntjac (an Asiatic deer) and the rat kangaroo (PTK_2). Both cell lines can be obtained from the American Type Culture Collection (Rockville, MD). The advantage of PTK_2 cells is that they remain flat throughout the cell cycle in monolayer culture, providing optimum features throughout cell division and thus facilitating the analysis of the distribution of antigens during mitosis. On the other hand, the Indian muntjac cells contain the simplest karyotype [$2N = 6$ (female), 7 (male)] and largest chromosomes of any mammal. Most chromosomes of the karyotype contain extensive centromere and kinetochore regions. The large size of these regions allows the distinction between the centromere and the kinetochore domain region (Earnshaw and Rattner, 1989) of metaphase chromosomes to be made at the light microscope level. This is usually not possible in other mammalian chromosomes. It is still unclear, however, whether it is possible to visualize specific staining of the kinetochore *plates* within the kinetochore domains by light microscopy.

Despite the morphological advantages provided by the choice of these special substrates, caution must be exercised when interpreting results obtained using cells of nonhuman origin. Several factors should be kept in mind when applying human antoantibodies to nonhuman substrates. First, the autoantibody may react with an epitope on an authentic homologous protein. The distribution of such homologous proteins may be the same or different in different species. [For example, CENP-B, the 80-kDa centromeric autoantigen, recognizes a 17-bp DNA sequence in human alphoid DNA

(Masumoto *et al.*, 1989). This sequence is also present in mouse minor satellite DNA (Wong and Rattner, 1988), but it is unknown whether the detailed distribution of CENP-B in mouse and human is similar.] Second, the auto-antibody may recognize a cross-reactive epitope on a protein otherwise un-related to the human autoantigen. In this case, the pattern cannot be easily correlated with that seen on the human substrate.

When tissue culture cells are to be used as substrates, cells from stock cultures can be plated onto microscope slides placed within a petri dish or within a temporary chamber mounted on the slide. Slides with chambers can be obtained commercially (Lab-Tek tissue culture chamber/slides, Miles Scientific). Alternatively, cells may be plated directly onto coverslips. The 18 × 18 mm size is usually sufficient and fit into 35 × 10 mm plastic dispos-able petri dishes. Cells are usually plated at 50% confluency at least 24 hours prior to use. For the accumulation of mitotic cells, anti-mitotic agents such as Colcemid (0.01 μg/ml for 2–12 hours) can be added to the culture prior to use. Care must be taken, however, since many cells types round up during division and are easily dislodged from the culture surface.

2. TISSUE SECTIONS

Tissue sections are often used in the early stages of autoantibody charac-terization. Cryopreserved organ tissues, which have proved to be the best tissue substrates, are available from commercial suppliers (Inter Medico, Markham, Ontario, Canada) or can be prepared fresh in the laboratory. Several methods have been reported in the literature; a basic procedure is as follows. Microscope slides are coated with a 0.25% aqueous solution of so-dium metasilicate (Sigma) by immersing the slide in the solution for 10 minutes and then allowing the slides to air-dry. The slides may then be stored for several days in an air-tight box. The tissue material is excised and placed in an aluminum foil capsule containing OCT (Cryostat embedding medium OCT, Ames Co; or Tissue Tek II OCT, Miles Laboratories). The capsule is then quick-frozen in liquid nitrogen or acetone dry ice. The capsules can then be stored at $-70°$C in a box containing desiccant. To produce sections, the capsules are mounted in a cryostat and allowed to equilibrate for 30 minutes. Four-micron sections are then cut and mounted on coated slides and allowed to air-dry. The slides may be stored at $-20°$C. To process the slides, the tissue sections are first fixed in acetone for 10 minutes and allowed to air-dry. The sections are then reacted with the antibody as described below.

3. ISOLATED MITOTIC CHROMOSOMES AND NUCLEI

In special cases, one may be interested in looking at isolated components such as nuclei and chromosomes. There are several common nuclear isolation protocols that gives rise to nuclear suspensions that can be deposited directly on coverslips or microscope slides (see for example, Davis *et al.,* 1986).

A simple procedure for obtaining a chromosome suspension is as follows. Mitotic cells obtained from logarithmically growing cultures are collected, typically by selective mitotic detachment, and pelleted at 1000 *g* for 5 minutes and resuspended in 0.5 ml of growth media without fetal calf serum. (We use a table-top centrifuge, such as a Beckman TJ-6, with 15-ml conical centrifuge tubes.) The cells are then diluted 1:3 in medium containing 0.1–1% Nonidet NP-40 (BDH, Poole, England) that has been made up in growth medium without fetal calf serum. Lysis is monitored by placing a drop of the solution on a slide and observing it in a light microscope. One should be able to observe the release of the chromosomes. Care must be taken in transferring the sample, since the chromosomes are susceptible to mechanical shear. In lieu of growth media, other chromosomal buffers can be substituted as required by the study. The goal is to maintain morphology. The chromosome suspension is then deposited on coverslips as described above. This procedure provides a lawn of mitotic chromosomes, but will also work equally well for preparations of nuclei if interphase cells are used for starting material.

B.   Techniques for Indirect Immunofluorescence of Chromosomes and Nuclei

1.   PROTOCOLS FOR INDIRECT IMMUNOFLUORESCENCE

Many autoantibodies of interest have been identified because they recognize interesting structural patterns upon immunofluorescence. This section is devoted to the description of procedures used to localize autoantigens in chromosomes and nuclei.

Indirect immunofluorescence (IIF) methods are very sensitive to experimental conditions, and it is important to bear in mind the following considerations when developing a research plan in order to minimize effects of experimenter bias.

First, it is important to design the experiment to answer the desired question properly. If, for example, it is simply desired to know if an antigen is chromosomal or cytoplasmic, the method of choice will be IIF, using some form of permeabilized, fixed cells as substrate. If, on the other hand, ultrastructural localization of the antigen to a particular chromosomal domain (such as the kinetochore) is required, it will be necessary to perform the much more laborious procedure of immunoelectron microscopy. This is because, while it is possible to *detect* objects below the limit of resolution of the light microscope by fluorescence microscopy, this technique cannot be used to locate the antigens to better than ~0.2 μm resolution. In the chromosomes, the outer kinetochore plate is separated from the centeromeric chromatin by only 0.025–0.04 μm (Rieder, 1982), clearly below the resolving power of the microscope.

If it is important to observe details of cell architecture at the same time as antibody labeling, use of an enzymatic detection system may be required. The enzymatic detection systems are also compatible with simultaneous visualization in parallel with other types of labeling procedures (such as autoradiography; Heck and Earnshaw, 1986).

All immunolocalization protocols are designed to show us what we want to see, and may therefore not always absolutely reflect the true distribution of antigens *in vivo*. Any immunolocalization protocol reflects a "competition" between detection of antigen and structural preservation of the cell. Many antigens (such as those recognized by ACA) are sensitive to chemical fixation procedures, and are undetectable under conditions where cell structure is optimally preserved. Thus, compromises must be struck with the degree of fixation employed. It is therefore important when characterizing a novel autoantigen that several different fixation protocols be employed in order to arrive at an optimal and representative procedure.

Similarly, some antigens are present in both soluble and bound forms. If cells are permeabilized prior to fixation, the soluble forms may be lost. It is therefore important in the initial characterization of a novel antigen to use fixation procedures like cold methanol or acetone that work by precipitating proteins and permit detection of otherwise soluble forms of antigen. If soluble pools exist, permeabilization prior to fixation will permit an enhanced visualization of bound antigen, but it is important not to exclude the soluble pool from the ultimate map of the distribution of the given antigen.

The physiological state of the substrate cells has a profound effect on the distribution of a number of antigens. Many studies of chromosomal antigens use cells that have been arrested with Colcemid or other antimitotic drugs. However, recent results show clearly that the distribution of some mitotic spindle proteins changes on Colcemid treatment of cells, causing these proteins to become associated with centromeres (Compton *et al.*, 1991). It is therefore important to look both at blocked and cycling cells.

Finally, choice of system is critical. If, for example, one wishes to study the centromere region, an organism such as the Indian muntjac, which has enormous kinetochores, is a great advantage. At the same time (as discussed above), one should be aware that human autoantibodies primarily recognize human antigens, and that such antigens may not be distributed equivalently in all organisms (or even in all mammals).

The following procedure works well for detection of chromosomal antigens with autoantibodies. In all the protocols below, we suggest dilutions of antibody that work well in our laboratories. However, it is important to titer all reagents so that they are used at the maximum dilution that gives an acceptable signal. This maximizes the number of procedures that can be performed with each batch of reagent and minimizes background.

Basic protocol for Immunofluorescence Analysis of Chromosomal or Nuclear Proteins in Situ. All buffers are given in Table II. Other reagents are given in Table III. Note that after fixation, PBS and TBS may be used interchangeably. PBS is preferable prior to fixation with aldehydes.

1. Grow cell cultures *in situ* on glass coverslips in petri dishes or in microwell slides. Alternatively, nonadherent cells are adsorbed to

TABLE II

BUFFERS FOR INDIRECT IMMUNOFLUORESCENCE

Buffers[a] (10 × final concentration)	Stock solution	Add/1 liter final
10 × TBS		
100 mM Tris-HCl, pH 7.7	1 M	100 ml
1.5 M NaCl	5 M	300 ml
1% BSA	30% Pentex grade	33 ml
10 × PBS		
80.6 mM Na$_2$HPO$_4$(dibasic)	—	11.4 g
14.7 mM KH$_2$PO$_4$(monobasic)	—	2 g
1.37 M NaCl	5 M	274 ml
26.8 mM KCl	4 M	6.7 ml
4.92 mM MgCl$_2$	1 M	4.92 ml
6.8 mM CaCl$_2$	1 M	6.8 ml
10 × TEEN		
10 mM Triethanolamine-HCl, pH 8.5	1 M	10 ml
2 mM NaEDTA	100 mM, pH 9.0 [NaOH]	20 ml
250 mM NaCl	5 M	50 ml
10 × PIPES-MT		
800 mM PIPES, pH 7.2	1 M	800 ml
10 mM EGTA	0.5 M	20 ml
50 mM MgCl$_2$	1 M	50 ml
10 × PHEM		
450 mM PIPES, pH 6.9	—	136.1 g
450 mM HEPES	—	106.8 g
100 mM EGTA	0.5 M	200 ml
50 mM MgCl$_2$	1 M	50 ml
Add PMSF to 1 mM immediately before use		

[a] BSA, Bovine serum albumin; NaEDTA, sodium ethylenediaminetetraacetic acid; PIPES, piperazine-*N,N'*-bis(2-ethanesulfonic acid); HEPES, *N*-2-hydroxyethylpiperazine-*N'*-2-ethanesulfonic acid; EGTA, ethylene glycol bis(*β*-aminoethyl ether) *N,N'*-tetraacetic acid; PMSF, phenylmethylsulfonyl fluoride.

TABLE III

OTHER SOLUTIONS FOR INDIRECT IMMUNOFLUORESCENCE

3% Paraformaldehyde in $1 \times$ PBS-azide. (Made from solid paraformaldehyde.) Add paraformaldehyde to a 90% volume of double-distilled water (ddH$_2$O). Heat to near boiling with stirring on a hotplate in a fume hood. Add 10 N NaOH dropwise until solution clears. Add $10 \times$ PBS-azide and ddH$_2$O as necessary to the desired final volume. Cool and filter through a 0.2-μm filter. We routinely make formaldehyde solutions of up to 8% in this way

10% Triton X-100 (in ddH$_2$O). Dissolve, filter through a 0.2-μm filter, and store refrigerated

30% BSA (Sigma A-7284)

0.8% Sodium citrate in ddH$_2$O

Mounting medium. (Moviol is very convenient, because it hardens to a clear solid and does not require sealing around the edges of coverslips. The formula given here does not contain antifade agents, and will bleach fairly rapidly under intense illumination. The glycerol/ Dabco mixture is quite stable against fading, but requires the more tedious steps of sealing carefully around all coverslips with nail polish.)

Moviol (We use 10 mM Tris-HCl, pH 7.7, 150 mM NaCl, but any buffer will do.) For 4 ml: Add 0.4 g of Moviol solid + 1 ml ddH$_2$O to 1 g glycerol in 5-ml plastic tube. Stir thoroughly to mix, then incubate for 2 hours at room temperature. Add 2 ml of desired buffer (at 2 × the desired final concentration). Incubate overnight at 50°C. Remove undissolved solids by centrifugation at 5000 g for 15 minutes. Aliquot into glass tubes and store at either 4°C or −20°C

Glycerol/Dabco: Dissolve Dabco (1,4-diazabicyclo[2.2.2]octane; Aldrich D2,780-2) at 100 mg/ ml in 1 × PBS-azide + 25% glycerol. Seal around edges of coverslips with nail polish

---

Adhesio Slides (M & M Developments) as per the manufacturer's instructions. Cells may be blocked in mitosis with Colcemid or other drugs as required.

2. For cells grown *in situ*, carefully transfer to tissue culture hood and aspirate medium.

3. Wash cells in two changes of PBS, 2–5 minutes each (optional step).

4. Fix with 3% paraformaldehyde in PBS for 5–10 minutes at room temperature. See below for discussion of other fixation protocols. For processing of single slides, all incubations are carried out in a 9-cm petri dish. If multiple slides are processed, Coplin jars are used.

5. Wash coverslips three times for 2 minutes each with 1 × TBS buffer + 0.5% Triton X-100 + 0.1% BSA at room temperature. (When acetone or methanol fixations have been used as described below, the slides will have been air-dried. Therefore it is necessary to rehydrate the slide in PBS for approximately 5 minutes.)

6. The preparation is incubated with blocking solution to prevent unspecific reactions. (This is an optional step, see discussion below.)

7. Add autoantibody (dilute appropriately in 1 × TBS buffers) to cover-slips. Incubate for 30 minutes at 37°C. This is usually done in a closed box containing moistened paper towels or Kimwipes, which maintain the moisture in the chamber and prevent the preparation from drying out. It is essential to prevent drying throughout the procedure.

8. Wash three times for 2, 5, and 3 minutes at room temperature with 1 × TBS buffer.

9. Add preabsorbed biotinylated anti-human antibody (Vector Labs) di-luted 1 : 200 (or as appropriate) in 1 × TBS buffer. (A protocol for anti-body absorption is given below.) Incubate for 30 minutes at 37°C. If double labeling (for example, for tubulin) is required, add monoclonal or rabbit antibody recognizing the desired component.

10. Wash as in step 8.

11. Add streptavidin: Texas Red (BRL) diluted 1 : 800 (or as appropriate) in 1 × TBS buffer. Incubate for 30 minutes at 37°C. If double labeling is being performed, add the recommended dilution of fluorescein iso-thiocyanate (FITC)-conjugated anti-mouse or anti-rabbit antibody as appropriate.

12. Wash for 2 minutes with 1 × TBS buffer at room temperature.

13. Counterstain for DNA with 4′,6-diamidino-2-phenylindole (DAPI, 0.5 μg/ml in 1 × TBS buffer) for 5 minutes at room temperature.

14. Wash for 3 minutes with 1 × TBS buffer at room temperature.

15. Mount with either Moviol, glycerol/Dabco (Table III), or commercial mounting medium (see below).

Procedures for Preparation of Aqueous Mitotic Chromosome Spreads for Indirect Immunofluorescence. Substitute the following for steps 1–3 in the Basic Protocol, above.

1a. If spreads of cells grown *in situ* are required, gently pipette 2 ml of a hypotonic swelling solution (usually 0.8% sodium citrate or 75 m*M* KCl) into the petri dish (hypotonic swelling step).

1b. Incubate for 20–25 minutes at room temperature, then remove this solution very carefully by aspiration.

1c. Place coverslip or microwell slide on filter paper in petri dish and spin in a table-top centrifuge at maximum acceleration, stopping when a speed corresponding to ∼2000–2400 *g* has been reached. (This takes 1–2 minutes.) We use the Beckman TJ-6 table-top centrifuge, which has flat-bottomed steel buckets large to hold a 9-cm plastic petri dish. We place a piece of foam rubber in the bottom of the buckets as a cushion.

1d. Stop centrifuge, remove coverslips, and rapidly immerse coverslips in 1 × PBS-azide.

1e. Aspirate carefully and proceed with step 4 of the Basic Protocol.

The following alternative method is suitable for production of spreads of metaphase chromosomes from mitotic cells in suspension. Substitute for steps 1 and 2 in the Basic Protocol.

A small plexiglass chamber made from a hollow plexiglass rod or purchased commercially (for example, Polysciences Incu Rings, cat. no. 8412) is temporarily cemented to a glass slide or coverslip using melted paraffin wax.

1a. A solution containing the mitotic cells is then placed within the chamber and placed in table-top centrifuge operated at 3000 g for 1–2 minutes. Most centrifuge buckets can be adapted for this purpose. For example, in centrifuges that accommodate large ($\sim$ 11 by 11 cm) buckets, the coverslip with chambers can be taped to a glass microscope slide that is in turn taped to the centrifuge bucket. For centrifuges that can only accommodate small-diameter buckets, the coverslips with chambers can be placed directly at the bottom of the bucket using long forceps. The bottom of the bucket is covered with a rubber pad to absorb shock and vibration.

1b. After centrifugation the slides are removed and the chamber dislodged with a single-edge razor blade. The sample is then ready to be processed. The residual paraffin ring serves as a dam to hold small amounts of solution over the area containing the sample.

2. Aspirate carefully and proceed with step 3 of the Basic Protocol.

Procedures for Preparation of Methanol/Acetic Acid Chromosome Spreads for Indirect Immunofluorescence. Substitute the following for step 3 in the Basic Protocol.

3. Gently replace medium with 2 ml of a solution of 0.8% sodium citrate. Incubate for 20–25 minutes at room temperature.

4a. Add 2 ml of a solution of methanol/acetic acid (3:1). Incubate for 2 minutes at room temperature.

4b. Remove this solution by aspiration.

4c. Add 2 ml of methanol/acetic acid (3:1) for 5 minutes at room temperature.

4d. Aspirate the methanol/acetic acid.

4e. Insert the petri dishes and vigorously bang them on a paper towel on the bench top three times to remove most of the remaining fixative.

4f. Immediately blow the coverslip dry with an aquarium pump (Hagen 800 or 800D, using a cut 1-ml plastic pipette as nozzle), with the nozzle $\sim$ 1–3 inches above the coverslip. Move the nozzle over the surface rapidly and at random. (Process two or three slides simultaneously, depending on how many you can hold in your hand).

5a. *As soon as* the surface is dry, remove the coverslips from the petri dishes (to place forceps under the coverslips, flex the petri dishes by pressing down on a rubber stopper). Immediately immerse the coverslips in a 1 × solution of PBS-azide for 5 minutes at room temperature (or as long as it takes to process the rest of the coverslips).

5b. Wash coverslips three times for 1 minute each with 1 × TEEN buffer (Table II) + 0.1% Triton X-100 + 0.1% BSA at room temperature.

6. Add antibody (diluted appropriately in 1 × TEEN + 0.1% Triton X-100 + 0.1% BSA) to coverslips. Incubate for 30 minutes at 37°C.

7. Proceed to step 9 of the Basic Protocol.

2. VARIATIONS OF THE ABOVE PROTOCOLS

There are many ways of performing immunofluorescence analysis. The above three-stage procedure has proved useful for the detection of faint antibody labeling patterns, but it takes ~45 minutes longer than a simpler alternative, in which the human antibody is detected directly with fluorochrome-conjugated anti-human antibody. In this case, if simultaneous labeling of a second subcellular component is desired, the specific mouse or rabbit antibody is added concurrently with the autoantibody.

Depending on the details of the antigen to be studied, the main points of variation are as follows.

Fixation. Different antigens tolerate different degrees of fixation. Thus, for centromeric autoantigens (which are relatively sensitive to fixation), mild aldehyde treatment is used as described above. This is not optimal for microtubules, however, and if preservation of overall cell architecture is desired, then somewhat harsher fixations are needed. These commonly include higher percentages of formaldehyde in PBS for longer periods of time, mixtures of formaldehyde and glutaraldehyde (usually between 0.01 and 0.1% glutaraldehyde), other cross-linkers such as ethylene glycol bis(succinimidyl succinate (EGS, Pierce) (Gorbsky *et al.,* 1987), periodate:lysine:paraformaldehyde (McLean and Nakane, 1974), absolute methanol or acetone at −20°C, and anhydrous methanol at −20°C for 10 minutes followed by 100% acetone at −20°C for 10 minutes. In the case of methanol and acetone fixation, the slides are air-dried following fixation and can be stored at −20°C in a tightly sealed container for some time. Samples cannot be stored when other fixation protocols are used.

The general rule when characterizing an novel antigen is to try a variety of fixation protocols to determine the one that produces the best morphology commensurate with good antibody labeling. Comparison of results obtained with different fixation protocols is also important in revealing potential artifacts (such as redistribution of the epitope during processing, and loss of the epitope due to the fixation) that can arise at this stage. In studies where microtubule integrity is a concern, PIPES-MT buffer (Table II) is used to stabilize microtubules. Treatment with PIPES-MT buffer containing the following is typically used as a fixation protocol that preserves microtubules: 3.7% paraformaldehyde (for 10–30 minutes); 2% paraformaldehyde and 0.1% glutaraldehyde (for 20 minutes). Microtubules are also preserved with methanol or methanol–acetone fixation.

In some studies it may become apparent that the cytoplasm also contains the antigen under study. In this case, some workers use an extraction protocol to remove cytoplasmic antigen prior to fixation in order to obtain better images. For example, the localization of p34^{cdc2} cannot be clearly visualized until the cytoplasmic pool of this protein is removed (Rattner *et al.*, 1990). This extraction can be achieved by immersing the specimen in PHEM buffer (Table II) + 0.1% Triton X-100 for ≤60 seconds. The sample is then fixed and processed.

Permeabilization.   Detergent treatment is necessary to permit access of antibodies to the interior of aldehyde-fixed cells. Typically, 0.1% Trition X-100 or NP-40 is used. If desired, other detergents such as 0.1% digitonin or 0.1% saponin may be used, or the Triton/NP-40 concentration may be lowered to 0.01%–0.05%. Additional permeabilization steps are not needed after methanol or acetone fixations.

Blocking.   Nonspecific binding can result from electrostatic or hydrophobic interactions between the antibody and the substrate. These binding sites can be blocked by the use of proteins such as fetal calf serum, bovine serum albumin, or nonimmune sera. In general, these proteins are added at a concentration of 0.1–1.5% and incubated for 10–30 minutes at 37°C. The excess is removed by tilting the coverslip but the residual solution is not washed off. Rather, the first antibody is added directly to the sample. As noted above, this step is optional.

Primary Antibody.   Dilutions of primary antibody are prepared in the appropriate buffer (see Table II). We typically use either TBS or PBS for this purpose. Addition of protein at this stage will help to block nonspecific binding and stabilize the antibody solution. In addition, detergents can also be added to the primary antibody solution to facilitate penetration. These include 0.1% Tween 20, 0.01%–0.1% Triton X-100, or 0.1–0.05% NP-40.

When novel autoantigens are being characterized, it is essential to perform control incubations in parallel. These include elimination of the primary antibody, or substituting sera from individuals with no known autoimmune disease for the primary antibody. Care must be taken in the latter case to characterize the control sera by immunoblotting, since individuals with no known autoimmune disease may contain autoantibodies that are in some cases present at relatively high titers. When using antibodies to known antigens, it is possible in some cases to preabsorb with the antigen to be localized. This should abolish specific labeling.

Secondary Antibody and Detection System. Bound primary antibodies can be detected by a variety of detection systems. Each has its own characteristics and these features must be considered when choosing a detection system. In order to achieve the cleanest possible signal with minimal background due to nonspecific binding by the secondary antibody, it is best to adsorb the secondary antibody reagents with extracts made from the cell type to be examined. A protocol for this absorption is given below.

Avidin and Streptavidin. The avidin–biotin system for detection of bound primary antibody is preferred by most workers. This system offers great sensitivity, since the secondary anti-human antibody is multiply biotinylated, offering the opportunity for some direct amplification of the fluorescence signal when > 1 fluorochrome-tagged streptavidin molecules binds to it. It is possible to purchase biotinylated secondary antibodies, although kits for direct biotinylation of the primary antibody are also available [Clontech, Boehringer Mannheim]. Streptavidin is the reagent of choice for detection of biotinylated antibodies bound to nuclear and chromosomal components. It binds to biotin with K_d similar to that of egg while avidin ($K_d = 10^{-15}M$), but lacks the extremely charged amino acid composition that causes avidin to exhibit considerable nonspecific binding to chromatin.

Flurochromes. Instead of the three-step streptavidin: biotin system described in the Basic Protocol, many laboratories use a two-step immunodetection protocol. This protocol uses fluorochrome-conjugated secondary antibodies for detection of bound primary antibody. Antibodies to human immunoglobulin are commercially available conjugated to a variety of flurochromes, including fluorescein isothiocyanate (FITC), rhodamine B isothiocyanate (RITC), or Texas Red. Even though these antibodies are typically affinity-purified by the supplier, it is strongly recommended that they be absorbed with the cell type to be studied in order to minimize background.

Enzyme Conjugates. Enzyme conjugates offer the opportunity to view the reaction and the counterstain simultaneously. (This can also be done with immunofluorescence, if the immunolabeling is sufficiently strong, by leaving the phase-contrast illumination on while viewing the fluorescence.) Furthermore, the signal on slides prepared using enzyme conjugates has the advantage of being permanent. The method can be used on most substrate preparations including paraffin sections and only a conventional light microscope is required. In addition, material reacted with enzyme substrates, for example, peroxidase, can be used for both light and electron microscopy. The main disadvantage of this system is that the precipitate of the enzyme reaction may diffuse from the side of reaction. Horseradish peroxidase (HRP) and alkaline phosphatase conjugated to anti-human secondary antibodies are the most common enzymes used in immunocytochemistry. In the case of HRP, the enzyme is detected following reaction with one of several chromogens. These include 3,3'-diaminobenzidine (DAB; Sigma), which leaves a water- and alcohol-insoluble brown deposit at the reactions site; DAB in conjunction with nickel, which leaves a dark brown deposit at the reaction site; 3-amino-9-ethylcarbazole (AEC; Sigma), which leaves a water-insoluble red or brown red deposit at the reaction site; 4-chloro-1-napthol, which leaves a water-insoluble blue-black reaction product; and Hanker-Yates substrate (Sigma), which produces a blue-black product insoluble in water and alcohol. When using substrates with endogenous peroxidase activity, this activity can be inhibited prior to addition of peroxidase-coupled antibody by incubation in a solution of methanol nitroferricyanide containing 1% sodium and 1% acetic acid (Zymed). Substrates for alkaline phosphatase include Fast Blue, Fast Red, and new fuchsin.

Double Labeling. It is often informative to detect the positions of two antigens simultaneously. To do this, one must use primary antibodies from two different sources and secondary antibodies conjugated with labels that can be differentiated. Many combinations are possible. For example, it is possible to locate the bound autoantibodies with streptavidin: Texas Red and colocalize tubulin using a commercial monoclonal antibody to tubulin plus FITC-conjugated anti-mouse secondary antibody. Localization of DNA using DAPI stain (see below), provides a third level of simultaneous detection. Double labeling is also possible using the enzymatic detection systems. For example, one antibody can be detected with a peroxidase-conjugated secondary reagent while the other is detected with an alkaline phosphatase-conjugated reagent. If hydrogen peroxidase/AEC is used with peroxidase, a red deposit is produced. Use of AP-Blue alkaline phosphatase gives a blue color.

Colloidal Gold Labeling. Colloidal gold labeling has applications at both the light and electron microscope levels. At the light microscope level, colloidal gold labeled primary antibodies have been used on dewaxed or cryostat sections as well as 1-μm resin sections. Using a silver enhancing procedure, it is possible to visualize the gold label. This method is particularly useful when detection of low levels of antigen is required. In general, one should chose small particle sizes (1 nm; Amersham, Pelco) since they facilitate penetration. Gold-conjugated reagents are commercially available from a variety of sources and are generally available in 1 to 20-nm particle sizes.

Specific application of the colloidal gold detection system to electron microscopy is outside the scope of this review. See Cooke *et al.* (1990) for a suitable protocol.

Secondary Antibody Absorption. Despite the fact that many commercial antibodies are affinity-purified, these reagents often given variable levels of binding to cellular structures. Therefore, one should routinely absorb such reagents with extracts prepared from the cell of choice. The following protocol describes the preparation of a secondary antibody suitable for use with either human or chicken cells:

1. Grow the following cell cultures in spinner flasks: 1 liter each of HeLa and chicken MSB-1 cells blocked overnight with 0.1 μg/ml Colcemid, plus 1 liter each of HeLa and MBS-1 cells in exponential growth. HeLa cultures should be at $\sim 2 \times 10^5$/ml. MBS-1 cells grow well to $\sim 1 \times 10^6$/ml.

2. Harvest cells by centrifugation at 800 g for 3 minutes and wash in 100 ml/liter of culture of $1 \times$ PBS (Table II) + PI (Table IV).

3. Resuspend each liter in 9 ml of $1 \times$ PBS + PI. Sonicate on ice with 20 pulses of a Branson sonifier with microprobe at setting 5, 70% duty cycle. (Check for cell lysis after 10–15 pulses by phase-contrast microscopy.)

4. Add 9 ml of 8% formaldehyde in ddH$_2$O to each tube. Incubate with mixing in the cold for 1 hour.

5. Centrifuge fixed cell debris at 12,000 g for 30 minutes at 4°C.

6. Resuspend each pellet by sonication in 2 ml of quenching solution containing 0.1M Tris-HCl, pH 7.5, with 0.5 mg/ml NaBH$_4$ (prepared fresh!). Incubate for 5 minutes on ice.

7. Centrifuge cell debris in microfuge for 5 minutes at 4°C.

8. Quench twice more as in step 6 and 7.

9. Divide each liter's worth of cells in two and wash twice by sonication in 20 mM NaPO$_4$ buffer, pH 7.4, followed by centrifugation as in step 7.

10. Resuspend washed pellets in 0.25 ml of 20 mM NaPO$_4$ buffer, pH 7.4, +0.5 ml of carrier serum (normal goat serum or fetal bovine serum) +1 ml of commercial secondary antibody + PI minus PMSF.

11. Incubate overnight with mixing in the cold.

12. Centrifuge at 140,000 g_{av} for 1 hour at 4°C. We use the Beckman type 50ti rotor.

TABLE IV

SOLUTIONS FOR SECONDARY ANTIBODY ABSORPTION

PI (protease inhibitor cocktail) (final concentration)	Stock solution
0.1% Trasylol	10% in ddH$_2$O
1 mM PMSF	1 M in ethanol
1 mg/ml Chymostatin, leupeptin, antipain, pepstatin	1 mg/ml in ddH$_2$O (mixture of all four)

1 M NaPO$_4$ buffer, pH 7.4 (dilute to 20 mM for use)

1 M Tris-HCl, pH 7.5 (dilute to 0.1 M for use)

8% Paraformaldehyde in ddH$_2$O (made from solid paraformaldehyde). Add paraformaldehyde to ddH$_2$O. Heat to near boiling with stirring on a hotplate in a fume hood. Add 10 N NaOH dropwise until solution clears. Add ddH$_2$O as necessary to the desired final volume

13. Titer the supernatant using a known primary antibody, aliquot into 25- to 50-μl aliquots and store frozen at $-80°C$.

3.   DETECTION OF DNA

The choice of a counterstain for DNA detection will to a large degree depend on the detection system used with the secondary antibody, and also on the specific microscope application. DAPI (0.1 μg.ml) is commonly used because of the ease of staining, and because it is compatible with the simultaneous use of fluorescein and rhodamine/Texas Red. In addition to the staining method given in the Basic Protocol, DAPI (0.01 μg/ml), ethidium bromide (0.05 μg/ml), or 33258 Hoechst (50 μg/ml) can also be added directly to the mounting medium.

Some applications, such as confocal microscopy, require detection of DNA in either the green or red channels. This can be accomplished either with $10^{-4}M$ chromomycin A3 (White et al., 1987) (Calbiochem; fluorescein channel), 1 μg/ml propidium iodide (Sigma; rhodamine channel), or a 1 : 50 dilution of commercial monoclonal anti-DNA antibody (Chemicon). If staining with propidium iodide is desired, it is necessary to remove the endogenous RNA by digestion with heat-treated RNase A (at 50 μg/ml in the final antibody incubation for 30 minutes at 37°C).

4.   MOUNTING

When fluorochrome detection systems are used, the immunofluorescence is stable for several months if the specimen is mounted in an alcohol-based medium and stored at 4°C in the dark. The pH should reflect the maximum fluorescence emission of the fluorochrome. For example, for FITC, it is pH 8.5. There are several commercially available mounting media that meet these criteria, for example, Citifluor (Ted Pella, Reading, CA). A solution of 90% glycerol containing paraphenylenediamine (PPD) to inhibit fading is also widely used. Two alternative mounting media are described in more detail in Table III.

5.   PHOTOGRAPHY

IIF images can vary in intensity and are subject to fading. For these reasons, photography should be carried out using a high-speed, fine-grain film. Common black and white films used are IlFord XPI-400, IlFord HP5, and Kodak T-Max 400. We use Kodak Ektar 125 film when color photographs are required, and Kodak Ektachrome 400 for color slides.

IV.   Identification of Nuclear and Chromosomal Antigens Recognized by Autoantisera

Immunoblotting

The first step toward the identification of the antigens recognized by an autoimmune serum is immunoblotting. This is assumed by most people to be a standard procedure, but this clearly is not the case for all antigens. In the case of ACA, we found that the bulk of sera recognize three antigens, CENP-A (17 kDa), CENP-B (80 kDa), and CENP-C (140 kDa) (Earnshaw *et al.*, 1986; Earnshaw and Rothfield, 1985), with many of the sera also recognizing CENP-D (50 kDa) (Kingwell and Rattner, 1987). However, many other laboratories have detected only CENP-A, CENPs A and B, or CENPs A and C in their blots. This does not seem to be due to differences in the antibodies, since antibodies from those laboratories that were sent to us gave the normal pattern in immunoblots. Thus, immunoblotting protocols are not necessarily straightforward and may require some troubleshooting.

The procedure we routinely use for preparation of crude chromosomes for immunoblotting is given below.

Procedure for Isolation of a Crude Chromosome Suspension.   Block a 500-ml culture of HeLa cells growing in RPMI 1640 in spinner culture [2×10^5 cells/ml] in mitosis by adding Colcemid at 0.1 μg/ml late the day before. (Add 1μl/ml of a 100 μg/ml stock.)

1. *Make solutions 1–4.* (The following values are for a 500-ml culture.) See Table V for buffer formulas.
    Solution 1: 5 ml of 10× buffer A + 90 ml of H_2O. Store at room temperature.
    Solution 2: 5 ml of 10× buffer A + 45 ml of H_2O + 500 μl of 10% digitonin. Chill on ice.
    Solution 3: 20 ml 10× solution 3 + 2.0 ml of 10% AMK + 180 ml of H_2O. Chill on ice.
    Solution 4: 1 ml 10× solution 4 + 0.1 ml of 10% AMX + 8.9 ml of H_2O. Chill on ice.
    Add TRAS and PMSF to solutions 1–3 just prior to use (final dilution for both is 1 : 1000). Add TRAS only to solution 4.
2. *Harvest cells by centrifugation.* Pour culture into two 500-ml flat-bottomed polycarbonate centrifuge tubes (Beckman). Spin at 800 *g* for 5 minutes. (We use a Sorvall RC-5 preparative ultracentrifuge.) Next, aspirate any bubbles from the surface with a Pasteur pipette, then tilt the tube and completely aspirate the rest of the supernatant.
3. *Swell the cells.* Gently resuspend cell pellet in 10 ml of solution 1. When

TABLE V

REAGENTS FOR ISOLATION OF CRUDE MITOTIC CHROMOSOMES

10 × final concentration	Stock solution	
10 × Buffer A		Add/250 ml final volume
150 mM Tris-HCl, pH 7.4	1 M	37.5 ml
800 mM KCl	4 M	50 ml
20 mM K-EDTA, pH 7.4	100 mM	50 ml
7.5 mM Spermidine	30/70 mM	25 ml
3.0 mM Spermine		
10 × Buffer 3		Add/500 ml final volume
50 mM Tris-HCl, pH 7.4	1 M	25 ml
20 mM KCl	4 M	2.5 ml
20 mM K-EDTA, pH 7.4	0.1 M	100 ml
3.75 mM Spermidine	75 mM	25 ml
10 × Buffer 4		Add/50 ml final volume
50 mM Tris-HCl, pH 7.4	1 M	2.5 ml
20 mM KCl	4 M	0.25 ml
3.75 mM Spermidine	75 mM	2.5 ml

Colcemid (Sigma), 100 μg/ml. Filter sterile. Store 1-ml aliquots at $-20°$C
10% Digitonin (Sigma) in DMSO. Store at room temperature
10% AMX (Ammonyx Lo; Onyx Chemical Co., Jersey City, NJ)
RNase A, 5 mg/ml in 10 mM sodium acetate, pH 4; boil for 7 minutes; store at 4°C
0.1% SDS in H_2O
Two Wheaton dounce homogenizers: 15 ml with pestle A (the tighter one), plus 40 ml with pestle B

all tubes are resuspended, pour in solution 1 to 45 ml for each of the two 50-ml tubes. Let cells swell at room temperature for 5 minutes, then spin at 800 g for 5 minutes in a table-top centrifuge at room temperature.

4. *Lyse the cells.* Aspirate the supernatant with a Pasteur pipette as above. Add 100 μl of RNase A to each pellet. Vigorously pipette 10 ml of solution 2 onto the pellet, disrupting it. Pipette the solution up and down twice. Dounce with a total of 10 vigorous strokes (Wheaton Dounce tissue grinder with 'B' pestle). Pour into a 15-ml tube. From this point on, all operations are performed on ice.

The above steps, from aspiration of the solution 1 to the first dounce stroke, should take ≤ 30 seconds or the yield will drop significantly. This is because the detergent permeabilizes the cells and the chromosomes and cytomatrix begin to clump together.

5. *Remove nuclei and debris.* Spin 15-ml tube at 250 g for 5 minutes at 4°C. We use a Beckman TJ6 table-top centrifuge for this and all subsequent centrifugations. Carefully remove supernatant from over the pellet

of nuclei and unlysed cells. Transfer supernatant into 50-ml round-bottomed centrifuge tubes. We use Sorval flanged polycarbonate tubes (03146).

6. *Spin down chromosomes.* Add ~40 ml solution 3 to each tube. Pour into the 40-ml dounce and homogenize with four gentle strokes. Spin at 1300 g for 20 minutes at 4°C. Carefully aspirate the supernatant. The pellet should look like a ring. Let the tube sit on ice ~2–3 minutes then gently tap tube with finger to resuspend the chromosomes. Resuspend chromosomes in 0.5 ml of solution 4 by repipetting gently with a cut blue pipette tip to minimize shear.

7. *Run chromosomes on SDS gel.* Add $CaCl_2$ to 2 mM and micrococcal nuclease to 50 μg/ml. Digest on ice for 20 minutes. Add 0.5 ml of 3x Laemmli sodium dodecyl sulfate (SDS) sample buffer per 500 ml of starting culture. Boil and sonicate. Electrophorese on two single-slot 16-cm-wide SDS gels.

Experimental Procedure for Immunoblotting.   (For buffers, see Table VI.)
   1. Run an SDS-polyacrylamide gel containing the antigen of interest.

TABLE VI

Solutions and Reagents for Immunoblotting

	Stock solution	Add/final volume
Transfer buffer (to make 4 liters)		
12.1 g Sigma 7-9 (cheap Tris)		
57.7 g Aldrich 98% glycine		
800 ml methanol		
3.2 liters H_2O		
20 ml 20% SDS		
GB buffer (5 × final concentration) (to make 500 ml final)		
250 mM Triethanolamine-HCl, pH 7.4	1 M	125 ml
500 mM NaCl	5 M	50 ml
10 mM K-EDTA, pH 7.4	0.1 M	50 ml
2.5% Triton X-100	10%	125 ml
0.5% SDS	20%	12.5 ml
PTX buffer (10 × final concentration) (to make 1 liter final)		
100 mM $NaPO_4$, pH 7.5	0.5 M	200 ml
1.5 M NaCl	5 M	300 ml
10 mM Na EGTA	0.1 M	100 ml
2% Triton X-100	10%	200 ml
(1 mM NaN_3, optional)	10%	6.5 ml
PTX/BSA (to make 100 ml final)		
10 ml 10 × PTX		
13.3 ml 30% BSA (Sigma A-7284)		
H_2O to 100 ml		
[125]I-Labeled protein A		

2. Electrophoretically transfer the proteins from this gel to nitrocellulose paper. You can transfer too vigorously. Significant amounts of CENP-B begin to pass through nitrocellulose after 2–3 hours under the these blotting conditions, so the length and current of the blotting procedure may need to be optimized for particular antigens. We recommend starting with 5 hours at 23 watts in transfer Buffer. When characterizing novel antigens, it is very important to transfer both the stacking and resolving gels, since very large antigens or antigens that have not been fully solubilized in the SDS sample buffer may fail to enter the resolving gel.

3. Stain the nitrocellulose with Ponceau S (Sigma). If desired, blot it dry. (Nitrocellulose may be stored at room temperature between sheets of filter paper for several years prior to use.) If markers are present, mark their position by pricking the filter with a hypodermic needle.

4. Incubate with PTX/BSA buffer for 20 minutes at room temperature with shaking. This blocking step saturates nonspecific binding sites for proteins on the nitrocellulose. Other protocols use Tween or Blotto (powdered milk) for the blocking step. We found that use of powdered milk decreased the binding of autoantibodies to the CENP-C autoantigen, and thus have continued to use BSA despite its greater cost. We use a reciprocating shaker (New Brunswick), although orbital and tilting shakers are also often used.

5. Add antibodies to desired dilution in the PTX/BSA buffer. The dilution may range from undiluted for some affinity-purified antibodies to 1 : 20,000 or more for autoimmune sera.

6. Incubate for 3 hours to overnight. For many antibodies, the longer you incubate after 3–4 hours, the greater problems you may have with background.

7. Wash five times for 3 minutes each with GB buffer. Use enough volume so the the nitrocellulose moves back and forth in the container.

8. Rinse once briefly with 1x PTX buffer (to lower the SDS concentration).

9. Add the minimum amount of PTX/BSA buffer sufficient to cover the nitrocellulose paper completely.

10. Incubate for 5 minutes at room temperature.

11. Add ^{125}I-labeled protein A to desired amount (typically 150 cpm/μl). This is a point of some debate. One of us (WCE) prefers the radioactive label because it permits numerous exposures to be taken of the same immunoblot. Thus, one can take two different exposures in experiments where two lanes on the same gel contain greatly different levels of antigen. Others who find it unpleasant or inconvenient to use radioactivity prefer to use enzymatic markers. It is important to note that, in many cases, the chemical substrates for these reactions have their own attendant dangers. (For example, DAB is a potential carcinogen.)

12. Incubate for 30 minutes at room temperature.
13. Wash five times for 3 minutes with GB buffer.
14. Dry thoroughly, mark the positions of marker proteins and the boundary between the stacking and resolving gels with radioactive pen, mount on stiff backing covered with plastic wrap, and expose to film. We make a radioactive pen by removing the tip from a Pilot felt-tip pen with forceps or pliers and injecting some "tired" $[^{35}S]$-methionine into the ink pad using a 26-gauge needle on a 1-ml syringe. This pen can be regenerated many times (until it eventually runs out of ink).

The major difference between autoantisera and experimental sera that have been raised against purified antigens is that preimmune sera are not available for the autoantisera, since people do not typically predict when they are about to become ill. [Two exceptions to this rule have been found, for individuals who were being monitored for other reasons in a clinic before developing ACA on the one hand (Earnshaw and Rothfield, 1985) and Sm on the other (ter Borg et al., 1988).] As a result, it is often not clear which of the antigens recognized by antibody on immunoblots is responsible for the observed IIF pattern. Furthermore, it should not be assumed that even if a single antigen is recognized by blotting, this is responsible for the patterns observed in IIF. Antigens seen by immunoblotting have been previously denatured by SDS while those seen in IIF have not. The unpublished lore of monoclonal antibody production is full of examples of antibodies that work by fluorescence and not by immunoblotting, and vice versa.

The formal connection between results obtained by immunoblotting and those obtained by IIF is attained by the use of affinity-purified antibodies. Affinity-purification procedures are standard for experimental antibodies, but are somewhat more tricky for autoantibodies for two reasons. First, the autoantigens may not be previously characterized proteins and may, in fact, be extremely minor components of nuclei and chromosomes. As a result, purification of the antigen and conjugation to a chromatography resin may not be possible. Second, some autoantibodies bind extremely tightly to their antigens. Thus, for example, we found that the following buffers were unable to dissociate antibodies to CENP-B from the chromosomal protein when it had been electrophoretically transferred to nitrocellulose: glycine buffer, pH 2.8 (Olmsted, 1981), 1 or 2 M acetic acid, 1 M NH$_4$OH, 8 M urea, 4 M MgCl$_2$, or 4 M guanidine-HCl. The buffer that eventually did work (3 M ammonium thiocyanate) combines both high pH and chaotropic properties. One protocol developed for antibody affinity purification from bands on electroblots is given below.

Procedure for Affinity Purification of Autoantibodies from Antigens Immobilized on Nitrocellulose.   The following protocol was used for affinity purification of anti-centromere antibodies from chromosomal proteins following SDS-PAGE and electroblotting to nitrocellulose. The method is suitable for purification of small amounts of high-affinity antibody from trace amounts of impure antigen. The best success will be achieved with sera that have a high titer to begin with. (All stock solutions are listed in Tables IV and VII).

1. Run an SDS gel containing the antigen of interest in a single wide slot.
2. Electroblot the proteins from this gel to nitrocellulose paper.
3. Stain the nitrocellulose with Ponceau S. Blot it dry.
4. Cut a strip from one side of the blot and process by the *Immunoblotting Protocol* given above through step 8. Use PTX buffer without azide, since azide inhibits horse radish peroxidase. Then proceed with peroxidase detection as below. Prepare the DAB solution during the washes (step c), but don't add the H_2O_2 until immediately before use. *Wear gloves for the subsequent steps. DAB is a potential carcinogen and should be disposed of following your laboratory protocol for hazardous material.*
   a. Incubate in PTX/BSA blocking buffer for 5 minutes.
   b. Add peroxidase-coupled secondary antibody (TAGO; dilution: ~1 : 1000). Incubate for 30 minutes to 1 hour.
   c. Wash with PTX (five times for 2 minutes each).
   d. Wash once with 50 mM Tris-HCl, pH 7.4, for 3 minutes.
   e. Add the H_2O_2 to the DAB solution (3.33 μl of commercial 3% H_2O_2 per milliter of DAB solution).

TABLE VII

SOLUTIONS AND REAGENTS FOR AFFINITY PURIFICATION OF AUTOANTIBODIES

	Stock solution	Add/1 liter final
10 × PBS/azide buffer (10 × final concentration)		
100 mM NaPO$_4$, pH 7.5	0.5 M	200 ml
1.5 M NaCl	5 M	300 ml
10 mM Na-EGTA	0.1 M	100 ml
0.1% NaN$_3$	10%	10 ml

Elution solution (make fresh just prior to use). In a 15-ml plastic tube, carefully weigh out 0.18 g of NH$_4$OH using a Pasteur pipette. Add 2.92 g of potassium thiocyanate (Sigma) (easiest if weighed out on a waxed weighing paper). Add H_2O to 8 ml and mix to dissolve the potassium thiocyanate. Add 333 μl of 30% BSA (Sigma A-7284). Mix and add H_2O to 10 ml. Chill on ice.

DAB solution (made fresh just prior to use). *Wear Gloves. DAB is a potential carcinogen.* 50 mg of diaminobenzidine (DAB) is dissolved in 100 ml of 50 mM Tris-HCl, pH 7.4. Immediately before use; add 333 μl of commercial 3% H_2O_2. (This should be ≤2–3 months old.)

Sephadex G-25 swollen and equilibrated in 0.1 M Tris-HCl, pH 6.8, 50 mM NaCl, 1% BSA

Ponceau S, 0.2% in 3% TCA

f. Add the complete DAB solution to the blot and incubate for 15 seconds to 2 minutes (or as necessary).

g. Wash once with H_2O (quick rinse).

h. Stop the reaction by washing with 10% TCA for 2 minutes.

i. Wash twice for 2 minutes with H_2O.

j. Dry and photograph if desired.

5. Align peroxidase-stained strip with rest of blot and, using a razor blade, carefully excise the immunopositive band. (If the band is a minor component, use the general trend of the bands seen by Ponceau staining to keep your cut parallel to the band.)

6. Dice the excised strip with the razor blade into ~2-mm squares.

7. Place the squares in a 10-ml round bottomed plastic tube.

8. Add 5 ml of PTX/BSA and incubate with the diced nitrocellulose for 20 minutes to block the strips.

9. Add antibody (1:250–1:500 for anti-centromere autoantibodies) and incubate overnight with shaking at room temperature.

10. In the morning, prepare the required number of spin columns for desalting of the antibody solution (see Fig. 2).

11. Rinse the nitrocellulose with PTX (five times for 2 minutes each).

PACKING THE BLUE TIPS   **ASSEMBLING THE SPIN COLUMN**

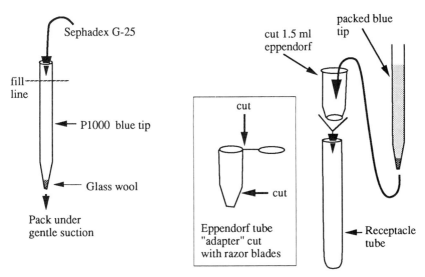

FIG. 2.   How to prepare the spin column for rapid removal of ammonium thiocyanate from affinity-purified antibodies.

(During these rinses, prepare the ammonium thiocyanate elution solution.)

All subsequent steps will be performed at 4°C (place a styrofoam container with ice on the shaker).

12. Rinse again twice for 2 minutes each with PBS/azide at 4°C.
13. Carefully aspirate away all liquid.
14. Add 1 ml of elution solution to each tube. Set shaker velocity so that the nitrocellulose fragments move briskly.
15. Place one set of the spin columns in a table-top centrifuge (such as the Beckman Accuspin) and spin for 1 minute at 1000 *g*.
16. At end of spin, remove the eluate from the bottom of the receptable tube by aspiration.
17. Draw off the ammonium thiocyanate elution solution from the nitrocellulose with a micropipette, and pipette 500 *μ*l into each of two blue tips.
18. Add another 1 ml of elution solution to the nitrocellulose and continue to shake briskly.
19. Set timer for 10 minutes.
20. Spin the blue tips with elution solution at 1100 *g* for 1 minute.
21. At the end of the spin remove the desalted antibodies from the receptable tube with a Pasteur pipette. Store on ice in a polyallomer tube.
22. Two minutes before the timer sounds, repeat steps 15 and 16.
23. After timer sounds, repeat steps 17 and 20.
24. Rinse strips several times with PTX. Store them frozen in PTX/BSA. They can be reused multiple times and, in fact, appear to work *better* upon reusing.
25. Dialyze the eluted antibodies versus PBS-azide overnight at 4°C with several changes.

These affinity-purified antibodies will work in immunoblots at dilutions of 1 : 5–1 : 20. For immunofluorescence, try a range from undiluted to 1 : 10. If the antibodies do not work, try shortening the elution times or try a less harsh elution buffer such as glycine, pH 2.8 (Olmsted, 1981).

V.   cDNA Cloning of Chromosomal Autoantigens

Autoantibodies have proved to be very useful for cloning nuclear and chromosomal autoantigens; however, in many cases, this cloning presents special problems. The problems do not arise primarily in the initial stages of obtaining immunopositive clones from cDNA libraries in *λ*gt11. This is done by standard means (Snyder *et al.,* 1988; Young and Davis, 1983a,b). Instead, the problems arise both from the nature of expression vector cloning and from the nature of the autoantibodies themselves.

It is essential to remember that expression vector cloning clones *epitopes*, not antigens. Thus, any polypeptide containing an epitope recognized by the serum used for screening will be recognized as a positive clone, regardless of whether the epitope is present in the bona fide protein, as a fortuitous component of an unrelated protein, or as a spurious expression artifact. In most cDNA libraries, only one-sixth of the clones contain cDNAs inserted in the correct orientation and reading frame. When this is coupled with the fact that many immunoreactive epitopes need be only six amino acid residues long (Harlow and Lane, 1988), clearly the potential for encoding a spurious epitope is not negligible. Thus, validation that cDNA clones encode the bona fide antigen is an essential part of any expression vector cloning study. Unambiguous validation can be extremely difficult if nothing is known about the antigen aside from its reactivity with the antiserum used for cloning.

Typically, when one prepares an antibody to a chromosomal protein, the protein has been purified to some extent during preparation of the immunogen. This implies some knowledge of the properties of the antigen. Also, if sufficient quantities of the antigen are available for standard immunization protocols, these are often sufficient to obtain some amino acid sequence information. In contrast, many autoantibodies detect otherwise unknown proteins on immunoblots and by immunofluorescence. The question therefore becomes, when a cDNA clone is isolated, how does one verify that this is the bona fide cDNA encoding the protein recognized in cells by the autoantibody?

As a start, Northern (RNA) blotting with the cloned cDNA can be used to ask if the mRNA length makes sense for a protein of the size known from immunoblots. It is important to remember, however, that many antigens migrate anomalously in SDS-PAGE. [For example, CENP-B, which runs in SDS-PAGE with a M_r of 80 kDa, is actually at 65-kDa protein (Earnshaw *et al.,* 1987b).] Second, if the autoantigen is amenable to two-dimensional gel analysis, the pI predicted from the deduced sequence should roughly match that observed on two-dimensional gels. If either of these criteria is far off, then any subsequent demonstration that the cDNA is a valid clone must be extremely convincing.

Validation of the cDNA clones

1. PROCEDURES THAT DO NOT WORK

A number of workers have claimed to validate clones based on the demonstration that antibody affinity-purified from the recombinant phage plaques rebinds to the antigens of interest in immunoblots and/or immunofluorescence. As is clear from the above discussion, this procedure only demonstrates that the correct *epitope* has been cloned.

Other workers have claimed to validate clones based on the ability of antibodies raised against the bacterially expressed cloned protein to recognize the endogenous protein. This is also circular logic, since by definition it is known that the cloned protein contains at least one correct epitope. Therefore, regardless of the nature of the cloned protein (i.e., correct or spurious), if that epitope elicits antibodies, then the resultant antibodies will recognize (among other things) the endogenous protein.

2.   PROCEDURES THAT WORK FOR ABUNDANT OR WELL-KNOWN PROTEINS

The easiest way to validate a clone is to find that its DNA sequence (or the polypeptide sequence deduced from it) is highly similar to the sequence of a homologous cDNA or protein that has already been characterized. For example, when autoantibodies were used to clone human topoisomerase I, the validity of the cDNA was readily determined when the sequence analysis revealed extensive similarity to the previously cloned and sequenced yeast enzyme (D'Arpa et al., 1988). A second way to validate cDNA clones is to obtain limited protein sequence from the purified antigen and demonstrate that this sequence is encoded by the cloned cDNA. Clearly, this only works for antigens that one can purify in sufficient amounts for microsequencing.

Alternatively, if the cDNA encodes a fairly abundant antigen, then the clone can be validated by hybrid-selected translation of mRNA (Jagus, 1987). This method was used to validate the first clone of the human La autoantigen (Chambers and Keene, 1985). Briefly, the cDNA adsorbed to a filter was used to purify the complementary mRNA by hybridization. The mRNA was then eluted and translated *in vitro*, giving a protein product of the right length. The identity of this protein product was confirmed by comparison with the endogenous protein by partial proteolytic mapping (Cleveland et al., 1977). Hybrid-selected translation only works for autoantigens expressed as fairly abundant mRNAs. This method was unsuccessful for CENP-B (W. C. Earnshaw and D. W. Cleveland, unpublished).

3.   PROCEDURES THAT WORK FOR POORLY CHARACTERIZED AUTOANTIGENS

When an otherwise unknown cloned autoantigen is present in sufficiently small amounts that the above methods do not work, an immunological approach may be used that both validates the clones and provides valuable reagents for future experiments. This approach depends on obtaining poly-clonal (and monoclonal if needed) antibodies to the fusion proteins. These antibodies are then used to demonstrate that the bacterially expressed cloned proteins share a minimum of three independent epitopes with the endogenous protein. Since an epitope is typically composed of ≥ 6 amino acid residues, the probability that two otherwise unrelated proteins would share three independent epitopes is negligible.

Procedure for Production of Antibodies to Cloned Proteins Expression. First, cDNA inserts are expressed in *Escherichia coli* using an appropriate expression vector. A variety of vectors have been used for this, including pT7-7 [which expresses the cloned species as a fusion with six residues of the T7 gene 10 product (Tabor and Richardson, 1985)], pATH [trpE fusion—fusion moiety 32 kDa (Koerner *et al.,* 1990)], pRX [trpE fusion—fusion moiety 18 amino acids (Rimm and Pollard, 1989)], pUCX [β-galactosidase—fusion moiety 14 amino acids (D. W. Cleveland and W. C. Earnshaw, un-published)]; and λgt11 lysogens [β-galactosidase—fusion moiety 113 kDa (Young and Davis, 1983b)]. Trial and error will indicate which of these vectors will give the highest level of expression of a stable fusion protein.

Fusion proteins are readily obtained by subcellular fractionation if they form inclusion bodies (Rothfield *et al.,* 1987), or as bands excised from SDS gels. We typically attempt to fractionate the bacteria prior to electrophoresis, since if the fusion protein is insoluble, simple low-speed centrifugation can give a significant enrichment. If the expressed protein is soluble, we typically use a gel band from a whole-cell lysate. The latter gives an antigen preparation that is potentially contaminated by bacterial proteins of similar size, and therefore must be depleted of antibodies to bacterial antigens by extensive absorption prior to use. Antigens are typically injected into rabbits for production of polyclonal antibodies. Note that it is *extremely important* that you obtain preimmune serum from the rabbits prior to injection, and that you make sure that the rabbits are not expressing antibodies against human proteins. We find that $< 50\%$ of the rabbits that we obtain are suitable for injection!

Polyclonal antibodies raised by the following procedure against CENP-B fusion proteins target numerous epitopes along the polypeptide.

Preparation of SDS Gel Bands for Injection
1. Run your antigen on an SDS-polyacrylamide gel.
2. Rinse gel three times for 10 minutes each with ddH$_2$O to remove SDS.
3. Stain gel with aqueous Coomassie Blue (Table VIII) for 1 hour.
4. Destain gel in 1x Laemmli running buffer minus SDS (Table VIII) until you can clearly see your band.
5. Cut the band out of gel using a fresh razor blade. (Note: pressing straight down works much better than drawing the blade along the gel as polyacrylamide rips easily.)

TABLE VIII

STOCK SOLUTIONS FOR IMMUNIZATION WITH SDS GEL BANDS

	Stock solution	Add/final volume
PBS for tissue culture (10 × final concentration)		1 liter final
64.6 mM Na$_2$HPO$_4$ (dibasic)	—	11.5 g
14.7 mM KH$_2$PO$_4$ (monobasic)	—	2 g
1.37 M NaCl	5 M	274 ml
26.8 mM KCl	4 M	6.7 ml
Laemmli gel running buffer without SDS		
(10 × final concentration)		4 liters final
0.25 M Trizma base	121.2 g	
1.92 M glycine (Aldrich)	576.8 g	

Aqueous Coomassie Blue. Add 0.1% Coomassie Brilliant Blue R to 1 × Laemmli running buffer (without SDS)

6. Place the band in a centrifuge tube and store at −80°C *or* proceed to next step.
7. Grind band to fine powder that will easily pass through a needle for injection. Place ceramic mortar and pestle in a foam ice bucket and pour liquid N$_2$ on it until it cools to −190°C (liquid N$_2$ stops boiling). Add a few milliliters of liquid N$_2$ to the bottom of the mortar, then add your gel band. Grind until it is a fine powder. (You cannot grind too much, but you can grind too little.)
8. Prepare your samples for injection. Using a small piece of cold weighing paper, scrape into a centrifuge tube. (Note: gel will become impossible to handle when it thaws!) You can rinse the last little bit from the sides of the mortar with a little liquid N$_2$. It will settle into a cake on the bottom. When all antigen is in the tube, add an equal volume of 1 × PBS (Table VIII). Do not use PBS with azide or EDTA! Take half of the sample and put it into a tube labeled "initial injection." Divide the remainder in half, putting each in tubes labeled "boost."
9. Injection. We inject at multiple sites (subscapular, subcutaneous, and intramuscular) using antigen mixed 50:50 with Freund's complete adjuvant (first injection) or incomplete adjuvant (subsequent injections). Injections are made on day 0, 30, 60, etc. Bleeds (from ear veins) are performed on days 40, 70, etc.

4. DEMONSTRATION OF MULTIPLE EPITOPES

Demonstration that cloned and endogenous antigens share multiple epitopes is accomplished by immunoblotting with bacterially expressed proteins encoded by subcloned regions of the cDNA. Several nonoverlapping regions of the polypeptide are expressed in E. *coli* and shown to bind auto-antibody or experimentally raised antibody by immunoblotting. Antibodies affinity-purified from these subcloned proteins are then tested for binding to the cellular protein. A clone is considered validated if ≥ 3 shared epitopes can be demonstrated.

As controls, antibodies affinity-purified from one subcloned fragment are tested for their ability to recognize proteins expression by the other non-overlapping subcloned fragments. This is necessary to demonstrate that the epitopes on different fragments are distinct.

If the polyclonal antibodies do not recognize multiple regions along the cloned protein, an alternative approach is to generate a library of monoclonal antibodies to the cloned protein. The monoclonals are then tested for binding to independent epitopes in competition binding experiments (Kiehart *et al.*, 1984).

5. FURTHER CRITERIA FOR CLONE VALIDATION

One straightforward method for confirmation of the above results is to perform partial proteolysis on bacterially expressed proteins and compare this to the map from the cellular protein. This is done by immunoblotting, and has the advantage that this is not necessary to obtain pure protein prior to the proteolysis. Only the subset of the proteolytic fragments bearing epitopes is detected by the antibody.

The bacterially expressed protein is excised from an SDS gel, as is the appropriate region of a gel of chromosomes. The proteins are then cleaved with either α-chymotrypsin (Worthington) or V8 protease (Miles Labora-tories) (Cleveland *et al.*, 1977), or BNPS-skatole (Sigma) (Detke and Keller, 1982) (cleaves at Trp) and subjected to a second round of SDS-PAGE. Anti-genic fragments are detected by immunoblotting. This has been found to work well for both topoisomerase I and the centromeric autoantigens (Shero *et al.*, 1986).

VI. Conclusions

A major goal of modern cell biology has been to understand the functions of the nucleus and chromosomes through the identification and characterization of their component proteins. The identification of autoantibodies to these components has greatly facilitated this process. Perhaps one of the best ex-amples of the use of these antibodies has been their application to the study of the mammalian centromere. Autoantibody probes have allowed the identification of a family of centromere-specific proteins, many of which have now been localized to specific domains within the centromere by light and electron microscopy. In addition, some of these proteins have been sequenced and their interaction with specific DNA sequences determined (Pluta et al., 1990; Rattner, 1991). Studies like these have not only provided the first clues into functional organization of the centromere, but have established basic techniques and approaches applicable to other systems. These have been summarized in this chapter. As antibody probes continue to facilitate the identification and characterization of additional components, we can expect to begin to see the emergence of a more complete picture of nuclear and chromosomal organization and its relationship to specific functions.

ACKNOWLEDGMENTS

The methods described in this chapter come from our laboratory protocols. These have been developed over the years as a result of the advice of many friends. A partial list of those who have contributed either buffer recipes or methods includes B. Burke, C. Cooke, J. De Mey, W. Dunn, M. Fritzler, L. Gerace, B. Halligan, M. Heck, A. Hubbard, D. Kiehart, J. Kilmartin, B. Kingwell, U. Laemmli, R. Laskey, L. Liu, D. Murphy, and J. Sedat. Experiments from the authors' laboratories were supported by NIH grants GM 30985 and 35212 to WCE and grants from NSERC and the Canadian Cancer Society to JBR.

REFERENCES

Almendral, J. M., Huebsch, D., Blundell, P. A., Macdonald-Bravo, H., and Bravo, R. (1987). Cloning and sequence of the human nuclear protein cyclin: Homology with DNA-binding proteins. Proc. Natl. Acad. Sci. U.S.A. **84**, 1575–1579.

Alspaugh, M. A., and Maddison, P. (1979). Resolution of the identify of certain antigen-antibody systems in systemic lupus erythematosus and Sjögrens syndrome: An interlaboratory collaboration. Arthritis Rheum. **22**, 796–798.

Arana, R., and Seligmann, M. (1967). Antibodies to native and denatured deoxyribonucleic acid in systemic lupus erythematosus. J. Clin. Invest. **46**, 1867–1882.

Billings, P. B., Hoch, S. O., White, P. J., Carson, D. A., and Vaughan, J. H. (1983). Antibodies to the Epstein-Barr virus nuclear antigen and to rheumatoid arthritis nuclear antigen identify the same polypeptide. Proc. Natl. Acad. Sci. U.S.A. **80**, 7104–7108.

Bischoff, F. R., Maier, G., Tilz, G., and Ponstingl, H. (1990). A 47-kDa human nuclear protein recognized by antikinetochore autoimmune sera is homologous with the protein encoded by RCC1, a gene implicated in onset of chromosome condensation. Proc. Natl. Acad. Sci. U.S.A. **87**, 8617–8621.

Bustin, M., Reisch, J., Einck, L., and Klippel, J. H. (1982). Autoantibodies to nucleosomal proteins: Antibodies to HMG-17 in autoimmune diseases. Science **215**, 1245–1247.

Ceppelini, R., Polli, E., and Celada, F. (1957). A DNA reacting factor in serum of a patient with lupus erythematosus. Proc. Soc. Exp. Biol. Med. **96**, 572–574.

Chambers, J. C., and Keene, J. D. (1985). Isolation and analysis of cDNA clones expressing human lupus La antigen. Proc. Natl. Acad. Sci. U.S.A. **82**, 2115–2119.

Cherney, B. W., McBride, O. W., Chen, D., Alkhateb, H., Bhatia, K., Hensley, P., and Smulson, M. E. (1987). cDNA sequence, protein structure and chromosomal location of the human gene for poly(ADP-ribose) polymerase. *Proc. Natl. Acad. Sci. U.S.A.* **84**, 8370–8374.

Cleveland, D. W., Fisher, S. G., Kirschner, M. W., and Laemmli, U. K. (1977). Peptide mapping by limited proteolysis in sodium dodecyl sulfate and analysis by gel electrophrosis. *J. Biol. Chem.* **252**, 1102–1106.

Compton, D. A., Yen, T. J., and Cleveland, D. W. (1991). Identification of novel centromere/kinetochore associated proteins using monoclonal antibodies generated against human mitotic chromosome scaffolds. *J. Cell Biol.* (in press).

Cooke, C. A., Bernat, R. L., and Earnshaw, W. C. (1990). CENP-B: A major human centromere protein located beneath the kinetochore. *J. Cell. Biol.* **110**, 1475–1488.

Costa, O., Tchouatcha-Tchouassom, J. C., Roux, B., and Monier, J. C. (1986). Anti-H1 histone antibodies in systemic lupus erythematosus: Epitope localization after immunoblotting of chymotrypsin-digested H1. *Clin. Exp. Immunol.* **63**, 608–613.

D'Arpa, P., Machlin, P. S., Ratrie, H., Rothfield, N. F., Cleveland, D. W., and Earnshaw, W. C. (1988). cDNA cloning of human topoisomerase I: Catalytic activity of a 67.7 kDa carboxyl-terminal fragment. *Proc. Natl. Acad. Sci. U.S.A.* **85**, 2543–2547.

Davis, L. G., Dibner, M. D. and Battey, J. F. (1986). "Basic Methods in Molecular Biology," p. 48. Elsevier, New York.

Dedman, J. R., Lin, T., Marcum, J. M., Brinkley, B. R., and Means, A. R. (1980). Calmodulin: Its role in the mitotic apparatus. *In* "Calcium-Binding Proteins: Structure and Function" (F. L. Siegel, E. Carafoli, R. H. Kretsinger, D. H. MacLennan, and R. H. Wasserman, eds.), pp. 181–188. Elsevier/North-Holland, New York.

Detke, S., and Keller, J. M. (1982). Comparison of the proteins present in HeLa cell interphase nucleoskeletons and metaphase chromosome scaffolds. *J. Biol. Chem.* **257**, 3905–3911.

Deutscher, S. L., Harley, J. B., and Keene, J. D. (1988). Molecular analysis of the 60-kDa human Ro ribonucleoprotein. *Proc. Natl. Acad. Sci. U.S.A.* **85**, 9479–9483.

Douvas, A. S., Achten, M., and Tan, E. M. (1979). Identification of a nuclear protein (Scl-70) as a unique target of human antinuclear antibodies in scleroderma. *J. Biol. Chem.* **254**, 10514–10522.

Earnshaw, W. C., and Rattner, J. B. (1989). A map of the centromere (primary constriction) in vertebrate chromosomes at metaphase. *In* "*Aneuploidy: Mechanisms of Origin*" (B. Vig and M. Resnick, eds.), pp. 33–42. Alan R. Liss, New York.

Earnshaw, W. C., and Rothfield, N. (1985). Identification of a family of human centromere proteins using autoimmune sera from patients with scleroderma. *Chromosoma* **91**, 313–321.

Earnshaw, W. C., Bordwell, B., Marino, C., and Rothfield, N. (1986). Three human chromosomal autoantigens are recognized by sera from patients with anti-centromere antibodies. *J. Clin. Invest.* **77**, 426–430.

Earnshaw, W. C., Machlin, P. S., Bordwell, B., Rothfield, N. F., and Cleveland, D. W. (1987a). Analysis of anti-centromere autoantibodies using cloned autoantigen CENP-B *Proc. Natl. Acad. Sci. U.S.A.* **84**, 4979–4983.

Earnshaw, W. C., Sullivan, K. F., Machlin, P. S., Cooke, C. A., Kaiser, D. A., Pollard, T. D., Rothfield, N. F., and Cleveland, D. W. (1987b). Molecular cloning of cDNA for CENP-B, the major human centromere autoantigen. *J. Cell Biol.* **104**, 817–829.

Eilat, D., Steinberg, A. D., and Schechter, A. N. (1978). The reaction of SLE antibodies with native, single stranded RNA: Radioassay and binding specificities. *J. Immunol.* **120**, 550–557.

Fritzler, M. J. (1986). Autoantibody testing: Procedures and significance in systemic rheumatic diseases. *Methods Achiev. Exp. Pathol.* **12**, 224–260.

Fritzler, M. J., Ali, R., and Tan, E. M. (1984a). Antibodies from patients with mixed connective tissue disease react with heterogeneous nuclear ribonucleoprotein or ribonucleic acid [hnRNP/RNA]. *J. Immunol.* **132**, 1216–1222.

Fritzler, M. J., Valencia, D. W., and McCarty, G. A. (1984b). Speckled pattern antinuclear antibodies resembling anti-centromere antibodies. *Arthritis Rheum.* **27**, 92–96.

Gorbsky, G. J., Sammak, P. J., and Borisy, G. G. (1987). Chromosomes move poleward in anaphase along stationary microtubules that coordinately disassemble from their kinetochore ends. *J. Cell Biol.* **104**, 9–18.

Guldner, H. H., Lakomek, H.-J., and Bautz, F. A. (1984). Human anti-centromere sera recognise a 19.5 kD non-histone chromosomal protein from HeLa cells. *Clin. Exp. Immunol.* **58**, 13–20.

Guldner, H.-H., Szostecki, C., Vosberg, H.-P., Lakomek, H.-J., Penner, E., and Bautz, F. A. (1986). Scl 70 autoantibodies from scleroderma patients recognize a 95 kDa protein identified as DNA topoisomerase I. *Chromosoma* **94**, 132–138.

Guldner, H.-H., Netter, H. J., Szostecki, C., Lakomek, H.-J., and Will, H. (1988). Epitope mapping with a recombinant human 68-kD (U1) ribonucleoprotein antigen reveals heterogeneous autoantibody profiles in human autoimmune sera. *J. Immunol.* **141**, 469–475.

Hannestad, K., and Stollar, B. D. (1978). Certain rheumatoid factors react with nucleosomes. *Nature (London)* **275**, 671–673.

Hardin, J., and Thomas, J. O. (1983). Antibodies to histones in systemic lupus erythematosus: Localization of prominent autoantigens on histones H1 and H2B. *Proc. Natl. Acad. Sci. U.S.A.* **80**, 7410–7414.

Hargraves, M. M., Richmond, H., and Morton, R (1948). Presentation of two bone marrow elements: The "tart" cell and the "LE" cell. *Mayo Clin. Proc.* **23**, 25–28.

Harlow, E., and Lane, D. (1988). "Antibodies. A Laboratory Manual." Cold Spring Harbor Laboratory, Cold Spring Harbor, New York.

Heck, M. M. S., and Earnshaw, W. C. (1986). Topoisomerase II: A specific marker for proliferating cells. *J. Cell Biol.* **103**, 2569–2581.

Hildebrandt, S., Weiner, E., Senécal, J.-L., Noell, S., Daniels, L., Earnshaw, W. C., and Rothfield, N. F. (1990). The IgG, IgM, and IgA isotypes of anti-topoisomerasse I and anticentromere autoantibodies. *Arth. Rheum.* **33**, 724–727.

Hoffmann, A., Heck, M. M. S., Bordwell, B. J., Rothfield, N. F., and Earnshaw, W. C. (1989). Human autoantibody to topoisomerase II. *Exp. Cell. Res.* **180**, 409–418.

Holman, H. R., and Kunkel, H. G. (1957). Affinity between the lupus erythematosus serum factor and cell nuclei and nucleoprotein. *Science* **126**, 162–165.

Jagus, R. (1987). Hybrid selection of mRNA and hybrid arrest of translation. In "Guide to Molecular Cloning Techniques" (S. L. Berger and A. R. Kimmel, eds.), pp. 567–572. Academic Press, Orlando, Florida.

Kiehart, D. P., Kaiser, D. A., and Pollard, T. D. (1984). Monoclonal antibodies demonstrate limited structural homology between myosin isozymes from Acanthamoeba. *J. Cell Biol.* **99**, 1002–1014.

Kingwell, B., and Rattner, J. B. (1987). Mammalian kinetochore/centromere composition: A 50 kDa antigen is present in the mammalian kinetochore/centromere. *Chromosoma* **95**, 403–407.

Koerner, T. J., Hill, J. E., Myers, A. M., and Tzagoloff, A. (1990). High-expression vectors with multiple cloning sites for construction of trpE-fusion genes: pATH vectors. In "Methods in Enzymology" (submitted).

Li, X., McNeilage, L. J., and Whittingham, S. (1989). Autoantibodies to the major nucleolar phosphoprotein B23 define a novel subset of patients with anticardiolipin antibodies. *Arthritis Rheum.* **32**, 1165–1169.

Lischwe, M. A., Ochs, R. L., Reddy, R., Cook, R. G., Yeoman, L. C., Tan, E. M., Reichlin, M., and Busch, H. (1985). Purification and partial characterization of a nucleolar scleroderma antigen (Mr = 34,0000, pI 8.5) rich in Ng, Ng-dimethylarginine. *J. Biol. Chem.* **260**, 14304–14310.

McKeon, F. D., Tuffanelli, D. L., Fukiyama, K., and Kirschner, M. W. (1983). Autoimmune response directed against conserved determinants of nuclear envelope proteins in a patient with linear scleroderma. *Proc. Natl. Acad. Sci. U.S.A.* **80,** 4374–4378.

McKeon, F. D., Tuffanelli, D. L., Kobayashi, S., and Kirschner, M. W. (1984). The redistribution of a conserved nuclear envelope protein during the cell cycle suggests a pathway for chromosome condensation. *Cell (Cambridge, Mass.)* **36,** 83–92.

McLean, I. W., and Nakane, P. K. (1974). Periodate-lysine-paraformaldehyde fixative: A new fixative for immunoelectron microscopy. *Histochem. Cytochem.* **22,** 1077–1083.

Martin, L., Cusano, R., Kingwell, B., Fritzler, M. J., and Rattner, J. B. (1991). Autoantibodies to chromosomal proteins in rheumatic disease. *J. Clin. Lab. Immunol.* (in press).

Masumoto, H., Masukata, H., Muro, Y., Nozaki, N., and Okazaki, T. (1989). A human centromere antigen (CENP-B) interacts with a short specific sequence in alphoid DNA, a human centromeric satellite. *J. Cell Biol.* **109,** 1963–1973.

Matsumoto, K., Moriuchi, T., Koji, T., and Nakane, P. K. (1987). Molecular cloning of cDNA coding for rat proliferating cell nuclear antigen (PCNA)/cyclin. *EMBO J.* **6,** 637–642.

Maul, G., French, B. T., van Venrooij, W. J., and Jimenez, S. A. (1986). Topoisomerase I identified by scleroderma 70 antisera: Enrichment of topoisomerase I at the centromere in mouse mitotic cells before anaphase. *Proc. Natl. Acad. Sci. U.S.A.* **83,** 5145–5149.

Meliconi, R., Bestagno, M., Sturani, C., Negri, C., Galavotti, V., Sala, C., Facchini, A., Ciarrocchi, G., Gasbarrini, G., and Astaldi-Rocotti, G. C. B. (1989). Autoantibodies to DNA topoisomerase II in cryptogenic fibrosing alveolitis and connective tissue disease. *Clin. Exp. Immunol.* **76,** 184–189.

Mitchison, T. J., and Kirschner, M. W. (1985). Properties of the kinetochore *in vitro.* I. Microtubule nucleation and tubulin binding. *J. Cell Biol.* **101,** 755–765.

Moroi, Y., Peebles, C., Fritzler, M. J., Steigerwald, J., and Tan, E. M. (1980). Autoantibody to centromere (kinetochore) in scleroderma sera. *Proc. Natl. Acad. Sci. U.S.A.* 77, 1627–1631.

Muller, S., Briand, J.-P., and Van Regenmortel, M. H. V. (1988). Presence of antibodies to ubiquitin during the autoimmune response associated with systemic lupus erythematosus. *Proc. Natl. Acad. Sci. U.S.A.* **85,** 8176–8180.

Olmsted, J. B. (1981). Affinity purification of antibodies from diazotized paper blots of heterogeneous protein samples. *J. Biol. Chem.* **256,** 11955–11957.

Palmer, D. K., and Margolis, R. L. (1987). A 17-kD centromere protein (CENP-A) copurifies with nucleosome core particles and with histones. *J. Cell Biol.* **104,** 805–815.

Palmer, D. K., O'Day, K., Le Trong, H., Charbonneau, H., and Margolis, R. L. (1991). Purification of the centromeric protein CENP-A and demonstration that it is a centromere specific histone. *Proc. Natl. Acad. Sci. U.S.A.* **88,** 3734–3738.

Pettersson, I., Hinterberger, M., Mimori, T., Gottlieb, E. and Steitz, J. A. (1984). The structure of small nuclear ribonucleoproteins. Identification of multiple protein components reactive with anti-(U1)ribonucleoprotein and anti-Sm autoantibodies. *J. Biol. Chem.* **259,** 5907–5914.

Pluta, A. F., Cooke, C. A., and Earnshaw, W. C. (1990). Structure of the human centromere at metaphase. *Trends Biol. Sci.* **15,** 181–185.

Price, C. M., McCarty, G. A., and Pettijohn, D. E. (1984). NuMa protein is a human autoantigen. *Arthritis Rheum.* **27,** 774–779.

Query, C. C., and Keene, J. D. (1988). A human autoimmune protein associated with U1 RNA contains a region of homology that is crossreactive with retroviral p30gag antigen. *Cell (Cambridge, Mass.)* **51,** 211–220.

Rattner, J. B. (1991). The structure of the mammalian centromere. *BioEssays* **13,** 51–56.

Rattner, J. B., Kingwell, B. G., and Fritzler, M. J. (1988). Detection of distinct structural domains within the primary constriction using autoantibodies. *Chromosoma* **96,** 360–367.

Rattner, J. B., Lew, J., and Wang, J. H. (1990). p34cdc2 kinase is localized to distinct domains within the mitotic apparatus. *Cell Motil. Cytoskeleton* **17,** 227–236.

Rekvig, O. P., and Hannestad, K. (1980). Human antibodies that react with both cell nuclei and plasma membranes display specificity for the octamer of histones H2A, H2B, H3 and H4 in high salt. *J. Exp. Med.* **152,** 1720–1733.

Rieder, C. L. (1982). The formation, structure and composition of the mammalian kinetochore and kinetochore fiber. *Int. Rev. Cytol.* **79,** 1–58.

Rimm, D. L., and Pollard, T. D. (1989). New plasmid vectors for high level synthesis of eukaryotic fusion proteins in *Escherichia coli. Gene* **75,** 323–327.

Robbins, W. C., Holman, H. R., Deicher, H. R., and Kunkel, H. G. (1957). Complement fixation with cell nuclei and DNA in lupus erythematosus. *Proc. Soc. Exp. Biol. Med.* **96,** 575–579.

Rodriguez-Sanchez, J. L., Gelip, C., Juarez, C., and Hardin, J. A. (1987). Anti-NOR 90: A new autoantibody in scleroderma that recognized a 90-kDa component of the nucleolus-organizing region of chromatin. *J. Immunol.* **139,** 2579–2584.

Rothfield, N., Whitaker, D., Bordwell, B., Weiner, E., Senecal, J.-L., and Earnshaw, W. C. (1987). Detection of anticentromere antibodies using cloned autoantigen CENP-B. *Arthritis Rheum.* **30,** 1416–1419.

Rothfield, N. F., and Stollar, B. D. (1967). The relation of immunoglobulin class, pattern of anti-nuclear antibody and complement-fixing antibodies to DNA in sera from patients with systemic lupus erythematosus. *J. Clin. Invest.* **46,** 1785–1794.

Shero, J. H., Bordwell, B., Rothfield, N. F., and Earnshaw, W. C. (1986). High titers of auto-antibodies to topoisomerase I (Scl-70) in sera from scleroderma patients. *Science* **231,** 737–740.

Snyder, M., Elledge, S., Sweetser, D., Young, R. A., and Davis, R. W. (1988). 1gt11: Gene isolation with antibody probes and other applications. *In* "Methods in Enzymology" (R. Wu and L. Grossman, eds.), Vol. 154, pp. 107–128. Academic Press, Orlando, Florida.

Solden, M. H. L., Van Eekelen, C. A. G., Habets, W. J. A., Vierwonden, G., van de Putte, L. B. A., and van Venrooij, W. J. (1982). Anti-nuclear matix antibodies in mixed connective tissue disease. *Eur. J. Immunol.* **12,** 783–786.

Stanford, D. R., Rohleder, A., Neiswanger, K., and Wieben, E. D. (1987). DNA sequence of a human Sm autoimmune antigen. The multigene family contains a processed pseudogene. *J. Biol. Chem.* **262,** 9931–9934.

Stetler, D. A., Rose, K. M., Wenger, M. E., Berlin, C. M., and Jacob, S. T. (1982). Antibodies to distinct polypeptides of RNA polymerase I in sera from patients with rheumatic autoimmune disease. *Proc. Natl. Acad. Sci. U.S.A.* **79,** 7499–7503.

Stollar, B. D., Levine, L., and Marmur, J. (1962). Antibodies to denatured deoxyribonucleic acid in lupus erythematosus serum. II. Characterization of antibodies in several sera. *Biochim. Biophys. Acta* **61,** 7–18.

Tabor, S., and Richardson, C. C. (1985). A bacteriophage T7 RNA polymerase/promoter system for controlled exclusive expression of specific genes. *Proc. Natl. Acad. Sci. U.S.A.* **82,** 1974–1078.

Takasaki, Y., Deng, J.-S., and Tan, E. M. (1981). A nuclear antigen associated with cell pro-liferation and blast transformation. Its distribution in synchronized cells. *J. Exp. Med.* **154,** 1899–1909.

Tan, E. M. (1982). Autoantibodies to nuclear antigens (ANA): Their immunobiology and medicine. *Adv. Immunol.* **33,** 167–240.

Tan, E. M. (1989). Antinuclear antibodies: Diagnostic markers for autoimmune diseases and probes for cell biology. *In* "Advances in Immunology" (F. J. Dixon, ed.), pp. 93–151. Academic Press, San Diego, California.

Tan, E. M., Robinson, J., and Robitaille, P. (1976). Studies of antibodies to histones by im-munofluorescence. *Scand. J. Immunol.* **5,** 811–818.

ter Borg, E. J., Horst, G., Hummel, E., Jaarsma, D., Limburg, P. C., and Kallenberg, C. G. M. (1988). Sequential development of antibodies to specific Sm polypeptides in a patient with systemic lupus erythematosus: Evidence for independent regulation of anti-double-stranded DNA and anti-Sm antibody production. *Arthritis Rheum.* **31,** 563–1566.

Thomas, J. O., Wilson, C. M., and Hardin, J. A. (1984). The major core histone antigenic determinants in systemic lupus erythematosus are in the trypsin-sensitive regions. *FEBS Lett.* **169,** 90–96.

White, J. G., Amos, W. B., and Fordham, M. (1987). An evaluation of confocal versus conventional imaging of biological structures by fluorescence light microscopy. *J. Cell Biol.* **105,** 41–48.

Wieben, E. D., Rohleder, A. M., Nenninger, J. M., and Pederson, T. (1985). cDNA cloning of human autoimmune nuclear ribonucleoprotein antigen. *Proc. Natl. Acad. Sci. U.S.A.* **82,** 7914–7918.

Wong, A. K. C., and Rattner, J. B. (1988). Sequence organization and cytological localization of the minor satellite of mouse. *Nucleic Acids Res.* **16,** 11645–11661.

Yamanaka, H., Willis, E. H., Penning, C. A., Peebles, C. L., Tan, E. M., and Carson, D. A. (1987). Human autoantibodies to poly(adenosine diphosphate-ribose) polymerase. *J. Clin. Invest.* **80,** 900–904.

Young, R. A., and Davis, R. B. (1983a). Efficient isolation of genes by using antibody probes. *Proc. Natl. Acad. Sci. U.S.A.* **80,** 1194–1198.

Young, R. A., and Davis, R. B. (1983b). Yeast polymerase II genes: Isolation with antibody probes. *Science* **222,** 778–782.

Chapter 7

Meiotic Chromosome Preparation and Protein Labeling

C. HEYTING

Department of Genetics
Agricultural University
NL-6703 HA Wageningen, The Netherlands

A. J. J. DIETRICH

Institute of Human Genetics
University of Amsterdam
NL-1105 AZ Amsterdam, The Netherlands

I. Introduction

The prophase of the first meiotic division appears an ideal experimental system for the analysis of the functional organization of the nucleus. The

177

chromatin as well as the nuclear matrix and lamina undergo a series of precisely timed and well-defined structural changes to fulfil the apparently equally well-defined functions of genetic exchange and chromosome segregation. The synaptonemal complexes (SCs) look particularly inviting: These structures, which are formed between homologous chromosomes during meiotic prophase, undergo a series of morphological alterations which closely correlate with the successive rearrangements of chromatin.

The assembly of SCs involves the following steps (von Wettstein *et al.,* 1984):

1. After premeiotic S phase a thin, proteinaceous axis is found along each chromosome (leptotene).
2. The axes of the homologous chromosomes are aligned and transversal filaments are formed between them; the axes become the lateral elements (LEs) of the SCs (zygotene).
3. A central element (CE) is formed between the LEs on the transversal filaments (zygotene and pachytene).

Recombination is probably mediated by small, electron-dense structures, the recombination nodules (RNs) which appear between the lateral elements during zygotene and pachytene. At the beginning of diplotene the SCs disassemble: the CE, the transversal filaments, and the LEs disappear, and the two recombinant chromosomes of each bivalent segregate (diakinesis, metaphase I, and anaphase I). Although the correlation between the morphological alterations of SCs and the successive chromatin rearrangements of meiotic prophase is obvious, it is not clear which functions SCs fulfil precisely and which factors influence the timing of their assembly and disassembly. For the elucidation of these questions, it is important to analyze the composition of SCs in successive stages of meiotic prophase and to study the function of the separate SC components. Several investigators have noted the need for a good SC-isolation procedure, and at least three groups have published about their efforts to isolate SCs (Walmsley and Moses, 1981; Gorach *et al.,* 1985; Heyting *et al.,* 1985). In the next section, we describe in detail the SC-isolation procedure we use.

II.   Isolation of Synaptonemal Complexes

The outline of the isolation protocol is shown in Fig. 1. There are no suitable stops in the procedure, and the isolation should be completed within one day. This can be accomplished if sufficient preparations are made in advance.

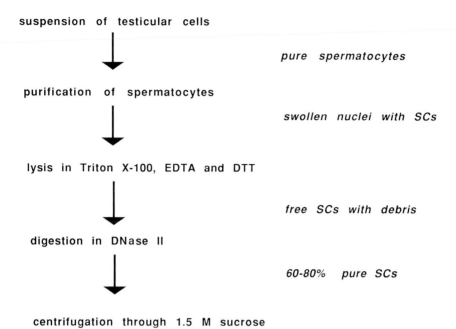

suspension of testicular cells

pure spermatocytes

purification of spermatocytes

swollen nuclei with SCs

lysis in Triton X-100, EDTA and DTT

free SCs with debris

digestion in DNase II

60-80% pure SCs

centrifugation through 1.5 M sucrose

FIG. 1. Outline of the SC isolation procedure.

A. Protocol for the Isolation of Synaptonemal Complexes

Solutions

1. Spermatocyte isolation medium (SIM), 2 liters [104 mM NaCl, 45 mM KCl, 1.2 mM MgSO$_4$, 0.6 mM KH$_2$PO$_4$, 0.1% (w/v) glucose, 6 mM sodium lactate, 1 mM sodium pyruvate, 60 mg/liter penicillin G (Serva, cat. no. 31749), 100 mg/liter streptomycin sulfate (Serva, cat. no. 35500)]. Adjust the pH to 7.3 with HCl and filter sterilize. Adjust the pH to 7.3 with HCl and filter sterilize.

2. SIM with 0.1% bovine serum albumin (BSA), 2 liters. Dissolve 2 g of BSA fraction V (Boehringer cat. no. 735108) in 2 liters of SIM, adjust the pH to 7.3 with NaOH, and filter sterilize.

3. SIM with 1% BSA, 100 ml. Adjust the pH to 7.3 with NaOH and filter sterilize.

4. SIM with 400 μg/ml DNase I (Boehringer, cat. no. 104159). Aliquot the solution in 1.5-ml portions. Freeze in liquid nitrogen and store at $-20°$C.

5. 2 × SIM without CaCl$_2$ and MgSO$_4$, 40 ml [208 mM NaCl, 90 mM KCl, 12 mM Na$_2$HPO$_4$, 1.2 mM KH$_2$PO$_4$, 0.2% (w/v) glucose, 12 mM sodium lactate, 2 mM sodium pyruvate, pH 7.3 with HCl].

6. 27% Percoll in SIM, 60 ml. Mix 16.2 ml of Percoll (Pharmacia), 13.8 ml of demineralized water, and 30 ml of 2 × SIM (solution 5). Dialyze overnight against 1 liter of SIM, adjust the pH to 7.3 with HCl, and the refractive index (N_{D20}) to 1.3400 with SIM or Percoll. Filter through 0.45-μm filter.

7. 8 mM Tris-HCl, 1 mM phenylmethylsulfonyl fluoride (PMSF), pH 7.0, 100 ml. The PMSF is added a few hours before use from a 1 M stock solution in dimethyl sulfoxide (DMSO). Filter the solution through a 0.2-μm filter as soon as the PMSF has dissolved.

8. Lysis mix, 50 ml. [8 mM Tris-HCl, 5 mM ethylenediaminotetraacetic acid (EDTA), 0.5 mM dithiothreitol (DTT), 1% (v/v) Triton X-100, 1 mM PMSF, pH 7.0]. The PMSF is added a few hours before use (see above), and the solution is then filtered through a 0.2-μm filter. The DTT is added immediately before use from a 0.5 M stock solution.

9. 1.5 M sucrose in solution 7, 10 ml. Filter through a 0.2-μm filter.

10. 2.4 M sucrose in solution 7, 10 ml. Filter through a 0.2-μm filter.

11. DNase II solution. Dissolve 12,000 units DNase II (Sigma, cat. no. D5275) in 1 ml of solution 7. Aliquot in portions of 60 μl, freeze in liquid N_2, and store at $-20°C$.

12. Aprotinine solution. Dissolve 2 mg of aprotinine (Boehringer, cat. no. 9815) in 1 ml of solution 7. Aliquot in portions of 100 μl, freeze in liquid N_2, and store at $-20°C$.

13. Trypsin inhibitor solution. Dissolve 10 mg of trypsin inhibitor (Sigma, cat. no. T9003) in 1 ml of solution 7, freeze in liquid N_2, and store at $-20°C$.

14. 4% (w/v) paraformaldehyde (PFA). Add 2 g of PFA (Merck, no. 4005) to 40 ml of demineralized water, heat to 70°C while stirring, and add dropwise 1 N NaOH until the solution becomes clear. Add 5 ml of 0.1 M sodium phosphate buffer, pH 7.0, cool to 20°C, adjust the pH to 7.3 with NaOH and the volume to 50 ml. Filter through a 0.2-μm filter, and cool the solution to 4°C. This solution has to be made freshly.

15. 1 M PMSF in DMSO; store at $-20°C$.

We prepare preweighed aliquots of collagenase (20 mg; Sigma, cat. no. C0130), trypsin (10 mg; Worthington, cat. no. 3703), trypsin inhibitor (20 mg; Sigma, cat. no. T9003), and DNase I (200 mg; Boehringer, cat. no. 104159) and store these in small vials at $-20°C$.

Equipment.   Testicular tubules and cells tend to stick to glass. Therefore, handle the cells in plastic pipettes, containers, etc., and siliconize the glass surfaces with which they come into contact. Otherwise, no special equipment is required, apart from a high-speed centrifuge (e.g., a Beckman J2.21) with a swing-out rotor in which 15- or 30-ml tunes can be centrifuged at 15,000 g (e.g., Beckman JS13 or 13.1), and an angle rotor in which 30-ml tubes can be spun at at least 11,400 g (e.g., Beckman JA20). In addition, and elutriator rotor is required (Beckman JE6 or JE6.1), or, if such a rotor is not available, a unit gravity sedimentation chamber according to Tulp *et al.* (1980). This sedimentation chamber is not available commercially, but Tulp *et al.* describe its construction in detail.

Purification of Spermatocytes

Testicular Cell Suspension (see Romrell *et al.,* 1976)

1. Three male 37-day-old Wistar rats are killed with CO_2, and the testes are removed, decapsulated, and transferred to a 100-ml plastic Erlenmeyer flask with 20 ml of SIM and 20 mg of collagenase. Avoid squeezing the tubules. Incubate for 55 minutes in a waterbath shaker at 32°C. (During this incubation step, add PMSF to solutions 7 and 8. Adjust the pH of the dialyzed Percoll solution to 7.3 and the N_{D20} to 1.3400 with SIM or Percoll, and filter the Percoll solution through a 0.45-μm filter. Install the elutriator rotor and put aliquots of solutions 4, 11, 12, and 13 from the freezer on ice.)

2. During the collagenase incubation, the testes will fall apart into tubules and interstitial cells. To remove the interstitial cells, pour the tubule suspension into a plastic 50-ml conical tube (e.g., Greiner, cat. no. 227261), add SIM to 50 ml, mix, and allow the tubules to sediment at 1 g. Wash the tubules twice by resuspending them in SIM, sedimenting them at 1 g, and removing the supernatant.

3. Resuspend the tubules in 20 ml of SIM with 10 mg of trypsin (freshly dissolved) and 50 μl of DNase I (solution 4). Pour the tubules back into the plastic Erlenmeyer flask and incubate for 15 minutes at 32°C in a waterbath shaker.

4. Add 20 mg of trypsin inhibitor, 20 ml of SIM with 1% BSA (solution 3) and 50 μl DNase I (solution 4). Dissociate the tubules by gently pipetting them about 100 times through the mouthpiece of a plastic 10-ml pipette (e.g., Nalgene, cat. no. 3610). The suspension should become monodisperse; if not, continue by pipetting through the (narrower) tip of the pipette.

5. Pour the suspension through a nylon mesh (60-μm pore size), pre-wetted with SIM, into a plastic 50-ml tube. Adjust the volume to 50 ml with SIM, mix, and take a sample to count cells. From the testes of three 37-days-old rats one obtains 1 to 1.2×10^9 cells.

6. Centrifuge the cells for 5 minutes at 200 g. Remove the supernatant and carefully resuspend the pellet, by first adding 200 μl of solution 4, shaking gently, and then slowly adding SIM to a final volume of 7.5 ml; mix well by pipetting with a plastic Pasteur pipette.

Elutriation of Testicular Cells

1. About 10^9 cells can be separated in one run in a JE6 rotor. The

separation is performed in SIM with 0.1% BSA (solution 2), which is precooled to 10°C. When the cells are loaded the rotor is spinning at 2500 rpm at 10°C. We monitor the elutriation of cells at 360 nm (Grootegoed *et al.*, 1977). As soon as the OD_{360} decreases, the rotor speed is decreased or the flow increased so that cells with increasingly higher sedimentation coefficients are elutriated.

We use the following scheme:

Step	Flow (ml/min)	Speed (rpm)
1	15	2500
2	15	2000
3	15	1800
4	20	1800

After step 4, elutriation is continued until the OD_{360} has dropped to (almost) zero. This takes about 7 minutes.

2. Subsequently, the elutriated cells are collected in plastic 50-ml tubes while the flow is increased from 20 to 35 ml/min in about 2 minutes. The cells are collected until the OD_{360} has dropped again to (almost) zero. This takes about 6 minutes. In total, 200 to 250 ml of elutriate is collected.

3. Centrifuge the collected cells for 5 minutes at 200 *g*. Add 50 μl of solution 4 to each pellet, resuspend the cells in SIM, collect them in one 50-ml tube, add SIM to 50 ml, mix, and take a sample to count cells (from the testes of three 37-day-old rats one obtains about 1.3 × 10^8 cells of which 90–93% are spermatocytes).

4. Centrifuge the cells for 5 minutes at 200 *g*.

5. Dissolve 200 mg of DNase I and 20 mg of trypsin inhibitor in 58 ml of the Percoll solution.

6. Remove the supernatant carefully, and resuspend the cells in 58 ml of Percoll with DNase I and trypsin inhibitor. Mix well and check whether the N_{D20} is still 1.3400; adjust if necessary.

7. Divide the cell suspension between two siliconized 30-ml Corex glass tubes and spin for exactly 20 minutes at exactly 10,000 rpm ($= 11,400$ g_{av}) in a JA21 rotor at 20°C. Stop without brake.

8. Remove the Percoll till level a (see Fig. 2) and take the spermatocytes out (between a and b in Fig. 2; avoid the small cell clump below b, it contains spermatogania).

9. Transfer the spermatocytes to a 50-ml tube, add SIM to 50 ml, mix, and take a sample to count cells (the yield will be about 1.2 × 10^8 cells of which about 98% are spermatocytes).

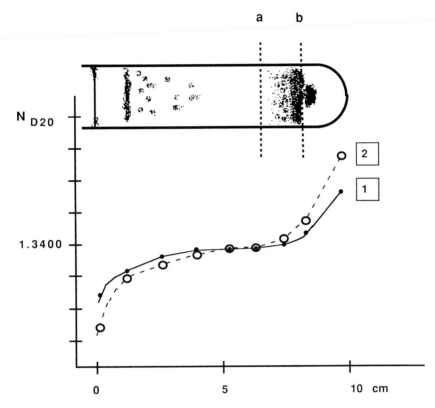

FIG. 2.   Purification of rat spermatocytes in a density gradient of Percoll. The cells were mixed with 27% Percoll in a balanced salt solution, and were separated on the basis of density differences while the density gradient was being formed (see Section II,A, Elutriation of Testicular Cells). At the top of the figure is shown how the cells, collected from the elutriator, are distributed through the centrifuge tube after 20 minutes of centrifugation at 11,400 g_{av}. Horizontal axis: distance from the top of the gradient; vertical axis: N_{D20} (one interval corresponds to an increase of the N_{D20} of 0.0010). Curve 1 shows the N_{D20} at different positions in the gradient after centrifugation for 10 minutes and curve 2 after centrifugation for 20 minutes at 11,400 g_{av}. Ninety eight to 100% of the cells between a and b are pachytene or diplotene spermatocytes. The layer at the top of the gradient contains up to 50% spermatocytes, plus spermatids, spermatogonia and nonspermatogenetic cells; the cell clump below b contains a mixture of spermatocytes and spermatogonia.

10. Centrifuge for 10 minutes at 200 g. Remove the supernatant, loosen the pellet with 50 μl of solution 4, and resuspend the cells in 18 ml of SIM. Mix well, and divide the cells in 2-ml portions between nine conical 10-ml polystyrene tubes. Add SIM to about 8 ml per tube and centrifuge at 200 g.

11. Add 50 μl of 0.5 M DTT to 50 ml of lysis mix (solution 8).

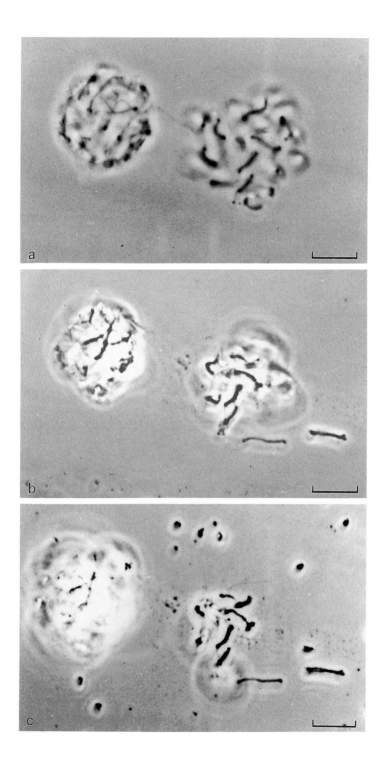

Lysis of Spermatocytes

1. Remove all supernatant carefully.
2. Loosen the nine pellets after addition of 2 μl of solution 4 per pellet.
3. Add to the wall of each tube (*not* to the pellet): 2 μl of 1 *M* PMSF in DMSO (solution 15), 5 μl of trypsin inhibitor (solution 13), and 5 μl of aprotinine (solution 12).
4. Treating the tubes one by one, quickly add 5 ml of lysis mix per tube, so that the spermatocytes are blown apart before they can stick to each other. Cap the tube, turn it upside down once, and put it on ice.
5. Incubate on ice for 1 hour. (A small sample can be viewed by phase contrast.)
6. Spin the swollen nuclei down by centrifuging for 5 minutes at 750 g at 4°C.

Digestion of Spermatocyte Nuclear Matrices with DNase II

1. Remove all supernatant carefully. The nuclei will stick together in a slimy pellet and stick to the wall of the tube.
2. To each tube, add 2 ml of solution 7, 5 μl of solutions 12 and 13, 2 μl of solution 15, and 3.3 μl of solution 11.
3. Cap the tubes and incubate them at 20°C on a horizontal mixing apparatus (like Denley spiramix) at 20°C. Every 20 minutes, add fresh DNase II (3.3 μl of solution 11 per tube). The pellets gradually fall apart and finally disappear entirely. Continue the incubation for 10 minutes after the pellet has dissolved (see Fig. 3). The DNase II incubation will take about 75 minutes (during this incubation make the sucrose step gradients as described below, and make solution 14).

Purification of the SCs

1. Collect the contents of the nine tubes in a plastic 50-ml tube and

FIG. 3. Phase-contrast appearance of nuclear matrices of lysed spermatocytes shortly after addition of DNase II. Spermatocytes, lysed for 60 minutes in lysis mix (See Section II,A, solution 8), were put onto a microscope slide, covered with a coverslip, and photographed with phase-contrast optics. The lysis mix was replaced by 8 m*M* Tris-HCl by adding this buffer to one side of the coverslip, and pulling it past the cells by means of a thick filter paper at the other side. Subsequently, a solution containing 83 Kunitz units DNase II per milliliter of this buffer was pulled past the cells, and the cells were photographed 1, 2, and 3 min after addition of DNase II (a, b, and c, respectively). First, the SCs appear to thicken, probably because chromatin collapsed onto the axes (compare the left cell in a and b). As the chromatin is digested by DNase II, the SC-like structures become thinner (compare the left cells in b and c), and SCs with adhering material are liberated from the nuclear matrix (c, right cell). After prolonged digestion, the SC-like structures become too thin for phase-contrast observation; in electron micrographs "clean" SCs can be observed (Heyting *et al.*, 1985). Bar, 15 μm.

centrifuge for 2 minutes at 60 g. Load the supernatant, except 1 ml above the pellet, onto two sucrose step gradients.

2. To make step gradients, pipet into each of two siliconized Corex glass tubes 2 ml of Fluorinert FC77 (3 M; St. Paul) and 3 ml of 1.5 M sucrose (solution 9). Insert a layer of 400 μl of 2.4 M sucrose (solution 10) between the Fluorinert and the 1.5 M sucrose layer. Load 8.5 ml of supernatant onto each of these step gradients.

3. Centrifuge for 50 minutes at 10,000 rpm in a Beckman JS13.1 rotor at 20°C. The SCs are now in the 2.4 M sucrose and on the Fluorinert–2.4 M sucrose interface. (This is just visible as some fine dust on the interface.)

4. Remove the supernatant plus 1 ml of the 1.5 M sucrose layer. Dry the walls of the tube carefully with paper towels, and remove the remainder of the 1.5 M sucrose layer.

5. Collect the 2.4 M sucrose layer; avoid mixing the 2.4 M sucrose with the Fluorinert. Wash the Fluorinert surface twice with 800 μl of solution 7, and mix these well with the 2.4 M sucrose, to obtain the final SC suspension.

6. Mix a sample of 10 μl of the SC suspension with 90 μl of solution 7 and 100 μl of solution 4 and put it on ice for 15 minutes. Put droplets of 30 μl of this mixture onto an agar filter for ultrastructural analysis (see Section III,A, Protocol for Agar Filtration).

7. Store the SC suspension in liquid N_2.

B.   Comments on the Synaptonemal Complex Isolation Protocol

1.   WHY PURIFY SPERMATOCYTES?

From the reports of at least three groups (Walmsley and Moses, 1981; Ierardi et al., 1983; Raveh and Ben-Zeev, 1984), it appeared difficult to detach SCs from the nuclear matrices of meiocytes. We therefore looked for conditions by which the nuclear matrix, but not the SCs, could be dissociated. We considered two possibilities: (1) treatment of nuclei with successively low salt, DNase I (and Mg^{2+}), high salt, and again DNase I (and Mg^{2+}) (Kaufmann et al., 1981), or (2) treatment of nuclei with Triton X-100, EDTA, and DNase II (an endonuclease which does not require Mg^{2+}) (Galcheva-Gargova et al., 1982). Both treatments were reported to leave the nuclear laminae intact. We chose the second treatment because we had observed, by phase-contrast microscopy, that any treatment of spermatocyte nuclei with Mg^{2+} or Ca^{2+} led to a higher contrast of their substructures; this effect was not reversed by EDTA. We considered this as an indication of irreversible aggregation of macromolecules, and speculated that this might be one of the reasons why SCs were so difficult to detach from Mg^{2+}-treated nuclei. We found (Heyting et al., 1985) that SCs can stand treatment with EDTA and DNase II. Since spermatocytes are exceptional in that they do not possess a nuclear lamina (Fawcett, 1966; Stick and Schwartz, 1982), contamination of isolated SCs with nuclear laminae can be prevented if very pure fractions of spermatocytes are used as starting material.

The procedures which we use for the purification of rat spermatocytes are essentially those of Romrell et al. (1976) (the preparation of the testicular cell suspensions) and Bucci et al. (1986) (the procedures of cell separation). The protocol in Section II,A for purification of spermatocytes describes the separation conditions as they are adapted to our requirements: a high yield of very pure spermatocytes in a good physiological condition. This is the reason SIM is used as the isolation medium. It is a modification of the testis isolation medium of Dietrich et al. (1983) in which spermatocytes can survive for a long time; SIM has a higher KCl/NaCl ratio than most balanced salt solutions (cf. Grootegoed et al., 1977). The isolation is medium is supplemented with pyruvate and lactate, which are essential energy sources for spermatocytes (Jutte et al., 1981). Elutriation is performed in the presence of 0.1% BSA to inhibit possible traces of trypsin and to stabilize the cells; it is carried out at 10°C to avoid shortage of oxygen while the cells are packed in the separation chamber of the elutriator rotor. The protocol gives the optimal conditions for the purification of spermatocytes in this medium (SIM with 0.1% BSA) and 10°C. The collected cell fractions will consist mainly of late pachytene and diplotene spermatocytes; early spermatocytes (leptotene/zygotene and early pachytene) are underrepresented in these fractions. This can be analyzed in Giemsa-stained preparations of the cell fractions (Oud and Reutlinger, 1981). Smaller amounts of cells (up to 2×10^8) can be separated in a Staput instead of an elutriator (Heyting et al., 1985; Tulp et al., 1980).

The cell separation in Percoll is based on density differences among cell types. After centrifugation for 20 minutes at 11,400 g_{av} in a JA21 rotor, a rather flat gradient is formed (Fig. 2), which allows the separation of cells which only slightly differ in density. We could separate more cells (up to 1.5×10^8 per 30-ml tube) and achieve a better separation if we mixed the cells with the Percoll and performed gradient formation and cell separation in a single run at 11,400 g if we layered the cells onto a performed gradient and separated them at a low g force (1,000 g). The damage caused to the cells by the high g forces will probably have little effect on the quality of the isolated SCs, because the cells are lysed immediately after the separation in Percoll.

2. Treatment with DNase II

If nuclei, isolated in the absence of divalent cations, are pelletted, they will stick together and cannot be dispersed. DNase II digestion of such a pellet of 1.3×10^8 nuclei takes several hours. We circumvented this problem by distributing the cells over a large numbers of tubes, so that a large number of small pellets instead of a single large pellet are digested. The conditions (8 mM Tris-HCl, pH 7.0) are far from optimal for DNase II, which has an optimum at pH 4.6. However, we could not dissociate SCs from spermatocyte matrices at this low pH, and we suspect from phase-contrast microscopical observations that this low pH may also cause aggregation of nuclear matrix components. At pH 7, a large amount of DNase II is needed for digestion. At temperatures above 25°C the enzyme may denature and precipitate; this has to be avoided because the denatured enzyme tends to stick to the SCs and copurifies with them. A small amount of Triton X-100 (0.005%) in the DNase II digestion medium helps to prevent this.

3. Morphology and Yield of Isolated SCs

Most isolated SCs have the appearance of late pachytene and diplotene SCs (Heyting *et al.,* 1985); this corresponds to the composition of the spermatocyte fractions from which they were isolated. We have also tried to purify spermatocytes from earlier stages of meiotic prophase, and to isolate SCs from these. However, in lysates from fractions consisting of about 50% zygotene and early pachytene cells and about 50% spermatogonia, we found only very few recognizable SCs, and most of these were early pachytene SCs or small paired segments (unpublished observations). It is possible that the (unpaired) axial elements of zygotene SCs cannot stand the isolation procedure.

Also from late pachytene and diplotene spermatocytes a large fraction of the SCs is lost during the isolation procedure (Heyting *et al.,* 1985): From 1.3×10^8 spermatocytes we obtain about 5×10^8 SCs, which is about 20% of the theoretical yield (See Table 1 in Heyting *et al.,* 1985). The missing SCs are not found in the 1.5 M sucrose layer or in the supernatant of the step gradient. It is possible that part of the SCs was dissociated into unrecognized parts during the DNase II digestion.

The morphology of the isolated SCs has been described in detail by Heyting *et al.* (1985). Most substructures that can be discerned in SCs in sections or surface-spread spermatocytes are also present in the isolated SCs, with the exception that the lateral elements of isolated SCs appears less dense than in sections, that recombination nodules are very rare, and that the central element is lacking from most isolated SCs. Thus, it is possible that only part of the SC proteins are present in the SC preparations.

III. Analysis of Synaptonemal Complex Preparations

A. Ultrastructural Analysis

Protocol for Agar Filtration. We analyze the SC preparations ultra-structurally by means of the agar filtration technique (Fukami and Adachi, 1965; Woldringh *et al.*, 1977) (Fig. 4). The procedure is as follows:

1. Pour agar plates, consisting of 2% agar (Difco bacto agar, cat. no. 0140-01) and 0.7% NaN$_3$ in water, in petri dishes.

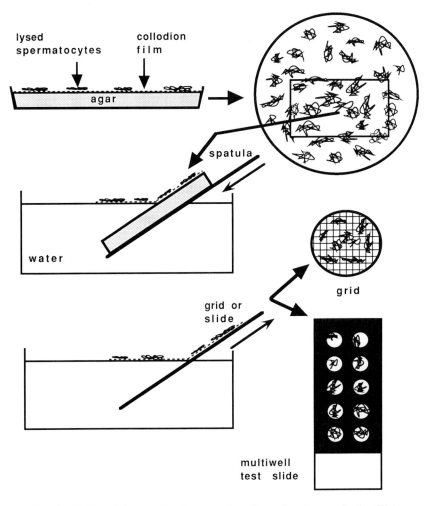

FIG. 4. Outline of the agar filtration procedure. For explanation, see Section III,A.

2. Dry the plates overnight upside down, like roof tiles (lid off), at room temperature.

3. Cool the agar plates one by one for 2 minutes at $-20°C$, and spread 5 ml of a 0.4% collodion solution rapidly over the plate; let the plates drain upside down, at an angle of 30°, on moistened filter paper in a hood. [The 0.4% collodion solution is obtained by dilution of a 4% stock solution (Merck cat. no. 2644) in isoamylacetate (Merck no. 1231).] As the solvent evaporates, a collodion film forms on the agar surface, with tiny holes where condense drops were present; this film is the agar filter. The size and number of the holes can be influenced by changing the humidity in the hood, or the duration of the cooling of the agar plate at $-20°C$.

4. The plates with agar filters can be stored for about 2 months at 4°C.

5. Put droplets of 30 μl of fixed SCs (see Section II,A, Purification of the SCs) onto the agar filter and wait until the fluid has been sucked through the holes into the dry agar. This takes several hours; the plate can be left overnight at room temperature (lid on).

6. Cut out the collodion film together with the agar by means of a scalpel, and float the film off (Fig. 4).

7. The film can be picked up on a grid by means of a metal transfer loop. Drain the grid on its side on filter paper.

8. Dry the grid for 20 minutes at 37°C. Stain in 1% uranyl acetate for 20 minutes at room temperature; drain, destain in water for 10 seconds, drain again, and dry the grid for 20 minutes at 37°C. Electron micrographs of thus obtained preparations can be found in Heyting *et al.* (1985).

Comments on the Agar Filtration Technique.   Agar filtration is an ideal technique for the ultrastructural analysis of suspensions (Fukami and Adachi, 1965; Woldringh *et al.,* 1977). Because the fluid is sucked through the holes into the dry agar before the droplets can evaporate, the particles in the suspension do not aggregate, but spread evenly on the filter.

Agar filtration has also been important for the selection of the most suitable conditions of spermatocyte lysis for the isolation of SCs. For this purpose, we simply put suspensions of unfixed lysed spermatocytes onto the agar filter, and further treated them according to the agar filtration protocol. Electron micrographs of agar filtrates of spermatocytes, lysed in the mixture which we finally selected (solution 8), show nicely spread nuclear matrices without any indication of aggregation (Fig. 5). Such agar filtrates are also useful for the screening of hybridoma supernatants (see below). If *isolated* unfixed SCs are subjected to agar filtration, their morphology is not well preserved (not shown). It is possible that unfixed isolated SCs cannot stand direct contact with the collodion.

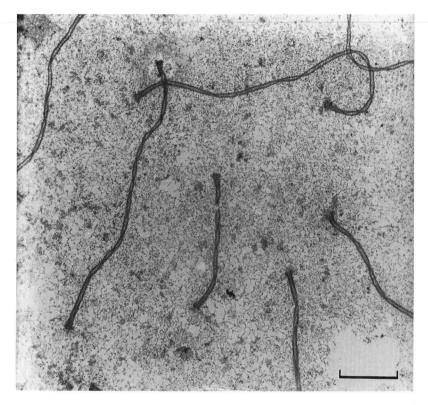

Fig. 5.   Synaptonemal complexes in a pachytene spermatocyte, lysed in a mixture containing Triton X-100, EDTA, and DTT (see Section II,A, Lysis of Spermatocytes). Bar, 4 μm.

B.   Light Microscopical Analysis

For a light microscopical inspection of the result of an effort to isolate SCs, one can make use of fluorescence microscopy (Fig. 6).

Protocol

1. Mix the final suspension with an equal volume of 4% PFA (solution 14). Fix for 15 minutes at room temperature.
2. Transfer the suspension to a vial with a flat bottom (e.g., Greiner, cat. no. 205111), with a coverslip on the bottom.
3. Centrifuge for 15 minutes at 1700 *g* in a swing-out rotor.

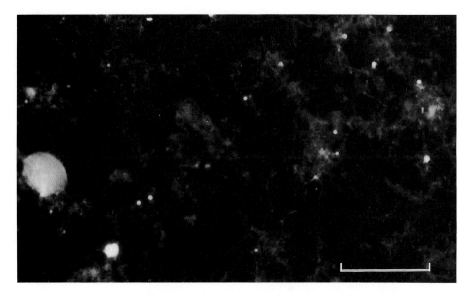

FIG. 6.   Light microscopical analysis of SC purification after staining with FITC. The spermatocyte nuclear matrices have not been completely digested by DNase II (see Section II,A, Digestion of Spermatocyte Nuclear Matrices with DNase II). Note that parts of the SCs are still captured in presumed remnants of nuclear matrices and note the material still adhering to part of the SCs. Bar, 20 μm.

4. Remove the supernatant and wash the coverslips in, successively, 70% and 100% ethanol. Air-dry the coverslips, and then wash twice in fluorescein isothiocyanate (FITC) buffer (150 mM NaCl, 50 mM Na$_2$CO$_3$, pH 9.5). All washing steps are performed by adding the washing fluid, centrifugation for 5 minutes at 1700 g, and removing the supernatant. Avoid touching the coverslips.

5. Add 2 ml of 20 μg/ml FITC (Sigma, cat. no. F7250) in FITC buffer to each vial, and incubate for 30 minutes in the dark; wash twice in FITC buffer.

6. Mount the coverslips with 50% glycerol (v/v) in FITC buffer on object slides.

Figure 6 shows a micrograph of such a preparation. We took this picture while we were optimizing the SC isolation procedure. The SCs in this particular preparation are not free from adhering material, and part of the SCs appear still to be clustered in presumed remnants of spermatocyte nuclei. The split LEs of diplotene SC can be easily discerned.

IV. Immunocytochemical Labeling of Synaptonemal Complex Proteins

A. Immunization

For the identification and localization of SC components, antibodies specifically recognizing the separate SC proteins are indispensable. We chose to prepare monoclonal anti-SC antibodies for the following reasons: (1) we had not been able to find conditions to dissociate SC proteins and to purify them in a more or less native conformation; and (2) the amount of SC proteins that we could isolate was limited (see Section II); therefore, immunopurification of specific antibodies from a polyclonal antiserum did not seem feasible.

We have performed nine immunizations for the production of anti-SC monoclonal antibodies (Mabs), with, respectively, native SCs (3 ×), PFA-treated SCs (3 ×), SCs treated with 0.3% sodium dodecyl sulfate (SDS) (2 ×), and fraction of high-molecular-weight SC proteins, obtained after dissociation of SCs in SDS and size-fractionation by means of HPLC (1 ×). The immunizations with native SCs were by far the most successful ones, and yielded 40 of the 47 clones which fulfilled our selection criteria (see Section IV,B). The most successful immunization protocol involved one injection with 5×10^8 SCs ($=$ about 50 μg) followed by four injections with 2.5×10^8 SCs. The interval between the injections was 2 weeks, and the fusion of the mouse spleen cells with the SP2 mouse myeloma cells was performed 75 hours after the last injection (Heyting et al., 1987). The SCs were pelleted, washed twice with PBS, resuspended in 150 μl of PBS, and sonicated until the pellet had fallen apart. (The SC pellet is very hard to disperse.) The suspension was mixed by sonication with 150 μl of complete (first injection) or incomplete Freund's adjuvant and was injected intraperitoneally.

We also prepared two polyclonal anti-SC antisera by immunization of rabbits by the same protocol, except that the SCs were injected subcutaneously and intramuscularly. This yielded two strong anti-SC antisera which mainly recognized SC proteins of 30 and 33 kDa, respectively. Apparently, these SC components are relatively strong antigens, at least in rabbits.

B. Screening of Hybridomas

For the selection of hybridomas producing anti-SC Mabs, we needed a screening test which fulfilled two requirements: (1) it should require little antigen, and (2) the test should match as closely as possible the experiments in which the Mabs were going to be used. For this purpose, we developed an immunocytochemical test on agar filtrates of lysed spermatocytes. The procedure is as follows (Fig. 4).

1. Lyse 6×10^6 purified spermatocytes (see section II,A Purification of Spermatocytes) in 5 ml of lysis mix with protease inhibitors, exactly as described in Section II,A, Lysis of Spermatocytes.
2. Put the tubes one ice for 75 minutes.
3. Mix the suspension by gently turning the tube upside down once (not more).
4. On each of a series of agar filters (see Section III,A), cover a square of 5×5 cm with 350 μl of the suspension. Spread the cells with a bent siliconized glass rod, without touching the agar filter.
5. Leave the agar filters overnight at room temperature (lid on).
6. Using a scalpel, cut out two pieces of 2×5 cm of filter (plus agar).
7. Using a spatula, lift the filter plus agar from the plate, and float the filter off on the surface of deionized water, filtered through a 0.2-μm filter.
8. Allow each filter to spread on the surface for at least 2 minutes; pick up each filter on an epoxy-coated multiwell testslide (Cel-Line Inc., cat. no. 10–103; cleaned overnight in ethanol/diethyl ether 1 : 1 and wiped dry with paper towels).
9. Let the slides drain and dry overnight in slide racks.
10. Wrap the slides in aluminum foil and store in liquid N_2; the slides are stable for months.
11. Before use, allow the slides to warm to room temperature while still in the foil.

These slides can be used to screen hybridomas for the production of anti-SC antibodies by incubating them with the hybridoma supernatants. The binding of the anti-SC antibodies can be visualized by the standard indirect immunoperoxidase technique with DAB and H_2O_2 as substrates (see Heyting et al., 1987). After DAB-staining, the slides are not mounted, but they are screened with a dry 40 × phase-contrast objective. The DAB precipitate shows up with high contrast, so that even weak staining of SCs can be discerned. Phase contrast should *not* be used to demonstrate that certain antigens make up part of SCs, because nuclear substructures with a high contrast cannot be discriminated with certainty from structures which have been weakly stained with DAB. We discarded all hybridomas which, after repeated subcloning, did not produce a detectable signal in the above-described immunocytochemical test with *bright-field* illumination (for examples, see Heyting et al., 1987, 1989).

Our second criterion for the selection of hybridoma clones was the ability of their Mabs to recognize antigens on Western blots of SCs (Heyting et al., 1987, 1989). For this purpose, we separate the proteins of 2.5×10^8 SCs (2.5 μg) on a 7–18% linear gradient SDS-polyacrylamide slab gel with a single 10-cm-wide slot. The proteins from the entire gel are transferred to nitrocellulose by electroblotting. The best results with respect to transfer and immunoreactivity are obtained if the gels are frozen for at least 17 hours at

−20°C, thawed at 4°C, and subjected to Western blotting in a carbonate buffer at pH 9 (Dunn, 1986), for 24 hours at 5 V/cm at a temperature below 35°C. The intensive cooling of the blotting apparatus required is accomplished by immersing it on ice on a stirring motor and pumping ice-cold water through a cooling coil (Dunn, 1986). Results using this technique are described in Heyting *et al.* (1988). Strips of nitrocellulose filters obtained in this way were used for the screening of hybridoma supernatants (Heyting *et al.*, 1989).

Four classes of hybridoma clones have thus been selected which produce Mabs recognizing polypeptides of, respectively 190, 125–120, 65–55, and 30 + 33 kDa. The 190, 125–120, and 30 + 33 kDa antigens have been localized ultrastructurally on the SCs (Heyting *et al.*, 1987, 1989; Moens *et al.*, 1987). The 65–55 kDa antigens, recognized by Mabs of 2 of the 47 clones selected, are not confined to SCs. This result shows the advantage as well as the limitation of the selection procedure: All selected clones except two produce Mabs which recognize SC-specific antigens (Heyting *et al.*, 1988; Offenberg *et al.*, in press). However, clones producing Mabs recognizing SC components which are *not* SC-specific will not pass the immunocytochemical test.

C. Labeling of Synaptonemal Complex Proteins in Cells and Tissues

Immunocytochemical labeling of SC proteins has been performed on isolated spermatocytes (Heyting *et al.*, 1987, 1989; Moens *et al.*, 1987; Dresser, 1987; Haaf *et al.*, 1989) and on frozen sections of testes and ovaries (Heyting *et al.*, 1988; Demartino *et al.*, 1980; Dietrich *et al.*, in preparation). Labeling of tissue sections provides information about the distribution of the antigen in various cell types, and also within cells; labeling of lysed or surface-spread spermatocytes allows a more detailed analysis of individual SCs and of whole SC complements.

1. LABELING OF TISSUE SECTIONS

We have attempted to label SC proteins immunocytochemically in sections of frozen as well as paraffin-embedded testes of the rat. We never obtained any signal on sections of paraffin-embedded testes, neither with polyclonal nor with monoclonal antibodies, and neither with tissue fixed in PFA, nor with tissue fixed by freeze substitution (Heyting *et al.*, 1983). Several (but not all) anti-SC Mabs and both polyclonal antisera reacted detectably with frozen sections of the testis. Cross-linking fixatives reduced (0.5–4% PFA) or abolished (0.1–2% glutaraldehyde) the antigenicity. A good morphological conservation and a reasonable strong signal can be obtained with frozen sections that were fixed postsectioning in 1% PFA, 10 mM sodium phosphate buffer, pH 7.3, for 30 minutes at 4°C; a strong signal and a reasonable morphological conservation can be achieved with frozen sections, fixed for 45 minutes at −20°C in acetone.

For visualization of antibody binding to sections we prefer indirect immunofluorescence above indirect immunoperoxidase staining with DAB and H_2O_2: The background staining of sections with DAB is relatively high, and the only mildly fixed frozen sections are damaged considerably during the exposure to DAB and H_2O_2.

The protocol for the immunofluorescence labeling of SC proteins in frozen sections of the testis is as follows:

1. Slices of rat testes, 2 mm thick, are quickly frozen in liquid N_2, wrapped in aluminum foil, and put at −18°C for at least 2 hours. Subsequently, 8- to 10-μm-thick sections are cut at −18°C and collected on glycerin–chromalum-coated slides.

2. The sections are quickly dried by means of a hairdryer without a heater, wrapped in aluminum foil, and stored in liquid N_2 or −80°C; they retain their antigenicity at least for months. Before immunocytochemical staining, the slides are allowed to warm to room temperature while still in the foil.

3. Subsequently, they are fixed for 45 minutes at −20°C in precooled dehydrated acetone, or for 30 minutes at 4°C in precooled 1% PFA in 10 mM sodium phosphate buffer, pH 7.3.

4. After four washes in PBS (140 mM NaCl, 10 mM sodium phosphate, pH 7.3), the slides are incubated for 15 minutes in blocking buffer (3% nonfat dry milk, 0.1 mM PMSF, and 10% normal goat serum in PBS; adjust the pH to 7.3 with NaOH, centrifuge for 30 minutes at 50,000 g, and use the supernatant). Follow with an incubation for 45 minutes at room temperature in hybridoma supernatant (diluted 1:1 to 1:20 in blocking buffer), three washes of 5 minutes each in PBS, incubation for 15 minutes in blocking buffer and for 45 minutes in a goat anti-mouse–FITC conjugate, diluted in blocking buffer, three washes of 5 minutes each in PBS, a short rinse in 150 mM NaCl, 50 mM sodium carbonate, pH 9.5, and mounting in 50% glycerol in the same buffer. (Be careful: The acetone-fixed sections are very vulnerable to pressure on the coverslip.)

2. LABELING OF ISOLATED SPERMATOCYTES

For light as well as electron microscopical analysis, immunocytochemical labeling can be performed on (1) surface-spread spermatocytes, (2) agar filtrates of lysed spermatocytes, and (3) spermatocytes, "dried down" onto glass or plastic film in the presence of a high concentration of sucrose.

Light Microscopical Analysis. The indirect immunoperoxidase technique, applied to agar filtrates of lysed spermatocytes, is a very easy and reliable technique for light microscopical analysis (see Sections III,A and IV,B). In contrast to frozen sections, agar filtrates show little background and a good conservation of morphological detail, at least at the light microscope level, after the immunoperoxidase incubation and exposure to DAB plus H_2O_2 (see Heyting *et al.*, 1987, 1989). Immunofluorescence staining of agar filtrates yields preparations of equivalent quality. It is also possible to label SC proteins immunocytochemically by the immunoperoxidase or the immunofluorescence technique in surface-spread or dried-down spermatocytes, but the signal will be less intense than in the agar filtrates.

Electron Microscopical Analysis. It is necessary to attach the cells to a thin support film and to carry out the staining protocol without losing the fragile film for electron microscopical analysis. In agar filtrates, the agar filter itself serves this purpose. For dried-down or surface-spread cells, the film is produced on microscope slides as support, as is described in detail below.

Immunogold Staining of SCs in Surface-Spread Spermatocytes (Moens *et al.*, 1987). The immunogold staining procedure allows an accurate localization of antigen; according to this technique, binding of an anti-SC antibody is visualized by means of a second antibody, conjugated to gold particles with a diameter of 1–20 nm (Slot and Geuze, 1981; Geuze *et al.*, 1981). A problem which we initially encountered with this technique was low labeling intensity: Probably, the relatively large gold particles hamper the second antibody in reaching its target. Treatment of the preparations with DNase I or with detergents improves access to the antigens, but also causes loss of morphological detail at the ultrastructural level. The following protocol is an empirically determined compromise which allows sufficient conservation of morphology and a reasonable strong immunocytological signal.

Preparation of Slides

1. Spray slides with Windex and rub with a paper towel. Rinse in running tap water and in distilled water, and wipe dry. Suspend the slide in a solution of 0.3% polystyrene (Falcon petri dish 3080) in chloroform, and then pull up just above the solution for 5 seconds to let excess plastic drain in the saturated chloroform vapor. The timing alters the thickness of the film. When the film is dry, test one slide to ensure that the film can be removed. Scratch the film along the edges of the slide and float off on a clean water surface by slowly emerging the slide, starting at the far edge. In reflected light, a good film looks silvery (about 60 nm in thickness). A thick film (about 90nm) has a golden hue.

2. Surface-spreading of spermatocytes is performed as described in [4], this volume. The spread cells are picked up on plastic-coated slides instead of plain glass, and are transferred immediately to a solution containing 4%PFA and

0.1–0.15% SDS adjusted to pH 8.2 with NaOH. The SDS concentration is critical; test a small series of SDS concentrations in the indicated range. The cells are fixed in this solution for 3 minutes at room temperature, and then for another 3 minutes in 4% PFA, pH 8.2, without SDS. Subsequently, they are washed for 10 seconds in 0.4% Photo-Flo (Eastman Kodak Co.) and dried. (The slides can be wrapped in aluminum foil and stored at −20°C for weeks.)

3. Take slides through the following solutions at room temperature: 10 minutes in PBS, 20 minutes in 1 μg/ml DNase I (Sigma, cat. no. D4527) in minimal essential medium (MEM) (Gibco, cat. no. 320.1570), 10 minutes in detergent (5 mM EDTA, 0.25% gelatin, 0.05% Triton X-100, in PBS), 10 minutes in PBS. Accumulate the slides in holding buffer [= antibody dilution buffer (see step 4), diluted 1:10 in PBS]. A small magnetic stirring bar in the bottom of the staining dish, arranged so that it does not touch the slides, improves the efficiency of washing.

4. Shake excess holding buffer from the slide and add 60 μl of the appropriate dilution of the first antibody in antibody dilution buffer (3% BSA, 10% normal goat serum, 0.05% Triton X-100). Rock the slide to spread the solution over the slide. Put the slide cell-side down on a strip of Parafilm, and incubate for 1 hour at room temperature and then overnight at 4°C.

5. Flood the slides with PBS and gently lift off the Parafilm without disturbing the plastic film. Wash for 10 minutes in PBS, 10 minutes in detergent, 10 minutes in PBS, and 10 minutes in holding buffer. Shake off excess holding buffer and add the immunogold-conjugated second antibody, diluted in antibody dilution buffer.

6. Again, place the slides face down on Parafilm and keep them at 37°C for 45 minutes. Fifteen-nanometer gold (Amersham, cat. no. RPN422) is the most easily visible with the electron microscope. Five-nanometer gold (Amersham, cat. no. RPN420) is difficult to see but has the advantage of precise localization and a better access to the antigen. The two can be combined. A still better access to the antigens is presumably provided by 1-nm gold, but this has to be "enhanced" to a larger size (see below). (Suppliers of gold-conjugated secondary antibodies usually supply their preferred protocol.)

7. Flood the slides with PBS before they are picked up, and wash several times in PBS. After a final rinse in 4% Photo-Flo in water, adjust to pH 7 with dilute NaOH and let the slides dry. Score the plastic film along the edges and float off on a clean water surface. Place clean electron microscope grids on the floating film. Put Parafilm on top of the film with grids and lift up after a few seconds. Let the grids dry overnight. The next day, the individual grids can be postfixed in glutaraldehyde or osmium tetroxide, or stained with 1% uranyl acetate in water for 5 to 10 minutes. If the grids have not been thoroughly dried, the osmium tetroxide or uranyl acetate solution may remove the gold grains.

8. For 1-nm gold, rinse the slides in water twice for 5 minutes each. Flood with a 1:1 mix of the enhancer and initiator solutions of the Intense BL kit (Amersham, cat. no. RPN491). After no more than 5 minutes, wash several times in distilled water. Rinse in 4% Photo-Flo, dry the slide, and float off the plastic film as above. After the washes, the slides may be postfixed in 1% glutaraldehyde. If the "enhance" tends to bead, some Photo-Flo can be added to improve wetting of the slide. Examples of immunogold staining of surface-spread spermatocytes can be found in Moens *et al.* (1987) and Heyting *et al.* (1987, 1989).

Agar Filtration of Spermatocytes for Immunogold Staining. See Section III,A, Protocol for Agar Filtration, and Fig. 4. The agar filtrates are fixed and labeled as has been described above for surface-spread cells. An example of an immunogold-stained SC in an agar filtrate can be found in Heyting *et al.* (1987).

Drying Down Testicular Cells for Immunogold Staining. Put large drops (about 200 μl) of 0.15 M sucrose onto polystyrene-coated slides (see Section IV,C,2), then 20 μl of a suspension of mechanically dispersed testicular cells in MEM is added per drop. The sucrose drop is then mixed and spread with a bent Pasteur pipette to cover a surface of 2 × 2 cm; avoid touching the plastic film. Air-dry the slides (this takes about 2 hours), and rinse twice for 1 minute in deionized water. Then fix the slides and proceed as has been described above for surface-spread cells. Figure 7 shows an example of an immunogold-stained SC in a dried-down spermatocyte.

Of the three techniques tested for the immunocytochemical analysis for SCs in isolated spermatocytes, *surface spreading* provides the best conservation of ultrastructural detail and the highest intensity of immunogold labeling; however, this technique involves relatively large losses of cells. *Drying down* allows a high cell yield and a reasonable intensity of immunogold labeling; this technique is the best compromise for electron microscopical analysis if selective loss of cells has to be avoided. *Agar filtration* also allows a high cell yield, but provides an even lower intensity of immunogold labeling than drying down. However, it is an easy, reliable technique, which allows a very intense labeling of SCs by the immunoperoxidase technique; for light microscopical analysis it is the method of choice.

V. Applications and Concluding Remarks

The isolation of SCs is an essential step for the biochemical analysis of the role of these structures in meiotic chromosome behavior. The preparation of polyclonal as well as monoclonal anti-SC antibodies is another precondition

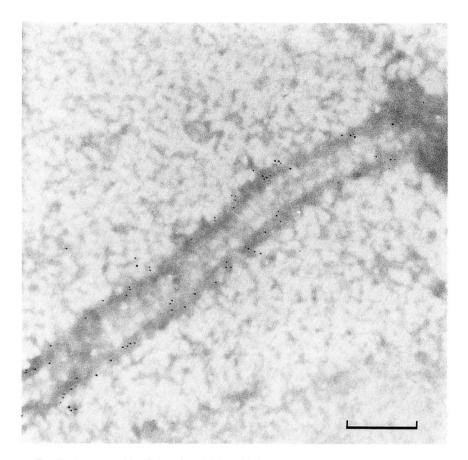

Fig. 7.   Immunogold staining of an SC in a dried-down spermatocyte (see Section IV,C,2). The first antibody was Mab II52F10, which recognizes SC proteins of 30 and 33 kDa; the second antibody was goat anti-mouse IgG, conjugated to 15-nm gold particles. Bar, 0.5 μm.

to be fulfilled for such an analysis. The antibodies can now be used to analyze the assembly and disassembly of SCs by immunocytochemical anaylsis; the relation between the presence or absence of chromosome homology and assembly of SCs can be studied in detail; changes in the polypeptide composition and/or the pattern of SC protein modifications in the course of meiotic prophase can be analyzed; and the genes coding for SC components can be isolated by screening expression libraries with the now available antibodies. If sufficient homology exists between SC genes of rats and of lower eukaryotes, a detailed analysis of the role of SCs and of their components in the regulation of meiotic chromosome behavior comes within reach.

ACKNOWLEDGMENTS

We thank P. B. Moens (York University, Toronto, Canada), H. H. Offenberg, R. J. Dettmers, and P. de Boer (Agricultural University, Wageningen, The Netherlands) for fruitful cooperation, discussion, and advice; T. de Vries for performing immunogold staining of SCs in dried-down spermatocytes in the course of his undergraduate study; and A. C. G. Vink and E. J. W. Redeker for expert technical assistance.

REFERENCES

Bucci, L. R., Brock, W. A., Johnson, T. S., and Meistrich, M. L. (1986). Isolation and biochemical studies of enriched populations of spermatogonia and early primary spermatocytes from rat testes. *Biol. Reprod.* **34,** 195–206.

Demartino, C., Capanna, E., Nicotra, M. R., and Natali, P. G. (1980). Immunochemical localization of contractile proteins in mammalian meiotic chromosomes. *Cell Tissue Res.* **213,** 159–178.

Dietrich, A. J. J., Scholten, R., Vink, A. C. G., and Oud, J. L. (1983). Testicular cell suspensions of the mouse *in vitro. Andrologia* **15,** 236–246.

Dietrich, A. J. J., *et al.* (in preparation).

Dresser, M. E. (1987). The synaptonemal complex and meiosis: An immunocytochemical approach. *In* "Meiosis" (P. B. Moens, ed.), pp. 245–274. Academic Press, Orlando, Florida.

Dunn, S. D. (1986). Effects of the modification of transfer buffer composition and the renaturation of proteins in gels on the recognition of proteins on Western blots by monoclonal antibodies. *Anal. Biochem.* **157,** 144–153.

Fawcett, D. W. (1966). On the occurrence of a fibrous lamina on the inner aspect of the nuclear envelope in certain cells of vertebrates. *Am. J. Anat.* **119,** 129–146.

Fukami, A., and Adachi, K. (1965). A new method of preparation of a self-perforated micro plastic grid and its application. *J. Electron Microsc.* **14,** 112–118.

Galcheva-Gargova, A., Petrov, P., and Dessev, G. (1982). Effect of chromatin decondensation on the intra-nuclear matrix. *Eur. J. Cell Biol.* **28,** 155–159.

Geuze, H. J., Slot, J. W., Van der Ley, P. A., and Scheffer, R. C. T. (1981). The use of collodial gold particles in double labeling immunoelectron microscopy of ultrathin frozen sections. *J. Cell Biol.* **89,** 653–662.

Gorach, G. G., Safronov, V. V., Kolomiets, O. L., Dadashev, S. Y., and Bogdanov, Y. F. (1985). Biochemical and ultrastructural analyses of a synaptonemal complex in mammalian spermatocytes. *Tsitologya* **27,** 1347–1352.

Grootegoed, J. A., Grolle-Hey, A. H., Rommerts, F. F. G., and Van der Molen, H. J. (1977). Ribonucleic acid synthesis *in vitro* in primary spermatocytes, isolated from rat testis. *Biochem. J.* **168,** 23–31.

Haaf, T., Mackeus, A., and Schmid, M. (1989). Immunocytogenetics. II. Human autoantibodies to synaptonemal complexes. *Cytogenet. Cell. Genet.* **50,** 6–13.

Heyting, C., Van der Laken, C. J., Van Raamsdonk, W., and Pool, C. W. (1983). Immunohistochemical detection of O^6-ethyldeoxyguanosine in the rat brain after *in vivo* applications of N-ethyl-N-nitrosourea. *Cancer Res.* **43,** 2935–2941.

Heyting, C., Dietrich, A. J. J., Redeker, E. J. W., and Vink, A. C. G. (1985). Structure and composition of synaptonemal complexes, isolated from rat spermatocytes. *Eur. J. Cell Biol.* **36,** 307–314.

Heyting, C., Moens, P. B., Van Raamsdonk, W., Dietrich, A. J. J., Vink, A. C. G., and Rederker, E. J. W. (1987). Identification of two major components of the lateral elements of synaptonemal complexes. *Eur. J. Cell Biol.* **43,** 148–154.

Heyting, C., Dettmers, R. J., Dietrich, A. J. J., Redeker, E. J. W., and Vink, A. C. G. (1988). Two major components of synaptonemal complexes are specific for meiotic prophase nuclei. *Chromosoma* **96**, 325–332.

Heyting, C., Dietrich, A. J. J., Moens, P. B., Dettmers, R. J., Offenberg, H. H., Redeker, E. J. W., and Vink, A. C. G. (1989). Synaptonemal complex proteins. *Genome* **31**, 81–87.

Ierardi, L. A., Moss, S. B., and Bellvé, A. R. (1983). Synaptonemal complexes are integral components of the isolated mouse spermatocyte nuclear matrix. *J. Cell Biol.* **96**, 1717–1726.

Jutte, N. H. P. M., Grootegoed, A., Rommerts, F. F. G., and Van der Molen, H. J. (1981). Exogenous lactate is essential for metabolic activities in isolated rat spermatocytes and spermatids. *J. Reprod. Fertil.* **62**, 399–405.

Kaufmann, S. H., Coffey, D. S., and Shaper, J. H. (1981). Considerations in the isolation of rat liver nuclear matrix, nuclear envelope and pore complex lamina. *Exp. Cell Res.* **132**, 105–123.

Moens, P. B., Heyting, C., Dietrich, A. J. J., Van Raamsdonk, W., and Chen, Q. (1987). Synaptonemal complex antigen location and conservation. *J. Cell Biol.* **105**, 93–103.

Offenberg, H. H., Dietrich, A. J. J., and Heyting, C. (1991). Tissue distribution of the two major components of synaptonemal complexes of the rat. *Chromosoma* (in press).

Oud, J. L., and Reutlinger, A. H. H. (1981). Chromosome behavior during early meiotic prophase of mouse primary spermatocytes. *Chromosoma* **83**, 395–407.

Raveh, D., and Ben-Zeev, A. (1984). The synaptonemal complex as part of the nuclear protein matrix of the flour moth, *Ephestia kuhniella*. *Exp. Cell Res.* **153**, 99–108.

Romrell, L., Bellvé, A. R., and Fawcett, D. W. (1976). Separation of mouse spermatogenic cells by sedimentation velocity. *Dev. Biol.* **49**, 119–131.

Slot, J. W., and Geuze, H. J. (1981). Sizing of protein A collodial gold probes for immunoelectromicroscopy. *J. Cell Biol.* **90**, 533–542.

Stick, R., and Schwartz, H. (1982). The disappearance of the nuclear lamina during spermatogenesis. An electron microscopic and immunofluorescence study. *Cell Differ.* **11**, 235–243.

Tulp, A., Collard, J., Hart, A. A. M., and Aten, J. A. (1980). A new unit gravity sedimentation chamber. *Anal. Biochem.* **105**, 246–256.

Walmsley, M., and Moses, M. J. (1981). Isolation of synaptonemal complexes from hamster spermatocytes. *Exp. Cell Res.* **133**, 405–411.

Wettstein, D. von, Rasmussen, S. W., and Holm, P. B. (1984). The synaptonemal complex in genetic segregation. *Annu. Rev. Genet.* **18**, 331–413.

Woldringh, C. L., De Jong, M. A., Van den Berg, W., and Koppes, L. (1977). Morphological analysis of the division of two *Escherichia coli* substrains during slow growth. *J. Bacteriol.* **131**, 270–279.

Chapter 8

Distribution of Chromosomal Proteins in Polytene Chromosomes of Drosophila

ROBERT F. CLARK, CYNTHIA R. WAGNER,
CAROLYN A. CRAIG, AND SARAH C. R. ELGIN

Department of Biology
Washington University
St. Louis, Missouri 63130

I. Introduction

In recent years, our understanding of gene regulation has been facilitated by the identification and characterization of proteins that interact with DNA (Johnson and McKnight, 1989). Proteins specifically associated with either euchromatin or heterochromatin have also been identified and characterized, increasing our knowledge of differential DNA packaging (Cartwright *et al.,*

METHODS IN CELL BIOLOGY, VOL. 35

1983; Gross and Garrard, 1988). Techniques that allow us to establish the pattern of interaction between a given protein and the genome as a whole are obviously useful for these studies.

In *Drosophila*, many of the chromosomal proteins have been studied using the technique described in this chapter, that of immunolocalization of proteins on polytene chromosomes (Silver and Elgin, 1976; Rodriguez-Alfageme *et al.*, 1976; Jamrich *et al.*, 1977). In the larval stages of *Drosophila melanogaster*, many of the nuclei undergo endoreduplication, successive rounds of DNA replication without concurrent mitosis and cytokinesis. Since the replicated chromatids remain tightly aligned, the result is a set of polytene chromosomes. The salivary gland nuclei of the third instar larvae produce the largest polytene chromosomes, with an octopus-like structure as seen in Figs. 7 and 8: the fused centromeres (the chromocenter) of the four chromosomes form the body, and the arms of the chromosomes form the tentacles. The number of rounds of replication varies among tissues but is generally reproducible for any given tissue. The DNA within the chromosome is not replicated equally: in general, both the α- and β-heterochromatin are under-replicated, with the α-heterochromatin (made up largely of satellite DNA) undergoing few, if any, rounds of replication. In contrast, euchromatic regions can undergo as many as 10 rounds of endoreduplication.

A distinct "banding" pattern is observed for each chromosome arm (except for the Y, which does replicate in polytene nuclei), in large part due to differential packaging. Some of the conspicuous constrictions represent regions of underreplication (Spierer and Spierer, 1984); aside from these few sites, the bands and interbands generally have the same level of polyteny and differ only in the degree of compaction of the DNA (Spierer and Spierer, 1984). A detailed map of each chromosome arm was drawn by Bridges in 1935, and a photographic representation of each arm has also been presented (Lefevre, 1976). More recently, the technique of *in situ* hybridization to the polytene chromosomes has allowed a physical map to be determined, which correlates a specific band (or region within a band) on the chromosome with the location of a specific gene or repetitive sequence (Gall and Pardue, 1969). The techniques described in [1–5], this volume, are used for these localizations.

During the process of endoreduplication, the protein components of the chromosomes are "amplified" along with the DNA. By using an antibody specific for a given protein, this extensive amplification of both DNA and protein allows one to obtain a "total map" of a protein's distribution on the polytene chromosomes. Phase-contrast optics allow the unstained chromosome to be seen, which is a necessity for this technique; ultraviolet optics are used to localize the proteins tagged with fluorescent antibodies. (If a viewer wishes merely to visualize the chromosomes, aceto-orcein stain and a light microscope are adequate.) A microscope with 400 × magnification will allow one to identify individual bands within the polytene chromosome. With the advent of optical sectioning and three-dimensional reconstruction (see [10], this volume), the resolution one can obtain is greater.

The nuclei containing polytene chromosomes are in interphase. Both DNA replication and RNA transcription occur in these nuclei, as demonstrated by the incorporation of radiolabeled precursors (Bonner and Pardue, 1976). One can often readily observe a change in the structure of the polytene chromosome when a region is subject to high levels of transcriptional activity. This phenomenon, known as puffing, is a decondensation of the chromatids at that region. As development proceeds in the third instar larvae, a highly regulated and sequential pattern of puffing activity occurs. The release of ecdysone in late third instar larvae causes a rapid induction of approximately six early puffs. These puffs continue to increase in size for about 4 hours, after which time they regress. After formation of the early puffs, one sees a large number of late puffs appearing (Ashburner, 1972). The regression of the early puffs, as well as the induction of the late puffs is dependent on protein synthesis, unlike the induction of the early puffs, which is unaffected if protein synthesis is blocked. The induction of high levels of transcription at a small set of loci by heat shock is also manifest in the development of puffs at those sites. The "puffing" phenomenon is useful since it allows one to follow the distribution of a specific protein during the process of gene activation at known loci. Both ecdysone and heat shock will induce changes in gene expression of intact glands, allowing one to capture more quickly the induced response.

II. Squashing and Staining Protocols

We present here a technique for determining the *in situ* distribution of chromosomal proteins in polytene chromosomes. Emphasis is placed on a procedure using formaldehyde fixation followed by squashing in acetic acid (Silver and Elgin, 1976; Silver *et al.*, 1977); similar techniques have been developed by others (Rodriguez-Alfageme *et al.*, 1976; Jamrich *et al.*, 1977; Zink and Paro, 1989). This two-step fixation procedure produces polytene chromosome spreads of good morphology and antigen reactivity that retain essentially a full complement of chromosomal proteins. Polytene chromosomes are reacted with antibodies specific for a particular chromosomal protein. A fluorescent or enzyme-linked secondary antibody directed against the primary antibody is then incubated on the chromosomes. Finally, the chromosomes are viewed and photographed with either a fluorescence or bright-field microscope. Using this technique, proteins can reproducibly be localized on polytene chromosomes to the resolution of individual polytene bands.

A. Fixation of Polytene Chromosomes

The ideal polytene chromosomes for immunofluorescent microscopy should have the following characteristics: (1) the chromosome arms should be well spaced and very flat; (2) the chromosomes should exhibit good morphology, with the majority of the bands distinct and easily identifiable; and (3) no nonchromosomal components should be adhering to or lying over the chromosomes. The fixation and squashing procedures described here will, with practice, give chromosomes that reproducibly exhibit these characteristics in at least a portion of the spread.

The best polytene chromosomes are obtained from third instar larvae that have been grown with minimal crowding. The larvae should be fat and just crawling up out of the medium. Salivary glands are dissected from the larva as exhibited in Fig. 1. Larvae are carefully removed from bottles with dissection forceps (Dumont forceps) and washed in microwells in Cohen buffer

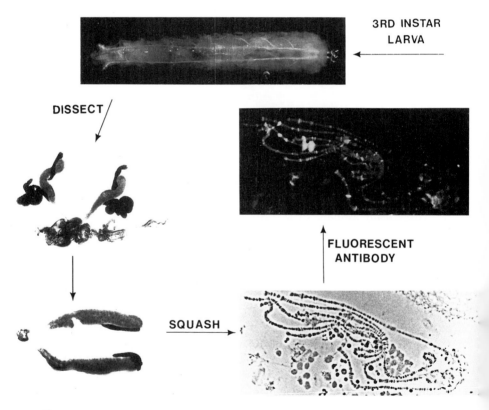

FIG. 1.   Step-by-step method of isolating, squashing, and staining polytene chromosomes from *Drosophila*. [From Elgin (1983) with permission.]

(Cohen and Gotchel, 1971). A larva is then placed in a new microwell, and, using two pairs of dissection forceps, pulled apart while grasping the head and rear of the larva. The salivary glands are then dissected away from the other head parts and the fat body. The paired salivary glands are clear or slightly milky and somewhat cucumber-shaped. They usually have associated cream-colored fragments of fat body. Care should be taken not to damage the salivary glands at this stage. If necessary, adhering fat body can be removed during acetic acid fixation (see below). The excised glands are incubated in Cohen buffer for 8–10 minutes. Pretreatment of glands with this buffered nonionic detergent solution causes the dissolution of cytoplasmic membrane structures. Longer incubations result in poor morphology of the chromosomes. The glands are then transferred to formaldehyde fixative and incubated for 25 minutes. Shorter incubations can be used, but may result in loss of chromosomal proteins during subsequent manipulations, while longer incubations result in chromosomes that are extremely difficult to spread well. The glands are then transferred to 45% acetic acid and incubated for 10–60 minutes. Extraneous tissue (fat body) can be removed using forceps during this incubation. Acetic acid fixation is necessary to achieve good phase morphology of the spread polytene chromosomes. Phase morphology of the chromosomes will deteriorate if they are left any longer than 1 hour. This procedure will be referred to as "the formaldehyde fixation technique." This technique minimizes extraction of chromosomal protein (less than 1%) while the proteins generally retain good antigenicity (Silver and Elgin, 1976). The histones are nicely fixed with this technique; many other proteins require less fixation.

Reagents. All stock solutions should be kept at 4°C unless otherwise indicated in the right-hand column.

	Final concentration	Volume/5 ml	Stock solution
Cohen buffer	10 mM $MgCl_2$	50 µl	1 M
	25 mM sodium glycerol		
	3-phosphate	120 µl	1 M ($-20°$C), pH 7
	3 mM $CaCl_2$	150 µl	0.1 M
	10 mM KH_2PO_4	250 µl	0.2 M
	0.5% NP40	250 µl	10%
	30 mM KCl	750 µl	0.2 M
	160 mM sucrose	1 ml	0.8 M ($-20°$C)
	H_2O	2.43 ml	
Can be kept at 4°C for 2–3 days			
Formaldehyde fixative	0.1 M NaCl	100 µl	5 M
	2 mM KCl	50 µl	0.2 M
	10 mM $NaPO_4$	50 µl	1 M, pH 7
	2% NP40	1 ml	10%
	2% formaldehyde	270 µl	37%
	H_2O	3.55 ml	
Must be made fresh daily. Use 37% formaldehyde stock within 6 months of purchase.			

Conventional squashing techniques without prior formaldehyde fixation result in extraction of most of the histones and approximately 10% of the nonhistone proteins (particularly lysine-rich proteins) from the chromatin (Silver and Elgin, 1976). Occasionally, it may be necessary or desirable to prepare polytene chromosomes that have not been fixed with formaldehyde prior to squashing. Many nonhistone chromosomal proteins are neither extracted nor perturbed by acetic acid. In this protocol, salivary glands are excised in acetic acid/formaldehyde squashing solution (45% acetic acid, 3.7% formaldehyde) and incubated for 10–40 minutes. This solution gives better preservation than 45% acetic acid alone (Holmquist, 1972), even though extraction with acetic acid is much more rapid than is fixation by formaldehyde (Dick and Johns, 1968; Chalkley and Hunter, 1975). Glands fixed in this manner are squashed in the same solution. This "conventional fixation technique" results in polytene chromosomes of excellent band morphology and good antigenicity of many proteins, although many other proteins are lost. An example using this fixation procedure is seen in Fig. 6. The staining pattern observed with a particular antiserum using a particular fixation technique is influenced by the differential extractability and the differential accessibility of the protein in question. The same results are often seen for nonhistone proteins with the two protocols, although this is not always the case. Interpretation of results obtained with a given fixation protocol depends on the amino acid composition (i.e., lysine content) and solubility of the antigen, as well as on the reactivity of the antiserum used.

Acetic Acid/Formaldehyde Squashing Solution

Final concentration	Volume/5 ml	Stock solution
45% Acetic acid	2.25 ml	100%
3.7% Formaldehyde	0.25 ml	37%
H$_2$O	2.25 ml	
Must be made fresh daily.		

B. Squashing of Polytene Chromosomes

The salivary glands are now ready for squashing. A drop of 45% acetic acid is placed in the center of a clean siliconized coverslip [coverslips are siliconized as described in Sambrook *et al.* (1989)]. The fat body and any miscellaneous larval tissue have been carefully removed from the salivary glands during the incubation in acetic acid. The salivary gland is transferred with forceps to the drop of 45% acetic acid on the coverslip. A glass slide is cleaned with 95% ethanol and dried, and the middle of the slide is placed on the coverslip. The salivary gland nuclei are broken open by picking up the slide and by moving the coverslip back and forth by means of gentle tapping of the edges of the slide between the thumb and forefinger. The salivary gland cells and nuclei will split open, and the arms of the chromosomes will spread out. Flipping the slide over so that the coverslip is now on top of the slide, the chromosome spreading is monitored by viewing under a phase-contrast microscope at 400×. If moving the coverslip did not spread the chromosome arms sufficiently, the slide is placed on the bench (coverslip up) and the coverslip is tapped with the eraser end of a pencil. When satisfactory spreading of the chromosome arms has occurred, the chromosomes are flattened by very firmly applying thumb pressure to the coverslip. A folded tissue should be placed between the thumb and the coverslip to prevent grease from getting on the slide. The coverslip must not move under pressure, or stretching and breaking of the chromosome arms will result. Too much pressure will result in fragmented chromosomes. A little practice is necessary to achieve the desired results. Care must also be taken not to allow the squash to dry out, or the subsequent staining of the chromosomes will be poor. The slide is then quickly submerged in liquid nitrogen using hemostatic forceps. When the bubbling stops, the slide is removed from the liquid nitrogen and the coverslip is flicked off with a razor blade. Care must be taken not to scrape the slide and thus remove the chromosomes. Before the specimen thaws, the slide is immediately immersed in Tris-buffered saline (TBS). If the slides are not to be used within 2–4 hours, they can be stored in slide storage medium at −20°C for up to a few weeks.

This squashing procedure, like many other techniques, is easy to learn, but not so easy to master. Everyone improves their results with a few weeks practice. If the specimen is allowed to dry at any time during the procedure, the polytene chromosomes will stain poorly, giving a stain outlining the chromosomes rather than uniform staining of the chromosome bands (Fig. 2). The squashes are most vulnerable after flattening, as the liquid layer is very thin. If the 45% acetic acid begins to recede from the edge of the coverslip, the specimen may be too dry for good staining. Therefore, during the squashing procedure all steps should be performed as quickly as caution allows. A troubleshooting guide is provided at the end of this section to help solve different problems.

Reagents

	Final concentration	Volume	Stock solution
10× TBS	0.2 M Tris-HCl	200 ml	1 M (pH 8)
	17% NaCl	170 g	
	Tween-20 (Sigma)	10 ml	100%
	H_2O	Bring to 1 liter	
Dilute 10-fold in H_2O before use.			
Slide storage medium	67% Glycerol	335 ml	100%
	33% PBS	165 ml	PBS
		500 ml	

One liter of phosphate-buffered saline (PBS) is made with 20 g of NaCl, 0.5 g of KCl, 0.5 g of KH_2PO_4, and 1.45 g of Na_2HPO_4.

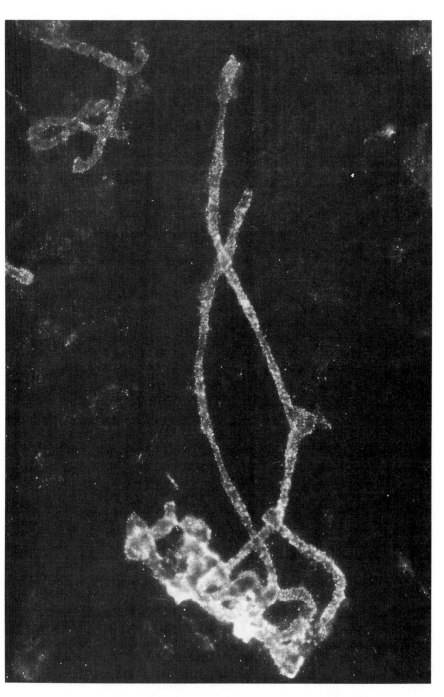

Fig. 2. An example of the chromosomal outlining seen when the slide is allowed to dry or when the chromosomes are refractory (not flat enough). Chromosomes were stained with a mouse monoclonal antibody, C1A9, as the primary antibody, and with fluorescein-conjugated rabbit anti-mouse IgG as the secondary antibody.

C. Antibody Staining Procedure

To localize the positions of chromosomal proteins on the polytene chromosomes, the chromosome spreads are reacted with a primary antibody against a particular protein, followed by reaction with a fluorescent or other (i.e., enzyme-linked) secondary antibody against the primary antibody. Only immunofluorescent labeling will be described in detail here. A variety of other secondary labeling techniques can be adapted for use with this protocol.

The use of fluorescently labeled antibodies to localize antigenic molecules precisely was pioneered by Coons in the early 1940s (Coons *et al.,* 1941, 1942). The indirect staining technique takes advantage of the fact that antibody molecules are themselves capable of acting as antigens for subsequent antibodies. Since 5–10 secondary antibodies can bind to each primary antibody (Goldman, 1968; Sternberger, 1974; Williams and Chase, 1976), the fluorescence of the original signal is much increased. In the procedure described here, polytene chromosome spreads are labeled using the indirect staining technique, essentially as performed in Weller and Coons (1954), with only minor modifications. This procedure was originally developed and described by Silver and Elgin (1976, 1978).

All antibodies used for immunofluorescent microscopy of polytene chromosomes must be evaluated by several criteria. First, both primary and secondary antibodies must be titered to correct dilutions. Too high a concentration of either may lead to extensive nonspecific labeling, while too high a dilution will result in loss of signal. We typically use dilutions ranging from 1 : 50 to 1 : 5000. Second, the appropriate controls must be carried out. Many animals (i.e., rabbits, mice, etc.) carry antibodies that react with chromosomal proteins in their preimmune serum. Virtually all sera will carry antibodies that react with the chromosomes generally, and some will carry antibodies that give very specific patterns. An example is given in Fig. 3. Fortunately, nonspecific antibodies are frequently of low titer, and the background signal will be lost on dilution. This must be established for every animal used. Both polyclonal and monoclonal antibodies should be tested for specific binding to *Drosophila* chromosomal proteins by techniques such as Western blotting or immunoprecipitation to establish their specificity. However, even if the antibody under study does not label nuclear proteins on Western blots, it may still react to polytene chromosomes, and vice versa. A control without primary antibody should also be used to check every batch of secondary antibodies, as these may label polytene chromosomes nonspecifically.

To obtain immunofluorescent labeling of polytene chromosomes, the following protocol has been used. Slides with polytene chromosome squashes (used immediately or removed from storage) are washed with gentle agitation twice for 7 minutes each in cold TBS. Antiserum (or monoclonal antibody)

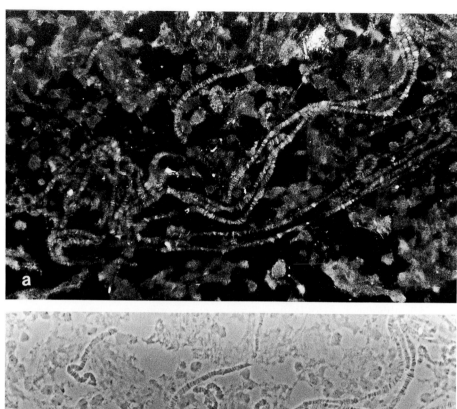

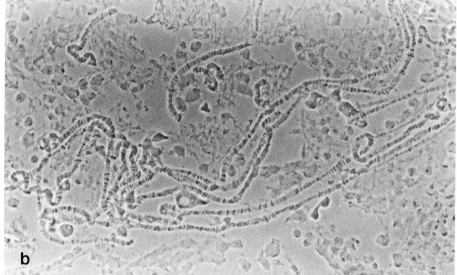

FIG. 3.   An example of cytoplasmic fluorescent staining due to nonspecific binding. Chromosomes were stained with a rabbit preimmune serum as the primary antibody and with fluorescein-conjugated goat anti-rabbit IgG as the secondary antibody. (a) Phase contrast; (b) fluorescence. [From Silver and Elgin (1978) with permission.]

specific for the chromosomal protein of interest is diluted appropriately in TBS. The addition of blocking agents to the solution can be helpful in reducing nonspecific labeling. Fetal calf or goat serum (10%), 5 mg/ml γ-globulin, and/or 5% nonfat milk (all final concentrations) can significantly reduce background without altering antibody labeling patterns. The bottom and top of the slide (except for the squash area) are quickly dried, and the slide is placed horizontally in a humidity chamber. The humidity chambers we use are plastic boxes having a lid with a tight seal. Wet paper towels are placed at the bottom; glass rods on clay holders are positioned to keep the slides above the towels and perfectly horizontal. One hundred to 200 μl of diluted antiserum (with blocking agents as needed) is then applied immediately to the squash area. The squash must be kept wet at all times. The slides are incubated with the primary antibodies for 15 to 120 minutes at room temperature.

After incubation of the slide with primary antiserum, the antiserum is rinsed from the slide with cold TBS in a squirt bottle, and the slide is then washed in cold TBS with gentle agitation in a rack three times for 5 minutes each. The slides are wiped dry except for the squash area and placed back in the humidity chamber. One hundred to 200 μl of labeled secondary antibody, diluted as appropriate in TBS (with blocking agents as needed), is applied to the squash area on the slide, and the slide is incubated for 15–120 minutes at room temperature as described above. The slides must be kept away from bright light when reacted with fluorescent antibody, or the fluorescent markers will fade. This slides are then washed with cold TBS as before, and wiped dry except for the squash area. One drop of mounting solution is immediately placed on the squash, and a nonsiliconized coverslip is placed over it. Excess solution is blotted from the slide and, using a tissue to prevent fingerprints, the coverslip is pressed hard onto the squash without letting the coverslip slide. After excess solution is again blotted from the slide, clear fingernail polish is used to seal the squash and prevent drying of the specimen. The polytene chromosome squash is now ready for viewing and photography using the appropriate fluorescence microscope. We generally use a Leitz orthoplan microscope at 600 $\times$ with an epifluorescence attachment and exposure meter built in, and photograph using 35 mm black and white Tri-X pan film (Kodak). It is generally useful to include some nonchromosomal debris in the picture to establish background levels of staining.

Mounting Solution

Final concentration	Volume/100 ml	Stock solution
90% Glycerol	90 ml	100%
0.1 M Tris-HCl	10 ml	1 M (pH 7)
0.2% n-Propyl gallate	0.2 g	

Pulverize *n*-propyl gallate with a mortar and pestle before weighing, and allow it to dissolve by stirring overnight in the solution. *n*-Propyl gallate inhibits the loss of fluorescence during viewing.

D.   Troubleshooting Guide

In this section, a troubleshooting guide is given. For each problem given here, there is at least one possible explanation for the problem and a means to help correct it.

A. *Small polytene chromosomes*
Small and/or overcrowded larvae, or larvae harvested prior to last round of replication.
Correction: Harvest larvae at the time when they are the fattest and crawling up the side of the bottle.

B. *Poor chromosome morphology (i.e., indistinct bands under phase contrast, smeared staining pattern)*
Incubation too long in Cohen buffer, incubation not long enough in formaldehyde fixative, or incubation too long in 45% acetic acid.
Correction: Make sure incubation periods are correct, or try to vary them according to the properties of the protein under study.

C. *Polytene chromosomes don't spread well*
1. Incubation too long in formaldehyde fixative.
   Correction: Incubate in formaldehyde fixative for 25 minutes or less.
2. Spreading incomplete.
   Correction: Constantly monitor squashing using a phase-contrast microscope; tap repeatedly with the pencil eraser until chromosome arms are well spread.
3. Squashed too gently.
   Correction: Apply more pressure in final flattening.

D. *Too much background material*
1. Poor washing of glands in Cohen buffer.
   Correction: Incubate glands in buffer for at least 8 minutes.
2. Dirty coverslips and/or slides.
   Correction: Clean all glass with 95% ethanol, dry with lens paper and/or tissue, handle glass at edges, and squash with tissue.

E. *Broken polytene chromosomes*
1. Squashed too hard or allowed coverslip to slide.
   Correction: Constantly monitor squashing under phase-contrast microscope, and do not attempt to further spread well-spread chromosomes. Hold coverslip while tapping with eraser to prevent sliding.

2. Incubation not long enough in formaldehyde fixative, or too long in 45% acetic acid.
   Correction: See B1.

F. *Refractile polytene chromosomes*
   1. Squashed too gently.
      Correction: See C3.
   2. Allowed chromosomes to dry.
      Correction: Keep squash area wet at all times. Use a small humidifier on the bench during squashing steps if air is very dry, as can occur during freezing winter weather.
   3. Allowed chromosomes to thaw in air after liquid nitrogen freezing.
      Correction: Immerse slides immediately in TBS after specimen is fully frozen.

G. *Staining pattern outlines the chromosome*
   Allowed chromosomes to dry.
   Correction: See F2.

H. *Stretched-out chromosomes*
   1. *Allowed coverslip to slide during flattening.*
      Correction: See E1.
   2. Inadequate formaldehyde fixation.
      Correction: See B1.

I. *No intact chromosomes in squash area*
   1. Squashed gland floated out from under the coverslip.
      Correction: Use less 45% acetic acid on coverslip.
   2. Squashed too hard or allowed coverslip to slide during squashing.
      Correction: See E1.
   3. Can't find chromosomes in squash area.
      Correction: Mark position of squashed gland on the edges of the slide with a waterproof marker.
   4. Scraped off chromosomes with razor blade.
      Correction: Only flick the edge of the coverslip with the razor blade after freezing the slide.
   5. Chromosomes fell off after frozen on slide.
      Correction: Keep slides in slide storage medium if not used within 2–4 hours.

J. *High levels of labeling of nonchromosomal debris*
   1. Antiserum contains antibodies that cross-react with cellular components other than the desired protein.
      Correction: Affinity purify or preabsorb the antiserum.

2. No blocking agents.
   Correction: Add 10% calf or goat serum, 5 mg/ml γ-globulin, and/or 5% nonfat milk to both primary and secondary antibodies.
3. Antibody dried onto chromosome squash.
   Correction: Use humidity chamber, make sure slides are perfectly level, and use 200 μl of antibody solutions.
4. Titer of antibody too high.
   Correction: Decrease antibody titer.

K. *Weak or absent signal on chromosomes*
1. Antibody doesn't react to antigen on chromosomes.
   Correction: Try variations in fixation protocol. If antigens denatured by acid, can use the technique described in Hill *et al.* (1989).
2. Overfixation in formaldehyde.
   Correction: See C1.
3. Faded fluorescent label.
   Correction: After reaction to fluorescent-conjugated antibody, keep slides in the dark as much as possible. *n*-Propyl gallate is used in the mounting medium to prevent bleaching.
4. Titer of antibody too low.
   Correction: Increase antibody titer.

III. Applications and Results

Although the primary use of the immunofluorescent localization technique is to map the distribution of one protein on the polytene chromosomes, there are many extensions of this technique. The two variations discussed here are the immunofluorescent localization of more than one protein simultaneously and immunofluorescent localization of proteins in whole-mount salivary glands. We then provide two examples of studies where the determination of protein distribution patterns has been particularly helpful in analyzing the function of a chromosomal protein.

A. Double-Labeling Technique

It is often useful to determine to what degree the patterns of different proteins superimpose. To determine the *in situ* locations of more than one protein simultaneously on polytene chromosomes, two criteria must be met: (1) multiple secondary antibodies with different fluorochromes (i.e., fluorescein-conjugated, rhodamine-conjugated) must be available, and (2) to prevent cross-reaction of these secondary antibodies, the primary antibodies must either have been made in different species (i.e., rabbit, mouse, rat, goat) or have been modified differently so that the appropriate secondary antibodies recognize this modification (e.g., biotinylation). The indirect immunofluorescent staining method is the same, except that multiple primary antibodies are reacted to the chromosomes squashes at the same time and, following washing, multiple secondary antibodies are then reacted to these squashes. Using different filters on the immunofluorescence microscope, the labeling patterns of the different antibodies are discerned.

An example of this technique is shown in Fleischmann *et al.* (1987). To localize RNA polymerase II and a 69-kDa nonhistone chromosomal protein (P69) simultaneously on polytene chromosomes, a rabbit anti-RNA polymerase II antibody and a mouse anti-P69 antibody were reacted to polytene squashes, followed by reaction of these squashes with a fluorescein-conjugated goat anti-rabbit antibody and a Texas Red-conjugated goat anti-mouse antibody. When using the appropriate Leitz filters during immunofluorescent microscopy, the location of RNA polymerase II fluoresced green, while that of P69 fluoresced red (Fig. 4). The results show that P69 is associated with the inactive loci, those not associated with RNA polymerase II. Little cross-reaction was evidenced here. Many such combinations of primary and secondary antibodies are possible.

B. Immunofluorescence of Whole-Mount Salivary Glands

Another variation of the immunofluorescent staining of polytene chromosomes is the immunofluorescent labeling of whole-mount salivary glands (James *et al.*, 1989). Since the localization of proteins on chromosomes in an intact salivary gland is necessary to provide three-dimensional information, a procedure for the staining of intact salivary glands has been developed. Intact glands are dissected into Cohen buffer and sequentially fixed with the formaldehyde fixative and 45% acetic acid for the same times as previously described (see Section II,A). After dissection, glands are moved from one solution to the next with forceps. The fixed glands are twice rinsed with TBS in a microfuge tube with the bottom cut out and replaced with nylon mesh. The glands are then incubated in a microwell in a 500-μl solution of primary antiserum for 1 hour at room temperature. The glands are then washed twice again in TBS, incubated in 500 μl of fluorescent-conjugated secondary antibody for 30 minutes at room temperature, and washed twice again in TBS. The glands are then mounted as previously described, but without any squashing. The polytene nuclei so obtained are intact, with the chromosomes tightly packaged (see Fig. 5). They retain their *in vivo* structure with the chromosomes polarized in "Rabl" orientation, i.e., with the chromocenter oriented toward

63 BC 67B 68C 70B 71EF 74EF 75B

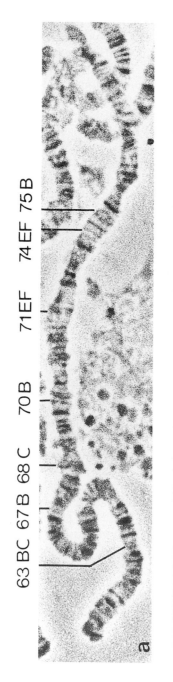

63 BC 67B 68C 70B 71EF 74EF 75B

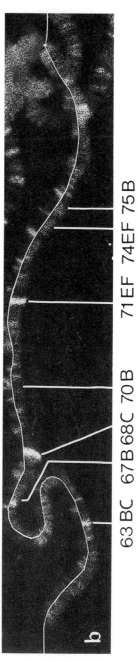

Fig. 4. Polytene chromosome arm 3L, double labeled with antibodies against P69 and RNA polymerase II. (a) Phase contrast; (b) fluorescence. The distribution pattern of P69 is shown on the upper half of the chromosome and the distribution pattern of RNA polymerase II is seen on the lower half of the chromosome. [From Fleischmann *et al.* (1987) with permission.]

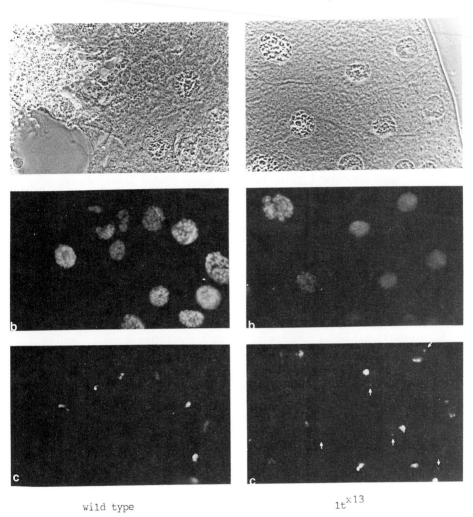

wild type   1t[x13]

FIG. 5. Immunofluorescent staining pattern using a monoclonal antibody against HP1 in intact third instar larval salivary glands from wild-type and T(2; 3)It[x13] stocks. The rearrangement removes a small fragment of heterochromatin from the chromocenter and places it near the end of chromosome arm 3R (see Fig. 8). (a) Phase contrast; (b) staining with DAPI; (c) fluorescence. [From James *et al.* (1989) with permission.]

one side of the outer surface of the nuclei and the telomeres oriented toward the other side. (Ellison and Howard, 1981; Foe and Alberts, 1985). For additional information on the three-dimensional analysis of polytene nuclei, see [10], this volume.

C. Topoisomerase I

Topoisomerase I is an enzymatic nonhistone chromosomal protein that introduces a transient single-stranded break in the DNA, changing the linking number by one. This "nicking" of the DNA is thought to be necessary in order to relieve torsional strain on the DNA as transcription occurs. Antibodies against topoisomerase I were used to determine the distribution patterns of this protein on the polytene chromosomes before and after heat shock. (Fig. 6). The protein is widely distributed at discrete loci throughout the length of the chromosome arm. After heat shock, one sees an accumulation of topoisomerase I at the heat-shock loci (puffs), strongly suggesting that topoisomerase I is associated with actively transcribing regions of the genome. The contrast in staining before and after heat shock suggests that topoisomerase I is recruited to these loci as part of the activation process. The nucleolus also shows high concentrations of topoisomerase I; this is evident before and after heat shock, as ribosomal DNA transcription continues under these conditions (see Fleischmann *et al,* 1984). The results shown here are qualitative; however, using the method of Rykowski ([10], this volume), quantitative results could be obtained.

Subsequent analyses used camptothecin, an inhibitor which stabilizes the topoisomerase I–DNA covalent intermediate that forms during relaxation of torsionally strained DNA. This allows the sites of topoisomerase I interaction along the chromatin fiber to be mapped, and has shown that topoisomerase I nicks the DNA primarily within the actively transcribed region (Gilmour and Elgin, 1987; Stewart and Schutz, 1987). Again, an abrupt transition is observed. Little or no topoisomerase I nicking is seen when the gene is inactive, but extensive nicking is seen after heat shock. Genetic analysis of topoisomerase I mutants in yeast has shown that the gene is not essential (Uemura and Yanagida, 1984); topoisomerase II apparently can substitute for topoisomerase I. In a double mutant (containing a temperature-sensitive allele of topoisomerase II), transcription from ribosomal DNA is clearly reduced (Brill *et al,* 1987). Thus, the results originally obtained from a cytological observation have been fully confirmed by the higher resolution biochemical studies and genetic analysis subsequently performed.

D. Heterochromatin Protein 1

Another excellent example of the use of immunofluorescence microscopy of polytene chromosomes to help determine the function of an unknown protein

---

FIG. 6. Distribution of topoisomerase I on non-heat-shocked and heat-shocked chromosome arms 3R and 3L. Chromosomes were obtained from a third instar larva grown at 25°C (a and d) and from a larva heat-shocked at 37°C for 20 minutes prior to dissection (b, c, e, and f). (a–c) Chromosomal arm 3L; (d–f) chromosomal arm 3R; (a, b, d, and e) fluorescent images; (c and f) phase-contrast images of b and e. [From Fleischmann *et al.* (1984) with permission.]

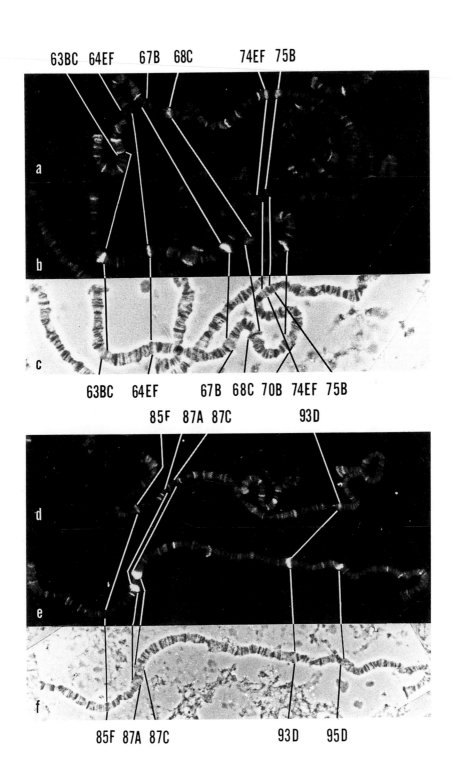

63BC 64EF 67B 68C 74EF 75B

63BC 64EF 67B 68C 70B 74EF 75B

85F 87A 87C 93D

85F 87A 87C 93D 95D

comes from studies of heterochromatin protein 1 (HP1). A 1–2 *M* potassium isothiocyanate protein extract of *Drosophila melanogaster* embryonic nuclei that contained a 19-kDa protein was used to produce monoclonal antibodies (James and Elgin, 1986). Four different monoclonal antibodies were obtained that recognized a protein highly concentrated at the chromocenter (Fig. 7). Since the chromocenter is composed of the α- and β-centric heterochromatin of the four chromosomes, the antigen has been designated HP1, as it is predominantly associated with heterochromatin (James *et al.*, 1989). The genomic and cDNA clones encoding this protein have been sequenced, and the gene has been mapped to cytological position 29A (James and Elgin, 1986; Eissenberg *et al.*, 1990), a region where a dominant suppressor of position-effect variegation had earlier been mapped (Sinclair *et al.*, 1983). Position-effect variegation occurs when a gene normally packaged in euchromatin is placed next to heterochromatin; that gene becomes inappropriately repressed

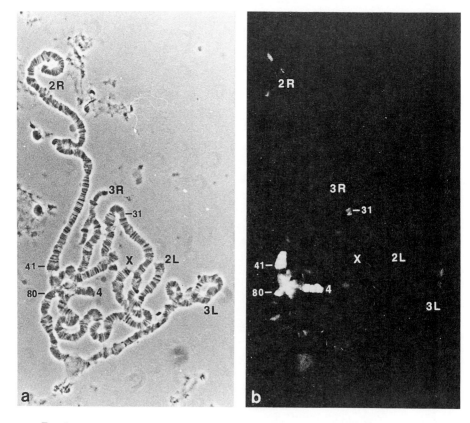

FIG. 7.   Distribution pattern of HP1 on polytene chromosomes. Note the intense staining of the chromocenter, the fourth chromosome, and regions 31, 41, and 80. (a) Phase contrast; (b) fluorescence. [From James *et al.* (1989) with permission.]

in some cells of a tissue, giving rise to a mosaic or variegated expression pattern (for review, see Spofford, 1976). Mutations that give rise to a dominant suppression of position-effect variegation are candidates to be structural proteins of chromatin or their modifiers (for reviews, see Tartof *et al.*, 1989; Eissenberg, 1989). Characterization of known suppressor of variegation [Su(var)] mutations at this locus has confirmed that a mutation in the HP1 gene results in suppression of position-effect variegation (Eissenberg *et al.*, 1990); HP1 therefore must contribute to heterochromatin structure.

The distribution pattern of HP1 is shown in Fig. 7. The chromocenter is the most strongly staining region. Stain extends into all of the banded region 41 at the base of chromosome 2R and all of the banded interval 80 at the base of 3L. The telomeres frequently exhibit staining, especially notable at 100EF at the end of 3R, and at the ends of X and 2R. Although a few sites within the chromosome arms may exhibit weaker staining, region 31 on 2L is consistently strongly stained. This region may be of special interest since it contains several Su(var) loci (Sinclair *et al.*, 1983). For the most part, however, the basis for the presence of HP1 in the chromosomal arms is unknown; there is no significant overlap between the staining regions in the arms and the sites of intercalary heterochromatin (Zhimulev *et al.*, 1982; Bolshakov *et al.*, 1985). Staining of polytene chromosomes with a polyclonal antibody prepared against a C-terminal peptide derived from the HP1 gene sequence shows a similar pattern of labeling to that of the monoclonal antibodies (R. F. Clark, C. Craig, and S. C. R. Elgin, unpublished observations).

Another prominent site of labeling is the fourth chromosome, which displays a banded pattern throughout its polytenized regions. In order to test the autonomous nature of this staining, immunofluorescent microscopy was performed on a translocation mutant [T(3; 4)f/In(3L)P], where a region of approximately seven bands of the fourth chromosome is translocated to position 65D1-2 on chromosome 3. This translocation appears as an unpaired "buckle" within the inversion loop in polytene chromosome squashes, and is stained by the monoclonal antibody C1A9, showing that the association of the fourth chromosome with HP1 is not dependent on its proximity to the chromocenter (James *et al.*, 1989).

The application of immunofluorescent microscopy of polytene chromosome spreads is again put to good use when examining rearrangements of the heterochromatin to determine whether the association of HP1 with heterochromatin is autonomous. In the mutant T(2; 3)It$^{\times 13}$, chromosome arm 2L is broken within the heterochromatin at 2LH37, between the *light* gene and the chromocenter, and joined to chromosome arm 3R at map position 97D2 (Wakimoto and Hearn, 1990). In Fig. 8 one observes that the heterochromatic portion of 2L translocated to 3R is strongly stained with the monoclonal antibody C1A9 (James *et al.*, 1989). The staining of the translocated heterochromatin is also seen with immunofluorescent microscopy on whole-mount

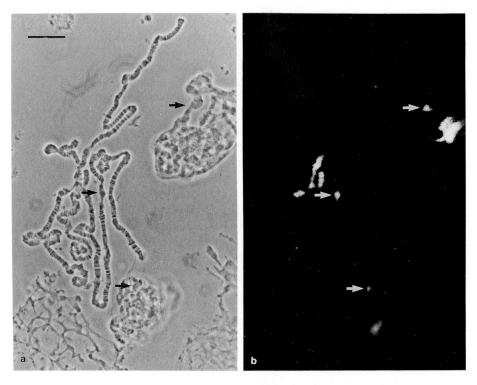

FIG. 8.   Distribution pattern of HP1 using the chromosomal rearrangement T(2; 3)It[×13]. Staining of the β-heterochromatin at the junction of 2L and 3R (arrows), distant from the chromocenter, is evident in both the spread chromosomes and the relatively intact nuclei. (a) Phase contrast; (b) fluorescence. [From James *et al.* (1989) with permission.]

salivary glands (Fig. 5). These results strongly support the fact that HP1 associates with heterochromatin irrespective of its proximity to the chromocenter.

The distribution pattern of HP1 protein does not mimic that of any known satellite DNA. The HP1 protein pattern does approximate the *in situ* hybridization pattern of clone Dr. D, a member of a family of middle repetitious sequences characterized by Miklos *et al.* (1988). Dr. D hybridizes extensively with the chromocenter of polytene chromosomes, in addition to the fourth chromosome and a few uncharacterized sites in the chromosome arms. However, even though the distribution patterns of HP1 and Dr. D are similar, there is no evidence as yet that HP1 is a DNA-binding protein. HP1 does share extensive homology to the Polycomb protein, a negative regulator of many homeotic genes (Zink and Paro, 1989), in a region which we designate as the "HP1/Pc box" (Paro and Hogness, 1991). Distribution patterns of HP1 and Polycomb are completely nonoverlapping, suggesting that the common box denotes a common functional domain unrelated to distribution signals. The possibility that this domain is a conserved element utilized in generating highly condensed chromatin structures is under further investigation.

In conclusion, a technique for determining the *in situ* distribution of chromosomal proteins in polytene chromosomes has been presented here. A few applications of this technique have been given in detail. Virtually any *Drosophila* chromosomal protein can be localized to specific chromosomal loci using this method or its variants. In addition, using the method described by Rykowski ([10], this volume), chromosomal proteins can be localized to polytene chomosomal bands with even higher resolution. Eventually, combining genetic, biochemical, and molecular biology techniques, the complete three-dimensional structure of the polytene chromosomes will be discerned.

ACKNOWLEDGMENTS

We thank the Elgin laboratory for critical reading of the manuscript, and Gerhard Fleischmann for the double-labeling photograph. This work was supported by NIH grant GM 31532 and HD 23844 to S. C. R. Elgin, and NIH grant GM 13347 to R. F. Clark.

REFERENCES

Ashburner, M. (1972). Puffing patterns in *Drosophila melanogaster* and related species. *In* "Results and Problems in Cell Differentiation, Volume 4, Developmental Studies on Giant Chromosomes" (W. Beermann, ed.), pp. 101–151. Springer, New York.

Bolshakov, V. N., Zharkikh, A. A., and Zhimulev, I. F. (1985). Intercalary heterochromatin in *Drosophila*. II. Heterochromatic features in relation to local DNA content along the polytene chromosomes of *Drosophila melanogaster*. *Chromosoma* **92,** 200–208.

Bonner, J. J., and Pardue, M. L. (1976). The effect of heat shock on RNA synthesis in *Drosophila* tissues. *Cell (Cambridge, Mass.)* **8,** 43–50.

Brill, S. J., DiNardo, S., Voelkel-Meiman, K., and Sternglanz, R. (1987). Need for DNA topoisomerase activity as a swivel for DNA replication for transcription of ribosomal RNA. *Nature (London)* **326,** 414–416.

Cartwright, I. L., Keene, M. A., Howard, G. C., Abmayr, S. M., Fleischmann, G., Lowenhaupt, K., and Elgin, S. C. R. (1983). Chromatin structure and gene activity: The role of nonhistone chromosomal proteins. *Crit. Rev. Biochem.* **13,** 1–86.

Chalkley, R., and Hunter, C. (1975). Histone-histone propinquity by aldehyde fixation of chromatin. *Proc. Natl. Acad. Sci. U.S.A.* **72,** 1304–1308.

Cohen, L. H., and Gotchel, B. V. (1971). Histones of polytene and non polytene nuclei of *Drosophila melanogaster*. *J. Biol. Chem.* **246,** 1841–1848.

Coons, A. H., Creech, H. J., and Jones, R. N. (1941). Immunological properties of an antibody containing a fluorescent group. *Proc. Soc. Exp. Biol. Med.* **47,** 200–202.

Coons, A. H., Creech, H. J., Jones, R. N., and Berliner, E. (1942). The demonstration of pneumococcal antigen in tissues by the use of fluorescent antibody. *J. Immunol.* **45,** 159–170.

Dick, C., and Johns, E. W. (1968). The effect of two acetic acid containing fixatives on the histone content of calf thymus deoxyribonucleoprotein and calf thymus tissue. *Exp. Cell Res.* **51,** 626–632.

Eissenberg, J. C. (1989). Position effect variegation in *Drosophila*: Towards a genetics of chromatin assembly. *BioEssays* **2,** 14–17.

Eissenberg, J. C., James, T. C., Foster-Hartnett, D. M., Hartnett, T., Ngan, V., and Elgin, S. C. R. (1990). Mutation in a heterochromatin-specific chromosomal protein is associated with suppression of position-effect variegation in *Drosophila melanogaster. Proc. Natl. Acad. Sci. U.S.A.* **87,** 9923–9927.

Elgin, S. C. R. (1983). Chromatin structure and gene expression in *Drosophila. Anatom. Rec., (Suppl.)* **1,** 103–143.

Ellison, J. R., and Howard, G. C. (1981). Non-random position of the A-T rich DNA sequences in early embryos of *Drosophila virilis. Chromosoma* **83,** 555–561.

Fleischmann, G., Pflugfelder, G., Steiner, E. K., Javaherian, K., Howard, G. C., Wang, J. C., and Elgin, S. C. R. (1984). *Drosophila* DNA topoisomerase I is associated with transcriptionally active regions of the genome. *Proc. Natl. Acad. U.S.A.* **81,** 6958–6962.

Fleischmann, G., Filipski, R., and Elgin, S. C. R. (1987). Isolation and distribution of a *Drosophila* protein preferentially associated with inactive regions of the genome. *Chromosoma* **96,** 83–90.

Foe, V. E., and Alberts, B. M. (1985). Reversible chromosome condensation induced in *Drosophila* embryos by anoxia; visualization of the interphase nuclear organization. *J. Cell Biol.* **100,** 1623–1636.

Gall, J., and Pardue, M. L. (1969). Formation and detection of RNA-DNA hybrids in cytological preparations. *Proc. Natl. Acad. Sci. U.S.A.* **63,** 378–383.

Gilmour, D. S., and Elgin, S. C. R. (1987). Localization of specific topoisomerase I interactions within the transcribed region of active heat shock genes by using the inhibitor camptothecin. *Mol. Cell. Biol.* **7,** 141–148.

Goldman, M. (1968). "Fluorescent Antibody Methods." Academic Press, New York.

Gross, D. S., and Garrard, W. T. (1988). Nuclease hypersensitive sites in chromatin. *Annu. Rev. Biochem.* **57,** 159–197.

Hill, R. J., Watt, F., Wilson, C. M., Fifis, T., Underwood, P. A., Tribbick, G., Geysen, H. M., and Thomas, J. O. (1989). Bands, interbands and puffs in native *Drosophila* polytene chromosomes are recognized by a monoclonal antibody to an epitope in the carboxy-terminal tail of histone H1. *Chromosoma* **98,** 411–421.

Holmquist, G. (1972). Transcription rates of individual polytene chromosome bands: Effects of gene dose and sex in *Drosophila. Chromosoma* **36,** 413–452.

James, T. C., and Elgin, S. C. R. (1986). Identification of a nonhistone chromosomal protein associated with heterochromatin in *Drosophila melanogaster* and its gene. *Mol. Cell. Biol.* **6,** 3862–3872.

James, T. C., Eissenberg, J. C., Craig, C., Dietrich, V., Hobson, A., and Elgin, S. C. R. (1989). Distribution patterns of HP1, a heterochromatin-associated nonhistone chromosomal protein of *Drosophila. Eur. J. Cell Biol.* **50,** 170–180.

Jamrich, M., Greenleaf, A. L., and Bautz, E. K. F. (1977). Localization of RNA polymerase in polytene chromosomes of *Drosophila melanogaster. Proc. Natl. Acad. Sci. U.S.A.* **74,** 2079–2083.

Johnson, P. F., and McKnight, S. L. (1989). Eukaryotic transcriptional regulatory proteins. *Annu. Rev. Biochem.* **58,** 799–839.

Lefevre, G. (1976). A photographic representation and interpretation of the polytene chromosomes of *Drosophila melanogaster* salivary glands. *In* "The Genetics and Biology of *Drosophila*" (M. Ashburner and E. Novitski, eds.), Vol. 2a, pp. 31–66. Academic Press, London.

Miklos, G. L. G., Yamamoto, M., Davies, J., and Pirrotta, V. (1988). Microcloning reveals a high frequency of repetitive sequences characteristic of chromosome 4 and the β-heterochromatin of *Drosophila melanogaster. Proc. Natl. Acad. Sci. U.S.A.* **85,** 2051–2055.

Paro, R., and Hogness, D. S. (1991). The polycomb protein shares a homologous domain with a heterochromatin-associated protein of *Drosophila*. *Proc. Natl. Acad. Sci. U.S.A.* **88**, 263–267.

Rodriguez-Alfageme, C., Rudkin, G. T., and Cohen, L. H. (1976). Locations of chromosomal proteins in polytene chromosomes. *Proc. Natl. Acad. Sci. U.S.A.* **73**, 2038–2042.

Sambrook, J., Fritsch, E. F., and Maniatis, T. (1989). "Molecular Cloning: A Laboratory Manual." Cold Spring Harbor Laboratory, Cold Spring Harbor, New York.

Sliver, L. M., and Elgin, S. C. R. (1976). A method for determination of the *in situ* distribution of chromosomal proteins. *Proc. Natl. Acad. Sci. U.S.A.* **73**, 423–427.

Silver, L. M., and Elgin, S. C. R. (1978). Immunological Analysis of Protein Distributions in *Drosophila* Polytene Chromosomes. *In* "The Cell Nucleus" (H. Busch, ed.), Vol. 5, pp. 215–262. Academic Press, New York.

Silver, L. M., Wu, C. E. C., and Elgin, S. C. R. (1977). Immunofluorescent techniques in the analysis of chromosomal proteins. *In* "Methods in Chromosomal Protein Research" (G. Stein, J. Stein, and L. Kleinsmith, eds.), pp. 151–168. Academic Press, New York.

Sinclair, D. A. R., Mottus, R. C., and Grigliatti, T. A. (1983). Genes which suppress position effect variegation in *D. melanogaster* are clustered. *Mol. Gen. Genet.* **191**, 326–333.

Spierer, A., and Spierer, P. (1984). Similar levels of polyteny in bands and interbands of *Drosophila* giant chromosomes. *Nature (London)* **307**, 176–178.

Spofford, J. B. (1976). Position-effect variegation in *Drosophila*. *In* "The Genetics and Biology of *Drosophila*" (M. Ashburner and E. Novitski, eds.), Vol. 2A, pp. 955–1018. Academic Press, Orlando, Florida.

Sternberger, L. A. (1974). "Immunocytochemistry." Prentice-Hall, Englewood Cliffs, New Jersey.

Stewart, A. F., and Schutz, G. (1987). Camptothecin-induced *in vivo* topoisomerase cleavages in the transcriptionally active tyrosine aminotransferase gene. *Cell (Cambridge, Mass.)* **50**, 1109–1117.

Tartof, K. D., Bishop, C., Jones, M., Hobbs, C. A., and Locke, J. (1989). Towards an understanding of position effect variegation. *Dev. Genet.* **10**, 162–176.

Uemura, T., and Yanagida, M. (1984). Isolation of type I and II DNA topoisomerase mutants from fission yeast: Single and double mutants show different phenotypes in cell growth and chromatin organization. *EMBO J.* **3**, 1737–1744.

Wakimoto, B. T., and Hearn, M. G. (1990). The effects of chromosome rearrangements on the expression of heterochromatic genes in chromosome 2L of *Drosophila melanogaster*. *Genetics* **125**, 141–154.

Weller, T. H., and Coons, A. H. (1954). Fluorescent antibody studies with agents of Varicella and Herpes zoster *in vitro*. *Proc. Soc. Exp. Biol. Med.* **86**, 789–794.

Williams, C. A., and Chase, M. W. (1976). Antigen-antibody reactions *in vitro*. *In* "Methods in Immunology and Immunocytochemistry" Vol. 5. Academic Press, New York.

Zhimulev, I. F., Semeshin, V. F., Kulichkov, V., and Belyaeva, E. S. (1982). Intercalary heterochromatin in *Drosophila*. I. Localization and general characteristics. *Chromosoma* **87**, 197–228.

Zink, B., and Paro, R. (1989). *In vivo* binding pattern of a trans-regulator of homoeotic genes in *Drosophila melanogaster*. *Nature (London)* **337**, 468–471.

Chapter 9

The Use of Monoclonal Antibody Libraries

H. SAUMWEBER

Universität zu Köln
Institut für Entwicklungsphysiologie
D-5000 Köln 41, Germany

I. Monoclonal Antibodies as Versatile Tools for Dissection of Nuclear Structure

The extraordinary structural and functional complexity of cell nuclei has taxed the ingenuity of chromatin researchers over the past 50 years. Two-dimensional gel electrophoretic analysis of nuclear proteins reveals several hundred polypeptides varying in abundance (e.g., Peterson and McConkey, 1976), and of course innumerable minor protein components will be missed

METHODS IN CELL BIOLOGY, VOL. 35

by this type of analysis. Given this level of complexity, it would be impossible to assign functions to more than a small number of these proteins by classical biochemical techniques. Thus, alternative approaches are required. The full power of genetic in the functional analysis of structural and regulatory nuclear proteins has become evident during the past decade, particularly in yeast and *Drosophila* (see Part V, this volume). On the other hand, *in vitro* transcription systems, in combination with a variety of *in vivo* and *in vitro* binding assays, have provided sensitive tests for the study of transcription factors from various species and tissues (reviewed by Johnson and McKnight, 1989). However, using both approaches only a certain fraction of nuclear proteins may be identified. Such screens for transcription factors and for proteins binding to specific DNA sequences may well fail to detect important factors, whose function depends on a more complex chromatin architecture. The fact that soluble extracts polymerize correctly initiated transcripts at about 10 nucleotides/hour, which is perhaps 0.01% of the transcription rate observed *in vivo* (Manley *et al.* 1980), indeed implies that functionally important proteins are not detected by current assays. The productive use of genetic approaches on the other hand depends on phenotypes that can be conveniently scored in screens for mutations. Mutant phenotypes for most nuclear proteins are *a priori* unknown or unpredictable; in the case of redundant proteins with similar function, mutations may not provoke any phenotypic change.

An alternative approach uses monoclonal antibody libraries to dissect molecularly the cell nucleus. The immune systems of small mammals are capable of producing more than 10^7 different antibody molecules, each with different ligand-binding properties. Clones of B lymphocytes, whose size and number are subject to a complex network of controls (Jerne, 1976), each produce a single kind of antibody. Expansion of particular clones may be stimulated by introducing as few as 10^{11} foreign molecules (i.e., 1 ng of a 20,000 MW polypeptide) into the system (Luben *et al.,* 1982). Provided that antibody-producing clones can be induced with sufficient frequency, the hybridoma technology (Köhler and Milstein, 1975) should allow one to propagate any desired clone producing a single monoclonal antibody indefinitely outside its original progenitor. This is done by fusion of spleen cells or cells from other lymphatic organs with one of the available myeloma cell lines. Repeated cloning of the resulting hybridomas allows one to establish permanent lines that secrete the desired antibody [many recent improvements of the monoclonal antibody technique may be found Langone and Vunakis, (1986)]. The use of crude nuclear protein fractions as the primary immunogen should result in generation of a large number of such clonal lines, each producing an antibody specific for only one of the proteins (epitopes) present in the original mixture: a monoclonal antibody library. An optimal library would contain monoclonal antibodies to all nuclear proteins of a given species. However, since the immune system is selective, this goal can never be reached. The proteins which are most antigenic will be overrepresented in the library regardless of their absolute amounts in the nucleus. On the other hand, immunological tolerance phenomena will reduce the probability of obtaining antibodies against conserved epitopes, which in consequence will be under-represented. One could argue that this would be an unavoidable disadvantage of this approach, but methods can be designed to obtain antibodies against even the most conserved nuclear proteins, such as the core histones (Bustin, 1989). Furthermore, the selectivity of the system can represent an advantage. As a result of tolerance phenomena, the system may select a tissue-specific subset of chromatin proteins, which include members that play a role in regulating genes that are expressed in a tissue-specific manner. This would provide us with powerful tools for the study of gene function in eukaryotes. It is likely that the mechanism of selection operative here will be different from that of other possible approaches, and thus we may pick up impor-tant nuclear proteins which escape detection by other means.

In this article I briefly review the areas in which this approach has been used and discuss the range of results obtained. I concentrate on work in which monoclonal antibody libraries have been used as a general approach to detect, classify, and investigate previously uncharacterized nuclear proteins. I do not list all monoclonal antibodies which have been made against nuclear proteins and I only briefly summarize work aimed at the production of monoclonal antibodies against specific proteins from an enriched but still heterogeneous nuclear fraction. Although most of the antibodies found are useful as molec-ular markers, it is inherent to the monoclonal library approach that at the beginning almost nothing is known about the proteins investigated. In a few selected cases, these functions have begun to be elucidated by a combination of molecular and genetic methods. However, in reviewing this field I often found myself confined to a mere listing of the molecular and cytological data presently known. Most existing libraries have been made using animal pro-teins, and much has been undertaken in *Drosophila melanogaster* and sev-eral species of amphibia. Recently, Mazzolini *et al.* (1989) have established 330 lines following immunization with crude nuclear extracts of *Nicotiana tabacum*. This is the first time that this technique has been applied to plants. The cell lines they described secreted monoclonal antibodies to conserved histone H1 epitopes, and some of them detected H1 in all higher eukaryotes. It is more than 10 years since we first used this approach for nuclear proteins of *Drosophila* (Saumweber *et al.* 1980). It now seems an appropriate occasion to review how this field had developed.

II.   Monoclonal Antibody Libraries to *Drosophila* Nuclear Proteins

A.   Summary of Available Libraries of Monoclonal Antibodies to *Drosophila* Nuclear Proteins

Drosophila melanogaster provides an attractive system in which to apply the monoclonal antibody library approach, in view of the possibility of carrying out cytological mapping of protein distribution on interphase polytene chromosomes (Silver and Elgin, 1976; Plagens *et al.,* 1976; see [8], this volume). Evolutionarily, *Drosophila* is considerably distant from mouse and rat, the animals of choice for the production of monoclonal antibodies. A further advantage of *Drosophila* lies in the wealth of information available on its genetics. The genome is small enough to make it feasible to obtain a complete physical map in the foreseeable future, and this will greatly facilitate genetic and functional analysis of any given gene product.

Saumweber *et al.* (1980) prepared nuclear protein extracts from the *Drosophila melanogaster* cell line Kc which they used as immunogens, either as total nuclear protein or in the form of complex protein fractions step-eluted from hydroxyapatite columns with increasing concentrations of salt, urea, and guanidinium chloride. In their monoclonal antibody library of 755 cell lines, they found 61 lines whose antibodies specifically stained nuclear structures in *Drosophila* polytene cells. The antibodies detect 31 different nuclear proteins (H. Saumweber unpublished). Kuo *et al.* (1982) used 3 to 6 hour *Drosophila* embryo nuclei as antigens, which were injected separately as a urea-soluble fraction containing mainly nonhistone proteins and an insoluble fraction which contained the histones and a minor amount of nonhistone proteins. Their antibody library of 170 cell lines resulted in 32 monoclonal antibodies whose antigens were predominantly located in nuclei of *Drosophila* tissue culture cells. There is no information available on how many different nuclear proteins can be detected by these antibodies. Frasch (1985) used 0 to 16 hour *Drosophila* embryo nuclear proteins to establish a library of 1412 cell lines. The nuclei had been previously digested by RNase to deplete RNA-binding proteins, and chromatin from these nuclei was further separated on sucrose gradients at 450 mM salt. Proteins dissociating from chromatin under these conditions, including histone H1 and nonhistone proteins, were collected from the top of the gradient and were further separated on QAE-Sephadex by step-elution with increasing salt and guanidinium chloride. The fractions running at the position of nucleosomal chromatin were separated on a Biorex column into a fraction containing mainly the core histones and a nonhistone protein fraction tightly bound to chromatin. One hundred and thirty-six cell lines were identified whose antibodies detected specific nuclear proteins. The antibodies identified 39 different nuclear proteins. Garzino *et al.* (1987) solubilized proteins from a crude nuclear preparation from 0 to 15 hours *Drosophila* embryos by dialysis against 0.15 and 0.3 M salt and used the high-speed centrifugation supernatants of the solubilized proteins for immunization. From 610 established cell lines, they selected 25 monoclonals which were specific for nuclear proteins. The antibodies detect at least 9 different nuclear proteins.

To identify a positive hybridoma line, most authors (including myself) often relied on one or the other version of a solid-phase assay (immunoradiometric assay, IRMA; enzyme-linked immunosorbent assay, ELISA) using crude nuclear fractions as antigens. However, due to the complex composition of the antigen mixture and our initial ignorance on the properties and concentrations of the antibodies, this assay most often resulted in an overestimate of the number of positive cell lines. Thus, in many experiments only 10% of the "ELISA prositive" hybridoma lines finally turned out to secrete antibodies positive in immunostaining and/or Western blotting on the same tissue (H. Saumweber and M. Frasch unpublished observation). We have therefore abandoned this assay as an initial screening method. Instead, we first screen for immunoglobulin (Ig) production (and Ig concentration) of the hybridoma lines, assaying culture supernatants $1000 \times$ diluted in phosphate-buffered saline (PBS) in a conventional ELISA. Known concentrations of mouse IgG in cell culture medium are treated in parallel to give a standard curve. Ig-producing lines are then tested by indirect immunofluorescence on polytene chromosomes, on fixed tissue culture cells, and on Western blots with the protein mixture which had been used as an antigen. Application of the proteins to be tested in one large gel slot allows one to cut the transfer filter after blotting in vertical stripes, which can be assayed with different antibodies independently. Thus, from one slab gel we can assay up to 50 different hybridoma supernatants at a time.

B. Monoclonal Antibodies to Components of the Nuclear Envelope and the Nucleolus

Eight lines in the library of Saumweber *et al.* (1980) secreted antibodies that detected components of the nuclear envelope in direct immunofluorescence on tissue culture cells. The antibodies detected the same polypeptides (Risau *et al.*, 1981; Frasch *et al.* 1988) which were later identified as the *Drosophila* lamins (Smith *et al.*, 1987). Twenty-one monoclonal antibodies in the library of Frasch (1985) also detect the lamin proteins and at least one of 13 nuclear envelope-specific antibodies produced by Kuo *et al.* (1982) detects the same antigen. Thus, this represents an example of a highly immunogenic protein. Using these antibodies, the fate of the *Drosophila* lamins during early development has been described (Fuchs *et al.*, 1983; Smith *et al.*, 1987; Frasch *et al.*, 1988). These antibodies also have been used to map quantitatively the discontinuous three-dimensional network of the nuclear lamina in diploid nuclei both in fixed embroys (Paddy *et al.*, 1990) and recently also *in vivo* (M. Paddy, Department of Biochemistry and Biophysics, University of California, San Francisco, personal communication). The antibodies recently allowed the isolation of the *Drosophila* lamin gene. Its amino acid sequence displays features diagnostic for lamins of higher eukaryotes (Gruenbaum *et al.*, 1988).

Two high-molecular-weight nuclear envelope antigens of unknown function were detected in the screen of Frasch (Frasch, 1985; Frasch *et al.*, 1988). These antibodies detect several high-molecular-weight polypeptides on Western blots. Similar observations have been made with monoclonal antibodies to a family of glycoproteins in the nuclear pore complex of rat liver (Snow *et al.*, 1987; see below). Four cell lines produced antibodies detecting a protein of unknown function that was specifically localized in the nucleolus (H. Saumweber, unpublished). These antibodies are useful as reagents to localize extra nucleoli in rDNA transformants (Karpen *et al.*, 1988), and as cellular markers in studies of genetic mosaics. In the latter case, an unstable ring-X chromosome is crossed into a *bobbed* (bb) background. X(bb)/O cells in mosaic animals can be easily identified, since in the absence of nucleoli these antibodies stongly stain the nuclei as a whole (Zusman and Wieschaus, 1987; Zusman *et al.*, 1987).

C.   Antibodies to Chromosomal Proteins

The majority of nuclear proteins detected by the *Drosophila* antibody libraries are as yet of unknown function. The use of the antibodies as specific markers allows one to classify them according to stage-and tissue-specific expression. Moreover, the superior cytological resolution of polytene chromosomes allows a classification according to differences in chromosomal distribution which cannot be recognized in diploid cells. Many of the structural features of polytene interphase chromosomes are likely to be typical for chromatin of diploid cells. Biochemical criteria based on nuclear fractionation allow further useful distinctions. Based on this preliminary information, interesting proteins have been singled out for detailed molecular and genetic analyses. In the following I use this classification scheme as a framework for discussing the results that have been obtained using monoclonal libraries in *Drosophila*.

1. Proteins in Condensed Chromosomal Regions

Although most of the dry mass of chromosomes is present in condensed regions, surprisingly few proteins have been detected specifically in bands on polytene chromosomes. Only one M_r 28,000 nonhistone protein which is present in condensed chromosomal bands was detected by antibodies from the library of Saumweber *et al.* (1980); antibodies to the same protein have also been found by Frasch (1985). Using antigens stripped from chromatin with 450mM salt, Frasch (1985) obtained six cell lines that secrete antibodies specific for *Drosophila* histone H1, which is predominantly present in condensed bands. The same was true for histone H2A. Of the 16 independently obtained histone H2A antibodies, all but one cross-reacted with bovine histone H2A. These results show that even using crude protein mixtures as immunogens one can readily obtain monoclonal antibodies to conserved proteins. A highly immunogenic protein of M_r 68,000 was described by Frasch (1985) as predominantly present in bands. During early development, this protein is preferentially found in nuclei of the nervous system. The set of 30 monoclonal antibodies obtained to several different epitopes allowed the isolation of cDNA coding for this protein. The amino acid sequence shows homology to the repeated motifs of Regulator of Chromatin Condensation (*RCC1*), a DNA-binding protein which appears to be involved in cell cycle control which may also be required for normal gene expression and chromatin structure (Ohtsubo *et al.*, 1991; Frasch, 1991). Kuo *et al.* (1982) also reported that some of their antigens were present in most bands; however, a heterochromatin-associated nonhistone chromosomal protein like the HP1 protein described in the previous section (Eissenberg *et al.*, 1990) has not been detected by antibodies of the general *Drosophila* monoclonal antibody libraries.

2. Proteins in Decondensed Chromosomal Regions

Many more different proteins have been found in decondensed regions on polytene chromosomes. Saumweber *et al.* (1980) obtained several antibodies to a protein of M_r 130,000 present in most, if not all, interbands on polytene chromosomes; however, this protein was not detected in the even more decondensed chromosomal puffs. The M_r 130,000 protein is also detected in diploid nuclei and may serve as a structural component for the maintenance of a regionally open chromation conformation. Kuo *et al.* (1982) have also obtained antibodies that bind to most interbands. Frasch (1985) reported a

M_r 67,000 protein with similar properties. Furthermore, a monoclonal antibody from this library that recognizes several bands on Western blots also detected proteins present predominantly in interbands. Using this antibody, staining was also seen in the chromocenter, where all centromeres meet, and in the centrosomal region (Frasch et al., 1986). This monoclonal antibody detects an epitope common to several proteins, since it was possible to generate a polyclonal serum specific for the M_r 185,000 centrosomal protein only. The monoclonal antibody allowed us to clone and further characterize the cDNA coding for this centrosomal protein (Whitfield et al., 1988). Frasch (1985) furthermore described two proteins of M_r 38,000 and 82,000 which were present in a subset of interbands only, but not in puffs.

A relatively large group of proteins present in a subset of decondensed regions including known puffs (the so called puff-specific proteins) has been found. Since these loci are known to be active in transcription, some effort has been invested in the investigation of the role of these proteins. Many of the puff-specific proteins first described by Saumweber et al. (1980) have now been characterized as RNA-binding proteins (Risau et al. 1983; Schuldt et al. 1989). Some of these proteins are also bound to specific lampbrush loops of the Y chromosome in Drosophila primary spermatocytes (Glätzer, 1984; Bonaccorsi et al., 1988). These antibodies have been used to study the evolutionary conservation of the cognate antigens. One of them reacts with HeLa proteins (Hügle et al., 1982), another detects proteins present on lampbrush loops of Xenopus laevis oocytes (C. Dreyer, MPI f. Entwicklungsbiologie Tübingen, personal communication, 1983). However, 18 further proteins are detected in Drosophila only (Kabisch et al. 1982; H. Saumweber, unpublished). The distribution of these RNA-binding proteins on polytene chromosomes is in no case coincident with that of RNA polymerase II, neither in normal salivary gland cells (Kabisch and Bautz, 1983), nor under heat-shock conditions (Dangli and Bautz, 1983; Dangli et al., 1983). We have studied the fate of these proteins throughout development. At early stages, they are located as maternal components in the cytoplasm before migrating into the nuclei at specific times during the blastoderm stage, when transcription starts in the embryo (Dequin et al., 1984; Frasch, 1985). A class of puff-specific proteins, later found associated with nucleosomal chromatin, shows a similar repartitioning (see below). Later, these proteins occur in nuclei of all tissues tested so far. cDNA clones coding for some of these proteins have been isolated and sequenced (Hovemann et al., 1991; K. H. Glätzer, Institut für Genetik, Universität Düsseldorf, personal communication, 1989). They show homology to the RNA-binding domain (RNP1, RNP2 consensus) and/or repeated motifs such as glycine-rich sequences, typical for RNA-binding proteins (Bandziulis et al., 1989). A role for these proteins in processing or specific RNA packaging has been discussed (Risau et al. 1983). Some of the puff-specific proteins described by Kuo *et al.* (1982) may belong to this class of proteins. Recent analysis of hnRNP proteins from mammals has revealed a class of 24 poly-peptides in the range of M_r 34,000–120,000, some of which may be the mammalian counterparts of these *Drosophila* proteins (Choi and Dreyfuss, 1984a; Pinol-Roma *et al.* 1988; see below).

In a search for antibodies to DNA-binding proteins, Frasch (1985) first eliminated these highly antigenic RNA-binding proteins before immunization, as has been mentioned above (Section II, A). Most of the 20 puff-specific proteins he then found are still bound to nucleosomal chromatin after RNase digestion and low-salt extraction of digested nuclei (Frasch 1985; Frasch and Saumweber, 1989). The chromosomal binding of two of these proteins, of M_r 82,000 and M_r 66,000, has been investigated in some detail (Frasch and Saumweber, 1989; Saumweber *et al.* 1990). Both are bound within 2.6 kb upstream of the *Sgs-4* gene at times when this gene is actively transcribed. It was shown that the binding of one of these proteins (M_r 66,000) is dependent on the presence of a 52-bp element in the enhancer region of the *Sgs-4* gene (Saumweber *et al.,* 1990). DNA coding for both proteins has been isolated and sequenced (Besser *et al.* 1990; Wieland *et al.,* 1991). The M_r 66,000 antigen is a basic protein rich in charged amino acids. It shows no significant homology to any known nuclear protein. The M_r 82,000 protein is a basic protein with homology to single-strand nucleic acid-binding proteins (Besser *et al.* 1990). It contains two RNA-binding domains each including the two RNP consensus motifs (Bandziulis *et al.,* 1989). Since the protein still binds to chromatin after RNase digestion, RNA binding per se apparently does not mediate the chromatin binding. DNA coding for another puff-specific protein detected by antibodies in this library has been cloned recently. Following heat shock, this protein binds specifically to several of the known heat-shock loci. By UV cross-linking (Gilmour and Lis, 1986), it could be shown that it is bound to DNA immediately downstream of the heat-shock transcripts at 87A. This DNA-binding protein contains one RNA-binding domain in the amino-terminal half and stretches of simple amino acid repeats in the carboxy-terminal half (Champlin *et al.,* 1991).

A major step in elucidating the function of these proteins involves isolation of mutations in their structural genes. Surprisingly, the M_r 82,000 protein has been found to be a product of the *nonA* gene, a gene discovered on the basis of its function in the visual system and in male courtship behavior (Hotta and Benzer, 1970; Heisenberg, 1971; Kulkarni *et al.,* 1988; Jones and Rubin, 1990; Besser *et al.,* 1990). Mutational alterations of this chromosomal protein may result in the malfunction of particular neurons. Here, the monoclonal library approach converges with genetics and molecular biology to help to elucidate problems to which none of the individual approaches has been able to provide satisfactory answers.

3. Stage-and Tissue-Specific Proteins

Although the antibodies described in the previous sections have most often turned out to be specific for *Drosophila* proteins, none of them is stage or tissue specific, with the exception of the M_r 68,000 protein preferentially detected in neuronal cells (see above). According to the data from Garzino et al. (1987), the 25 nuclear antigens they recovered were expressed stage specifically, and 9 of them were expressed in specific tissues. Several monoclonal antibodies specifically detected antigens in nuclei of the gut and associated structures. The gene for one of these, a M_r 66,000 polypeptide, has been isolated recently by screening an expression library. The *modulo* protein, so-called because of its characteristic array of charged regions with basic amino and carboxy termini, may bind to DNA (Krejci et al., 1989). The M_r of the other gut-specific antigens has not yet been reported, but judging by the developmental time at which these are first detectable in the embryo, they may represent at least two different antigens. Another of their monoclonal antibodies detected two polypeptide bands of M_r 250,000 and 45,000 preferentially expressed in the ventral nerve cord, similar to the M_r 68,000 neural-specific protein found by Frasch (1985; see above).

III. Monoclonal Antibody Libraries to Nuclear Proteins of Amphibia

A. Monoclonal Antibodies to *Xenopus* Proteins

In an attempt to identify maternal factors important in the early development of *Xenopus*, Dreyer et al. (1981, 1982, 1985) produced a library of several hundered clonal lines secreting antibodies to proteins of the oocyte nucleus or germinal vesicle (GV) of both *Xenopus laevis* and *Xenopus borealis*. The antibodies have proved to be valuable analytical tools with which to follow the fate of nuclear proteins at different stages of development on two-dimensional gels, comparing their patterns in *Xenopus laevis* and *Xenopus borealis*. With one exception, all proteins could be immunological detected in both species (Dreyer and Hausen, 1983; for a survey of the molecular data on 13 polypeptide antigens refer to Fig. 1 in Dreyer et al., 1985). One of the previously known antigens detected by this approach was nucleoplasmin, a protein representing about 10% of whole germinal vesicle protein, which is thought to bind histones and to be involved in the formation of nucleosomes during cleavage (Laskey et al., 1978). Antibodies to proteins of M_r 110,000 and 100,000, also known as N_1 and N_2 (Bonner 1975) and considered to complex histones H3 and H4 in the oocyte (Kleinschmidt and Franke, 1982; Kleinschmidt et al., 1986), are also contained in the library. The *Xenopus borealis* N_1 could be followed throughout development, and it was demonstrated that it is specifically expressed in the germ line (Wedlich et al., 1985). Monoclonal antibodies to a protein of M_r 175,000 with similar properties have been obtained from a monoclonal library to *Pleurodeles* GV proteins (Abbadie et al., 1987; see below).

Some of these antibodies, including one to a DNA-binding nucleolar protein of M_r 86,000 have been used to follow the fate of the antigens during oocyte maturation (Hausen et al., 1985) and development (Wedlich and Dreyer, 1988). Several antigens present in the GV were followed up to the adult stage. Their cellular distribution changes dramatically during development, but no major alterations have been detected in the proteins themselves (Dreyer et al., 1983). During GV breakdown at maturation, these proteins are released into the cytoplasm of the animal hemisphere to be reaccumulated again in nuclei at specific stages of development (Dreyer et al., 1981, 1982, 1983, 1985; Dreyer, 1987). Some proteins reenter the nuclei at early cleavage, others at the blastula stage or later. Similar observations have been made for *Drosophila* nuclear proteins (Dequin et al., 1984; Frasch, 1985, see above). Two of these proteins were found later to be localized in nuclei of specific tissues. A protein of M_r 46,000 was observed, both cytologically and by two-dimensional gel analysis, in the nuclei of the pharyngobranchial tract, the gut, the pancreas, and some cells of the kidney, and a M_r 80,000 polypeptide was found predominantly in the nuclei of the central nervous system. The CNS-specific protein probably binds to DNA and is found on many lampbrush loops of *Xenopus* (C. Dreyer, MPI f. Entwicklungsbiologie Tübingen, personal communication, 1991). The gene coding for this protein has been cloned, and the conceptual protein predicted from the cDNA sequence shows the presence of several zinc fingers (Etkin et al., 1990).

Immunization with a crude low-speed pellet from nuclear homogenates of *Xenopus laevis* oocytes has provided several monoclonal antibodies to nucleolus-specific proteins (Schmidt-Zachmann et al., 1984, 1987; Hügle et al., 1985a,b). These include an acidic M_r 185,000 protein found in the dense fibrillar component of nucleoli of oocytes and somatic cells (Schmidt-Zachmann et al., 1984), which is also found in residual nucleolar structures of transcriptionally inactive amphibian erythrocytes. A constitutive nucleolar protein of M_r 38,000, also detected in nucleoplasmic particles (Schmidt-Zachmann et al., 1987), whose amino acid sequence shows close homology to the histone-binding protein nucleoplasmin, is related to the nucleolar protein B23 (Prestayko et al., 1974; Chan et al., 1986a,b; see below) and is probably involved in nucleolar storage and pre-rRNA assembly of ribosomal proteins. The ribosomal protein S1 and ribocharin, a M_r 40,000 protein present in the granular component of the nucleolus and in 65S particles of the nucleoplasm containing 28S rRNA, are also detected by these monoclonal antibodies (Hügle *et al.,* 1985a,b). The latter protein may be involved in preibosomal assembly.

B.   Monoclonal Antibodies to Nuclear Proteins of Urodeles

Lacroix *et al.* (1985) established a monoclonal antibody library to GV antigens from *Pleurodeles waltl.* Of the 71 hybridoma lines initially isolated, 10 gave a positive reaction when tested on lampbrush chromosomes. Many of these antibodies detected homologous antigens in other urodele species, and have been used to identify homologous lampbrush loops (Ragghianti *et al.,* 1988). A M_r 80,000 protein was found to be present on most loops, except the "M" and "S (spheres)," structures which have a dense matrix. This protein was also detected in presumptive RNP particles in the nucleoplasm, and during development it is present in nuclei of many differentiated tissues. In mature oocytes and in cleavage embryos up to a short time before midblastula transition, the protein is present in the embryo, but not detectable in nuclei, again similar to the RNA-binding proteins in *Drosophila* described above (Dequin *et al.,* 1984; Abbadie *et al.,* 1987). A very similar RNP protein has been found in the newt by Roth and Gall (1987; see below). A cDNA coding for the carboxy-terminal part of this protein allowed the synthesis of a protein capable of migrating into the oocyte nucleus and binding to lampbrush loops (Roth and Gall, 1989). A monoclonal antibody detecting a protein in many lampbrush loops which was later found to occur mainly in the cytoplasm of somatic cells was also contained in the monoclonal library of Lacroix *et al.* (1985). Three monoclonal antibodies detected antigen(s) specific for a subset of about 30 lampbrush loops including M. In favorable cases, it could be shown that these antigens were specific for one of several transcription units in a particular loop. In sections of the ovary, embroys, and larvae the presence of these antigens could not be demonstrated. Another monoclonal antibody detected a protein of M_r 104,000, specifically located in lampbrush structures M and S. The latter structures are poor in RNA and are located near loci carrying the histone genes. Antibodies specific for condensed chromosomal structures have also been obtained. A M_r 270,000 protein present on the chromomeres and on the loop axis of lampbrush chromosomes was detected in nuclei of many tissues, with some enrichment in endodermal tissues (Abbadie *et al.,* 1987). After breakdown of the GV at oocyte maturation, this protein was present in the animal hemisphere cytoplasm, like many of the proteins described in *Xenopus* (Hausen *et al.,* 1985). From there it gradually entered nuclei around midblastula transition, and was found to be present on condensed chromosomes during mitosis.

Roth and Gall (1987) established a monoclonal library to proteins from GVs of the newt *Notophtalmus viridescens* which was screened on lampbrush chromosome preparations. Of the 535 hybridoma lines tested, 12 secreted antibodies specific for nucleolar antigens, 8 were specific for molecules present in the telomere and centromere granules, 10 antibodies detected antigens in the majority of lampbrush loops and 11 detected specific lampbrush loops. The M_r 90,000 protein found in most loops, very similar to the M_r 80,000 protein found in *Pleurodeles*, has already been discussed above. Another protein present in most loops has a M_r of 120,000. Two antibodies were highly specific for antigens in the "sequential labelling loops" and/or the "giant loops." The authors expressed their disappointment that these most interesting antibodies have so far not allowed them to characterize their cognate antigens molecularly, since they fail to react on Western blots. In addition, most antibodies turned out to be highly species specific. Of the 200 monoclonal lines against *Xenopus* or *Notophthalmus* GV proteins which had been tested, only 4 showed detectable cross-reaction between these species (Roth and Gall, 1987).

IV. Monoclonal Antibody Libraries to Nuclear Proteins from Birds and Mammals

Kane and co-workers (1982) established a monoclonal antibody library to urea-soluble red cell nuclear proteins from 14-day-old chickens with a view to the isolation of tissue-specific gene regulatory factors. They reported a series of fusion experiments in which 60–80% of the monoclonals obtained were directed against tissue-specific nuclear proteins. The 29 lines reported detect up to 18 different nuclear proteins. At least 7 of them turn out to be red blood cell specific: 2 of them are specific for proteins of the definitive erythrocyte lineage only (M_r 101,000 + 98,000; M_r 90,000), and 5 are specific for proteins which are also present in 5-day-old chicken red blood cell nuclei (M_r 88,000, M_r 85,000, M_r 70,000, M_r 60,000, M_r 50,000). The investigation of these potentially interesting antigens has been discontinued.

Vanderbilt and Anderson (1982, 1983; Anderson *et al.*, 1983) also established a library of monoclonal antibodies to chicken nuclear proteins. Following immunization with total chromatin from hen oviduct, 20 of the 25 monoclonal antibodies obtained were specific to nuclear proteins of this tissue. Unfortunately, most of these antigens were not characterized on Western blots. An antibody to an oviduct-specific protein of M_r 180,000 showed increased binding to oviduct chromatin following hormone stimulation, as did most of the other oviduct-specific antibodies (but not the nonoviduct-specific antibodies; Anderson et al., 1983). Of 207 hybridoma lines producing monoclonal antibodies to hen erythrocyte chromatin, 86% were specific to this tissue (Vanderbilt and Anderson, 1983). Of 77 antigens detected by immunoblotting, 4 were studied in more detail. Although these proteins are minor components, each representing between 0.001 and 0.12% of total chicken erythrocyte protein, multiple monoclonals to each of them have been obtained from separate fusions. A M_r 78,000 protein was present in all tissues examined; the other three were detectable in erythrocytes and reticulocytes only (M_r 87,000, M_r 93,000–23,000, M_r 87,000–210,000). The M_r 93,000–230,000 polypeptides probably orginate from a M_r 230,000 precursor, whose processing is developmentally regulated in vivo (Vanderbilt and Anderson, 1983).

Turner (1981) prepared several monoclonal antibody lines by immunizing with human liver chromatin. Of 266 cell lines tested, only 2 secreted antibodies specific to nuclear proteins. One antibody showed a speckled nuclear staining pattern, whereas the other antibody detected a protein which was more evenly distributed in nuclei and was also detectable on condensed metaphase chromosomes. Both antibodies (and other antibodies to cytoplasmic proteins obtained from the same fusions) showed cross-reaction to rat liver proteins.

Babin and Anderson (1988) established a monoclonal library to chromatin from a rat hepatoma line. Using whole chromatin, 95 of the 97 clones obtained secreted antibodies reactive with normal rat liver chromatin; 44 of these reacted with the same antigen. Immunization with chromatin depleted of this antigen by immunoabsorption resulted in the production of 34 clones whose antibodies showed preferential binding to tumor chromatin relative to normal liver chromatin. Together, these 36 tumor-specific lines defined 6 nuclear antigens, 4 of them preferentially and 2 of them exclusively detected in tumor cells. On closer inspection, each of these antibodies showed a characteristic pattern of reactivity to several transplantable hepatomas and transformed cell lines. One of the tumor-specific nuclear proteins of M_r 350,000 was localized to nucleoli. The other one, of M_r 350,000 + 290,000, showed a granular pattern of staining throughout the nucleus. Interestingly, the latter was selectively enriched in a chromatin fraction released after mild micrococcal nuclease digestion, but not following RNase digestion, suggesting that the protein is bound to active chromatin. In vitro binding studies suggested that this antigen is bound to DNA (Babin and Anderson, 1988). The antigens preferentially associated with tumors could be detected on Western blots only by omitting β-mercaptoethanol. A M_r 260,000 protein was homogeneously distributed in the nuclei and in vitro showed binding to DNA. A protein found at the nuclear envelope, with a M_r 95,000 in the presence of β-mercapto-ethanol, had a M_r of 500,000 in its absence, pointing to a protein structure consisting of several subunits. The same was observed for a protein of M_r 105,000, which showed a homogeneous nuclear staining of variable intensity and which had a M_r of 300,000 in the absence of β-mercaptoethanol. This complex could be resolved into several polypeptides of M_r 60,000, 55,000, 40,000, and 34,000, and it is not clear to which of the polypepetides the antibody binds. The antigen *in vitro* binds to DNA and is located in granular form inside nuclei and in some cytoplasmic areas adjacent to nuclei.

Monoclonal antibody-secreting lines to mammalian nuclear proteins have been established by several authors following the immunization of nuclear subfractions enriched for the component of interest. I only briefly review here the general results obtained and the reader is referred to the original papers.

Several investigators have isolated antibodies specific for constituents of the nuclear envelope. Monoclonal antibodies against lamina proteins have been successfully prepared following immunization with whole nuclei or crude nuclear fractions in several laboratories (e. g., Lehner *et al.*, 1985). Monoclonal antibodies against a family of related glycoproteins of the nuclear pore complex (nucleoporins) have been obtained by Davis and Blobel (1986, 1987) following immunization with Triton X-100-treated rat liver nuclei. Cross-reacting proteins were identified in the yeast nuclear envelope, and their genes have been isolated (Davis and Fink, 1990). Snow *et al.* (1987) isolated mono-clonal antibodies to eight glycoproteins specific for the nuclear pore complex following immunization with a salt- and detergent-extracted nuclear pore complex/lamina fraction from rat liver nuclei. By the same approach, Senior and Gerace (1988) isolated antibodies to an integral inner nuclear membrane protein, which may attach the nuclear membrane to the nuclear lamina. They also isolated antibodies to gp210, a conserved membrane-spanning glycopro-tein of the nuclear pore complex formerly known as gp190 (see Greber *et al.*, 1990). Park and co-workers (1987) also isolated a monoclonal antibody spe-cific for a glycoprotein found in nuclear pores following immunization with nuclear envelopes from Chinese hamster ovary cells.

Several monoclonal antibodies have been prepared against insoluble pro-teins of the nuclear interior. Following immunization with a salt- and urea-extracted HeLa chromatin, Bhorjee *et al.* (1983) isolated several antibodies to "nuclear matrix-like" antigens and one specific for nuclear region identi-fied as the chromocenter. Immunization with "nuclear matrices" from mouse or bovine lymphocytes resulted in four monoclonals specific for nonlamin "nuclear matrix" antigens which detected, to varying degrees, antigens in the nuclear envelope and in the nuclear interior of mouse 3T3 fibroblasts (Chaly *et al.*, 1984). Two of them also reacted with plant nuclear protein. The other two showed cross-reactivity with *Drosophila* antigens (Chaly *et al.*, 1986).

Several monoclonal antibodies to nucleolar and nuclear proteins of HeLa cells have been established by Busch and co-workers (Zweig et al., 1984; Freeman et al., 1985, 1986; Black et al., 1987) following immunization with purified nucleoli or extracted nucleolar proteins. Some of these antibodies detect antigens specific for proliferating cells (Zweig et al., 1984; Yaneva et al., 1985; Black et al., 1987) and one was directed against a M_r 145,000 protein specifically present in nucleoli of malignant tumor lines (Freeman et al., 1986). Several monoclonal antibody lines have been obtained to the higher antigenic nucleolar protein C23, also known as nucleolin (Lapeyre et al., 1987; Freeman et al., 1985). C23 shows binding to both rDNA-containing chromatin and rRNA transcripts, a property reflected in its predicted amino acid sequence (Lapeyre et al., 1987).

Monoclonal antibodies have proved to be particular useful tools in the analysis of hnRNA- and snRNA-associated proteins. Autoimmune antisera of the SLE type have been known to contain antibodies specific for Sm antigens present on snRNPs. Such antibodies have also been obtained from clonal lines established from autoimmune mice by Lerner et al. (1981). Following immunization with RNP particles from chicken and screening with mouse RNP complexes, Leser et al. (1984) established several cell lines that secrete monoclonal antibodies against the hnRNP core group proteins. Dreyfuss et al. (1984) used purified complexes of polyadenylated RNA with protein, obtained after UV-cross-linking in intact cells, to generate monoclonal antibodies to several hnRNA-associated proteins. hnRNP-specific monoclonal antibodies have also been obtained following immunization of hnRNP proteins eluted from ssDNA cellulose (Pinol-Roma et al. 1989). Using these antibodies, hnRNP complexes have been purified and their composition characterized in detail (Choi and Dreyfuss, 1984a, 1984b; Pinol-Roma et al., 1988). Some of the component proteins may play an important part in splicing reactions (Choi et al., 1986). cDNA clones coding for these proteins have been obtained (Nakagawa et al., 1986; Swanson et al., 1987). Lutz et al. (1988) recently also reported the establishment of a monoclonal library to purified hnRNP particles.

V.   Conclusions

The examples cited above illustrate the versatility and the widespread use of the monoclonal library approach in the analysis of nuclear structures and nuclear subfractions. The utility of monoclonal antibodies as specific labels to allow the *in situ* detection of the target antigens by microscopic techniques or to follow their distribution in biochemical fractionation procedures in the absence of functional tests is obvious. Immunopurification from crude extracts greatly helps to enrich and purify nuclear proteins present in minor amounts, and immunoprecipitation is a powerful method with which to copurify (protein or nucleic acid) components specifically associated with the antigen under study (Gilmour and Lis, 1986; Pinol-Roma et al., 1988). Screening of expression libraries using antibody probes eventually leads to the isolation of DNA coding for the antigen (Young and Davis, 1983) and, in systems where a genetic analysis is feasible, this allows one to isolate mutant alleles and to study their phenotypic effects in the intact animal (Eissenberg et al., 1990; Besser et al., 1990). Thus, the monoclonal antibody approach is nowadays tightly linked to other molecular and genetic approaches, and an entry at any point should allow one to use the advantages of any of the other methods.

The immune system is selective for highly immunogenic epitopes, which may dominate the response. Any system under study contains such components. However, it is intriguing that, in nearly all cases, a class of nuclear proteins associated with hnRNA has been obtained. The reason for their immunogenicity is not known. It may be that they are less conserved, that they occur as small particles better suited for antigen presentation to the immune system, or that they contain complexed RNA (e.g., Stollar and Ward, 1970; Bustin, 1989). It has become clear that the immune system is often selective for species-specific, tissue- and stage-specific determinants. This may be of advantage in some cases, such as in the establishment of tumor-specific nuclear markers (Freeman et al., 1986; Babin and Anderson, 1988). This selectivity may be circumvented by selecting monoclonal antibody secreting cells with an antigen only distantly related to the one used for immunization (e.g., Leser et al., 1984). One also can overcome these problems by protein fractionation or immunoabsorption of domaint antigens (Frasch, 1985; Babin and Anderson, 1988). We note that antibodies to almost all major classes of nuclear proteins have been obtained by the monoclonal library approach and, as demonstrated, monoclonal antibodies can be obtained to even most conserved nuclear proteins known, the histones.

However, although many attempts have been made to obtain them, to my knowledge specific transcription factors and regulatory proteins of developmental interest have not yet been identified by this approach. This is specially clear in *Drosophila*, where many such molecules have been identified by a combination of classical and molecular genetics (e.g., recent reviews by Manseau and Schüpbach, 1989; Levine, 1988; Ingham, 1988). Why this is so is not understood at the moment. These proteins are "seen" by the immune system and both specific polyclonal and monoclonal antibodies have been obtained. The limited amount of antigen per cell nucleus cannot be the only problem, since many of the proteins detected by the monoclonal antibody library approach are present in even smaller quantities per cell (H. Saumweber, unpublished). This finding possibly reflects the next level of selectivity imposed by the immune system which will have to be overcome by redesigning the immunization and screening methods. Despite this restriction, the monoclonal antibody library approach has, in recent years, given us many specific and powerful tools for the elucidation of complex biological functions.

Acknowledgments

I thank Dr. P. A. Hardy for critical reading of the manuscript. Part of this work was funded by the Deutsche Forschungsgemeinschaft SFB 243.

References

Abbadie, C., Boucher, D., Charlemagne, J., and Lacroix, J. C. (1987). Immunolocalization of three oocyte nuclear proteins during oogenesis and embryogenesis in pleurodeles. *Development* **101**, 715–728.

Anderson, J. N., Vanderbilt, J. N., Bloom, K. S., and Germain, B. J. (1983). Effects of steroid hormones on chicken oviduct chromatin. *In* "Gene Regulation by Steroid Hormones" (A. K. Roy and J. H. Clark, eds.), pp. 17–59. Springer-Verlag, New York.

Babin, J. K., and Anderson, J. N. (1988). Isolation and analysis of hepatoma nuclear proteins using monoclonal antibodies. *Cancer Res.* **48**, 5495–5502.

Bandziulis, R. J., Swanson, M. S., and Dreyfuss, G. (1989). RNA-binding proteins as developmental regulator. *Genes Dev.* **3**, 431–437.

Besser, H. von, Schnabel, P., Wieland, C., Fritz, E., Stanewsky, R., and Saumweber, H. (1990). The puff-specific *Drosophila* protein Bj6, encoded by the no-on transient *A* gene, shows homology to RNA-binding proteins. *Chromosoma* **100**, 37–47.

Bhorjee, J. S., Barclay, S. L., Wedrychowski, A., and Smith, A. M. (1983). Monoclonal antibodies specific for tight-binding human chromatin antigens reveal structural rearrangements within the nucleus during the cell cycle. *J. Cell Biol.* **97**, 389–396.

Black, A., Freeman, J. W., Zhou, G., Busch, H. (1987). Novel cell cycle-related nuclear proteins found in rat and human cells with monoclonal antibodies. *Cancer Res.* **47**, 3266–3272.

Bonaccorsi, S., Pisano, C., Puoti, F., and Gatti, M. (1988). Y. chromosome loops in *Drosophila melanogaster*. *Genetics* **120**, 1015–1034.

Bonner, W. M. (1975). Protein migration into nuclei II: Frog oocyte nuclei accumulate a class of microinjected oocyte nuclear proteins and exclude a class of oocyte cytoplasmic proteins. *J. Cell Biol.* **64**, 431–437.

Bustin, M. (1989). Preparation and application of immunological probes for nucleosomes. *In* "Methods in Enzymology" (P. M. Wassarman and R. D. Kornberg, eds.), Vol. 170, pp. 214–251. Academic Press, San Diego, California.

Chaly, N., Bladon, T., Setterfield, G., Little, J. E., Kaplan, J. G., and Brown, D. L. (1984). Changes in distribution of nuclear matrix antigens during the mitotic cycle. *J. Cell Biol.* **99**, 661–671.

Chaly, N., Sabour, M. P., Silver, J. C., Aitchison, W. A., Little, J. E., and Brown, D. L. (1986). Monoclonal antibodies against nuclear matrix detect nuclear antigens in mammalian, insect and plant cells: An immunofluorescence study. *Cell Biol. Int. Rep.* **10**, 421–428.

Champlin, D. T., Frasch, M., Saumweber, and Lis, J. T. (1991). Characterization of a *Drosophila* protein associated with boundaries of transcriptionally active chromatin. *Genes Dev.* (in press).

Chan, P. K., Aldrich, M., Cook, R. G., and Busch, H. (1986a). Amino acid sequence of protein B23 phosphorylation site. *J. Biol. Chem.* **261,** 1868–1872.

Chan, P. K., Chan, W. Y., Yung, B. Y. M., Cook, R. G., Aldrich, M. B., Ku, D., Goldknopf, I. L., and Busch, H. (1986b). Amino acid sequence of a specific antigenic peptide of protein B23. *J. Biol. Chem.* **261,** 14335–14341.

Choi, Y. D., and Dreyfuss, G. (1984a). Isolation of the heterogeneous nuclear RNA-ribonucleo-protein complex (hnRNP): A unique supramolecular assembly. *Proc. Natl. Acad. Sci. U.S.A.* **81,** 7471–7475.

Choi, Y. D., and Dreyfuss, G. (1984b). Monoclonal antibody characterization of the C proteins of heterogeneous nuclear ribonucleoprotein complexes in vertebrate cells. *J. Cell Biol.* **99,** 1997–2004.

Choi, Y. D., Grabowski, P. J., Sharp, P. A., and Dreyfuss, G. (1986). Heterogeneous nuclear ribonucleoproteins: Role in RNA splicing. *Science* **231,** 1534–1539.

Dangli, A., and Bautz, E. K. F. (1983). Differential distribution of nonhistone proteins from polytene chromosomes of *Drosophila melanogaster* after heat shock. *Chromosoma* **88,** 201–207.

Dangli, A., Grond, C., Kloetzel, P., and Bautz, E. K. F. (1983). Heat-shock puff 93 D from *Drosophila melanogaster*: Accumulation of an RNP-specific antigen associated with giant particles of possible storage function. *EMBO J.* **2,** 1747–1751.

Davis, L. I., and Blobel, G. (1986). Identification and characterization of a nuclear pore complex protein. *Cell (Cambridge, Mass.)* **45,** 699–709.

Davis, L. I., and Blobel, G. (1987). Nuclear pore complex contains a family of glycoproteins that includes p62: Glycosylation through a previously unidentified pathway. *Proc. Natl. Acad. Sci. U.S.A.* **84,** 7552–7556.

Davis, L. I., and Fink, G. R. (1990). The NUP1 gene encodes an essential component of the yeast nuclear pore complex. *Cell (Cambridge, Mass.)* **61,** 965–978.

Dequin, R., Saumweber, H., and Sedat, J. W. (1984). Proteins shifting from the cytoplasm into the nuclei during early embryogenesis of *Drosophila melanogaster*. *Dev. Biol.* **104,** 37–48.

Dreyer, C. (1987). Differential Accumulation of oocyte nuclear proteins by embryonic *Nuclei of Xenopus*. *Development* **101,** 829–846

Dreyer, C., and Hausen, P. (1983). Two dimensional gel analysis of the fate of oocyte nuclear proteins in the development of *Xenopus laevis*. *Dev. Biol.* **100,** 412–425.

Dreyer, C., Singer, H., and Hausen, P. (1981). Tissue specific nuclear antigens in the germinal vesicle of *Xenopus laevis* oocytes. *Wilhelm Roux Arch.* **190,** 197–207.

Dreyer, C., Scholz, E., and Hausen, P. (1982). The fate of oocyte nuclear proteins during early development of *Xenopus laevis*. *Wilhelm Roux's Arch. Dev. Biol.* **191,** 228–233.

Dreyer, C., Wang, Y. H., Wedlich, D., and Hausen, P. (1983). Oocyte nuclear proteins in the development of *Xenopus. In* "Current Problems in Germ Cell Differentiation" (A. McLaren and C. C. Wylie, eds.), pp. 329–351. Cambridge Univ. Press, Cambridge, England.

Dreyer, C., Wang, Y. H., and Hausen, P. (1985). Immunological relationship between oocyte nuclear proteins of *Xenopus laevis* and *Xenopus borealis*. *Dev. Biol.* **108,** 210–219.

Dreyfuss, G., Choi, Y. D., and Adam, S. A. (1984). Characterization of heterogeneous nuclear RNA-protein complexes *in vivo* with monoclonal antibodies. *Mol. Cell. Biol.* **4,** 1104–1114.

Eissenberg, J. C., James, T. C., Foster-Hartnett, D. M., Hartnett, T., Ngan, V., and Elgin, S. C. R. (1990). Mutation in a heterochromatin-specific chromosomal protein is associated with suppression of position effect variegation in *Drosophila melanogaster*. *Proc. Natl. Acad. Sci. U.S.A.* **87,** 9923–9927.

Etkin, L., Kloc, M., Reddy, B., and Miller, M. (1990). The role of maternally expressed genes during early *Xenopus* development. Meeting Abstract, 3rd Int. Xenopus Meeting, Les Diablerets.

Frasch, M. (1985). Charakterisierung Chromatinassoziierter Kernproteine von *Drosophila melanogaster* mit Hilfe monoklonaler Antikörper. Ph.D. Thesis, Eberhard Karls Universität, Tübingen, Germany.

Frasch, M. (1991). The maternally expressed *Drosophila* gene, encoding the chromatin-binding protein BJ1, is a homolog of the vertebrate gene *Regulator of Chromatin Condensation RCC1*. *EMBO J.* **10,** 1225–1236.

Frasch, M., and Saumweber, H. (1989). Two proteins from *Drosophila* nuclei are bound to chromatin and are detected in a series of puffs on polytene chromosomes. *Chromosoma* **97,** 272–281.

Frasch, M., Glover, D. M., and Saumweber, H. (1986). Nuclear antigens follow different pathways into daughter nuclei during mitosis in early *Drosophila* embryos. *J. Cell Sci.* **32,** 155–172.

Frasch, M., Paddy, M., and Saumweber, H. (1988). Development and mitotic behaviour of two novel groups of nuclear envelope antigens of *Drosophila melanogaster*. *J. Cell. Sci.* **90,** 247–263.

Freeman, J. W., Chatterjee, A., Ross, B. E., and Busch, H. (1985). Epitope distribution and immunochemical characterization of nucleolar phosphoprotein C23 using ten monoclonal antibodies. *Mol. Cell. Biochem.* **68,** 87–96.

Freeman, J. W., McRorie, D. K., Busch, R. K., Gyorkey, F., Gyorkey, P., Ross, B. E., Spohn, W. H., and Busch, H. (1986). Identification and partial characterization of a nucleolar antigen with a molecular weight of 145,000 found in a broad range of human cancers. *Cancer Res.* **46,** 3593–3598.

Fuchs, J. P., Giloh, H., Kuo, C., Saumweber, H., and Sedat, J. (1983). Nuclear structure: Determination of the fate of the nuclear envelope in *Drosophila* during mitosis using monoclonal antibodies. *J. Cell Sci.* **64,** 331–349.

Garzino, V., Moretti, C., and Pradel, J. (1987). Nuclear antigen differentially expressed during early development of *Drosophila melanogaster*. *Biol. Cell* **61,** 5–14.

Gilmour, D. S., and Lis, J. T. (1986). RNA polymerase II interacts with the promoter region of the noninduced hsp70 gene in *Drosophila melanogaster* cells. *Mol. Cell. Biol.* **6,** 3984–3989.

Glätzer, K. H. (1984). Preservation of nuclear RNP antigens in male germs cell development of *Drosophila hydei*. *Mol. Gen. Genet.* **196,** 236–243.

Greber, U. F., Senior, A., and Gerace, L. (1990). A major clycoprotein of the nuclear pore complex is a membrane-spanning polypeptide with a large lumenal domain and a small cytoplasmic tail. *EMBO J.* 1495–1502.

Gruenbaum, Y., Landesman, Y., Drees, B., Bare, J. W., Saumweber, H., Paddy, M. R., Sedat, J. W., Smith, D. E., Benton, B. M., and Fisher, P. A. (1988). *Drosophila* nuclear lamin precursor Dm_0 is translated from either of two developmentally regulated mRNA species apparently encoded by a single gene. *J. Cell Biol.* **106,** 585–596.

Hausen, P., Wang, Y. H., Dreyer, C., and Stick, R. (1985). Distribution of nuclear proteins during maturation of the *Xenopus* oocyte. *J. Embryol. Exp. Morphol.* **89,** (Suppl.) 17–34.

Heisenberg, M. (1971). Isolation of mutants lacking the optomotor response. *Drosophila Inf. Serv.* **46,** 68.

Hotta, Y., and Benzer, S. (1970). Genetic dissection of the *Drosophila* nervous system by means of mosaics. *Proc. Natl. Acad. Sci. U.S.A.* **67,** 1156–1163.

Hoveman, B., Dessen, E., Mechler, H., and Mack, E. (1991). The *Drosophila* snRNP associated protein P11, which specifically binds to heat shock puff 93D, reveals strong homology with hnRNP core protein A1 (manuscript in preparation).

Hügle, B., Guldner, H., Bautz, F. A., and Alonso, A. (1982). Cross-reaction of hnRNP-proteins of

HeLa cells with nuclear proteins of *Drosophila melanogaster* demonstrated by a monoclonal antibody. *Exp. Cell Res.* **142**, 119–126.

Hügle, B., Scheer, U., and Franke, W. W. (1985a). Ribocharin: A nuclear M_r 40,000 protein specific to precursor particles of the large ribosomal subunit. *Cell (Cambridge, Mass.)* **41**, 615–627.

Hügle, B., Hazan, R., Scheer, U., and Franke, W. W. (1985b). Localization of ribosomal protein S1 in the granular component of the interphase nucleolus and its distribution during mitosis. *J. Cell Biol.* **100**, 873–886.

Ingham, P. W. (1988). The molecular genetics of embryonic pattern formation in *Drosophila*. *Nature (London)* **335**, 25–34.

Jerne, N. K. (1976). The immune system: A network of lymphocyte interactions. *In* "The Immune System" (F. Melchers and K. Rajewsky, eds.), pp. 259–266. Springer, New York.

Johnson, P. F., and McKnight, S. L. (1989). Eukaryotic transcriptional regulatory proteins. *Annu. Rev. Biochem.* **58**, 799–839.

Jones, K. R., and Rubin, G. M. (1990). Molecular analysis of no-on-transient *A*, gene required for normal vision in *Drosophila*. *Neuron* **4**, 711–723.

Kabisch, R., and Bautz, E. K. F. (1983). Differential distribution of RNA polymerase B and non-histone chromosomal proteins in polytene chromosomes of *Drosophila melanogaster*. *EMBO J.* **2**, 395–402.

Kabisch, R., Krause, J., and Bautz, E. K. F. (1982). Evolutionary changes in non-histone chromosomal proteins within the *Drosophila melanogaster* group revealed by monoclonal antibodies. *Chromosoma* **85**, 531–538.

Kane, C. M., Cheng, P. F., Burch, J. B. E., and Weintraub, H. (1982). Tissue-specific and species-specific monoclonal antibodies to avian red cell nuclear proteins. *Proc. Natl. Acad. Sci. U.S.A.* **79**, 6265–6269.

Karpen, G. H., Schaefer, J. E., and Laird, C. D. (1988). A *Drosophila* rRNA gene located in euchromatin is active in transcription and nucleolus formation. *Genes Dev.* **2**, 1745–1763.

Kleinschmidt., J. A., and Franke, W. W. (1982). Soluble acidic complexes containing histones H3 and H4 in nuclei of *Xenopus laevis* oocytes. *Cell (Cambridge, Mass.)* **29**, 799–809.

Kleinschmidt, J. A., Dingwall, C., Maier, G., and Franke, W. W. (1986). Molecular characterization of a karyophilic, histone-binding protein: cDNA cloning, amino acid sequence and expression of nuclear protein N1/N2 of *Xenopus laevis*. *EMBO J.* **5**, 3547–3552.

Köhler, G., and Milstein, C. (1975). Continous cultures of fused cells secreting antibody of predefined specificity. *Nature (London)* **256**, 495–497.

Krejci, E., Garzino, V., Mary, C., Bennami, N., and Pradel, J. (1989). Modulo, a new maternally expressed *Drosophila* gene encodes a DNA-binding protein with distinct acidic and basic regions. *Nucleic Acids Res.* **17**, 8101–8115.

Kulkarni, S. J., Steinlauf, A. M., and Hall, J. C. (1988). The dissonance mutant of courtship song in *Drosophila melanogaster*: isolation, behaviour and cytogenetics. *Genetics* **118**, 267–285.

Kuo, C. H., Giloh, H., Blumenthal, A. B., and Sedat, J. W. (1982). A library of monoclonal antibodies to nuclear proteins from *Drosophila melanogaster* embryos. *Exp. Cell Res.* **142**, 141–154.

Lacroix, J. C., Azzouz, R., Boucher, D., Abbadie, C., Pyne, C. K., and Charlemagne, J. (1985). Monoclonal antibodies to lampbrush chromosome antigens of *Pleurodeles waltlii*. *Chromosoma* **92**, 69–80

Longone, J. J., and Vunakis, H. V., eds. (1986). Hybridoma technology and monoclonal antibodies. Section I: Production of hybridomas. *In* "Methods in Enzymology" (J. J. Langone and H. Van Vunakis, eds.), pp. 3–424. Academic Press, New York.

Lapeyre, B., Bourbon, H., and Amalric, F. (1987). Nucleolin, the major nucleolar protein of growing eukaryotic cells: An unusual protein structure revealed by the nucleotide sequence. *Proc. Natl. Acad. Sci. U.S.A.* **84**, 1472–1476.

Laskey, R. A., Honda, B. M., Mills, A. D., and Finch, J. T. (1978). Nucleosomes are assembled by an acidic protein which binds histones and transfers them to DNA. *Nature (London)* **275**, 416–420.

Lehner, C. F., Kurer, V., Eppenberger, H. M., and Nigg. E. A. (1985). The nuclear lamin protein family in higher vertebrates. *J. Biol. Chem.* **261**, 13293–13301.

Lerner, E. A., Lerner, M. R., Janeway, C. A., Jr., and Steitz, J. A. (1981). Monoclonal antibodies to nucleic acid containing cellular constituents: Probes for molecular biology and autoimmune disease. *Proc. Natl. Acad. Sci. U.S.A.* **78**, 2737–2741.

Leser, G. P., Escara-Wilke, J., and Martin, T. E. (1984). Monoclonal antibodies to heterogeneous nuclear RNA-protein complexes. *J. Biol. Chem.* **259**, 1827–1833.

Levine, M. (1988). Molecular analysis of dorsal-ventral polarity in *Drosophila. Cell (Cambridge, Mass.)* 52, 785–786.

Luben, R. A., Brazeau, P., Böhlen, P., and Guillemin, R. (1982). Monoclonal antibodies to hypothalamic growth hormone releasing factor with picomoles of antigen. *Science* **218**, 887–889.

Lutz, Y., Jacob, M., and Fuchs, J. P. (1988). The distribution of two hnRNP-associated proteins defined by a monoclonal antibody is altered in heat-shocked HeLa cells. *Exp. Cell Res.* **175**, 109–124.

Manley, J. L., Fire, A., Cano, A., Sharp, P. A., and Gefter, M. L. (1980). DNA-dependent transcription of adenovirus genes in a soluble whole-cell extract. *Proc. Natl. Acad. Sci. U.S.A.* **77**, 3855–389.

Manseau, L. J., and Schüpbach, T. (1989). The egg came first of course. *Trends Genet.* **5**, 400–405.

Mazzolini, L., Vaeck, M., and VanMontagu, M. (1989). Conserved epitopes on plant H1 histones recognized by monoclonal antibodies. *Eur. J. Biochem.* **178**, 779–787.

Nakagawa, T. Y., Swanson, M. S., Wold, B. J., and Dreyfuss, G. (1986). Molecular cloning of cDNA for the nuclear ribonucleoprotein particle C proteins: A conserved gene family. *Proc. Natl. Acad. Sci. U.S.A.* **83**, 2007–2011.

Ohtsubo, M., Yoshida, T., Seino, H., Nishitani, H., Clark, K. L., Sprague Jr., C. F., Frasch, M., and Nishimoto, T. (1991). Mutation of the hamster cell cycle gene *RCC1* is complemented by the homologous genes of *Drosophila* and *S. Cerevisae. EMBO J.* **10**, 1265–1273.

Paddy, M. R., Belmont, A. S., Saumweber, H., Agard, D. A. and Sedat, J. W. (1990). Interphase nuclear envelope lamins form a discontinuous network that interacts with only a fraction of the chromatin in the nuclear periphery. *Cell (Cambridge, Mass.)* **62**, 89–106.

Park, M. K., D'Onofrio, M., Willingham, M. C., and Hanover, J. (1987). A monoclonal antibody against a family of nuclear pore proteins (nucleoporins): O-linked N-acetylglucosamin is part of the immunodeterminant. *Proc. Natl. Acad. Sci. U.S.A.* **84**, 6462–6466.

Peterson, J. L., and McConkey, E. H. (1976). Non-histone chromosomal proteins from HeLa cells. *J. Biol. Chem.* **251**, 548–554.

Pinol-Roma, S., Choi, Y. D., Matunis, M. J., and Dreyfuss, G. (1988). Immunopurification of heterogeneous nuclear ribonucleoprotein particles reveals an assortment of RNA-binding proteins. *Genes Dev.* **2**, 215–227.

Pinol-Roma, S., Swanson, M., Gall, J. G., and Dreyfuss, G. (1989). A novel heterogeneous nuclear RNP protein with a unique distribution on nascent transcripts. *J. Cell Biol.* **109**, 2575–2587.

Plagens, U., Greenleaf, A. L., and Bautz, E. K. F. (1976). Distribution of RNA polymerase on *Drosophila* polytene chromosomes as studied by indirect immunofluorescence. *Chromosoma* **59**, 157–165.

Prestayko, A. W., Klomp, G. R., Schmoll, D. J., and Busch, H. (1974). Comparison of proteins of ribosomal subunits and nucleolar preribosomal particles from Novikoff hepatoma ascites cells by two-dimensional polyacrylamide gel electrophoresis. *Biochemistry* **13**, 1945–1951.

Ragghianti, M., Bucci, S., Mancino, G., Lacroix, J. C., Boucher, D., and Charlemagne, J. (1988). A novel approach to cytotaxonomic and cytogenetic studies in the genus *Triturus* using monoclonal antibodies to lampbrush chromosome antigens. *Chromosoma* **97**, 134–144.

Risau, W., Saumweber, H., and Symmons, P. (1981). Monoclonal antibodies against a nuclear membrane protein of *Drosophila*. *Exp. Cell Res.* **133**, 47–54.

Risau, W., Symmons, P., Saumweber, H., and Frasch, M. (1983). Nonpackaging and packaging proteins of hnRNA in *Drosophila melanogaster*. *Cell (Cambridge, Mass.)* **33**, 529–541.

Roth, M. B., and Gall, J. G. (1987). Monoclonal antibodies that recognize transcription unit proteins on newt lampbrush chromosomes. *J. Cell Biol.* **105**, 1047–1054.

Roth, M. B., and Gall, J. G. (1989). Targeting of a chromosomal protein to the nucleus and to lampbrush chromosome loops. *Proc. Natl. Acad. Sci. U.S.A.* **86**, 1269–1272.

Saumweber, H., Symmons, P., Kabisch, R., Will, H., and Bonhoeffer, F. (1980). Monoclonal antibodies against chromosomal proteins of *Drosophila melanogaster*. *Chromosoma* **80**, 253–275.

Saumweber, H., Korge, G., and Frasch, M. (1990). Two puff-specific proteins bind within the 2.5 kb upstream region of the *Drosophila melanogaster* Sgs-4 gene. *Chromosoma* **99**, 52–60.

Schmidt-Zachmann, M. S., Hügle, B., Scheer, U., and Franke, W. W. (1984). Identification and localization of a novel nucleolar protein of high molecular weight by a monoclonal antibody. *Exp. Cell Res.* **153**, 327–346.

Schmidt-Zachmann, M. S., Hügle-Dörr, B., and Franke, W. W. (1987). A constitutive nucleolar protein identified as a member of the nucleoplasmin family. *EMBO J.* **6**, 1881–1890.

Schuldt, C., Kloetzel, P. M., and Bautz, E. K. F. (1989). Molecular organization of RNP complexes containing P11 antigen in heat-shocked and non-heat-shocked *Drosophila* cells. *Eur. J. Biochem.* **181**, 135–142.

Senior, A., and Gerace, L. (1988). Integral membrane proteins specific to the inner nuclear membrane and associated with the nuclear lamina. *J. Cell Biol.* **107**, 2029–2036.

Silver, L. M., and Elgin, S. C. R. (1976). A method for determination of the *in situ* distribution of chromosomal proteins. *Proc. Natl. Acad. Sci. U.S.A.* **73**, 423–427.

Smith, D. E., Gruenbaum, Y., Berrios, M., and Fisher, P. A. (1987). Biosynthesis and interconversion of *Drosophila* nuclear lamin isoforms during normal growth and in response to heat shock. *J. Cell Biol.* **105**, 771–790.

Snow, C. M., Senior, A., and Gerace, L. (1987). Monoclonal antibodies identify a group oof nuclear pore complex glycoproteins. *J. Cell Biol.* **104**, 1143–1156.

Stollar, B. D., and Ward, M. (1970). Rabbit antibodies to histone fraction as specific reagents for preparative and comparative studies. *J. Biol. Chem.* **245**, 1261–1266.

Swanson, M. S., Nakagawa, T. Y., LeVan, K., and Dreyfuss G. (1987). Primary structure of human nuclear ribonucleoprotein particle C proteins: Conservation of sequence and domain structures in heterogeneous nuclear RNA, mRNA and pre-rRNA-binding proteins. *Mol. Cell. Biol.* **7**, 1731–1739.

Turner, B. M. (1981). Isolation of monoclonal antibodies to chromatin and preliminary characterization of target antigens. *Eur. J. Cell Biol.* **24**, 266–274.

Vanderbilt, J. N., and Anderson, J. N. (1982). Monoclonal antibodies to tissue specific, hormonally inducible chromosomal antigens in the chicken oviduct. *Fed. Proc.* **41**, 515.

Vanderbilt, J. N., and Anderson, J. N. (1983). Monoclonal antibodies to tissue-specific chromatin proteins. *J. Biol. Chem.* **258**, 7751–7756.

Wedlich, D., and Dreyer, C. (1988). Cell specificity of nuclear protein antigens in the development of *Xenopus* species. *Cell Tissue Res.* **252**, 479–489.

Wedlich, D., Dreyer, C., and Hausen, P. (1985). Occurrence of a species-specific nuclear antigen in the germ line of *Xenopus* and its expression from paternal genes in hybrid frogs. *Dev. Biol.* **108,** 220–234.

Whitfield, W. G. F. Millar, S. E., Saumweber, H., Frasch, M., and Glover, D. (1988). Cloning of a gene encoding an antigen associated with the centrosome in *Drosophila. J. Cell Sci.* **89,** 467–480.

Wieland, C., Mann. S., Besser, H., and Saumweber, H. (1991). The molecular structure of the puff-specific protein Bx42 from *Drosophila* (manuscript in preparation).

Yaneva, M., Ochs, R. McRorie, D. K., Zweig, S., and Busch, H. (1985). Purification of a 86–70 kDa nuclear DNA-associated protein complex. *Biochem. Biophys. Acata* **841,** 22–29.

Young, R. A., and Davis, R. W. (1983). Yeast RNA polymerase II genes: isolation with antibody probes. *Science* **222,** 778–782.

Zusman, S. B., and Wieschaus, E. (1987). A cell marker system and mosaic patterns during early embryonic development in *Drosophila melanogaster. Genetics* **115,** 725–736.

Zusman, S. B., Sweeton, D., and Wieschaus, E. (1987). Short gastrulation, a mutation causing delays in stage-specific cell shape changes during gastrulation in *Drosophila melanogaster. Dev. Biol.* **129,** 417–427.

Zweig, S.E., Rubin, S., Yaneva, M., and Busch, H. (1984). Production of a monoclonal antibody to a 94 KD/pI 6 human S phase specific nucleolar antigen. *Proc. Am. Assoc. Cancer Res.* **24,** 248.

Chapter 10

Optical Sectioning and Three-Dimensional Reconstruction of Diploid and Polytene Nuclei

MARY C. RYKOWSKI

Department of Anatomy and Arizona Cancer Center
University of Arizona
Tucson, Arizona 85719

I. Introduction

Cells, and processes that take place in and around cells, are three-dimensional (3D). Most of us who study cells, however, think and work more easily in two dimensions, so that samples, diagrams, and models are

253

usually constrained by the flatland of a two-dimensional surface. Cell biologists have explained a wide variety of cellular events without a full 3D treatment, but there remains a suspicion that we might be missing something. Happily, as curiosity rises, new optical microscopic and computational imaging techniques are being developed to observe and analyze samples in 3D and, in some cases, *in vivo* and in real time. Already this ability to see and measure cellular events in 3D has yielded surprising and tantalizing results, and as the technology becomes more widely accessible, the progress should be explosive.

The purpose of this article is to review and compile techniques for the observation and 3D analysis of nuclei, specifically the diploid and polytene nuclei of the fruit fly *Drosophila melanogaster*. Pioneering work in 3D microscopy was begun with *Drosophila* nuclei in the early 1980s and has progressed to the point that these may be the most completely understood nuclei from a cytological standpoint. Using film, video cameras, and, finally, sensitive charge-coupled device (CCD) cameras, first the polytene chromosomes then the diploid embryonic nuclei have yielded their secrets. Most recently, this 3D analysis has been extended to the study of living *Drosophila* embryos.

I focus on procedures for the OM-1 system, originally developed by J. W. Sedat, D. A. Agard, and colleagues at the University of California, San Francisco and the Howard Hughes Medical Institute, and now transplanted to several other laboratories. This system employs an IMT-2 inverted fluorescence microscope (Olympus, USA), a scientific grade, cooled CCD camera (Photometrics, Tucson, AZ), electronic shutters (Uniblitz, Vincent Associates, Rochester, NY), and motors (Compumotor, Petaluma, CA). The equipment is controlled by computer and specially designed software. Various computer systems have been employed; my laboratory uses a Microvax III (Digital Equipment Corporation, USA) and ZIP 3232 array processor (Mercury Computer Systems, Lowell, MA). In spite of my concentration on this particular microscope system, I would like to emphasize that many of the considerations for sample preparation and image quality presented below pertain equally for the CCD, video, and confocal microscope systems. Likewise, the attributes of the software required to analyze any set of 3D images will be very similar, no matter how the programs are actually implemented. For those interested only in image processing or some other subset of the techniques presented here, Section III,E contains some suggestions for alternatives to the complete microscope system that may fulfill the needs of some users.

A.   Structure of *Drosophila* Polytene Chromosomes

The enormous polytene chromosomes of *Drosophila melanogaster* allowed the first glimpse of the workings of interphase chromosomes, providing a test for models of gene structure and organization before electron microscopy or recombinant DNA methods were available (Beermann, 1972; Zhimulev *et al.*, 1981).

Polytene chromosomes are found in salivary glands and other tissues of larvae and adults and consist of a parallel arrangement of many copies of the euchromatic (mostly single-copy) portions of the chromosome. Each of the five major chromosome arms (1 or X, 2L and 2R, 3L and 3R) and the tiny fourth chromosome is joined to the chromocenter, a structure presumably derived from the centromere regions of the diploid chromosomes, so that a polytene chromosome resembles a spiny starfish. The familiar banded pattern of the chromosomes results from the alignment of condensed interphase units (called chromomeres) alternating with decondensed, interchromomeric sequences. The pattern of bands is highly reproducible from individual to individual and is unique for different regions of the chromosome. A comparison of polytene chromosomes from several *Drosophila* tissues (Hochstrasser and Sedat, 1987a) revealed that the structures were highly conserved at both the local (band to band correspondence) and global (chromosome–nuclear envelope attachment site conservation) levels. Some differences were seen, consistent with the idea that changes in banding pattern reflect tissue-specific differences in gene expression. For example, the bands corresponding to the heat-shock genes and those containing some developmentally regulated genes become "puffed" during times of high transcriptional activity (Ashburner and Berendes, 1978), providing compelling evidence that genes within bands decondense on activation. Demonstration of the actual organization of genes within bands and interbands has recently begun, using improved microscopic methods to determine the chromatin structure of specific DNA sequences within the band at high resolution (Kress *et al.*, 1985; Rykowski *et al.*, 1988).

The first studies on the 3D structure of *Drosophila* nuclei focused on the organization of polytene chromosomes of the salivary gland (Agard and Sedat, 1983) in order to determine what components of nuclear structure might be conserved, and hence important, for nuclear function. At 50 μm in diameter, these are the largest of the polytene nuclei, with each chromosome arm comprising 1000–2000 individual chromatids. In later work (Mathog *et al.*, 1984; Hochstrasser *et al.*, 1986; Hochstrasser and Sedat, 1987b; Mathog and Sedat, 1989), a number of basic attributes were determined: the chromocenter lies opposite the cluster of telomeres (so-called "Rabl" orientation); each of the chromosome arms coils, usually in a right-handed sense, from the chromocenter to the telomere; and each arm occupies its own sector of the nucleus, like the section of an orange.

With video microscopy of fluorescently labeled nuclei in whole glands, there was sufficient detail to identify the 20 numbered intervals along the arms, but the majority of the 60 or so individual bands in each interval were below the resolution of the method. Nevertheless, it could be concluded that no specific, reproducible intra- or interchromosomal associations occurred, except those between certain regions containing sequences called "intercalary heterochromatin" that have long been known to associate (so-called "ectopic pairing"). Further, no reproducible positional changes occured in either the heat-shock or ecdysone-inducible regions when they were induced to puff. Taken together, these observations suggest that, for the most part, the position of genes within the polytene nucleus is unlikely to affect their expression, and that changes in expression do not precipitate changes in the location of the genes, for example, movement toward the nuclear envelope.

Microscopic examination of less polytenized tissues (such as those in the gut) indicate (Hochstrasser and Sedat, 1987a) that chromosome arms are not always polarized in the Rabl orientation and, indeed, that they are not always connected at the chromocenter. The relative disorder in the nuclei of the gut cells, which experience extreme changes in cell shape during peristalsis, probably reflects the physical stresses on the polytenized chromosomes rather than a variant chromosome structure. The viability of these cells, however, suggests that the more common, organized, polytene structure may not be required for normal chromosome function, but may merely reflect the structure of the diploid nucleus from which the polytene nuclei were derived.

Important questions remain about the giant polytene chromosomes. For example, how are they formed; what signals their formation; what is the origin, structure, and function of the chromosomal fibers that mediate ectopic pairing; and what is the internal structure of the bands? Moreover, because they represent the diploid interphase chromatin structure in a macroscopic format, polytene chromosomes will remain useful for mapping the location of chromosomal proteins and DNA sequences.

B. Structure of Diploid Nuclei

Because of their small size the diploid cells of *Drosophila* were, until recently, largely passed over for cellular and subcellular studies in favor of other organisms. At the cellular blastoderm stage, each cell averages about 7 μm in diameter, with some diploid cells of later stages being even smaller. The egg itself is about 1/250 the volume of a *Xenopus* egg. *In vitro* fertilization of *Drosophila* eggs has not been possible, so synchronization of a large number of eggs is more difficult than with amphibian eggs. More recently, however, the close scrutiny of molecular and developmental biologists and improvements in microscopic techniques have made the embryo an irresistible target of cell biologists.

The syncytial blastoderm stage of the *Drosophila* embryo, during which a monolayer of nuclei lies directly under the surface of the embryo, presents an essentially flat sheet of nearly identical nuclei (reviewed in Campos-Ortega and Hartenstein, 1985). This sheet of nuclei is the product of 13 rapid, synchronous nuclear divisions. The last three divisions occur at progressively longer intervals and are characterized by shallow mitotic gradients originating near the poles and moving toward the center; these gradients may be required to accommodate the extended movements of anaphase in the crowded embryo subsurface. There is apparently no preferred direction of nuclear division, and some nuclei, having failed to complete mitosis successfully, are lost from the blastoderm layer without consequence (Minden et al., 1989; Sullivan et al., 1990). During the interphase of the fourteenth nuclear cycle, the blastoderm nuclei become segregated by plasma membranes, and gastrulation and germline elogation begin. During this phase, the embryo surface folds and stretches, moving cells into the interior and from the ventral to the dorsal side. The fourteenth mitosis is not synchronous, but rather occurs sequentially in patches of mitotic domains reminiscent of the pattern of the embryonic fate map, but at higher resolution (Foe and Odell, 1989). The exact position of the patches and the number of cells in each patch are slightly variable, but the shape, relative positions, and temporal order are similar from one embryo to another.

These attributes of the *Drosophila* embryo are important from several points of view. The absence of interstial plasma membranes, for example, has two happy consequences: images can be optically clear, and nuclei can be labeled by injection of molecules into living embryos at a site removed from the field of observation. The label is then incorporated into cortical structures and nuclei where they can be visualized *in vivo*, even after the nuclei are enclosed by plasma membranes (Minden et al., 1989). For example, fluorescently labeled histones have been incorporated into chromosomes to reveal chromosome structure throughout the nuclear division cycle, providing an unprecedented glimpse of the very early events of chromosome condensation (Hiraoka et al., 1989) and mitosis (Hiraoka et al., 1990b). The same technique was used to follow cells during the complex movements of gastrulation, permitting the first direct cell lineage analysis in the early *Drosophila* embryo (Minden et al., 1989). These studies and others (Karr and Alberts, 1986; Kellogg et al., 1988; Sullivan et al., 1990) have exploited the timing of nuclear and cell division (the synchrony of nuclear division in the early divisions, the shallow mitotic gradients in the later ones, and the mitotic patches in the cellular blastoderm) to study chromosome and cytoskeletal structure during the cell cycle, with a special emphasis on mitosis.

Attention has recently been focused on cell cycle mutations that affect embryos before and during blastoderm formation (reviewed in Glover, 1989). Molecular analysis of the *Drosophila* mutation *string* has revealed its homology with *cdc* 25 (Edgar and O'Farrell, 1989), a cell cycle regulatory gene previously identified in *Schizosaccaromyces pombe*; the gene encoding *Drosophila* cyclin A has been found to be homologous to cyclins previously identified in *S. pombe*, clam, and sea urchin (Lehner and O'Farrell, 1989). Staining of embryos, using monoclonal antibodies against cyclin A or *in situ* hybridization to the *string* messenger RNA, has produced a vivid picture of their spatial distribution in the mitotic domains of cycle 14 embryos. Moreover, because the timing of mitosis in each domain has been determined, the temporal pattern of expression of both these genes can be inferred from the spatial distribution of their respective signals. Another cell cycle mutation, *daughterless-abo-like*, has been characterized by direct observation of fluorescently labeled cytoskeletal components *in vivo* to reveal directly the spatial and temporal phenotypes at subcellular resolution (Sullivan *et al.*, 1990). Results indicate that the semilethality of this mutation derives from partial to complete failure of centrosome separation, resulting in faulty spindle attachment and abnormal segregation at mitosis.

These results demonstrate the generality of at least some aspects of cell cycle control over great evolutionary distance, and the potential of the *Drosophila* embryo system for examining the dynamics of cell cycle progression. The combination of the powerful genetic and molecular tools available in *Drosophila* with sensitive microscopic techniques should continue to yield insights into both the mechanics and regulation of diploid mitosis.

II. Sample Preparation

A. A Word about Materials

High-resolution optical microscopy requires sample preparation and optical technique that are fastidious in every detail. It is essential that chemicals and supplies be of the highest purity. In most cases, microscopic samples are small and only tiny amounts of reagents are required, so the additional cost of quality products is relatively little and, weighed against lost preparation time, insignificant.

The solvents used in these procedures are HPLC grade, the buffer salts reagent grade unless noted, and urea is ultrapure grade (Schwartz-Mann). Water quality is important as well and that used in the experiments below has a resistivity of 18 MΩ and is freshly prepared by mixed-bed resin (Milli-Q, Millipore, Inc.) purification of house-distilled water. Where available, detergents are freshly diluted from ampule-stored stocks (Surfact-amps, Pierce) to avoid peroxides, and all other detergent stocks are freshly diluted. Where available, other regeants are EM grade.

Microscope slides and cover glasses are Gold Seal (Clay Adams) and are cleaned before use with Milli-Q water and nonfluorescent lens paper. To minimize the presence of minute glass shards, frosted slides are not used. Glass and plasticware are clean and free of dust and other particles. All of the materials are kept in secondary wrappers in drawers or closed cabinets to prevent dust or chemicals from coming into contact with them.

B. Polytene Chromosome Methods: Isolation and Urea Stripping of Polytene Nuclei

There are probably a few questions remaining about the structure of polytene chromosomes in whole glands, but the less explored questions concern the structure of individual bands and such fine details as ectopic fibers. For this purpose, isolated nuclei, either intact or stripped with urea (Sedat and Manuelidis, 1977), are the ideal starting material. Microdissected chromosomes have been used for similar studies (Hill *et al.*, 1987), but the following method has the advantage of being easy for the novice to accomplish and probably involves less stress on the chromosomes.

Start with 3–5 bottles of third instar larvae at the crawling phase (enough to yield about 100 animals at the correct stage), about 2–3 hours before pupariation (or another developmental stage, if that is required). Any standard food may be used for culture. The chromosomes are usually more polytenized if the larvae have been cultured at 16–18°C, but, unlike chromosome "squashing," the success of this technique does not depend on the size of the chromosomes. Of course, the larger chromosomes are easier to "read" than the smaller ones, so I prefer to use the same larvae for both squashing and for nuclear isolation.

Using a sable water painting brush, remove as many larvae as possible from the wall of the culture bottle to a test tube containing 15% sucrose on ice and put the bottles back in the incubator for later use. The larvae should float and will become clean of the food fragments through their movements. Remove clean larvae, four at a time, to the depression slide containing cold buffer A [80 mM KCl, 20 mM NaCl, 0.5 mM EGTA, 2 mM EDTA, 15 mM PIPES buffer (pH 7.0), 15 mM β-mercaptoethanol, 0.5 mM spermidine, 0.2 mM spermine, (Belmont *et al.*, 1989)]. Holding the mouthhooks in one forceps and the tail in other, pull the larva apart and retrieve the salivary gland and the attached fat body. Remove any other fibrous tissue, but unless it is essential to examine uncontaminated salivary nuclei, the fat body can be left on to be removed in the process of nuclear isolation. After all the larvae are dissected, gather all four sets by the neck of the gland and transfer them to a drop (~50μl) of buffer A on a sheet of Parafilm that is in direct contact with the ice in the ice bucket. These steps are suggested to speed the process and to prevent tissue from being lost in a direct transfer to the centrifuge tube. Repeat this process until 75–100 animals have been dissected, going back to the culture bottles as required and starting additional buffer drops. This part of the procedure should take 1–2 hours with a little practice. Remove the glands to a 6 × 50 test tube with a capillary pipette (both previously heat-baked at 200°C for 2 hours). The glands can sit on ice for at least several hours without changes in morphology.

Allow the glands to settle to the bottom of the tube and remove as much buffer A as possible; add more buffer to rinse (volumes are not critical, fill the tube to about 5 mm from the top for all the rinses), settle again and remove top layer as before. Add buffer A with 0.1% digitonin to fill the tube about one-third full ($\sim 75\mu l$). Break up the glands by vortexing at the highest speed for 1 minute. Pellet for 10 seconds at low speed (20 g) in a clinical centrifuge at room temperature. The pellet will contain the unbroken glands and the supernatant the nuclei. Transfer the nuclei to a fresh tube, add more buffer A with digitonin to the unbroken glands and vortex as before; triturate using a drawn-out capillary to break up the resistant material and pellet. Repeat the steps until all of the glands are broken into nuclei. Combine the supernatants and pellet the nuclei at higher speed (200 g) for 5 minutes. Carefully remove the white film from the top (this contains the fat and membrane debris), and then the rest of the supernatant liquid. To check the supernatant for nuclei, add DAPI to a final concentration of approximately 0.1 $\mu g/ml$ and examine the nuclei using a fluorescence microscope equipped with a UV fluorescence filter set (excitation, 350 nm; emission, 450 nm). Centrifuge again at increased speed if necessary. Resuspend the loose pellet containing the nuclei in 100 μl of buffer A with 0.1% digitonin. The nuclei can be kept on ice in a refrigerator for several hours or several days, although some degradation can be seen after prolonged storage.

Just before use, add an equal volume of 0.5% Brij 58, mix, and spin at 200 g for 5 minutes. Remove the supernatant and suspend in a convenient volume of buffer A and use immediately. The nuclei will remain whole, and they can be observed directly. However, it is easier to see band morphology if the chromosome arms are freed of the nuclear envelope. To strip the nuclear envelope, put a drop of nuclei on a clean microscope cover glass of the appropriate size and thickness (22 × 40 mm, size 1 1/2, is usually a convenient size to start). Add equal volume of 1 M ultrapure urea and 0.1 $\mu g/ml$ DAPI in buffer A, and mix the two solutions by turning the cover glass and blowing on the drop with a drawn-out capillary. You should already have prepared a microscope slide with two 22 × 22 mm No. 1 cover glasses fastened with nail enamel at an appropriate distance to support the cover glass holding the specimen (see Fig. 1). Immediately after mixing the nuclei with the urea solution, pick up the cover glass with the prepared slide and seal it with the nail enamel. This should be done as quickly as possible to avoid evaporation. The nuclei can be observed for at least several hours, but some degradation may be

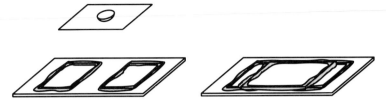

FIG. 1. Mounting three-dimensional samples. Two 22 × 22 mm cover glasses are fixed to a microscope slide with nail enamel. The specimen, in a drop of mounting medium, is placed on a 22 × 40 mm cover glass and picked up with the prepared slide. The large cover glass is sealed with nail enamel.

apparent after keeping the specimen overnight. Note that the suggested mounting medium in this case is buffer A, which has a low refractive index. As indicated below, the mismatch in refractive index will adversely affect resolution unless an immersion oil with a high index of refraction is used, or if a water immersion lens is used.

C.   Embryonic Chromosome Methods

Preparation of embryos is simple, but it is important to remember that mitosis is extremely sensitive to changes in environment. The preparation of the embryos should be as gentle and as fast as possible. For this reason, I prefer a bulk fixation procedure (Mitchison and Sedat, 1983) that involves dechorionation with bleach and devitelinization with a mixed solvent and heat shock. Hand dissection techniques are slow and laborious and, if not done proficiently, can induce artifacts like premature mitosis.

1.   COLLECTION OF EMBRYOS

In order to use bulk preparation methods, it is desirable to collect a large number of approximately synchronous embryos, at least 0.25 ml for the following protocol. The easiest way to get large volumes of eggs is to use a large population of flies (Elgin and Miller, 1978). If a population cage is not available, it is also possible to make minicages using 500–1000 flies and scale down the volumes and vessel sizes, but the recovered volume of eggs may be less than optimal due to sticking on glass surfaces. In either case, the population should be fed a rich paste of yeast and water at least a day or two prior to collection to ensure maximum egg yield.

Before collecting flies, the cage is fed using the same kind of food plate as will be used to collect in order to encourage the flies that have been holding eggs to lay them before the actual collection. Flies will hold their eggs to hatching if they find the food undesirable or are otherwise unhappy. Food plates and yeast paste are stored in a refrigerator away from chemicals or other unpleasant odors, the food is warmed to 25–27°C in a microwave or under a heat lamp before using it, and the plates are changed frequently, at least twice daily and more often on collection days. The fly room is kept moist with a humidifier if necessary. The flies lay better at night, so the fly room is on a reverse day–night schedule using an inexpensive darkroom red light during the day and incandescent lights on a timer (available from any hardware store) at night. Loud noises and constant vibrations are avoided (for example, the fly room is vacuumed only when the cages are being changed). The cage is maintained on a regular rotation so a new cage is started every 2 weeks. The cage is a cube of 18 inches per slide and is seeded with 175–200 ml of flies which were raised from 0.25–0.5 ml of eggs in each of 16 1-liter jars (Nalgene) filled to a depth of 1 inch with standard *Drosophila* food and about 10 ml of dry yeast. The first few days are not very productive, so the cage is changed on Friday to be ready for use on Monday.

For collection, a simple, high-agar food is preferred because it has a stiff surface that the eggs will not penetrate (88 g of agar, 360 ml of molasses, 2.5 liters of water; autoclave for 30 minutes, cool to 50°C, add 20 ml of ethyl acetate). Pour the food into shallow, rectangular plastic plates, about 6 × 8.5 inches, like the ones used in supermarket meat departments (obtained from a local restaurant supply house), or similar convenient dishes. For collection, spread a thin yeast paste in stripes over the collection plate; if the plate is completely covered with yeast, the flies may be discouraged from laying. The timing of collection will vary according to the life cycle time that is important and the size of the desirable window. For precellular blastoderm embryos, the collection in 1.5 hours with a 1-hour aging period to obtain embryos in nuclear division cycles 7–14.

The buffers and other solutions should be aerated to ensure that the embryos do not become anoxic, which leads to abnormal morphology (see below). If the number of embryos is very large, the solutions can be sparged with pure oxygen before use to avoid oxygen deficit. When the eggs are at the proper age, wash them off the plates with a fine paint brush and 0.7% NaCl, 0.1% Triton X-100 into a glass baking dish, then pour them through two stainless steel sieves, one 40 mesh to retain dead flies, the second 20 mesh ("Cellector" from Bellco), to catch the eggs. An inexpensive version can be made using Nitex of the appropriate mesh size glued over a plastic cylinder. Rinse the eggs to remove residual yeast and plunge them into 50% household bleach (2.5% sodium hypochlorite) in buffer A for 1.5 minutes to dechorionate them, stirring gently to keep them from falling to the bottom of the beaker. Collect the embryos in the 20 mesh sieve and plunge them into a beaker of

NaCl/Triton. Collect them in the 20 mesh sieve, and rinse with a stream of NaCl/Triton until the smell of bleach is gone. Try to remove as much detergent solution as possible before transfer to fixative, but don't allow the embryos to dry out.

The entire process, from the time the eggs are first rinsed until the tube is placed on the shaker should take no more than 7 minutes. In any case, embryos should be examined microscopically before use to be sure that the morphology is normal (see below).

2.   FIXATION OF EMBRYOS

The fixative solution should be prepared as follows: 1 ml of $10 \times$ buffer A, 8 ml of water, 10 ml of heptane equilibrated in a 25×250 mm screw-capped tube. It can be made up prior to collecting the embryos. Just before rinsing the embryos from the collection plates, add 1 ml of 37% formaldehyde freshly made from paraformaldehyde [1.8 g of paraformaldehyde (Polysciences) in 5 ml of water, add 70 μl of fresh 10 N NaOH, boil to dissolve, cool, and filter]. Mix well.

Transfer the dechorionated eggs from the sieve to the fixative: use the buffer A-saturated heptane to collect the embryos in one corner of the sieve (the heptane will not go through the wet sieve), and, using a Pastuer pipette, transfer at once to the test tube containing the fixative. Avoid transferring any residual aqueous phase, as the Triton X-100 will cause problems in subsequent steps. Lay the tube on its side and shake at about 350 rpm for 15–20 minutes.

Taking care to avoid transferring excess solvent, remove the embryos to a 25×250 mm screw-capped test tube containing 10 ml of heptane/10 ml of methanol which was previously cooled to 70°C, pack the tube on its side in crushed dry ice, and shake the vial for 10 minutes. Quickly transfer the tube to a 37°C water bath and shake until no more embryos sink. The fixed embryos can be used for immunofluorescence studies, DNA *in situ* hybridization analysis, or direct examination of the chromosomes. They may be stored in methanol for weeks or months in the refrigerator, although they are best used within a few days for optimum morphology.

Before use, the embryos should be rehydrated by passage through a methanol/buffer A series. They may be stained with DAPI and mounted under a supported cover glass as described for the polytene nuclei, used for *in situ* hybridization (Hiraoka *et al.,* 1990c), or prepared for indirect immunofluorescence (Fuchs *et al.,* 1983; Karr and Alberts, 1986; Hiraoka *et al.,* 1990b). Typically, a small number of embryos are preincubated with PBT (0.13 N NaCl, 7 mM Na$_2$HPO$_4$, 3 mM NaH$_2$PO$_4$, 0.1% Triton X-100) containing 10% normal goat serum, incubated with the appropriate dilution of the primary antibody in PBT, washed with PBT, and stained with fluorescently labeled secondary antibody. The embryos can be counterstained with DAPI in buffer A.

The embryos prepared as outlined above will be clustered in nuclear cycles 7 to 14, but it is nearly impossible to get them more synchronous. However, the density of the nuclei will reflect nuclear division cycle number, so if the microscope is equipped with a reticule, the embryos can be staged by counting the number of nuclei in a given area. In an area of 2200 μm^2, cycles 10, 11, 12, 13 and 14 will have 4–6, 7–10, 12–19, 27–44, and 60–84 nuclei, respectively (Foe and Alberts, 1983).

3. Assessing Nuclear Morphology

After every embryo preparation, the nuclear morphology should be examined to determine that the embryos were properly treated during dechorionation and fixation. There are treatments from which embryos can recover but which will cause temporary changes in nuclear architecture, principal among these being oxygen deprivation, or anoxia (Foe and Alberts, 1985). Incorrect salt, detergent, or other solvent concentrations can also adversely affect the chromosomes. In fact, it is important to bear in mind that the procedures used here are intended to preserve chromosome structure maximally; other variations may be necessary to preserve optimally actin, microtubules, or other components.

Anoxic embryos can have several characteristics, depending on how severe the oxygen deprivation was. Complete anoxia is difficult to accomplish accidently, but partial anoxia is common. Therefore, if there is any doubt the condition of the embryos, it is probably better to start over. In anoxia, the nuclei are swollen to about 1.5 times normal size, with the chromosomal material clinging to the nuclear envelope in small clumps rather than being dispersed throughout the nucleus. This is clear by focusing up and down through the nucleus. The most impressive sign of anoxia is premature chromosome condensation in which the individual chromosome arms become separate and highly condensed, lying in stripes along the nuclear envelope, in a Rabl orientation (i.e., telomeres toward the bottom of the nucleus, centromeres toward the top). Comparison with photographs can be helpful, but it may be more useful to prepare anoxic embryos intentionally, as described in Foe and Alberts (1985), and examine them in 3D.

Another good way to evaluate morphology is to find anaphase nuclei. The chromosomes of anaphase should be bundled at their centromeres but well separated at the telomeres. resembling a fan (see, for example, the nuclei labeled *in vivo* in Hiraoka *et al.,* 1989). Individual arms should be compact and relatively straight. Discard embryo preparations in which the anaphase chromosomes appear to have separated from the spindle or to be in several clumps along the mitotic axis. When the solvent is contaminated, or when the detergent concentration is suboptimal, the chromosomes of anaphase figures resemble the melted clocks in a Salvador Dali painting. If there is any doubt about the morphology, all solutions should be freshly prepared with special attention to the purify of the components.

III.   Image Acquisition and Analysis

A.   Introduction

Quantitative microscopy demands an integrated approach to the experiments. Success in the final product is guaranteed only when the entire process from the mounting of the specimens, through the choice of lens and camera system, to the final computational treatment of the data, is done with the end product in mind. Much has already been written in depth on the theory of image processing (Castleman, 1979; Agard, 1984; Agard *et al.,* 1989), the use of fluorescense in microscopy (Taylor *et al.,* 1986; Taylor and Salmon, 1989), and the use of the CCD camera for biological imaging (Hiraoka *et al.,* 1987, 1990a; Carrington *et al.,* 1989). Rather than duplicate that work, it is my goal to take an alternative approach for those who might prefer to start with a less mathematical treatment, or who would like a detailed step-by-step picture of how 3D imaging is done. In any case, the coverage here should be sufficient to understand subsequent discussion on obtaining images for detailed analysis.

1.   THE POINT SPREAD FUNCTION

When light passes through a lens, it is spread in a characteristic way. This means that no matter how fine the original light beam is, it will spread out to a finite size. It is this property of the lens that limits its resolution. Like resolution (see below, Section III,B,1), the amount of spreading is dependent on the wavelength of the light, the design of the lens, and the refractive index of the immersion medium (reviewed in Hiraoka *et al.,* 1990a).

The point spread can be observed by looking at a small fluorescent bead (Fluoresbrite beads, Polysciences) on a dark background in a conventional fluorescence microscope. Suspend a small volume of beads in 9 volumes of ethanol; drop 10 μl of this suspension on a cover glass and allow them to dry completely. Put a 2- to 4-μl drop of glycerol or mounting medium suitable for fluorescence microscopy on the cover glass and use a slide to pick it up. Seal the cover glass with nail enamel, let it dry, and mount the slide on the microscope. It is best to a high-numerical-aperture immersion lens for this experiments.

When the bead is precisely in focus, it looks like a solid disk of light, but above and/or below the plane of focus, it will expand into a set of rings, called Airy rings. If the bead is very tiny, less than half the resolution limit of the lens (0.1 μm; Pandex Laboratories), it acts like a true point light source and the pattern you observe represents the true point spread function of the microscope. The pattern observed for larger beads may resemble the point spread function, but it results from the superposition of the point spread from all the points of the object. A mathematical way of expressing this is that every point of light (represented by a δ function) is *convolved* with the point spread function to yield the observed image. Further, the general shape of the experimentally determined point spread function is quite different from the one calculated based on first principles (Hiraoka *et al.*, 1990a).

Ideally, the point spread function should be symmetric above and below the plane of focus but, more likely, the pattern on one side of focus is a distinct set of rings while the other side appears to contract into a smaller irregular disk. This is because the optical path length is not precisely correct, resulting in spherical aberrations that will reduce resolution. To achieve the best resolution, the problem ought to be corrected; a few solutions are suggested below, in Section III,B,2.

For the confocal microscope, the point spread function extends very little above and below the plane of focus, but it is finite, so resolution is still limited by the properties of the lens (Keller, 1989; Wilson, 1989). Thus, spherical aberration will affect the in-focus image of a confocal microscope to the same degree as that of a conventional one, so adjustments should be made in oil and lens choice to achieve maximum resolution.

2. THE FOURIER TRANSFORM

Many of the techniques in use for image processing depend on a mathematical tool called the Fourier transform. The reader is referred to Castleman (1979) and Hecht (1988) for a more complete treatment, but there are a few properties that are especially relevant to image work that I would like to point out.

A Fourier transform is an alternative way of representing a function. To simplify the explanation, I will first discuss a simple case of a periodic function and then talk about images. Every periodic function can be considered the sum of an infinite series of sinusoidal functions. Take for example the square wave in Fig. 2A. It can be described by the sum of a series of sine waves, the first three of which are drawn in Fig. 2B. The successive sums of these are shown in Fig. 2C, each one superimposed on the square wave. The shape of the curve

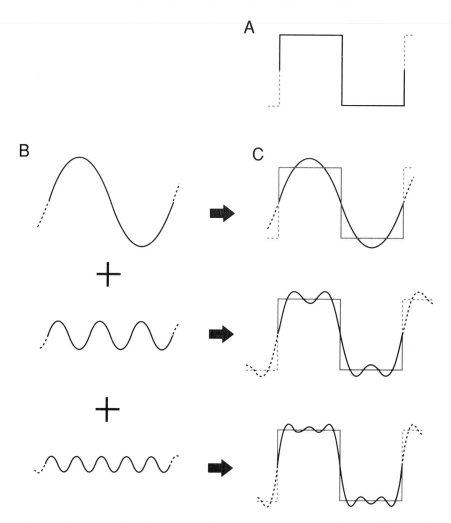

FIG. 2.   A Fourier series of a square wave. One cycle of the square wave function is shown in A. The successive terms of the series are represented in B and the cumulative sums are represented in C. See text for further explanation.

grows more and more like the square wave with every addition to the series. After adding an infinite series of sine and cosine functions, the series sum will look identical to (i.e., will converge to) the square wave. Because the component curves are simple periodic functions, they can be described by their frequency (how many cycles per unit of distance?), amplitude (what are the maximum and minimum values?), and phase (is this a sine or a cosine wave?).

In this way, it is possible to describe any function either the "normal" way, i.e., in the spatial domain, or by its component frequencies, phases, and amplitudes, i.e., in the frequency domain. The decomposition of the square wave function into its component frequencies and amplitudes constitutes "doing a Fourier series" of the function. The example of the square wave is a one-dimensional one, but the procedure can be extended to any number of dimensions. Likewise, it is possible to do Fourier series of continuous functions, like the square wave, which are defined at every point, as well as discrete functions, like the ones representing the values in a digitized image.

But how can an image be represented as a Fourier series, when almost no biological images are themselves periodic functions? The answer is that almost any function, even one that is not periodic, can be represented with a Fourier series. Take, for example, the function that describes the intensity of an image along a particular straight line (Fig. 3A). The function will be complicated, but it will be of finite length, running from, say, 0 to 10 cm. All that is necessary is to connect many copies of this function end-to-end, and we now have a function with period 10 cm. This complicated, but periodic, image function can be transformed and described as a set of simple periodic functions, even for a 3D image expanded in three dimensions. One consequence of doing this expansion is that, while the number of terms in the Fourier series is still infinite, the difference between successive terms in the series becomes very small. Another way to think about this is that the function representing the frequencies becomes a continuous line. Thus, instead of taking about the sum

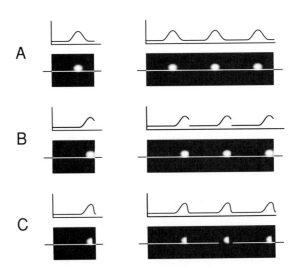

Fig. 3.   Fourier analysis of an image. See text for explanation.

of a series of functions, it is easier to describe the Fourier "series" as an integral, which is called a Fourier transform. This is the form encountered in image-processing texts, but is still helpful to think of it as a sum of a very large number of simple sinusoidal waves.

The Fourier transform is not the only kind of transform, but several of its properties make it well suited to our purposes. Most important is that the process of doing a Fourier transform is made easy by a computational technique called the fast Fourier transform, or FFT (Cooley and Tukey, 1965). Using this algorithm, as a part of a computer program, or programmed permanently into an FFT board for a computer, anyone can have access to the Fourier transform.

One might wonder what Fourier transforms have to do with image processing. The answer is that Fourier transforms have some very useful properties, especially with respect to the operation of convolution. Among them are

1. The Fourier transform of a Fourier transform yields the original function

$$\mathscr{F}^{-1}\{\mathscr{F}[f(x)]\} = \mathscr{F}^{-1}[F(x)] = f(x)$$

2. The Fourier transform of the convolution of two functions is the product of the Fourier transforms of the individual functions.

$$\mathscr{F}[f(x) * g(x)] = F(x) \cdot G(x)$$

These two properties together mean that the nasty-looking integral that describes the process of convolution can be changed into a simple multiplication: first the Fourier transforms of the two functions are determined separately, then the transforms are multiplied together, and finally the product is Fourier-transformed back again:

$$f(x) * g(x) = \mathscr{F}^{-1}[F(x) \cdot G(x)]$$

Convolution, as mentioned above, is a way to describe mathematically the process by which light is changed by passage through a lens to produce the observed image. That is, the function that describes the object is convolved with the point spread function to yield the observed image. If the point spread function can be determined accurately enough, it can be used to calculate and remove the contribution of the spreading function from a 3D image. Using a point spread function determined by collecting very accurate image data of very tiny beads using a CCD camera (see below), the point spread function for a 63 × lens was determined for the first time; this information is used in an image analysis program described below. Other processing techniques make use of convolution, including smoothing, filtering, and correlation techniques; the computations are often speeded tremendously by doing Fourier transforms. These are described in more detail in Castleman (1979) and Agard et al., (1989).

There are several things to notice about the short series illustrated in Fig. 2. The first is that the shape of the square wave is most closely approximated by the first curve in the series, the lowest frequency (and highest amplitude) curve, but the sharpness of the corners derives from the addition of the highest frequency (and lowest amplitude) terms. In fact, all of the terms but the first are of higher frequency than the original function. The sharper the feature is, the higher the frequency that is required to represent it. The frequencies contained in an ideal image will be limited by the point spread function of the lens. A real image, however, also contains random noise and may contain other nonrandom features: extraneous lines or "glitches." Some of these may cause the image to contain higher frequencies than the limit imposed by the point spread function. If the image data are properly oversampled (see below), the frequencies representing the true image can be separated to some extent from the noise using the appropriate "filters."

The next thing to notice is that the shape of the last composite curve in Fig. 2 deviates from the square wave more at the corners of the function than at the horizontal or vertical portions. Until an infinite number of terms has been added, some of this deviation will remain. The effect is called "ringing" and is a commonly encountered artifact of Fourier procedures, especially at the edges of images. Remember that to take the Fourier transform of an image, the original image first must be represented by a function that is repeated an infinite number of times in each direction, so it looks like a crystal of images. It is often the case that the intensity of the image at one edge is quite different from that at the other edge (Fig. 3B), so that, in the array of images, there is a discontinuity, much like the corner of the square wave function. To avoid this problem, images can be carefully cropped so the edges are all at the background level, or the edges can be smudged (Fig. 3C) using programs designed for that purpose. Ringing can sometimes be noticed at the edge of features within the image after image processing. When this happens the image might be said to be "overprocessed," meaning that, as a result of processing, the image now contains features it didn't have before. Except in cases where the high contrast is introduced for display purposes in a final image, this artifact should be avoided.

B.   Equipment

1.   CHOICE OF LENSES

Because it is the only light-gathering element of the system, the objective lens is the most important factor in determining the quality of the final image. The usual warnings for microscope use are even more imperative for quantitative microscopy: the lenses must be kept free of contaminating dyes and particles that will scratch them. When a new lens is purchased, it should be examined under a stereomicroscope for scratches and imperfections and thoroughly cleaned before use. The lens paper should be checked under fluorescent light to be certain it is not treated with fluorescent brighteners that can rub off on any surface and increase background in the 400 nm range.

Lens characteristics are usually coded on the objective: plan refers to a lens which is corrected so that the entire field will be in focus at the same time; achromat means that the lens is corrected for spherical and chromatic aberrations in blue and red; apochromat means that the lens is corrected for spherical and chromatic aberrations in dark blue, blue, green, and red light. All of the characteristics of the lenses are optimized for objects on the surface of the cover glass, which can lead to problems when doing 3D microscopy. Plan lenses are best for the image-processing technique called deconvolution (described below) because, for these lenses, the point spread function is the same over most of the field, i.e., it is nearly shift invariant.

The resolving power of the microscope in the plane of focus (x and y) is related to the numerical aperture (NA) of the lens. The numerical aperture is given by

$$NA = n_0 \sin \theta$$

where n_{01} is the refractive index of the immersion medium and θ is one-half the angle of the cone of light collected by the lens. Notice that NA is dimensionless. For a typical $60 \times$ oil lens, n_0 is 1.515, θ is about $67°$, and the NA is 1.4. The Raleigh resolution limit of the objective is given by

$$d = (0.61 \ \lambda)/NA$$

where d is the resolution and λ is the wavelength. Thus, the higher the NA, the better the resolution.

For 3D microscopy, the resolution in the direction of focus, i.e., the z direction, is as important to consider as the resolution in the plane of focus. The resolution limit in the direction of focus is given by the depth of field of the lens, and is related to the numerical aperture of the lens, the index of refraction of the immersion medium, and the wavelength of light (Taylor and Salmon, 1989):

$$D = \lambda \left/ \left(4 n_0 \sin^2 \frac{\theta}{2} \right) \right.$$

Substituting as above, the depth of field, hence the resolution in z, for this lens is about 0.31 μm. Keep in mind that the depth of field refers to the depth of truly in-focus material; out-of-focus material also contributes to the image as predicted from looking at the point spread function of the lens. Unless this out-of-focus contribution is removed, it will diminish the apparent resolution.

2. IMMERSION MEDIUM

Lenses are intended to be used with a particular refracting medium. If no immersion medium is specified, the lens is a "dry" or "air" lens and should not be brought into contact with anything. Likewise, the ideal thickness of the cover glass is usually indicated on the body of the lens; some lenses are intended to be used without a cover glass, for example, in the case of water lenses that can be used with physiological buffers. To minimize spherical aberration, it is important to match the index of refraction of the immersion and mounting media and to use the correct size of cover glass (Hiraoka *et al.*, 1990a). All of these elements contribute to the optical path length and must be matched to that of the lens to assure that the point spread function is as close to ideal as possible.

An oil immersion lens is commonly used for high-resolution work. Because of the high index of refraction of the oil, these lenses usually have the highest NAs. Immersion oil has several characteristics that will determine its performance, the most important of which are refractive index (actually, the refractive index measured at $20°C$ using 589 nm light) and dispersion (the variation in absorption and index of refraction as a function of wavelength). Some oils will autofluoresce at lower wavelengths, some will absorb light, etc. When experimenting with oils, be sure that they were intended for use as immersion media and that all the important properties are known.

If a specimen has previously been examined using a different oil on a different microscope, the old oil should be removed completely. To avoid dust and other contamination, the tip of the oil dropper should never touch any surface. Once the slide has oil on it, it should be put on the microscope, as dust is likely to collect in the oil, possibly endangering the lens. Be sure that no air is trapped in the oil as this will significantly degrade the image.

When working in 3D, it will be necessary to focus in areas of the specimen that are bathed in mounting medium, which is typically of lower refractive index than that of the immersion oil. This will result in a spherical aberration; this aberration can be partially corrected by increasing the refractive index of the immersion oil (Hiraoka *et al.*, 1990a). The optimum oil will vary, depending on the nature of the specimen and mounting medium, so tests of the point spread function should be done using beads mounted as the specimen will be, and using a series of oils of varying refractive index (Cargille Laboratories, Cedar Grove, NJ). Temperature should be rigorously maintained as well, since refractility is temperature dependent (keeping in mind that the microscope is an effective heat sink that requires hours or days to equilibrate after the room temperature is changed). The combination of mounting medium, cover glass, and immersion oil showing the most symmetrical point spread function should be used for data acquisition.

An alternative is to use a water lens without a cover glass on a specimen mounted in aqueous buffer. Although the NA of such a lens is lower than that of an oil immersion lens (NA = 1.2), the optical path of the microscope should be ideal without modifications (Hiraoka *et al.,* 1990a). If they should become available, high-NA lenses with correction collars could also be used to physically change the optical path length and correct the spherical aberrations caused by mismatched mounting media (Hiraoka *et al.,* 1990a).

3. AUXILIARY MAGNIFYING LENSES AND SAMPLING

In addition to the objective lens, many microscopes are equipped with supplementary magnifiers. This is "empty" magnification in that it does not increase resolution. However, this lens can be useful in attaining the best results after image analysis, because an accurate numerical solution depends on "oversampling" the image data (Castleman, 1979). That is, the image should be spread over enough detectors to see all the details.

This concept is illustrated in Fig. 4. To take an extreme case, if the image detector was a single photomultiplier tube ($n = 1$), we would only know how much light came from the entire image with no details. If the same area of the image were sampled with 16 smaller detectors, we would have a much better idea of where the light and dark areas of the image lie. However, we can recover maximum resolution only if there are more detectors (or, in our case, pixels) than we need to distinguish all the light and dark areas that the image contains. To obtain the best analysis, the data should be at least $2\times$ oversampled. In Fig. 4, $n = 256$, or 16 pixels/edge, allows us to just resolve the two diagonal lines.

To give an example, say the resolution limit of a $60\times$, 1.4 NA lens has been experimentally determined to be about 0.19 μ at 450nm, so for a DAPI image, there should be at least two pixels/0.19 μ. Direct measurement on OM-1 using a Ronchi reticule (Nikon) reveals that, using this lens, each pixel corresponds to 0.1126 μm of the specimen. This means that there are only 1.6 pixels/0.19 μ, so the data are not optimally oversampled. On the other hand, if a

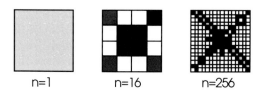

n=1 n=16 n=256

FIG. 4. A demonstration that the number of pixels/unit area will determine how much resolution is actually recovered by the imaging device.

1.5 × magnifier is inserted, the value increases to 1.5 × 1.6 = 2.4 pixels/0.19 μ, more than the minimum required for optimal results.

4. FLUORESCENCE FILTERS

When observing nuclear structure in fixed specimens of *Drosophila*, the chromosomes are usually stained with a single DNA-specific fluorescent dye, DAPI or Hoescht 33258, which are excited by light at about 350 nm and emit at about 450 nm. If other cellular components are to be visualized in addition to the chromosomes, other dyes must be used: fluorochrome-linked antibodies or lectins, for example. In order to localize each component separately, the light must be filtered to eliminate fluorescence from all but one fluorochrome at a time as the images are obtained. Filter sets are available for fluorescent microscopes through the microscope manufacturer, but these are typically either long- and short-pass filters or wide bandpass filters, which may allow light from several fluorophores to pass to the detector. The OM-1 microscope is equipped with highly selective interference filters (Marcus, 1988). (Omega Optical, Brattleboro, VT), with a band width of 10–40 nm, and the filters are antireflection coated to increase the efficiency of light transmission. In this way, three or four fluorophores can be visualized independently, without "cross-talk" between filters.

5. THE CHARGE-COUPLED DEVICE CAMERA

Precise determination of the point spread function for the microscope lens was first accomplished using charge-coupled device (CCD) camera. This measurement was essential to the use of the image analysis software (Hiraoka *et al.,* 1987, 1990a) described below. The design of the camera is discussed in detail in a number of excellent references (Kristian and Blouke, 1982; Aikens *et al.,* 1989).

The camera works using a silicon chip about 1 cm² (the size varies from one type to another) that is positioned behind a lens in the image plane, i.e., where the film would be. When the camera shutter is opened, light falls on the chip, ejecting electrons from the silicon. The electrons collect wherever the electrical potential is lowest. In order to get a "picture" on the chip, the area under the chip where the electrons collect is divided with barriers to form a square array of many thousands of "potential wells" (i.e., areas with low electrical potential surrounded by areas of high potential). The barriers are of two types. In one dimension, permanent channels are built, bounded by thin channel stops that prevent electrons from moving between channels. At right angles to the channels is a series of tiny polysilicon ribbons built into the underside of the chip that act as electrodes (Fig. 5). The ribbons are charged to produce a kind

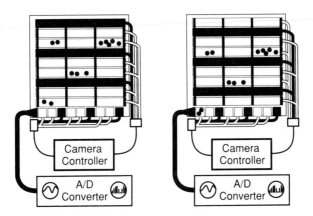

FIG. 5.   A diagram of a 3 × 3 pixel CCD chip showing two of three possible read-out cycles. The dark vertical lines represent the channel stops. The differently shaded horizontal rectangles represent the polysilicon ribbon-electrodes, three per pixel. The darkest are the highest potential, the lightest, the lowest potential. Electrons are represented by the small circles. The small squares along the lower edge of the chip are the collector. The camera controller changes the voltages applied to the electrodes so the potential well will move through the channels to the collector to be counted and digitized.

of "force field" to trap the electrons. The physical and the electrical barriers act together to create an array of square potential wells. While the chip is being exposed to light, each electron moves to the nearest potential well, so that the pattern of electrons in the chip, if they could be seen directly, would look just like the image. Each of the square potential wells corresponds to a single picture element, called pixel for short. A single chip can have 300 to several thousand pixels in each dimension.

Of course, the electrons collected in the chip cannot be seen directly; they must be removed from the chip in an orderly way and counted. This is done using the electrodes. There are three electrodes for every potential well: a high-, medium- and low-potential electrode, with the electrons from each well collecting at the lowest potential electrode. After the camera shutter closes, the potential applied to each electrode set is changed so that the "electrical boundaries" of each well move over by one electrode, shoving the electrons through the channel into the new potential well. At the edge of the chip, there is a collector that transfers the electron packets (one from each pixel) from all the channels in single file to be counted, one pack at a time. The process is repeated until all of the wells are empty and all the packets are counted. The electrons are counted using an analog-to-digital (A/D), converter and the numerical value corresponding to the number of electrons is transmitted to a computer.

The counting process is very efficient in that few electrons are left behind when the potentials change, which is essential with so many transfers. The accuracy and speed of the counting depend on the A/D converter, and are given in bits and hertz (Hz), respectively. Accuracy can be as high as 16 bits, or about 16,000 ($= 2^{14}$) distinguishable levels of electrons in each pixel, and speed can be as high as 500,000 Hz, or 500,000 pixels read per second. There is a trade-off between speed and accuracy, so that the fastest cameras are not the most accurate (about 12 bits). Even at the fastest speed, an image that has about 500 pixels on a side will need 0.5 second to read into memory, after which it will be read onto a magnetic disk to be saved.

The advantages of the CCD camera are many. First, CCDs count photons, so the response to light is linear until the potential wells fill up with electrons. Next, the chip does not bend or curl like film, nor is it scanned, as a video camera is, so it is dimensionally more stable than those two imagers. In addition, the CCD is sensitive to visible light (i.e., it has a high quantum efficiency in the wavelength range 400 to 1000 nm); 15–80% of the photons will result in electrons landing in a well, depending on the wavelength of the light and the type of chip. The chip can be thinned or coated to improve the quantum efficiency below 400 nm. If the chip is cooled (to a temperature of -50 to $-126°C$) to reduce thermal noise (electrons getting ejected from the silicon due to molecular motion) and/or special techniques are used in the A/D converter (Janesick *et al.*, 1989), the background noise can be very low, so that signal-to-noise ratio is good. CCD cameras have been expensive in the past, but newer models are priced comparably to some sensitive video cameras.

C. Obtaining the Image

1. CORRECTION FILES

Unprocessed CCD images are marred by small variations in intensity caused by the differential response of individual pixels. Before doing any data collection, the CCD has to be "calibrated" so that the raw data can be corrected. This is done by making a "correction file" that has a format similar to that of an image file, but instead of containing digitized intensity data, it contains pixel response data.

To make a correction file, it is necessary to produce an evenly illuminated microscope field using the same filters, lamp aperture, lens, and magnification as will be used when data are taken. Place 4 μl of the appropriate dye solution on the center of a 22 × 22 mm cover glass of the same thickness as the one you intend to use for the experiment (usually a size 1 1/2) and preferably from the same lot. Dye solutions should be as similar to ones being used in the experiment as possible, and made in the same mounting medium. Our laboratory typically uses 2% propyl gallate, 95% glycerol, and includes 30 mM Tris-HCl, pH 8.0, as some fluorescence is pH dependent.

DAPI will fluoresce only in the presence of DNA, so a different dye with a similar spectrum must be used. Start with coumarin 106-saturated glycerol/2% propyl gallate. Dilute as required, about 1/5 to 1/100. For fluorescein, Texas Red, rhodamine, and other dyes, start with a 1/10 dilution of antibody or avidin linked to the dye in mounting medium.

Pick up the drop with a clean slide. Turn the slide over and, before the dye solution has a chance to spread to the edges of the cover glass, seal the cover glass with nail polish to prevent dye from getting on the lens. Mount the slide, focus, and defocus by a small amount to minimize the effect of small dye aggregates in the solution. Take a single image to assess the signal intensity. Based on this value, calculate a series of exposure times that will cover the range of intensities in the samples. Most DAPI-stained samples will require 0.5–3 seconds exposure. Either a geometric series (i.e., 0.5, 1, 2, 4 seconds) or a linear series can be used.

After the exposures are taken and the software has calculated the correction file, the file must be evaluated. Slopes (expressed as ratios of the mean pixel values) usually range between 0.9 and 1.2 with correlation coefficients at least 0.998. By these criteria, there should be fewer than 1200 "bad" pixels out of 262,000. Some of the "bad" pixels are in fact consistently bad, having no response to light or being overly light-sensitive, but some are just pixels that received a much higher or lower than average dose of photons by chance. If there is a large number of bad pixels, it is probably because the dye sample was inhomogeneous or the lamp arc wandered, or perhaps stray light from the monitor or other light source in the room reached the camera. Be sure the lamp is warmed up and properly aligned and defocused before starting. Filter the dye solution before mixing it with glycerol, store the standard in the cold until use, prepare new dye standard solutions often, and work in a completely darkened room.

To evaluate the correction file visually, move the dye standard slide to a new field and take a corrected image using the correction file that was just made. This image should now be smooth, i.e., it should show only a small, random variation in intensity. "Scale" the image to display a narrow range of values. This will enhance any patterns in the image resulting from clumps of dye or other features on the slide that was used to make the correction file. If any occur, retake the picture over a third field to verify that the inhomogeneity was in the first rather than the second field chosen. The total range in intensities in the corrected image of the dye standard should be no more than 0.5–1% of the total full well signal.

2. MAKING OPTICAL SECTIONS

In order to get 3D image data, photographs of the object are taken at different levels of focus through the object. Each of the individual images is called an optical section and the sum of all of them is called a data set. The data set is obtained by moving the object just out of focus, taking a picture, storing the image on a computer disk, changing the focus (usually using a computer-controlled motor), and repeating the process until the object is once again out of focus on the other side. This tedious process can be automated by writing and executing a small computer program within the image system software, which is also responsible for formatting the data into a disk file that can be retrieved later. The steps are simple: measure the object roughly, calculate the number of sections required to achieve the desired resolution in z, place the object just out of focus, and direct the computer to take the pictures.

Measuring the Object. To write the data collection program, the diameter of the specimen must be measured. Put the center of the specimen in focus and change the focus by the estimated radius of the specimen (in micrometers) by moving the stage away from the objective lens, taking pictures to determine when the specimen is out of focus. For a diploid embryonic nucleus, this means focusing on the nucleus at its widest point and moving $2-3$ μm out of focus. Move the specimen up and down a few tenths of micrometers at a time until it is just out of focus, keeping track of the total number of micrometers traversed. When the first limit of focus is determined, define the other one in the same way: change the focus by a value equal to the estimated diameter of the object (in the case of a diploid nucleus, about 5 μm) and take a picture. Repeat until the correct level of focus is attained; the net distance traveled is the actual size of the specimen.

The Data Collection Program. In collecting data, it is safest to start with the objective lens close to the specimen and collect optical sections by moving the objective away from the specimen. This way, there will be no danger that a miscalculation of the number of required sections or the diameter of the object will cause the computer to push the objective through the specimen. Positive and negative values for the focus change commands will generally cause the specimen to be moved in opposite directions relative to the objective lens, and the "safe" direction should be committed to memory immediately.

The ideal number of optical sections per micrometer will be determined by a variety of factors: the desired resolution in z, desired time resolution for *in vivo* observations, the limits of safe exposure to the light (determined experimentally for each specimen and stain combination), the amount of storage space available on the disk, etc. As in the case of x and y, the z direction should be oversampled if possible, so collect two sections for each unit of the desired resolution. For analysis of fixed diploid embryonic nuclei, optical sections are obtained every 0.1–0.2 μm to obtain the optimal result after deconvolution. For live specimens, generally five sections through the nucleus (1 section/μm) is sufficient.

D. Image Treatment

1. IMAGE ANALYSIS

A 3D data set is usually treated with a series of computational procedures to gain the maximum value from it. These can be roughly divided into three types of procedures: processing, analysis, and interpretation. Processing refers to simple corrections for lamp flicker and bleaching, as well as cropping to make the most of disk space. After completion of the processing steps, the image appears very similar to the original.

Analysis refers to a computational image enhancement using one of three techniques, all of which make use of the experimentally determined point spread function to remove out-of-focus "contamination": a simple nearest neighbor (VAX-SIMPLE), an inverse filter, or full 3D deconvolution (DECON3D) scheme (Agard *et al.,* 1989). This step will yield an image with reduced contribution from out-of-focus material and with enhanced resolution; the methods are listed in order of speed and in reverse order according to the result. In the descriptions below, I refer to convolution, which is an operation in the spatial domain, but, in fact, the calculations are done in the frequency domain after Fourier transformation, and then back-transformed at the end.

Processing artifacts resulting from the analysis algorithms can be reduced by the application of some preliminary procedures to filter and smooth the image. Recall that Fourier transform methods can cause "ringing" around the strong features, especially ones at the edge of the image. Cropping should be done so the intensity of the pixels at the edge of the image are at the background level, or, failing that, that the edges have been blurred. Computational filters should also be applied to remove as much noise as possible because the processing software works on the assumption that the data are accurate. Typically, these two steps are incorporated into the analysis programs, but they can also be done in independent programs.

VAX-SIMPLE is a deblurring program that uses the point spread function to calculate the out-of-focus contamination from adjacent optical sections, on the premise that most such contamination results from objects that lie just beyond the focal plane. In my experience, this program works best when the interval between optical sections used in the computation is approximately 0.6–0.7 μm (M. C. Rykowski, unpublished). The results can be quite good, but they can suffer from residual background blur. Often, this is only a cosmetic problem in that the background can usually be distinguished from the objects of interest and further processing can sometimes remove it.

The next level of algorithm, the inverse filter method, is a single-step algorithm (as opposed to an iterative one) that uses a 3D point spread function and all the optical sections to calculate and remove the out-of-focus contamination. It takes more time than VAX-SIMPLE, with better results, but the final image can suffer from ringing near strong features. These can result in the loss of dim features that lie near strong ones. Agard et al., (1989) have suggested explanations for and possible solutions to the problem.

The very best results are obtained using an iterative program called DECON3-D that uses the full 3D point spread function and all the optical sections. This is done by actually blurring the observed image with the point spread function (i.e., convolving the point spread function and the observed image), and then subtracting the difference between the observed image and the blurred one from the observed image. This new image, or "guess" image, is blurred again and once again compared to the original observed image. The process is repeated many times, i.e., over many iterations, until the guess image looks just like the observed image when it is blurred with the point spread function, i.e., until the solution converges to a consistent solution. DECON3-D is helped by the addition of a "positivity constraint," that is, when any of the pixel values in the image becomes negative (a physical impossibility anyway), the value is changed to zero. It is also smoothed with a filter to prevent residual noise from causing problems. These extra steps prevent ringing and help the program to converge to a solution faster. Figure 6 shows a series of three optical sections, at 0.5-μm intervals, through a diploid nucleus in prophase. The upper three are the raw data, the lower have been treated with DECON3-D to improve resolution.

After completing the processing and analysis, the next step is image interpretation. Section data, whether optical or physical, whether they were obtained on a conventional, confocal, or electron microscope, are very hard to interpret in 3D. It is helpful to apply one or several programs to manipulate the stack of 2-dimensional images to appear 3D (Chen et al., 1989; Hiraoka et al., 1990b). One useful program, called PRISM, is used to draw stick figures through chromosomes or around nuclei, for example, to measure length or signal intensity, or to facilitate interpretation and display. Another group can be used to detect the edge of objects, or to create an artifical solid model based on the intensities of the objects.

Some programs combine optical section data to yield a computed projection that stimulates a 3D view through the image stack. The impression of depth can be enhanced by displaying two projections that are separated by 10–15°C of rotation, called stereo pairs. Because some people cannot "fuse" stereo pair images, even with special spectacles, etc., it can be useful to create a series of projections at closely spaced intervals, about every 2–3°. When

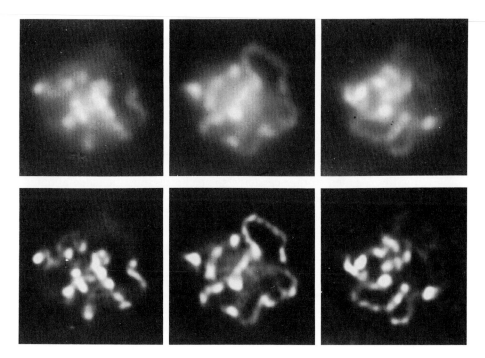

FIG. 6. Adjacent optical sections of a single *Drosophila* embryonic nucleus before (above) and after (below) image analysis with DECON3-D. A fixed *Drosophila* embryo in prophase of the nuclear division cycle was stained with the DNA-specific dye DAPI and optically sectioned at 0.25-μm intervals. The entire data set of XXX sections was used for the analysis, of which only three sections, separated by 0.5-μm intervals, are shown. (Photographs courtesy of Yasushi Hiraoka.)

shown in rapid succession, the set of projections will give the impression of rotating the objects in space. As some objects move in front of or behind others, the 3D effect is enhanced. A file generated by this kind of program is not truly 3D because information from the different optical planes has been combined to give the different views. To obtain a true 3D rotated data set, a different program is used. In some imaging systems, these steps can be done in real time using the optical section data, while in others, memory size or processor speed limitations will require that they be done first and viewed later.

2. WHAT TO SAVE

Image data files can be enormous: take, for example, a field that measures 40 μm on a side and is viewed at 450 nm with a 60 × NA 1.4 lens. If the image is oversampled by a factor of two (i.e., 0.08 μm/pixel), the minimum amount required to obtain the maximum resolution, the image will be 512 pixels on a side. If the data are 8 bit, the typical value from a video frame averager, a single optical section will be 256 kbytes. If the 3D image contains an optical section every 0.2–0.4 μm, again oversampled by two in the z direction, and is 10 μm thick, the whole image file will be 6–12 Mbytes. Because magnetic disk space is often limited, it is advisable to move data from the disks to some other medium: optical disk, 8 mm tape, etc. These devices are usually much cheaper per megabyte than hard disks, and usually only a little slower in transfers to and from memory or other disks, and so are essential if images are to be stored in digital form.

Even with expanded storage capacity, some choices must be made about what to store in order to avoid going adrift in a sea of data. In the beginning, the complete original image files should be saved, or, if only simple corrections were applied, the processed data, as it will be the starting material for any subsequent analysis steps. With more experience, the analysis programs can be applied immediately after data collection, using parameters that are known to yield acceptable results, so only the completed file needs be saved. For example, if the data were originally 12-bit data (usually written into 16-bit units) from a CCD camera, it may be reasonable to process it and store them as 8-bit data, thus using half the space of the original file. Similarly, a single projection or stereo pair may make the point adequately. Of course, a single photograph from the display screen will suffice for some applications.

E.   Other Alternatives

The OM-1 microscope system was designed to be a general-purpose tool for cell biology, but other alternatives can be used to obtain data for specific applications. For example, digitized images of polytene chromosomes taken with a video camera may not be as numerically accurate (8 bits) as they are from a CCD camera (12–14 bits) but they can be enhanced using a simple nearest-neighbor analysis scheme to give good results (Gruenbaum et al., 1984). Likewise, manual focusing can be done using the graduations on the fine-focus knob if >1 μm resolution in z is sufficient.

Image files can be transferred to a Macintosh computer using a public domain program called Kermit (Center for Computational Activities, Columbia University, New York, NY). Another public domain software program called Image, available from the National Technical Information Service (Springfield, VA; order number PB90-500687), can be used to manipulate images and do filtering, convolution, and FFT operations on a Macintosh. This program will read a number of standard image formats or can be customized to read virtually any nonstandard one. Image is not intended to handle 3D data sets, so it can be tedious to use, but the software source code is available for those who want to modify it for their own purposes. I have used a program called MacSpin (Abacus Concepts, Berkeley, CA) to create and manipulate simple stick figure models of objects from 3D image data by determining the coordinates of the objects in Image and transferring the coordinates to MacSpin; CAD packages that contain animation features will probably work as well.

As mentioned above, the cost of CCD cameras is decreasing, so that they are comparable in price to some top-of-the-range video cameras. Computer boards are available to obtain and display images directly on Macintosh computers, which, in combination with the Image program, can satisfy the needs of those who require accurate data but who cannot afford to buy separate imaging computers. In many cases, the software developed for OM-1 and other imaging systems is available in Fortran or C languages from the originators, often via electronic mail, and can be adapted to other systems by interested users.

IV. Summary

To build a coherent picture of mitosis and cell fates during blastoderm and through the complex movements of gastrulation, it will be important to localize and follow several markers simultaneously in live specimens, ideally in 3D, using high-resolution, specific, noninjurious staining and observation procedures. The study of early *Drosophila* development has already profited from the use of fluorescent labeling and low-light-level imaging of live embryos using a CCD camera. Chromosomes in fixed samples have been labeled using DNA-specific dyes, making the pattern of mitotic patches visible. *In vivo*, 3D microscopy of fluorescently tagged chromosomes, in conjunction with computerized image processing, has permitted the first direct cell lineage analysis in the early *Drosophila* embryo. Moreover, the techniques adapted to study *Drosophila* development have been used for analysis of *Drosophila* chromosome structure, mitosis, and cell cycle, and are general enough to be applied to a myriad of problems in cell biology.

"Optical sectioning" has always been used to scrutinize everything from onion roots to frog eggs, focusing up and down through the specimen, with the observer's brain responsible for the image processing. However, the volume of raw data generated by the high-resolution approach detailed above requires the use of sophisticated and adaptable computer systems to analyze and organize the results. Software designed to extract information from these complex images, either automatically or through an interactive approach, will become essential tools for cell and developmental biology. The brain of the experimenter remains the most important component in any image-processing system, but the support of technology will be essential.

ACKNOWLEDGMENTS

I thank John Sedat and David Agard for making the OM-1 system available to me. I am grateful to Yasushi Hiraoka for the photograph in Fig. 6. I would like to acknowledge my colleagues, Paul St. John, John Sedat, Andrew Belmont, Hans Chen, and Yasushi Hiraoka for helpful conversations; Paul St. John and Jean M. Wilson for critical reading of the manuscript; and Denise Bard for help in manuscript preparation. I was supported by grant NP-735 from the American Cancer Society during the preparation of the manuscript.

REFERENCES

Agard, D.A. (1984). Optical sectioning microscopy: Cellular architecture in three dimensions. *Annu. Rev. Biophys. Bioeng.* **13,** 191–219.

Agard, D. A., and Sedat, J. W. (1983). Three-dimensional architecture of a polytene nucleus. *Nature (London)* **302,** 676–681.

Agard, D. A., Hiraoka, Y., Shaw, P., and Sedat, J. W. (1989). Fluorescence microscopy in three dimensions. *Methods Cell Biol.* **30,** 353–377.

Aikens, R. S., Agard, D. A., and Sedat, J. W. (1989). Solid-state imagers for microscopy. *Methods Cell Biol.* **29,** 291–313.

Ashburner, M., and Berendes, H. D. (1978). Puffing of polytene chromosomes *In* "The Genetics and Biology of *Drosophila*" (M. Ashburner and T. R. F. Write, eds.), Vol. 2B, pp. 315–395. Academic Press, New York.

Beermann, W. (1972). Chromomeres and genes. *Cell Differ.* **1,** 1–33.

Belmont, A. A., Braunfeld, M. B., Sedat, J. W., and Agard, D. A. (1989). Large-scale chromatin structural domains within mitotic and interphase chromosomes *in vivo* and *in vitro*. *Chromosoma* **98,** 129–143.

Campos-Ortega, J. A., and Hartenstein, V. (1985). "The Embryonic Development of *Drosophila melanogaster*." Springer-Verlag, New York.

Carrington, W. A., Fogarty, K. E., Lifschitz, L., and Fay, F. S. (1989). Three-dimensional imaging on confocal and wide-field microscopes. *In* "The Handbook of Biological Confocal Microscopy" (P. Pawley, ed.), pp. 137–146. IMR Press, Wisconsin.

Castleman, K. R. (1979). "Digital Image Processing." Prentice-Hall, Englewood cliffs, New Jersey.

Chen, H., Sedat, J. W., and Agard, D. A. (1989). Manipulation, display, and analysis of three-dimensional biological images. *In* "The Handbook of Biological Confocal Microscopy" (P. Pawley, ed.), pp. 127–135. IMR Press, Wisconsin.

Cooley, J. W., and Tukey, J. W. (1965). An algorithm for the machine computation of complex fourier series. *Math. Comput.* **19,** 297–301.

Edgar, B. A., and O'Farrell, P. H. (1989). Genetic control of cell division patterns in the *Drosophila* embryo. *Cell (Cambridge, Mass.)* **57,** 177–187.

Elgin, S. C. R., and Miller, D. W. (1978). Mass rearing of flies and mass production and harvesting of embryos. *In* "The Genetics and Biology of *Drosophila*," Vol. 2a, pp. 112–121. Academic Press, New York.

Foe, V. E., and Alberts, B. M. (1983). Studies of nuclear and cytoplasmic behaviour during the five mitotic cycles that precede gastrulation in *Drosophila* embryogenesis. *J. Cell. Sci.* **61,** 31–70.

Foe, V. E., and Alberts, B. M. (1985). Reversible chromosome condensation induced in *Dro-*

sophila embryos by anoxia: Visualization of interphase nuclear organization. *J. Cell Biol.* **100**, 1623–1636.

Foe, V. E., and Odell, G. M. (1989). Mitotic domains partition fly embryos, reflecting early cell biological consequences of determination in progress. *Am. Zool.* **29**, 617–652.

Fuchs, J., Giloh, H., Kuo, C., Saumweber, H., and Sedat, J. (1983). Nuclear structure: Determination of the fate of the nuclear envelope in *Drosophila* during mitosis using monoclonal antibodies. *J. Cell Sci.* **64**, 331–349.

Glover, D. M. (1989). Mitosis in *Drosophila*. *J. Cell Sci.* **92**, 137–146.

Gruenbaum, Y., Hochstrasser, M., Mathog, D., Saumweber, H., Agard, D. A., Sedat, J. W. (1984). Spatial organization of the *Drosophila* nucleus: A three-dimensional cytogenetic study. *J. Cell Sci. (Suppl. 1)*, 223–234.

Hecht, E. (1988). "Optics," 2nd Ed., Addison-Wesley, Menlo Park, California.

Hill, R. J., Mott, M. R., and Steffensen, D. M. (1987). The preparation of polytene chromosomes for localization of nucleic acid sequences, proteins, and chromatin conformation. *Int. Rev. Cytol.* **108**, 61–118.

Hiraoka, Y., Sedat, J. W., and Agard, D. A. (1987). The use of a charge-coupled device for quantitative optical microscopy of biological structures. *Science* **238**, 36–41.

Hiraoka, Y., Minden, J. S., Swedlow, J. R., Sedat, J. W., and Agard, D. A. (1989). Focal points for chromosome condensation and decondensation revealed by three-dimensional *in vivo* time-lapse microscopy. *Nature (London)* **342**, 293–296.

Hiraoka, Y., Sedat, J. W., and Agard, D. A. (1990a). Determination of three-dimensional imaging properties of a light microscope system: Partial confocal behavior in epifluorescence microscopy. *Biophys. J.* **57**, 325–333.

Hiraoka, Y., Agard, D. A., and Sedat, J. W. (1990b). Temporal and spatial coordination of chromosome movement, spindle formation, and nuclear envelope breakdown during prometaphase in *Drosophila melanogaster* embryos. *J. Cell Biol.* **111**, 2815–2828.

Hiraoka, Y., Rykowski, M. C., Lefstin, J. A., Agard, D. A., and Sedat, J. W. (1990c). Three-dimensional organization of chromosomes studied by *in situ* hybridization and optical sectioning microscopy. *Proc. SPIE Int. Soc. Opt. Eng.* **1205**, 11–19.

Hochstrasser, M., and Sedat, J. W. (1987a). Three-dimensional organization of *Drosophila melanogaster* interphase nuclei. I. Tissue-specific aspects of polytene nuclear architecture. *J. Cell Biol.* **104**, 1455–1470.

Hochstrasser, M., and Sedat, J. W. (1987b). Three-dimensional organization of *Drosophila melanogaster* interphase nuclei. II. Chromosome spatial organization and gene regulation. *J. Cell Biol.* **104**, 1471–1483.

Hochstrasser, M., Mathog, D., Gruenbaum, Y., Saumweber, H., and Sedat, J. W. (1986). Spatial organization of chromosomes in the salivary gland nuclei of *Drosophila melanogaster*. *J. Cell Biol.* **102**, 112–123.

Janesick, J., Elliott, T., Fraschetti, G., Collins, S., Blouke, M., and Corrie, B. (1989), Charge coupled device pinning technologies. *Proc. SPIE* **1071**, 153–169.

Karr, T. L., and Alberts, B. M. (1986). Organization of the cytoskeleton in early *Drosophila* embryos. *J. Cell Biol.* **102**, 1494–1509.

Keller, H. E. (1989). Objective lenses for confocal microscopy. *In* "The Handbook of Biological Confocal Microscopy" (P. Pawley, ed.), pp. 69–77. IMR Press, Madison, Wisconsin.

Kellogg, D. R., Mitchison, T. J., and Alberts, B. M. (1988). Behaviour of microtubules and actin filaments in living *Drosophila* embryos. *Development* **103**, 675–686.

Kress, H., Meyerowitz, E. M., and Davidson, N. (1985). High resolution mapping of *in situ* hybridized biotinylated DNA to surface-spread *Drosophila* polytene chromosomes. *Chromosoma* **93**, 113–122.

Kristian, J., and Blouke, M. (1982). Charge-coupled devices in astronomy. *Sci. Am.* **297**, 67–74.

Lehner, C. F., and O'Farrell, P. H. (1989). Expression and function of *Drosophila* cyclin A during embryonic cell cycle progression. *Cell (Cambridge, Mass.)* **56,** 957–968.

Marcus, D. A. (1988). High-performance optical filters for fluorescence analysis. *Cell Motil. Cytoskeleton* **10,** 62–70.

Mathog, D., and Sedat, J. W. (1989). The three-dimensional organization of polytene nuclei in male *Drosophila melanogaster* with compound xy or ring x chromosomes. *Genetics* **121,** 293–311.

Mathog, D., Hochstrasser, M., Gruenbaum, Y., Saumweber, H., and Sedat, J. (1984). Characteristic folding pattern of polytene chromosomes *Drosophila* salivary gland nuclei. *Nature (London)* **308,** 414–421.

Minden, J. S., Agard, D. A., Sedat, J. W., and Alberts, B. M. (1989). Direct cell lineage analysis in *Drosophila melanogaster* by time-lapse, three-dimensional optical microscopy of living embryos. *J. Cell Biol.* **109,** 505–516.

Mitchison, T. J., and Sedat, J. (1983). Localization of antigenic determinants in whole *Drosophila* embryos. *Dev. Biol.* **99,** 261–264.

Rykowski, M. C., Parmelee, S. J., Agard, D. A., and Sedat, J. W. (1988). Precise determination of the molecular limits of a polytene chromosome band: Regulatory sequences for the notch gene are in the interband. *Cell (Cambridge, Mass.)* **54,** 461–472.

Sedat, J., and Manuelidis, L. (1977). A direct approach to the structure of eukaryotic chromosomes. *Cold Spring Harbor Symp. Quant. Biol.* **42,** 331–350.

Sullivan, W., Minden, J. S., and Alberts, B. M. (1990). *Daughterless-abo-like*, a *Drosophila* maternal-effect mutation that exhibits abnormal centrosome separation during the late blastoderm division. *Development* **110,** 311–323.

Taylor, D. L., and Salmon, E. D. (1989). Basic fluorescence microscopy. *Methods Cell Biol.* **29,** 207–237.

Taylor, D. L., Waggoner, A. S., Murphy, R. F., and Lanni, F. (1986). "Applications of Fluorescence in the Biomedical Sciences" (R. R. Birge, ed.), Alan R. Liss, New York.

Wilson, T. (1989). The role of the pinhole in confocal imaging systems. *In* "The Handbook of Biological Confocal Microscopy" (P. Pawley, ed.), pp. 99–113. IMR Press, Madison, Wisconsin.

Zhimulev, I. F., Belyaeva, E. S., and Semeshin, V. F. (1981). Informational content of polytene chromosome bands and puffs. *Crit. Rev. Biochem.* 303–340.

Part III. Identifying Specific Macromolecular Interactions

Chapter 11

Yeast Minichromosomes

SHARON Y. ROTH AND ROBERT T. SIMPSON

Laboratory of Cellular and Developmental Biology
National Institute of Diabetes and Digestive and Kidney Diseases
National Institutes of Health
Bethesda, Maryland 20892

I. Introduction

Episomal DNA is packaged into chromatin in yeast. Although yeast plasmids may lack centromeres and do lack telomeres, they are often referred to as minichromosomes, using nomenclature introduced for viral episomes in higher eukaryotes (Griffith, 1975). Yeast minichromosomes have provided a unique model system for study of many aspects of chromatin structure. This

289

chapter addresses the advantages and disadvantages of minichromosomes as an experimental system. First, we briefly review some of the questions regarding chromatin structure which have been successfully addressed using yeast minichromosomes. Then, general methods for their isolation are described, with an analysis of the strengths and weaknesses of each in regard to particular applications. Finally, we describe mapping techniques useful in determining nucleosome position.

A.   Analysis of Chromatin Structure Using Minichromosomes

Saccharomyces cerevisiae chromatin comprises nucleosomal subunits containing the four core histone molecules which are highly conserved among all eukaryotes (Fangman and Zakian, 1981). It differs from chromatin of other species in that it may not contain a lysine-rich linker histone (i.e., histone H1). The lack of an H1 histone and the high percentage of transcriptionally active chromatin in yeast is reflected in a higher degree of sensitivity to nuclease digestion relative to chromatin from other species. However, the composition of the nucleosome core particle is the same in yeast as in higher eukaryotes and therefore is an appropriate system for the study of many questions regarding the structure, placement, and functions of the nucleosome. Episomal plasmid systems such as those based on bovine papilloma virus allow parallel studies in higher eukaryotes (Cordingly *et al.*, 1987; Richard-Foy and Hager, 1987).

The question of whether nucleosomes have defined positions in chromatin has been extensively addressed using yeast minichromosomes (see Simpson, 1991, for review). These studies have shown that a variety of factors can influence nucleosome placement. Particular DNA sequences have a greater affinity for nucleosome formation than others, presumably due either to specific DNA–histone contacts or to a specific conformation of the DNA which favors folding into the nucleosome core particle (Thoma *et al.*, 1984; Thoma and Simpson, 1985). Nucleosome position may also be limited by sequences which exclude nucleosomes, forming boundaries to nucleosome formation or movement (Losa *et al.*, 1990; Thoma, 1986). Certain proteins also provide boundaries to nucleosome placement when bound to DNA, resulting in a statistical positioning of nucleosomes around their binding site (Fedor *et al.*, 1988). In addition, chromatin folding has been shown to influence nucleosome placement in yeast minichromosomes (Thoma and Zatchez, 1988).

One clear advantage to using a yeast minichromosomal system is the ability to manipulate the genetic background of the host cell readily. This is particularly advantageous in studies of the association of specific trans-acting factors with chromatin and the role of particular histones or histone domains in chromatin structure and function. The effect of binding of a particular factor on the chromatin structure of neighboring sequences, for example, can be addressed by simply introducing a plasmid containing the binding site for the factor into isogenic cells which either do or do not express the factor. This approach has been successfully used to address the effects of the $\alpha2$ repressor protein, which is involved in the determination of yeast mating type, on the placement of nucleosomes surrounding its operator sequence (Roth *et al.,* 1990). Comparison of the structure of minichromosomes containing the $\alpha2$ operator sequence isolated from α cells, which express $\alpha2$, and **a** cells, which do not express $\alpha2$, revealed that this protein organizes nucleosomes over sequences immediately abutting the operator (Fig. 1). Subsequent studies of genomic **a**-cell-specific genes confirmed that such organization also occurred in the chromosome (Shimizu *et al.,* 1991), further validating the use of minichromosomes as an experimental system to study these types of questions. The power of yeast genetics is also illustrated by the use of specific mutations in histone H4 (see [19], this volume, concerning mutations in chromosomal proteins) in these studies, to indicate that the organization of chromatin structure by $\alpha2$ requires the H4 amino-terminal tail (S. Roth, unpublished observations, 1991).

How nucleosomes affect the function of specific cis-acting DNA elements is another question successfully addressed by yeast minichromosomes. Specific sequences involved in autonomous replication (the ARS core element) are located in a nuclease-hypersensitive region near the edge of a precisely positioned nucleosome in TRP1/ARS1 plasmids (see below and Fig. 1). If these sequences must be in a nucleosome-free region to function, placement of a nucleosome over these sequences would be predicted to inhibit replication. Indeed, specific deletions which result in movement of a nucleosome over these sequences result in a lowered plasmid copy number (Simpson, 1990). Use of the $\alpha2$ operator as a second nucleosome positioning signal to shift the position of the nucleosome away from the ARS core sequence rescues the low-copy-number phenotype, further indicating that nucleosome position rather than specific deletions are responsible for the observed decrease in plasmid copy number.

Nucleosome position has also been implicated in the regulation of the acid phosphatase gene *PHO5*. Nucleosomes are positioned nonrandomly over upstream sequences when the *PHO5* gene is repressed, under conditions of high phosphate (Bergman and Kramer, 1983). By insertion of the *PHO5* gene into a minichromosome, sequences necessary for the maintenance of this ordered structure were defined and determined to also be involved in the regulation of *PHO5* expression by trans-acting factors (Bergman, 1986).

Yeast minichromosomes have been used to study various other types of questions as well, including a characterization of centromere structure (Bloom

A. TRP1/ARS1
   1453 bp

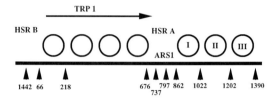

B. TALS
   1810 bp

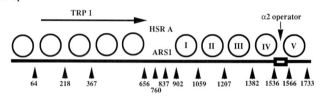

C. TRURAP
   2619 bp

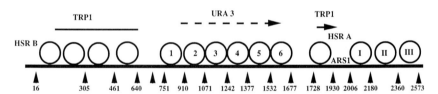

FIG. 1. Structure of TRP1/ARS1 and two derivative minichromosomes. The structure of (A) TRP1/ARS1 (Thoma *et al.,* 1984) is shown and compared to that of (B) TALS in α cells, (Roth *et al.,* 1990) and (C) TRURAP (Thoma, 1986). To aid in comparison, the three plasmids are shown in linear form, beginning with map site 1, which corresponds to an *Eco*RI restriction site in most TRP1/ARS1 derivatives. Major micrococcal nuclease cut sites in chromatin are indicated by arrowheads, and the inferred positions of nucleosomes are shown by circles. For simplicity, the numerous nuclease cut sites present in naked DNA are not shown. Nuclease-hypersensitive regions are also indicated (HSR A and HSR B). The locations of the TRP1 gene and the *URA3* gene are indicated by arrows. Notice that nucleosomes I, II, and III maintain their positions downstream of the ARS1 elements in all three minichromosomes. These positions are also maintained upon insertion of extra sequences between nucleosome I and II (not shown; see Thoma and Simpson, 1985), but are disrupted upon insertion of sequences into HSR B (Roth *et al.,* 1990). Also note that nuclease-hypersensitive regions are maintained upstream and downstream of TRP1 in both TRP1/ARS1 and TRURAP, and also flank the *URA3* gene in TRURAP. However, in TALS, sequences in HSR B are packaged into nucleosomes in α cells along with other sequences flanking the α2 operator. In **a** cells, which do not express α2, nucleosomes present on TALS are not maintained in stable positions (Roth *et al.,* 1990). Finally, notice that ARS1 elements are always located in a nuclease-sensitive region (HSR A), consistent with the finding that positioning of a nucleosome over the ARS1 core element inhibits its function (Simpson, 1990).

and Carbon, 1982) and a characterization of how yeast nucleosomes affect
the thermal untwisting of DNA relative to nucleosomes in other organisms
(Morse et al., 1987). The fate of nucleosomes during elongation of transcripts
by RNA polymerase II has also been addressed using specific minichromosomes carrying an inducible transcription unit (Pederson and Morse, 1989).
Analysis of repair rates of specific DNA lesions in different chromosomal
regions in minichromosomes has allowed study of the relationship between
transcription, nucleosome placement, and DNA repair (Smerdon et al., 1990;
Smerdon and Thoma, 1990). These studies, as well as many others, illustrate
both the utility and versatility of yeast minichromosomal systems.

B.   Structure of TRP1/ARS1 Plasmids

The TRP1/ARS1 plasmid YARp1 (Zakian and Scott, 1982) and its derivatives have been used extensively in studies such as those described above.
This plasmid was formed by circularization of a 1.45-kb EcoRI fragment of
yeast genomic DNA containing coding sequences for the TRP1 gene (N-5'-
phosphoribosyl-anthranilate isomerase), providing a selectable marker and
sequences sufficient for autonomous replication. TRP1/ARS1 plasmids are
maintained in high copy number (100–200 copies per cell) which, together
with their small size, facilitates isolation of these minichromosomes and
mapping of their chromatin structure. The original plasmid is packaged into
seven nucleosomes (Fig. 1). Four nucleosomes of limited stability are formed
in the region of the TRP1 gene, and three highly stable nucleosomes (I, II,
and III in Fig. 1A) are formed with sequences of unknown function (Thoma
et al., 1984). These nucleosomal domains are separated by two nuclease-
hypersensitive regions. Several derivatives of the original TRP1/ARS1 plasmid (Fig. 1) have now been characterized (Roth et al., 1990; Smerdon and
Thoma, 1990; Thoma, 1986; Thoma et al., 1984). Examination of how the
nucleosomal and hypersensitive regions present in TRP1/ARS1 are affected
by the insertion of heterologous sequences has been very informative as to
the nature of factors which determine nucleosome position. The structures
of these plasmids also serve as useful guidelines for the design of future minichromosomes. Knowledge of which regions of the plasmid might be expected
to be packaged as nucleosomes, for example, can greatly aid the choice of a
site of insertion or modification of sequences in the plasmid to contain defined
DNA sequence elements (see below).

C.   Minichromosomes and Trans-Acting Factors

The high copy number of TRP1/ARS1 plasmids not only facilitates structural
mapping, but also provides many targets for binding of trans-acting factors. If
such factors are sufficiently abundant and stable, minichromosomes can

provide an *in vivo* enrichment system for factors bound to a chromatin substrate. Insertion of a specific binding site into a yeast plasmid, then, provides a unique and powerful approach to the isolation of such factors, as well as a way to determine how they interact with or affect the chromatin structure of neighboring sequences as described above. In addition, resident features of the chromatin structure of a minichromosome can be useful in determining how the positions of nucleosomes or nuclease-sensitive regions affect the function of such factors.

One should be aware of particular problems which may arise in these kinds of studies. For example, if the binding site for a factor is itself packaged into a nucleosome in the plasmid, then the factor of interest may not be able to bind. While such occlusion may be interesting in terms of understanding the structure–function relationship of chromatin to that of the factor as described above, it will certainly frustrate attempts to coisolate the factor with the minichromosomes. Also, if the factor of interest is not present in amounts sufficient to bind all copies of the plasmid, then the structure of the bulk of the minichromosomes will not reflect the structure of the few minichromosomes actually bound by the factor. In such cases, use of single-copy plasmid containing a centromere (CEN) would be advantageous. In any case, it is important to determine what fraction of the plasmids is associated with the factor before structure can be related to function (or to binding). Such studies may be further complicated if the factor is unstable and easily lost during isolation procedures. Finally, because the small size of the TRP1/ARS1 plasmids limits room for nucleosome formation and placement, trans-acting factors affecting nucleosome position may be influenced by such constraints. Insertion of extra sequences to lengthen the plasmid is an obvious alternative in such a case. The plasmid might also be linearized and integrated into the genome, allowing analysis of its structure in a chromosomal context.

D. Topology of Minichromosomes

The circular nature of plasmids provides an advantage over structural studies of genomic chromosomal sequences in that it allows analysis of changes in supercoil density. The topology of chromatin is thought to be influenced by a number of factors, including wrapping of DNA around histone octamers (Morse and Simpson, 1988, for review) and transcription of genes by RNA polymerase II (Brill and Sternglanz, 1988). Each nucleosome, for example, imparts a change in linking number of about -1. Formation or dissociation of nucleosomes on a plasmid will change its superhelical density. Changes in linking number can also reflect changes in transcriptional activity, or even in the extent of posttranslational modification of histone molecules (Norton *et al.*, 1989). Analysis of the distribution of minichromosome topoisomers under different defined conditions can be very informative as to how these conditions affect the superhelicity of the chromatin DNA. Again, it is important to realize that structural studies of minichromosomes will reflect the structure of the bulk of the population, whereas functional studies may reflect the activity of only a few plasmids in the population.

E. Stability of Minichromosomes

Episomal plasmids in yeast are categorized by the type of origin of replication they possess and whether or not they contain a centromeric (CEN) element (Rose *et al.,* 1990, for review). ARS-containing plasmids (in the absence of a CEN) segregate unevenly during mitosis and are therefore unstable in the absence of selection. High-copy-number ARS plasmids, such as TRP1/ARS1, must therefore be maintained under selection in order to avoid loss. Unequal segregation might be problematic in some studies, since different cells in the population will contain different numbers of the plasmid molecules. Trans-acting factors associated with the plasmid chromatin, for example, might be limiting in some cells but not in others, complicating analysis of structure–function relationships. Plasmids containing a CEN element segregate stably, but the presence of a CEN reduces the copy number of the plasmid to one per cell.

Many yeast strains (designated cir$^+$) carry an endogenous plasmid, the 2μ circle (Broach, 1981), which is also packaged into nucleosomes. The 2μ plasmid is maintained at 20–50 copies per cell, and although it does not contain a CEN, it provides its own replication/segregation functions and is relatively stable (Som *et al.,* 1988). These plasmids offer an alternative to ARS plasmids for minichromosome studies. However, these plasmids have a high potential for homologous recombination (Volkert and Broach, 1986). Careful design of the minichromosome to avoid recombinogenic sequences and use of an appropriate host yeast strain can avoid these complications. Insertion of cis-acting Rep3 sequences from 2μ DNA to confer stability to ARS1-containing plasmids and transformation into a cir$^+$ strain (Simpson, 1990), or a strain expressing the trans-acting Rep1 and Rep2 factors from a genomic locus (Volkert and Broach, 1986), provides an attractive alternative to unstable TRP1/ARS1 plasmids.

II. Methods of Minichromosome Isolation

Various alternatives are available for the preparation of minichromosomes from yeast. We have found three types of procedures to be useful in the preparation of chromatin for the purposes of mapping nucleosome position

and analysis of factors associated with chromatin. These include nuclear isolation, biochemical isolation, and protein–nucleic acid affinity purification.
Each of these procedures has advantages and disadvantages which must be
weighed against the final purpose of the experiment in hand. As we describe
each procedure, we emphasize their strengths and weaknesses and give examples of experiments in which they have been used.

A.   Isolation of Nuclei

For many types of analyses, purified minichromosomes are not necessary.
Nucleosome mapping and footprint analysis of trans-acting factor binding,
for example, rely on the use of specific primers or probes which are not usually
influenced by background genomic sequences. Maintaining chromatin in a
concentrated state to avoid dissociation of nucleosomal and nonnucleosomal proteins is often more important than the purity of the final preparation
in these types of experiments (see Section III for a discussion of nucleosome
mapping). Rapid isolation is also a critical consideration to minimize degradation of the factor of interest. Isolation of nuclei without further purification provides a relatively rapid method of preparing a concentrated source
of total cellular chromatin. Nuclear isolation also provides good starting
material for further purification by biochemical or protein–nucleic acid affinity purification procedures.

The procedure described below is a slight modification of that developed by
Szent-Gyorgyi and Isenberg (1983). It can be completed within a few hours
and provides material which has been used successfully in many types of
mapping analyses. It differs from other procedures for the isolation of yeast
nuclei (Pederson *et al.*, 1986) in that it avoids the use of detergents and Percoll
gradients which might adversely affect the association of some proteins with
chromatin. An example of micrococcal nuclease-digested chromatin prepared
from nuclei isolated as described below is presented in Fig. 2.

Procedure

Buffers.   Add PMSF fresh to each before use.

   Sorbitol buffer: 1.4 M sorbitol; 40 mM HEPES [N-(2-hydroxyethyl)-
      piperidine-N-(2-ethanesulfonic acid], pH 7.5; 0.5 mM MgCl$_2$

   Ficoll buffer: 18% Ficoll 400 (Pharmacia); 20 mM PIPES [piperazine-
      N,N'-bis (2-ethanesulfonic acid)], pH 6.5; 0.5 mM MgCl$_2$

   Glycerol/Ficoll buffer: 7% Ficoll 400; 20% glycerol; 20 mM PIPES,
      pH 6.5; 0.5 mM MgCl$_2$

   Nuclease digestion buffer: 10 mM HEPES, pH 7.5; 0.5 mM MgCl$_2$; 0.05
      mM CaCl$_2$

1. Typically, 1 liter of culture in appropriate selective media is grown at
30°C with rapid shaking to a density of 0.7 to 1.0 as measured by absorbance at
600 nm (2–4 × 10^7 cells/ml).

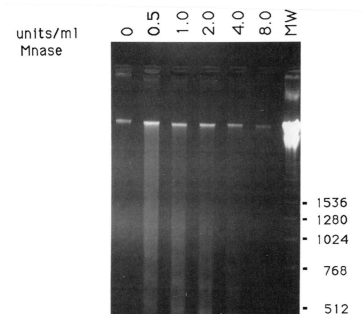

units/ml
Mnase

- 1536
- 1280
- 1024
- 768
- 512
- 256

FIG. 2. Micrococcal nuclease (Mnase) digestion of chromatin in isolated nuclei. Nuclei isolated as described in text were digested with increasing amounts of micrococcal nuclease as indicated. Digested DNA was then purified by standard procedures (Sambrook *et al.,* 1990) and resolved by electrophoresis in 1.2% agarose in 0.089 *M* Tris, 0.089 *M* boric acid, 0.002 *M* EDTA (TBE). Molecular weight markers (MW) consist of a partial digestion of a 5S ribosomal gene tandem repeat of 256 bp. Shown is the gel stained with ethidium bromide.

2. Protease inhibitors such as phenylmethylsulfonyl floride [PMSF (Boehringer Mannheim), final concentration 1 m*M*; diluted from a 100 m*M* stock made in isopropanol] may be added to the culture prior to harvest, and should be added fresh to all buffers and solutions throughout the isolation. Since PMSF will react with reducing agents, it should added to buffers prior to addition of dithiothreitol (DTT) or β-mercaptoethanol (BME).

3. Cells are harvested in a GSA rotor (Sorvall) or JA10 rotor (Beckman) at 5000 rpm (4100 g) for 5 minutes at room temperature.

4. Cells are collected and washed in sorbitol buffer containing 30 mM DTT or 10 mM BME and 1 mM PMSF, followed by centrifugation as above.

5. Cells are resuspended in 30 ml of the above sorbitol buffer and incubated at 30°C for 10 minutes. This step is optional, but has been found to increase the efficiency of spheroplast formation in some strains.

6. Cells are collected into a preweighed centrifuge tube by centrifugation at 3500 rpm (2000 g) in an HB4 rotor (Sorvall) for 5 minutes.

7. The cell pellet is weighed and resuspended in 4 volumes of sorbitol buffer containing 5 mM DTT or 2 mM BME and 1 mM PMSF. Zymolyase 100T (ICN) or oxalyticase (Enzogenetics) is added, and spheroplast formation is monitored carefully. The choice of enzyme preparation and concentration should be determined empirically for each strain since these enzymes can have quite different activities toward different strains. A concentration should be chosen which gives complete spheroplasting in less than 90 minutes. For most strains that we have used, 20–50 mg of Zymolyase 100T or 1–10 mg of oxalyticase per 3 g wet cell pellet gives adequate spheroplast formation in 60–90 minutes. Spheroplast formation should be monitored by cell wall degradation as assayed by the ability to destroy all cell morphology upon sliding a coverslip back and forth across the surface of a microscope slide. Complete cell wall degradation is essential for successful preparation of nuclei; hence, spheroplast formation is the most crucial step in this procedure. Because sorbitol is not utilized as a carbon source by yeast, cells undergo starvation during spheroplast formation. Many factors are unstable under these conditions, and a period of recovery in medium [YEPD (Rose *et al.*, 1990)] containing 1 M sorbitol may be required for regeneration of the factor of interest. Typically, 1 hour of regeneration time is sufficient, but the stability and synthesis of each factor should be monitored individually.

8. All following steps are done on ice or in a cold room. When spheroplast formation is complete, spheroplasts are diluted to 30 ml with cold sorbitol buffer without reducing agent but containing 1 mM PMSF and are collected by centrifugation at 3500 rpm (2000 g) in the HB4 swinging bucket rotor for 5 minutes. Spheroplasts are gently resuspended in the above buffer and recentrifuged. This wash is repeated once more to remove residual Zymolyase or oxalyticase.

9. Spheroplasts are resuspended in 20 ml of Ficoll buffer and homogenized with a motor-driven Teflon/glass homogenizer at 4°C. Cell lysis should be monitored microscopically.

10. The lysate is gently layered over 20 ml of glycerol/Ficoll buffer containing 1 mM PMSF and centrifuged at 11,500 rpm (20,000 g) in an HB4 rotor for 30 minutes.

11. The supernatant is carefully removed and the pellet is drained. The

pellet is then resuspended in 20 ml of Ficoll buffer by vortexing for 5 minutes at 4°C. The resuspended pellet is centrifuged at 4500 rpm (3000 g) in the HB4 rotor for 15 minutes.

12. The supernatant is carefully withdrawn and transferred to a fresh tube, avoiding the whitish layer which is formed just over the pellet. Centrifuge at 11,500 rpm (20,000 g) in the HB4 rotor for 25 minutes. Resuspend the nuclear pellet in appropriate buffer for nuclease digestion, etc.

B. Biochemical Purification of Minichromosomes

Pederson et al. (1986) have described a biochemical purification scheme for yeast minichromosomes. This procedure can yield highly purified plasmid chromatin and therefore may be the method of choice for isolation and analysis of nucleosomal and nonnucleosomal factors specifically associated with the minichromosome.

The biochemical purification scheme largely relies on size fractionation techniques (Dean et al., 1989; Pederson et al., 1986). Chromatin is eluted from isolated nuclei and then fractionated over a Sephacryl S-300 (Pharmacia) column to remove endogenous nucleases and other nonchromosomal proteins. Spheroplast lysates may also be fractionated in this manner and have been used successfully in chromatin mapping experiments without further purification. Chromatin eluted in the void volume of this column can be concentrated by isopycnic density gradient centrifugation using Nycodenz (Nyegaard, Oslo, Norway). Plasmid chromatin-containing fractions are then pooled and passed over a smaller Sephacryl column to remove the Nycodenz. The eluted chromatin is concentrated and further purified by sucrose gradient centrifugation.

Although very pure chromatin can be obtained by this procedure (Fig. 3), it requires several days to complete. A final yield of 10% of plasmid chromatin has been reported for this technique (Dean et al., 1989; Pederson et al., 1986). Dissociation of nucleosomal proteins has been observed. Knowledge of the stability and affinity of factors of interest under the various conditions employed in this procedure is imperative for isolation of chromatin containing the factors.

Procedure

1. Nuclei are isolated as described above (with appropriate scaling up of volumes, etc.) or by alternative procedures (Pederson et al., 1986) from 10 liters of cells grown to a density of 1.0–1.5 as measured by absorbance at 600 nm. The final nuclear pellet is resuspended in 8 ml of nuclei elution buffer (NEB) containing 0.5 mg/ml pepstatin A (Boehringer Mannheim) and 1 mM PMSF. (NEB is 200 mM NaCl; 5 mM MgCl$_2$; 10 mM PIPES, pH 7.3; 0.5 mM EGTA, 0.1% BME.) If nuclei are prepared by Percoll gradient centrifugation, they should be washed at least twice in NEB to remove residual Percoll.

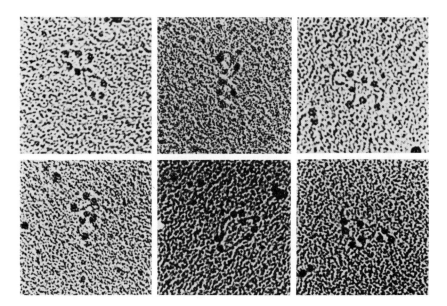

FIG. 3.   Electron micrographs of purified chromatin. A montage of six molecules visualized using electron microscropy is shown. Analysis of the number of nucleosomes per plasmid chromatin molecule in a larger sample showed the average to be 6.5. (From Pederson et al., 1986).

2. Nuclei are held on ice for 75 minutes with occasional mixing to elute plasmid chromatin, and then are collected by centrifugation for 5 minutes at 7500 rpm (9150 g) in an HB4 type rotor (Sorvall). The elution and collection are repeated twice more and the three eluates are combined. Residual nuclei and debris are removed by centrifugation at 11,000 rpm (19,600 g) for 10 minutes.

3. The eluate is then passed over a 200-ml Sephacryl S-300 column (Pharmacia) equilibrated in NEB. Plasmid chromatin elutes in the void volume.

4. Nycodenz is added to the column eluate to a concentration of 40% (w/v); the mixture is then centrifuged at 50,000 rpm (242,000 g) in a vertical rotor (VTi50, Beckman) for 36 hours. Following centrifugation, the gradients are fractionated into 1-ml fractions. A small portion, 80–100 μl, of each fraction is precipitated by the addition of 0.1 volume of 2.5 M sodium acetate and 2 volumes of ethanol. The precipitate is collected by centrifugation in a microfuge, subjected to digestion with RNase A at a final concentration of 0.5 μg/10 μl of 10 mM Tris, pH 7.5; 1 mM EDTA at 37°C for 30 minutes, and then analyzed by agarose gel electrophoresis using standard techniques (Sambrook et al., 1990) after addition of loading buffer.

5. Fractions containing the plasmid are pooled and passed over a second, 50-ml Sephacryl S-300 column, and again the void volume is collected.

6. The eluate is then loaded onto a preformed, 35-ml, 0.4–1.0 M linear sucrose gradient in 0.6 × NEB without BME and centrifuged at 50,000 rpm (242,000 g) in a VTi50 rotor for 80 minutes. Fractions are analyzed by agarose gel electrophoresis as above. Pooled fractions containing the plasmid can be stored in aliquots at −80°C if not used immediately.

C. Protein–Nucleic Acid Affinity Purification of Minichromosomes

The bacterial lactose (*lac*) repressor has a high affinity for its operator DNA, which can be exploited to purify DNA containing the *lac* operator sequence away from bulk genomic sequences. Use of a *lac* repressor–β-galactosidase fusion protein which possesses inducer and operator-binding properties of the repressor as well as galactosidase enzymatic activity has facilitated the application of protein–nucleic acid affinity to purification of DNA–protein complexes (Levens and Howley, 1985). Repressor binding and release can occur under a variety of buffer conditions. Both a galactosidase–*lac* repressor fusion protein and antibodies to β-galactosidase are now commercially available (Promega). Protein–nucleic acid affinity purification, then, can provide a gentle and widely adaptable method for the preparation of minichromosomes.

Dean *et al.* (1989) have described a protein–nucleic acid affinity purification procedure for yeast minichromosomes in some detail and have provided a clear discussion of its usefulness. The method consists of spheroplast formation followed by passive elution of the plasmid chromatin. The eluted chromatin is mixed with *lac* repressor–β-galactosidase fusion protein and subsequently with rabbit anti-β-galactosidase antibodies adsorbed to goat anti-rabbit immunobeads (Gibco Life Technologies). The beads are collected and washed to remove unbound chromatin. The bound plasmid chromatin is then released by the addition of IPTG (isopropylthio-β-D-galactopyranoside). A 20–25% yield of plasmid chromatin is typical for this technique (Dean *et al.,* 1989). Plasmid chromatin prepared by this method is shown in Fig. 4.

If this technique is used to purify a specific factor bound to chromatin, two potential problems must be considered. Although this procedure is quite gentle, the stability and affinity of the factor of interest for its target sequence should be monitored whenever possible under conditions used in the preparation in order to optimize conditions to minimize factor loss. Also, because the *lac* repressor has a certain affinity for nonspecific DNA, the minichromosome preparation may be contaminated with genomic chromatin to varying extents (20–50% of the total DNA; Dean *et al.,* 1989). Further purification

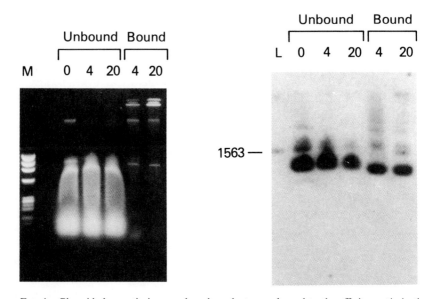

FIG. 4. Plasmid chromatin in a crude nulear eluate was bound to the affinity matix in the presence of increasing amounts of repressor–β-galactosidase fusion protein (4 and 20 μl at 0.5 mg/ml). After washing, the bound material was eluted with IPTG. Total nucleic acid in the bound and unbound fractions was electrophoresed in a 1% agarose gel and stained with ethidium bromide (left) and hybridized with a TRP1/ARS1 ^{32}P-labeled DNA probe (right) and autoradiography was performed. Lane M contains HaeIII digested $\phi \times 174$ DNA as size standards. Lane L contains linearized AT9 DNA. (From Dean et al., 1989).

by biochemical means may be necessary if proteins shared with bulk chromatin (such as histones) are the subject of interest.

We have observed that the chromatin structure of the plasmid can influence its recovery by this procedure (A. Dean and S. Roth, unpublished observations, 1988). Studies of the plasmid TALS best illustrate this point. TALS contains the binding site for the α-cell-specific protein α2. In α cells α2 binding to its operator results in the packaging of TALS into an extensive array of stably positioned nucleosomes as described above, one of which incorporates the lac operator sequence. In a cells, which do not express α2, nucleosomes on the plasmid are not organized into this stable nucleosomal array. Immunoaffinity purification of TALS works well with a cell lysates, but a low yield is obtained consistently from α cell lysates. The simplest explanation for this difference in yield is that the presence of a stable nucleosome over the lac operator in α cells inhibits lac repressor binding [in conflict with earlier studies, which suggested that the repressor could bind its operator in vitro even when present in a nucleosome (Chao et al., 1980)]. Consistent low yield, therefore, of a particular minichromosome may indicate that the lac operator is located in an inaccessible region of chromatin. Redesign of the minichromosome or an alternative isolation procedure may be required to facilitate purification under these conditions.

Procedure. A summary of the procedure developed by Dean *et al.* (1989) follows.

Buffers and Reagents. PMSF is added fresh to all buffers just prior to use.

Yeast wash buffer (YWB): 40 mM potassium phosphate, pH 7.5; 1 M sorbitol; 0.5 mM PMSF

Elution/binding buffer (EBB): 150 mM NaCl; 10 mM MgCl$_2$; 10 mM Tris-HCl, pH 7.4; 0.5 mM PMSF (filter sterilize and store at 4°C)

TEN + IPTG: 10 mM Tris-HCl, pH 7.4; 1mM EDTA; 150 mM NaCl; 2 mM PMSF; 2 mM IPTG (Gibco Life Technologies). This solution should be filter sterilized and stored at 4°C

lac repressor–β-galactosidase fusion protein: produced by *Escherichia coli* strain BMH 72-19-1 and purified as described by Fowler and Zabin (1983). This fusion protein is now commercially available from Promega

Rabbit anti-β-galactosidase antibodies: obtained from Cappel Laboratories. Repurified by affinity chromatography (Dean *et al.,* 1989). Anti–β-galactosidase antibodies are now available from Promega

Immunobeads: Goat anti-rabbit IgG immunobeads can be obtained from Gibco/Life Technologies and are prepared as described by Dean *et al.* (1989)

1. One liter of cell is grown to late log phase (absorbance at 600 nm of 1.0–1.5) in selective media.

2. Cells are harvested by centrifugation at 5000 rpm (4100 g) in a Beckman JA10 or Sorvall GSA-type rotor and washed twice with YWB.

3. Following washes, cells are resuspended for spheroplast formation in 50 ml of YWB containing 20 mM BME and 0.5 ml of 10 mg/ml Zymolyase 100T. Cells are incubated, slowly shaking (50–100 rpm), at 30°C for 30 minutes. Spheroplast formation is monitored by diluting cells 1 : 20 in 1% sodium dodecyl sulfate (SDS); the solution should become clear relative to a dilution in water as a control (absorbance at 600 nm in SDS is ≤0.1 of that in H$_2$O).

4. When ready, spheroplasts are collected by centrifugation in a disposable 50-ml conical tube at 3000 rpm (2300 g) for 5 minutes at 4°C.

5. Spheroplasts are resuspended gently in 30 ml of YWB at 4°C and recollected by centrifugation.

6. If desired, spheroplasts may be allowed a recovery period in medium containing 1 M sorbitol for 1 hour before continuing.

7. Washed spheroplasts are resuspended in 5 ml of EBB and are held on ice for 15 minutes.

8. Lysis is completed (as monitored microscopically) by homogenization with a motor-driven Teflon/glass homogenizer at the lowest speed and with the minimum number of strokes possible.

9. The lysate is held on ice for 2–4 hours with occasional mixing to elute the plasmid chromatin.

10. The lysate is transferred to microfuge tubes in 1-ml aliquots and spun for 20 minutes at 4°C at maximum speed (16,000 g) in a microfuge. The supernatant is transferred to fresh tubes on ice.

11. *lac* repressor–β-galactosidase fusion protein (20 μl) is added to each aliquot of the nuclear eluate, mixed, and held on ice for 15 minutes. The amount of fusion protein necessary to bind the plasmid in a given amount of nuclear eluate should be determined as described by Dean *et al.* (1989) to optimize specific versus nonspecific binding.

12. Four 50-μl aliquots of the rabbit anti-β-galactosidase antibody adsorbed to goat anti-rabbit immunobeads (as described by Dean *et al.,* 1989) are added to the eluate over a 30-minute period, and the mixture is incubated on ice for 1–1.5 hours following the last addition. Again, the amount of immunobeads required to bind a given amount of the fusion protein should be carefully titrated as described by Dean *et al.* (1989).

13. The eluate/bead mixture is spun briefly (30–60 seconds) in a microfuge. The supernatant (the unbound fraction) may be saved for later analysis.

14. The beads are washed twice in EBB at 4°C, and once in EBB containing 25–50 μg/ml poly[d(A–T)] (Boehringer Mannheim). The supernatant is removed completely after each wash.

15. Beads are suspended in 100–200 μl of TEN + IPTG and held on ice for 10 minutes, followed by centrifugation in a microfuge as above for 30 seconds at 4°C.

16. The supernatant, the bound and IPTG released fraction, may be stored up to 1 week at 4°C.

III.   Methods for Mapping Nucleosome Position

In minichromosomes, viral chromatin, and interphase genomic chromatin, the precise organization of DNA sequences relative to the histone octamer in the nucleosome is becoming of increasing interest in terms of the functional capacities of the DNA. Methods are available which allow one to ascertain whether nucleosomes are positioned precisely on a given DNA sequence or, at the opposite end of the spectrum of possible arrangements, are located totally randomly with respect to DNA sequence. Since many of the studies addressing positioning mechanisms and functional effects of positioning have been done in yeast minichromosomes, we review the methodology for mapping chromatin structure in this chapter. We consider the reagents used, the means of analysis of digestion patterns, and considerations important when interpreting the data as an inferred chromatin structure.

A.   Reagents

The most important single facet of a successful mapping experiment is the quality of the chromatin. Preparation of nuclei or minichromosomes as described above should ensure a good substrate for experiments in yeast. For other cell types, nuclear preparation is also necessary, to allow access of cutting reagents to the chromatin. High-purity nuclei *are not required* for mapping experiments; nuclei wherein rearrangement of the native chromatin structure has not occurred *are necessary*. The most useful criterion for nuclear quality is a micrococcal nuclease digestion nucleosome ladder. If a clear tandem repeat with low background between nucleosome oligomers is not evident on an ethidium bromide-stained gel, there is little point in proceeding with attempts to map the locations of nucleosomes on a sequence of interest. Note that an equally clear ladder may not be present for the sequence of interest when a blot of the gel is probed by hybridization; this may indicate a specific disruption of a regular chromatin structure for the probe region. What is important is preservation of a regular chromatin structure for bulk chromatin. A mandatory control for any method of analysis of nucleosome positioning is digestion of protein-free DNA with the same reagent used for chromatin studies.

Two nucleases, micrococcal nuclease and DNase I, and two chemical reagents, methidiumpropyl–EDTA–iron(II) complex (MPE) and hydroxyl radical, are the most useful DNA-cutting reagents for mapping nucleosome positions. Micrococcal nuclease preferentially cuts DNA in linker regions, making double-stranded scissions and thereby being the benchmark for definition of a nucleosome. The sequence preference of any nuclease creates problems in interpretation of results of mapping experiments (see below), and micrococcal nuclease is no exception to this rule. MPE was developed as an attempt to circumvent this sequence preference and has been largely successful in this goal (Hertzberg and Dervan, 1984). The methidium group is an intercalator which ensures that the reagent is located in the vicinity of DNA when the tethered EDTA–iron complex, with dissolved oxygen or hydrogen peroxide, generates hydroxyl radicals that cleave DNA. Single-strand scissions are the rule, but the physical localization of the radical generating complex leads to frequent nearby double-strand breaks. Presumably, intercalation within the nucleosome core particle is disfavored relative to intercalation in linker DNA, leading to the preferential cutting of chromatin DNA between nucleosomes by MPE. Our general preference is to use micrococcal nuclease for primary experiments. It is a large molecule and therefore more likely to sense gross features of chromatin structure. Refinement of results obtained with the nuclease can then be made with the MPE complex if necessary. In fact, recent results obtained with the nuclease have indicated that detection of nucleosome positioning with base pair precision is possible even with this large reagent (Shimizu *et al.,* 1991; see Fig. 6).

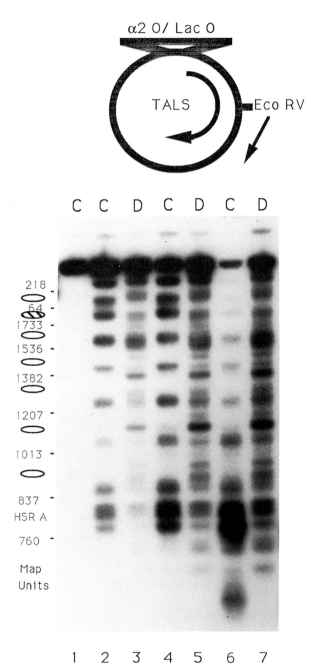

FIG. 5. Mapping of nucleosome position in TALS chromatin by indirect end labeling. Micrococcal nuclease cut sites were mapped in TALS chromatin (C) and DNA (D) isolated from α cells (as described by Thoma *et al.*, 1984) clockwise from the *Eco*RV restriction site at 385 map units. The probe consisted of a 230-bp *Eco*RV–*Hind*III TALS restriction fragment. Locations of cut sites were determined by comparison of individual fragment migration to that of molecular

In contrast to those reagents which preferentially cut between nucleosomes, DNase I and hydroxyl radical generated by a Fenton reaction occurring in solution (not preferentially in the neighborhood of chromatin DNA) cut chromatin DNA within the core particle as well (Hayes *et al.*, 1990). This leads to a DNA digestion pattern for single-stranded fragments which has a periodicity of about 10 nucleotides, arising from accessibility of the double helix to the enzyme or radical where it is most exposed to solvent and limited accessibility where it lies against the histone octamer. When a nucleosome is precisely positioned *in vivo*, this pattern will be strikingly evident (Shimizu *et al.*, 1991), just as it is for positioned nucleosomes *in vitro* (Simpson and Stafford, 1983).

B.   Methods

Nucleases are purchased from any of several commercial suppliers, dissolved in 10 mM Tris-Cl, pH 8, or water, aliquotted, and stored at $-20°$C. Nuclease digestions are performed at $37°$C in 10 mM HEPES, pH 7.5; 0.5 mM MgCl$_2$; 0.05 mM CaCl$_2$. Use of both calcium and magnesium reduces the sequence selectivity of DNase I. Detailed protocols for preparation and use of the MPE complex have been provided by Cartwright and Elgin (1989). The conditions for hydroxyl radical mapping are described by Hayes *et al.* (1990). A range of nuclease or reagent concentrations and times of digestion should be used in preliminary experiments. Some indication of the ballpark to work in can be gleaned from the primary literature. For yeast minichromosomes, micrococcal nuclease concentrations of 1–250 U/ml and DNase I concentrations of 0.01–2.5 U/ml for 10 minutes at $37°$C have been used successfully to map nucleosome positions. Naked DNA controls are digested with 5–100 times lower concentrations of nuclease. The goal is to obtain digests wherein the region of interest has been cut once per molecule. Some of the full-length parent fragment should remain when analyzed as described below for results to be meaningful. Different levels of digestion should be analyzed, since nucleosome positioning may not be stable during the course of the analysis (Thoma *et al.*, 1984). There is no substitute for varying conditions of digestion to optimize the display of cutting patterns (Fig. 5).

---

weight standards on the same gel (not shown). Inferred positions of nucleosomes are indicated by ellipses (⌒). Notice that these are between hypersensitive sites in chromatin which are separated by about 140 bp and that intervening sites which are exposed in naked DNA are protected. The striped ellipse indicates a region in which no cut sites are present in naked DNA. Therefore, no information about nucleosome placement in this region can be obtained from this map. Micrococcal nuclease concentrations used: 0 units/ml (lane 1), 2 units/ml (lane 2 and 3), 5 units/ml (lane 5), 10 units/ml (lanes 4 and 7), and 50 units/ml (lane 6). (Reproduced from Roth *et al.*, 1990.)

C.   Analysis

After chromatin DNA has been cut with one of the four reagents, a population of linear fragments is generated. The goal in mapping experiments is to define just where these cut sites are, relative to a known site in the primary sequence of the DNA. To achieve this goal, one needs to define an origin for the map—to create the beginning of a molecular ruler. Two good methods are currently used for such mapping, indirect end labeling (for low-resolution studies) and primer extension (for high-resolution experiments).

Indirect end labeling was introduced in 1980 by Wu for DNase I-hypersensitive sites and Nedospasov and Georgiev for micrococcal nuclease cutting sites in chromatin. DNA is purified from the chromatin digest and secondarily restricted with a restriction endonuclease have a cutting site near the region of interest. This leads to a population of molecules having one end defined by the restriction site and the other end created by the nuclease or chemical reagent cut site. This population is separated by agarose gel electrophoresis and blotted to a nitrocellulose (Schleicher and Schuell) or nylon (e.g., Gene Screen from DuPont or Duralon from Stratagene) membrane. The blot is then hybridized with a short (approximately 200 bp) radioactive probe which abuts the restriction site. Only those fragments which extend away from the restriction site will be labeled, leading to the equivalent of labeling the DNA at the restriction site *in vitro*, that is, an indirect end label. The distance from the restriction site is determined by comparison with molecular weight standards or internal restriction fragments, allowing location of the nuclease or reagent cutting sites. Detailed descriptions of the method and its background are provided by Wu (1989) and Nedospasov *et al.* (1989). An example of an indirect end label map which shows positioned nucleosomes on a yeast minichromosome containing the $\alpha 2$ operator is shown in Fig. 5.

An analogous method for genomic sequencing was developed by Church and Gilbert (1984) and adapted by others to high-resolution-analysis chromatin mapping. While seminal for extending mapping experiments to nucleotide-level resolution, this approach has largely been supplanted by primer extension mapping of cut sites, due to the ease of its application. Primer extension is similar in principle to indirect end labeling, but in this case, a short, end-labeled oligonucleotide is hybridized in solution to the digested, purified DNA. The complex is then extended using AMV reverse transcriptase, a modified T7 DNA polymerase, or *Thermus aquaticus* (*Taq*) DNA polymerase. The limitations and virtues of each enzyme are discussed by Hull *et al.* in [15], this volume. Extension terminates any given DNA molecule at the cutting site for the probing nuclease or chemical reagent. Separation of the resultant fragments on a sequencing gel allows determination of the cutting sites in the population of chromatin analyzed. A single cycle of primer extension is sufficient for analysis of digests of high-copy-number minichromosomes. Determination of cutting sites in single-copy yeast gene chromatin requires about 15 cycles of linear amplification of the primer extension products. An example of a primer extension map showing the location of nucleosomes on the promoter and beginning of the yeast genomic *STE6* gene is shown in Fig. 6. For the more complex genomes of larger eukaryotic cells, polymerase chain reaction amplification of the primer extension product using an oligonucleotide coupled to the probe cutting site may be necessary (Mueller and Wold, 1989). Since any sequence can be used as the origin for our ruler, one is not limited by the availability of a suitable restriction endonuclease site.

D. Interpretation

At the end of this analysis, one is left with a pattern of cutting sites for the probing reagent on chromatin and naked DNA. Inclusion of the naked DNA samples as controls is mandatory for this type of analysis, given the sequence selectivity of most, if not all, of the probing reagents. Sites which are cut in naked DNA but protected in chromatin are informative, as are the opposite, sites cut in chromatin but not in naked DNA. Sites which are cut in both experimental and control samples may be informative, while sites that were not cut in either sample cannot contribute to the analysis. Compare Fig. 5 (for data) and Fig. 1B (for interpretation) as an indication of the analytical process involved in inference of a chromatin structure from mapping experiments.

The operational definition of a positioned nucleosome is a region of chromatin spanning 146–200 bp flanked by micrococcal nuclease or MPE cutting sites, with protection of intervening sites which are exposed in naked DNA. While many other structures might lead to protection of this length of DNA from nuclease cutting, the sole case where both nuclease mapping of chromatin structure and morphological characterization of the minichromosome have been performed showed agreement between the two types of analysis, lending credence to the biochemical approach to determination of the structure of chromatin (c. f. Thoma *et al.,* 1984 and Pederson *et al.,* 1986). Extension of micrococcal nuclease or MPE mapping to experiments which show a 10-bp periodicity for cutting of chromatin DNA by DNase I or hydroxyl radical provides much firmer support for the conclusion that a positioned nucleosome is located in the region of interest. Recently, such experiments have been carried out using primer extension analysis of the structure of yeast minichromosomes and single-copy yeast genomic genes (Shimizu *et al.,* 1991).

The precision and nature of positioning of nucleosomes must be an element of the analysis of mapping experiments. The precision of positioning can only be judged by higher and higher resolution analysis. It is apparent that agarose gel indirect end label mapping over several thousand base pairs cannot define the location of a nucleosome with the precision that a sequencing gel primer extension experiment can, irrespective of the probing reagent employed. Differentiation of the mechanisms of positioning, statistical (Kornberg and Stryer, 1988), DNA sequence related (Thoma and Simpson, 1985) or enforced by nonhistone proteins (Roth *et al.,* 1990; Figs. 5 and 6), requires high-resolution analysis. The nature of positioning refers to rotational versus translational positioning. Rotational positioning reflects the orientation of the DNA double helix on the surface of the histone octamer—a uniformly bent DNA might adopt several translational positions on the octamer but maintain a constant rotational orientation, enforced by inherent bending of the nucleic acid. This positioning would be scored as positive in a DNase I or hydroxyl radical map but negative using micrococcal nuclease or MPE. On the other hand, a nucleosome which is statistically positioned near a nonhistone protein might have a translational position which is precise to ± 5 bp, but have completely random rotational positioning. This would be defined as a positioned nucleosome in MPE or micrococcal nuclease digests but would be scored as random in a DNase I or hydroxyl radical experiment. In contrast, a nucleosome which is absolutely precisely positioned translationally will also be precisely positioned rotationally; this has been observed for nucleosomes flanking the α2 operator in yeast cells (Shimizu *et al.,* 1991; Fig. 6).

Only recently has it been shown that the location of a cis-acting element within a nucleosome may be important in its function. The yeast plasmid replication origin is limited in its efficiency when located within 40 bp of the pseudodyad of a nucleosome although it is as efficient in the peripheral regions of the nucleosome as when located in linker DNA (Simpson, 1990).

FIG. 6. Chromatin structure of a single-copy yeast gene mapped by primer extension. The chromatin structure of the single-copy, **a**-cell-specific gene *STE6* was analyzed by primer extension mapping of micrococcal nuclease cut sites in chromatin (C) or DNA (D) isolated from α cells(α) or unrelated **a** cells (**a**). The position of the mRNA start site is indicated by the arrow, and the locations of the α2 operator (required for *STE6* transcriptional repression in α cells), putative TATA sequences, and the 5′ portion of the *STE6* coding region are shown (Wilson and Herskowitz, 1986). Numbering is relative to the AUG. X indicates a band corresponding to an artifactual primer extension stop, not a micrococcal nuclease cut site, as determined by examination of extension products of undigested, naked DNA (not shown). Naked DNA prepared from nuclei was also used in dideoxy sequencing reactions (G corresponds to a reaction terminating in ddC, A for that terminating in ddT, etc.). Protection of several nuclease cut sites present in naked DNA over a 142-bp region in chromatin from α cells, along with nuclease-hypersensitive sites at the ends of this protected region (at −166 and −24), indicates that a nucleosome is positioned in the promoter region (ellipse). Protected regions are also observed downstream, in the *STE6* coding region. These regions of protection are disrupted in **a** cells; several cut sites exposed in naked DNA are also exposed in the chromatin of these cells. Chromatin was digested with 20 units/ml micrococcal nuclease. DNA was digested with 0.5 units/ml micrococcal nuclease. Digested fragments were purified by standard procedures (Sambrook *et al.,* 1990), annealed to a ^{32}P-end-labeled primer, and amplified with 15 cycles of a linear polymerase chain reaction (94°C, 1 minute; 55°C, 2 minutes; 73°C; 3 minutes) using *Taq* polymerase (Cetus).

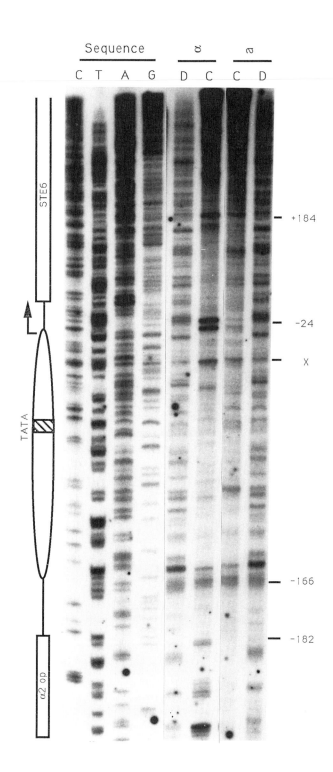

When located in a positioned nucleosome facing toward the solvent, the glucocorticoid response element is able to interact with the steroid receptor; another element which faces the histone octamer appears to interact less well (Perlmann and Wrange, 1988). These observations reinforce the need to use reagents which assess both the translational and the rotational position of nucleosomes, at a high-resolution level of analysis, when attempting to correlate the location of a nucleosome with function of DNA in chromatin.

IV. Closing Remarks

Much has been learned in recent years about specific factors involved in regulation of many genes, but the mechanisms by which these factors activate or repress transcription are not well understood. Although changes in chromatin structure have been correlated to changes in gene expression, the significance of the structural changes has been difficult to assess. Little is known, for example, about how specific trans-acting factors interact with chromatin or how specific cis-acting elements are influenced by the chromatin structure of neighboring sequences. Yeast minichromosomes provide an *in vivo* system in which these types of questions may be directly addressed. The ability to manipulate the genetic background of the host yeast cell together with the relative ease with which plasmid chromatin can be isolated and analyzed make yeast minichromosomes a very powerful system to study not only the structure of chromatin, but also the functional consequences of changes in chromatin structure.

REFERENCES

Bergman, L. W. (1986). A DNA fragment containing the upstream activator sequence determines nucleosome positioning of the transcriptionally repressed *PHO5* gene of *Saccharomyces cerevisiae*. *Mol. Cell. Biol.* **6**, 2298–2304.

Bergman, L. W., and Kramer, R. A. (1983). Modulation of the chromatin structure associated with derepression of the acid phosphatase gene of *Saccharomyces cerevisiae*. *J. Biol. Chem.* **258**, 7223–7227.

Bloom, K. S., and Carbon, J. (1982). Yeast centromere DNA is in a unique and highly ordered structure in chromosomes and small circular minichromosomes. *Cell (Cambridge, Mass.)* **29**, 305–317.

Brill, S., and Sternglanz, R. (1988). Transcription dependent DNA supercoiling in yeast DNA topoisomerase mutants. *Cell (Cambridge, Mass.)* **54**, 403–411.

Broach, J. R. (1981). The yeast 2μ circle. *In* "The Molecular Biology of Yeast *Saccharomyces*: Life Cycle and Inheritance" J. N. Strathern, E. W. Jones, and J. R. Broach, eds.), pp. 445–470. Cold Spring Harbor Laboratory, Cold Spring Harbor, New York.

Cartwright, I. L., and Elgin, S. C. R. (1989). Nonenzymatic cleavage of chromatin. *In* "Methods in Enzymology" (P. M. Wassarman and R. D. Kornberg, eds.), Vol. 170, pp. 359–369. Academic Press, San Diego, California.

Chao, M. V., Gralla, J. D., and Martinson, H. G. (1980). Lac operator nucleosomes I. Repressor binds specifically to operator within the nucleosome core. *Biochemistry* **19**, 3254–3260.

Church, G., and Gilbert, W. (1984). Genomic sequencing. *Proc. Natl. Acad. Sci. U.S.A.* **81**, 1991–1995.

Cordingly, M. G., Riegel, A. T., and Hager, G. L. (1987). Steroid dependent interactions of transcription factors with the inducible promoter of mouse mammary tumor virus *in vivo*. *Cell (Cambridge, Mass.)* **48**, 261–270.

Dean, A., Pederson, D. S., and Simpson, R. T. (1989). Isolation of yeast plasmid chromatin. *In* "Methods in Enzymology" (P. M. Wasserman and R. D. Kornberg, eds.), Vol. 170, pp. 26–41. Academic Press, San Diego, California.

Fangman, W. L., and Zakian, V. A. (1981). Genome structure and replication. *In* "The Molecular Biology of Yeast *Saccharomyces*: Life Cycle and Inheritance" J. N. Strathern, E. W. Jones, and J. R. Broach, eds.), pp. 27–31. Cold Spring Harbor Laboratory, Cold Spring Harbor, New York.

Fedor, M. J., Lue, N. F., and Kornberg, R. D. (1988). Statistical positioning of nucleosomes by specific protein binding to an upstream activating sequence in yeast. *J. Mol. Biol.* **204**, 109–127.

Fowler, A. V., and Zabin, I. (1983). Purification, structure, and properties of hybrid β-galactosidase proteins. *J. Biol. Chem.* **258**, 14354–14358.

Griffith, J. D. (1975). Chromatin structure: Deduced from a minichromosome. *Science* **187**, 1202–1203.

Hayes, J. J., Tullius, T. D., and Wolffe, A. P. (1990). The structure of DNA in a nucleosome. *Proc. Natl. Acad. Sci. U.S.A.* **87**, 405–409.

Hertzberg, R. P., and Dervan, P. B. (1984). Cleavage of DNA with methidiumpropyl-EDTA-iron(II): Reaction conditions and product analyses. *Biochemistry* **23**, 3934–3945.

Kornberg, R. D., and Stryer, L. (1988). Statistical distributions of nucleosomes: Non-random locations by a stochastic mechanism. *Nucleic Acids Res.* **16**, 6677–6690.

Levens, D., and Howley, P. M. (1985). Novel method for identifying sequence-specific DNA-binding proteins. *Mol. Cell. Biol.* **5**, 2307–2315.

Losa, R., Omari, S., and Thoma, F. (1990). Poly(dA)·Poly(dT) rich sequences are not sufficient to exclude nucleosome formation in a constitutive yeast promoter. *Nucleic Acids Res.* **18**, 3495–3502.

Morse, R. H., and Simpson, R. T. (1988). DNA in the nucleosome. *Cell (Cambridge, Mass.)* **54**, 285–287.

Morse, R. H., Pederson, D. S., Dean, A., and Simpson, R. T. (1987). Yeast nucleosomes allow thermal untwisting of DNA. *Nucleic Acids Res.* **15**, 10311–10330.

Mueller, P. R., and Wold, B. (1989). *In vivo* footprinting of a muscle specific enhancer by ligation mediated PCR. *Science* **246**, 780–786.

Nedospasov, S. A., and Georgiev, G. P. (1980). Nonrandom cleavage of SV40 DNA in the compact minichromosome and free in solution by micrococcal nuclease. *Biochem. Biophys. Res. Commun.* **92**, 532–539.

Nedospasov, S. A., Shakhov, A. N., and Georgiev, G. P. (1989). Analysis of nucleosome positioning by indirect end labeling and molecular cloning. *In* "Methods in Enzymology" (P. M. Wasserman and R. D. Kornberg, eds.), Vol. 170, pp. 408–420. Academic Press, San Diego, California.

Norton, V. G., Imai, B. S. Yau, P., and Bradbury, E. M. (1989). Histone acetylation induces nucleosome core particle linking number change. *Cell (Cambridge, Mass.)* **57**, 449–457.

Pederson, D. S., and Morse, R. H. (1990). Effect of transcription of yeast chromatin on DNA topology *in vivo*. *EMBO J.* **9**, 1873–1881.

Pederson, D. S., Venkatesan, M., Thoma, F., and Simpson, R. T. (1986). Isolation of an episomal yeast gene and replication origin as chromatin. *Proc. Natl. Acad. Sci. U.S.A.* **83**, 7206–7210.

Perlmann, T., and Wrange, O. (1988). Specific glucocorticoid receptor binding to DNA reconstituted in a nucleosome. *EMBO J.* **7**, 3973–3079.

Richard-Foy, H., and Hager, G. L. (1987). Sequence specific positioning of nucleosomes over the steroid-inducible MMTV promoter. *EMBO J.* **6**, 2321–2328.

Rose, M. D., Winston, F., and Hieter, P. (1990). "Methods in Yeast Genetics," pp. 34–43. Cold Spring Harbor Laboratory, Cold Spring Harbor, New York.

Roth, S. Y., Dean, A., and Simpson, R. T. (1990). Yeast α2 repressor positions nuclesomes in TRP1/ARS1 chromatin. *Mol. Cell. Biol.* **10**, 2247–2260.

Sambrook, J., Fritsch, E. F., and Maniatis, T. (1990). "Molecular Cloning: A Laboratory Manual," 2nd Ed. Cold Spring Harbor Laboratory, Cold Spring Harbor, New York.

Shimizu, M., Roth, S. Y., Szent-Gyorgi, C. and Simpson, R. T. (1991). Nucleosomes are positioned with base pair precision adjacent to the α2 operator in *Saccharomyces cerevisiae*. *EMBO J.* **10** (in press).

Simpson, R. T. (1990). Nucleosome positioning can affect the function of a cis-acting DNA element *in vivo*. *Nature (London)* **343**, 387–389.

Simpson, R. T. (1991). Nucleosome positioning: Occurrence, mechanism and functional consequences. *Prog. Nucleic Acid Res. Mol. Biol.* **40**, 143–184.

Simpson, R. T., and Stafford, D. W. (1983). Structural features of a phased nucleosome core particle. *Proc. Natl. Acad. Sci. U.S.A.* **80**, 51–55.

Smerdon, M. J., and Thoma, F. (1990). Site specific DNA repair at the nucleosome level in a yeast minichromosome. *Cell (Cambridge, Mass.)* **61**, 675–684.

Smerdon, M. J., Bedoyan, J., and Thoma, F. (1990). DNA repair in a small yeast plasmid folded into chromatin. *Nucleic Acids Res.* **18**, 2045–2051.

Som, T., Armstrong, K. A., Volkert, F. C., and Broach, J. R. (1988). Autoregulation of yeast 2μ circle gene expression provides a model for maintenance of stable plasmid copy levels. *Cell (Cambridge, Mass.)* **52**, 27–37.

Szent-Gyorgyi, C., and Isenberg, I. (1983). The organization of oligonucleosomes in yeast. *Nucleic Acids Res.* **11**, 3717–3736.

Thoma, F. (1986). Protein DNA interactions and nuclease hypersensitive regions determine nucleosome positions on yeast plasmid chromatin. *J. Mol. Biol.* **190**, 177–190.

Thoma, F., and Simpson, R. T. (1985). Local protein-DNA interactions may determine nucleosome positions on yeast plasmids. *Nature (London)* **315**, 250–253.

Thoma, F., and Zatchez, M. (1988). Chromatin folding modulates nucleosome positioning in yeast minichromosomes. *Cell (Cambridge, Mass.)* **55**, 945–953.

Thoma, F., Bergman, L. W., and Simpson, R. T. (1984). Nuclease digestion of circular TRP1 ARS1 chromatin reveals positioned nucleosomes separated by nuclease sensitive regions. *J. Mol. Biol.* **177**, 715–733.

Volkert, F. C., and Broach, J. R. (1986). Site-specific recombination promotes plasmid amplification in yeast. *Cell (Cambridge, Mass.)* **46**, 541–550.

Wilson, K. L., and Herskowitz, I. (1986). Sequences upstream of the *STE6* gene required for its expression and regulation by the mating type locus in *Saccharomyces cerevisiae*. *Proc. Natl. Acad. Sci. U.S.A.* **83**, 2536–2540.

Wu, C. (1980). The 5′ ends of *Drosophila* heat-shock genes in chromatin are sensitive to DNase I. *Nature (London)* **286**, 854–860.

Wu, C. (1989). Analysis of hypersensitive sites in chromatin. *In* "Methods in Enzymology" (P. M. Wasserman and R. D. Kornberg, eds.) Vol. 170, pp. 269–289. Academic Press, San Diego, California.

Zakian, V. A., and Scott, J. F. (1982). Construction, replication, and chromatin structure of TRPI RI circle, a multicopy synthetic plasmid derived from *Saccharomyces cerevisiae* chromosomal DNA. *Mol. Cell. Biol.* **2**, 221–232.

Chapter 12

Nucleosomes of Transcriptionally Active Chromatin: Isolation of Template-Active Nucleosomes by Affinity Chromatography

VINCENT G. ALLFREY AND THELMA A. CHEN

Laboratory of Cell Biology
The Rockefeller University
New York, New York 10021

I. Introduction

A. Alterations of Nucleosome Structure in Transcriptionally Active Chromatin

The isolation method to be described is based on the premise that the nucleosome is a dynamic structure, changing its conformation and composition to facilitate transit of the RNA polymerases along the DNA template, and that the changes in nucleosome structure associated with transcription are

315

rapid and reversible. They are mediated in part by postsynthetic modifications of the histones and other nucleosome-associated nonhistone proteins that relax the constraints on the DNA strand. The changes in nucleosome topology are augmented momentarily by the passage of the transcription complex through the DNA of the nucleosome core. The net effect, an "unfolding" of the nucleosome cores along the transcribed DNA sequences, alters the accessibility of thiol groups in histone H3 and in certain nonhistone proteins associated with active nucleosomes. This change in thiol reactivity of transcriptionally active nucleosomes, which has been shown to occur *in vivo* (Prior *et al.*, 1983), in intact nuclei (Johnson *et al.*, 1987; Chan *et al.*, 1988), and in isolated nucleosomes (Johnson *et al.*, 1983; Sterner *et al.*, 1987), can be exploited to permit their isolation by affinity chromatography on an organomercurial matrix (Allegra *et al.*, 1987; Sterner *et al.*, 1987; Chen and Allfrey, 1987; Walker *et al.*, 1990; Chen *et al.*, 1990; Boffa *et al.*, 1990).

Here we describe the method as applied to cultured murine fibroblasts (BALB/c 3T3 cells) for the study of nucleosome structural changes during the induction and repression of the c-*fos* and c-*myc* oncogenes.

This chromatographic separation of active and inactive nucleosomes depends at the outset on enzymatic methods for the release of nucleosomes from the interphase nucleus. It has long been recognized that the active regions of chromatin exist in an extended state in which the DNA is highly accessible to hydrolytic cleavage by DNase I (Weintraub and Groudine, 1976; Garel *et al.*, 1977), DNase II (Gottesfeld *et al.*, 1974; Gottesfeld and Butler, 1977), and micrococcal nuclease (Bellard *et al.*, 1977; Bloom and Anderson, 1978, 1979; Sanders, 1978; Dimitriades and Tata, 1980; Smith *et al.*, 1984).

Because sensitivity to DNase I is not tightly coupled to the timing or rate of transcription (Weintraub and Groudine, 1976; Garel *et al.*, 1977) and can extend into the nontranscribed DNA sequences flanking the gene (Bellard *et al.*, 1980; Lawson *et al.*, 1982), we have focused on nucleosome fractions generated by limited digestions of isolated nuclei with micrococcal nuclease (MNase). There is ample evidence that this enzyme selectively cleaves rapidly transcribed genes; with the release of nucleosomes and nucleosome oligomers (Bellard *et al.*, 1977; Bloom and Anderson, 1978, 1979; Sanders, 1978; Dimitriadis and Tata, 1980; Smith *et al.*, 1984). Of particular significance are observations showing that the micrococcal nuclease sensitivity of a given gene is related to the timing of its transcription, that the MNase-sensitive domain does not include the nontranscribed DNA sequences flanking the gene, and that the sensitivity to MNase is lost when transcription stops. These are significant advantages in development of a method to isolate transcriptionally active nucleosomes; but there are limitations and technical problems that require consideration.

Chief among these is the fact that the enzyme can cleave DNA within the nucleosome core in addition to its more effective attack on "linker" DNA sequences between the cores. Because the nucleosomes of transcriptionally active chromatin are released rapidly into a digestion buffer in which the enzyme remains active, a substantial fraction of the released nucleosomes may be degraded and lost as digestion is prolonged. This effect is particularly evident in the unfolded nucleosomes of rapidly transcribed DNA sequences, such as the ribosomal genes of *Physarum*, where MNAse released over 43% of the rDNA-containing nucleosomes in 2 minutes, but only 9% could be detected after a 10-minute digestion period (Johnson *et al.*, 1978).

The cutting of DNA sequences within the cores of active nucleosomes results in a blurring of the nucleosomal repeat-length ladder generated by MNase digestion, as observed, for example, in the heat-shock genes of *Drosophila* (Wu *et al.*, 1979), and the ovalbumin (Bloom and Anderson, 1982) and lysozyme (Stratling *et al.*, 1986). genes of the chicken. To minimize this intranucleosomal DNA degradation and favor the preferential release of the active nucleosomes, micrococcal nuclease digestions must be limited. In practice, conditions are chosen to release no more than 10% of the total nuclear DNA. It is especially important, when investigating changes in chromatin structure of a given gene under different experiment conditions, to control the digestions to achieve equivalent DNA release.

The lability of nucleosomes to endonuclease attack is affected by changes in the interactions between the core histones and the enveloping DNA strand. Nucleosomes of transcriptionally active genes are hyperacetylated while those of inactive genes are not (Johnson *et al.*, 1987; Allegra *et al.*, 1987; Sterner *et al.*, 1987; Hebbes *et al.*, 1988; Walker *et al.*, 1990; Boffa *et al.*, 1990; and reviews in Reeves, 1984; Matthews and Waterborg, 1985). This modification of the basic NH_2-terminal domains of the histones is known to increase the accessibility to nucleases of specific DNA sites within the nucleosome core (Simpson, 1978), and to facilitate unfolding of the nucleosome, as evidenced by a reduction in the negative linking number change per nucleosome (Norton *et al.*, 1989) and by electron spectroscopic imaging of the DNA (Oliva *et al.*, 1990; Lochlear *et al.*, 1991).

Histone acetylation is also a factor in altering nucleosome topology to reveal the previously shielded cysteinyl-SH groups of histone H3 (Bode *et al.*, 1980).

In order to maintain the unfolded state of the active nucleosomes during isolation of the nuclei, MNase digestion, and subsequent steps in the chromatographic separation, the deacetylation of the histones should be minimized. For this purpose, we employ 5 mM sodium butyrate as a deacetylase inhibitor (Riggs *et al.*, 1977; Boffa *et al.*, 1978); it is present from the time the cells are broken until the chromatographic separation of the nucleosomes is completed. [It should be noted, however, that hyperacetylation of the histones, although necessary, is not sufficient to establish the transcriptionally active state of the nucleosome core (Boffa *et al.,* 1990; Chen *et al.* 1991).]

B. Rationale for the Chromatographic Separation of Transcriptionally Active and Inactive Nucleosomes

1. REACTIVITY OF HISTONE H3 THIOLS AS A BASIS FOR SELECTIVE MODIFICATION OF TRANSCRIPTIONALLY ACTIVE NUCLEOSOMES

Studies of the ribosomal genes in *Physarum* established that nucleosomes in the transcribed regions of the rDNA "unfold" to reveal the thiol groups of histone H3, and that those groups could be derivatized with iodoacetamidofluorescein (Prior *et al.,* 1983) or with [^{3}H]iodoacetate (Johnson *et al.,* 1987). Two additional points were made: first, the H3 thiols were *not* reactive in the nucleosomes of the nontranscribed central "spacer" of the rDNA molecule; and second, when rRNA synthesis was suppressed, the thiol reactivity of nucleolar H3 was lost (Prior *et al.,* 1983).

It was subsequently shown that the unfolding of nucleosomes is not limited to ribosomal genes; the SH reactivity of histone H3 was also demonstrated in the active nonnucleolar chromatin of *Physarum* (Johnson *et al.,* 1987) and in transcriptionally active chromatin fractions of avian cells (Chan *et al.,* 1988).

2. AFFINITY CHROMATOGRAPHY OF THIOL-REACTIVE NUCLEOSOMES

The close correlation between the reactivity of histone H3 thiol groups and transcription suggested that the nucleosomes of active genes could be separated from inactive nucleosomes by affinity chromatography on an organomercurial column which had been developed for the purification of histone H3(Ruiz-Carrillo and Allfrey, 1973). Tests were carried out by passing the mixture of nucleosomes released during a limited micrococcal nuclease digestion of isolated nuclei through a short organomercurial column, washing the column to elute the unbound nucleosomes, and then displacing the bound nucleosomes with 10 m*M* dithiothreitol. The DNA sequence content of the unbound and bound nucleosome fractions was then compared by quantitative slot-blot hybridizations to a variety of ^{32}P-labeled DNA probes.

It was shown that liver nucleosomes retained by the mercury column were enriched in DNA sequences transcribed by hepatocytes (e.g., albumin and transferrin genes) but lacked DNA sequences not expressed in the liver but in the brain (e.g., preproenkephalin DNA) (Allegra *et al.,* 1987). Mercury-affinity chromatography of nucleosomes prepared from HeLa cells at successive stages in the cell cycle showed that the histone H2A and H4 genes were present in the Hg-bound nucleosome fraction during the S phase when histone mRNA synthesis is maximal, but those histone DNA sequences were shifted to the class of unbound nucleosomes in the G_2 phase when synthesis of the cognate histone mRNAs is curtailed (Sterner *et al.*, 1987).

A third and more rigorous test of the procedure was an analysis of the rapid kinetics of activation and repression of the c-*fos* and c-*myc* oncogenes when quiescent 3T3 cells are stimulated to reenter the growth cycle (Greenberg and Ziff, 1984). At every time point examined, the extent of retention of the oncogenic DNA on the column coincided closely with the extent of transcription of each oncogene, as determined by run-off transcription assays on the isolated nuclei (Chen and Allfrey, 1987).

3. RAPID REVERSIBILITY OF THE ACTIVE STATE WHEN TRANSCRIPTION CEASES

Transcription of the c-*fos* gene in serum-stimulated 3T3 cells reaches a peak in 15 minutes and then declines abruptly, reaching preactivation levels by 30 minutes. This rapid change in transcription is mirrored in the Hg-binding of the c-*fos* nucleosomes, which are fully retained at 15 minutes and not retained at 30 minutes (Chen and Allfrey, 1987). When transcription of c-*fos* and c-*myc* was blocked by α-amanitin, an inhibitor of RNA polymerase II, their nucleosomes failed to bind to the column 10 minutes after addition of the drug (Chen *et al.*, 1990). This is a clear indication that mercury-affinity chromatography can be used to monitor rapid changes in nucleosome structure during gene activation and repression. Significantly, there was no effect of α-amanitin on the Hg-binding of nucleosomes from the 28S ribosomal gene which is transcribed by amanitin-resistant RNA polymerase I (Chen *et al.*, 1990).

4. TWO TYPES OF MERCURY BINDING BY TRANSCRIPTIONALLY ACTIVE NUCLEOSOMES

Recent studies of the binding of nucleosomes to mercury columns have confirmed that a major mode of binding involves the formation of a covalent mercaptide bond between the organomercurial and the accessible thiol groups of histone H3 (Walker *et al.*, 1990). In yeast, an organism in which histone H3 has no cysteinyl residues, this type of nucleosome binding cannot occur, and there is no enrichment of transcriptionally active DNA sequences on the column. However, when the yeast histone H3 gene is altered by site-directed mutagenesis to encode a cysteinyl residue at position 110 of the polypeptide chain, transcriptionally active nucleosomes are recovered on the mercury column (Chen *et al.*, 1991).

Electrophoretic analyses of the proteins in the active nucleosomes of mammalian cells show the presence of many nonhistone proteins, some of which appear in stoichiometric proportions to the histones and also contain reactive SH groups. Two of these proteins were identified as high-mobility group proteins, HMG-1 and HMG-2 (Walker *et al.*, 1990). Because these HMG proteins have been shown to stimulate markedly RNA synthesis in chromatin (Stoute and Marzluff, 1982) and in reconstituted transcription systems using purified RNA polymerases II and III (Tremethick and Molloy, 1986), their role in the Hg-binding of the active nucleosomes had to be considered. The ease of extraction of nuclear HMG-1 and HMG-2 in 0.5 M NaCl suggested that nucleosomes retained on the mercury column because of their association with these thiol-reactive proteins (or other salt-extractable, thiol-reactive proteins) could be released by raising the ionic strength of the eluting buffer. When HeLa and 3T3 nucleosomes were applied to the mercury column and then eluted in 0.5 M NaCl, a substantial fraction of the Hg-bound nucleosomes was released, while the HMG proteins remained covalently linked to the column (Walker, *et al.*, 1990).

The nucleosomes released by 0.5 M NaCl were shown to contain actively transcribed DNA sequences and hyperacetylated histones (Walker *et al.*, 1990; Chen *et al.*, 1990), and their longer monomeric DNA lengths were consistent with shielding of linker DNA sequences by associated HMG proteins (Walker *et al.*, 1990). Subsequent treatment of the salt-washed column with 10 mM dithiothreitol released those nucleosomes that were covalently linked to the mercurated support through the thiol groups of histone H3 (Walker *et al.*, 1990).

In light of this evidence that active nucleosomes can be fractionated into two classes with different modes of binding to the mercury column, the original one-step elution procedure with 10 mM dithiothreitol has been modified to permit separation of the salt-labile nucleosome fraction from the nucleosomes with reactive H3 thiol groups. Both classes of Hg-bound nucleosomes are enriched in the c-*fos* and c-*myc* DNA sequences at the time of their transcription, and both are depleted of those sequences when their transcription is inhibited, but the classes differ significantly in their degree of enrichment. Comparisons of the salt-eluted and DTT-eluted nucleosome classses have shown that the specific c-*fos* or c-*myc* DNA concentration (expresssed as intensity of hybridization signal/microgram of DNA) can be 2- to 3-fold higher in the DTT-eluted nucleosomes than in the corresponding salt-eluted fraction (Chen *et al.*, 1990).

The two-step method for the separation of Hg-bound nucleosomes is described in detail in Section II,B.

II.   Methods

A.   Isolation of Nucleosomes from Serum-Stimulated Murine Fibroblasts

1.   CELL CULTURE AND TIMING OF c-*fos* AND c-*myc* EXPRESSION

BALB/c 3T3 cells (clone A31 murine fibroblasts) were obtained from the American Type Culture Collection (Gaithersburg, MD). The cells are grown in Dulbecco's modified Eagle's minimal essential medium (D-MEM) (Dulbecco and Freeman, 1959; Smith *et al.,* 1960; Gibco Laboratories, Grand Island, NY) containing 4500 mg/liter D-glucose 100 U/ml penicillin G, 100 μg/ml streptomycin sulfate, 10% newborn calf serum (NCS). The cells (1×10^6 in 25 ml of growth medium) are seeded in a sterile 150-cm^2 tissue-culture flask with a plug seal and canted neck (Corning Glass Works, Corning, NY). The cultures are incubated at 37°C in humidified air containing 10% CO_2. Cell numbers are amplified by taking subconfluent cultures, removing the growth medium, and washing the cells twice with Dulbecco's phosphate-buffered saline (D-PBS) (Dulbecco and Vogt, 1954) without $CaCl_2$. The cells are treated with a solution of 0.05% trypsin/0.53 mM EDTA (Gibco) until they become detached. Trypsinization is stopped by diluting the cell suspension with an excess of growth medium and pelleting the cells by centrifugation at 25°C for 7 minutes at 500 g. After two washings, the cells are resuspended in fresh growth medium and reseeded as described above. After 5–6 days of incubation, the cells reach confluency at ~1×10^7 cells per flask, and they deplete the medium of growth factors by 7–8 days. Note that when the cells are actively growing they appear elongated or spindle-shaped under an inverted light microscope, but quiescent cells are avoid or rounded.

Quiescent cell cultures are stimulated to reenter the growth cycle by removing the old medium by suction, and replacing it with fresh medium containing 20% (v/v) NCS, without antibiotics. The cultures are incubated under growth conditions for the appropriate time. Under these conditions, the maximal transcriptional activity of c-*fos* occurs at 15 minutes after serum stimulation; it then declines abruptly to prestimulation levels at 30 minutes (Greenberg and Ziff, 1984). Transcription of c-*myc* begins at about 30 minutes, peaks at 1–2 hours, and declines to preactivation levels in about 4 hours (Greenberg and Ziff, 1984). Similar results are obtained when purified platelet-derived growth factor, at a concentration of 3 units/ml in D-MEM, is used instead of serum.

It is recommended that the transcriptional activity of c-*fos* and c-*myc* (or other genes under investigation) be tested by nuclear run-on transcription assays (rather than Northern blotting) in order to assure that the genes are being expressed, or not expressed, in the short time frames of the experiment. This will allow a meaningful comparison of gene activity with the distribution of the cognate DNA sequences in the bound and unbound nucleosome fractions prepared by mercury-affinity chromatography.

The number of cells required for nucleosome chromatography depends on the goals of the experiment. If the intent is simply to compare the DNA sequence content of the Hg-bound and unbound nucleosome fractions, about 1×10^8 cells would be adequate (i.e., about 10 flasks of 3T3 cells). If the nucleosomes are to be analyzed for their histone and nonhistone protein contents, at least 3×10^8 cells are recommended. Similar requirements have been observed for other mammalian cell types (HeLa-S3, COLO 320, and rat liver cells).

2. ISOLATION OF CELL NUCLEI

All reagents used in the procedures to follow are of molecular biology quality. All solutions are sterilized by filtration through a 0.2-μm membrane, or by autoclaving stock solutions before making the final working buffers.

The cells are harvested at the appropriate times. The growth medium is removed quickly, and the cells are rapidly washed three times with ice-cold D-PBS without $CaCl_2$, and containing 5 mM sodium butyrate as a deacetylase inhibitor (Boffa et al., 1978). Cells are gently scraped off the flask in ice-cold D-PBS using a rubber policeman or disposable cell scraper (Falcon), and transferred to a sterile centrifuge tube with a sterile pipette. The flask is rinsed once with D-PBS to remove any remaining cells. The combined cell suspensions are centrifuged in a swinging-bucket rotor at 500 g for 5 minutes at 4°C.

To lyse the cells, the pellet is resuspended in buffer A [10 mM Tris-HCl, pH 7.4, 10 mM NaCl, 3 mM $MgCl_2$, 5mM sodium butyrate, 0.5% Nonidet P-40, 0.1 mM phenylmethylsulfonyl fluoride (PMSF), and 0.1 mM 1,2-epoxy-3-(p-nitrophenoxy)propane (EPNP)]. The protease inhibitors PMSF and EPNP are made up as 0.1 M stock solution in 95% isopropanol and 95% ethanol, respectively. They are stored at 25°C and added to the buffer just prior to use. After pipetting the suspension up and down for several cycles to disperse the cells, the suspension is placed on ice for 5 minutes and centrifuged in a swinging-bucket rotor at 500 g for 7 minutes at 4°C.

The nuclear pellet is resuspended in buffer A and the purified nuclei are collected by centrifugation. The purity of the nuclear preparation is assessed by taking a 100-μl aliquot of the suspension and adding 10 μl of 0.4% (w/v) Trypan Blue dye. A drop is placed in a hemocytometer and examined under the light microscope. If cytoplasmic debris is present, the nuclear pellet is resuspended in 50 mM Tris-HCl, pH 7.5, 25 mM KCl, 2.3 M sucrose, 5 mM MgCl$_2$, 5 mM sodium butyrate, 0.1 mM PMSF, 0.1 mM EPNP, and centrifuged at 120,000 g for 1 hour in a swinging-bucket (Beckman SW28) rotor.

3. Release of Nucleosomes by Limited Endonuclease Digestion

The nuclear pellet is resuspended in buffer B (10 mM Tris-HCl, pH 7.4, 25 mM KCl, 25 mM NaCl, 0.35 M sucrose, 5 mM butyrate, 5 mM MgCl$_2$, 0.1 mM PMSF, 0.1 mM EPNP) at a concentration of 1×10^7 nuclei/ml. Small aliquots are taken to determine the DNA concentration of the suspension, as follows: duplicate 25 to 50-μl aliquots are added to test tubes containing 1 ml of 0.1 N NaOH; the clumps are dispersed by sonication in a Branson 200 Cell Disruptor at a continuous-pulse setting of 3 for 10 seconds, and the A_{260} of each sample is measured. Based on the estimate of the total DNA in the total nuclear suspension, the nuclei are pelleted and resuspended in buffer B at a concentration of 20 A_{260} units/ml (~ 1 mg DNA/ml).

The nuclear suspension is transferred to a Dounce homogenizer fitted with a loose (type B) pestle and incubated in a waterbath until its temperature reaches 37°C. Micrococcal (staphylococcal) nuclease (Worthington Chemical) is added to a concentration of 10 U/ml, and the enzyme is activated by adding 1 M CaCl$_2$ to a final concentration of 0.5 mM. The reaction is timed from this point. To assure uniform distribution of the enzyme, the suspension is homogenized gently for four cycles. After 5 minutes, the reaction is stopped by placing the homogenizer in an ice-bath and quickly adding 0.1 M ethylenebis(oxyethylenenitrilo)tetraacetic acid (EGTA), pH 7.5, to a final concentration of 5 mM. After gentle mixing, to ensure that further endonuclease activity is blocked, the suspension is transferred to a centrifuge tube and centrifuged in a Beckman SW28 swinging-bucket rotor at 10,000 g for 20 minutes at 4°C. The supernatant, S1, containing the released nucleosomes, is analyzed for its DNA content by measuring the A_{260}/ml and by the Hoechst dye-binding assay (Cesarone et al., 1979). Under these conditions, the DNA released by endonuclease digestion of 3T3 cell nuclei is $10 \pm 2\%$ of the total DNA.

B. Chromatographic Separation of Active and Inactive Nucleosome Fractions

Techniques of Mercury-Affinity Chromatography

a. Column Preparation. The structure of the organomercurial linked to an agarose matrix is shown in Fig. 1. The mercury atom is Hg^{II}, and the organic linker increases its accessibility for thiol-binding. The Hg^{II} affinity matrix is

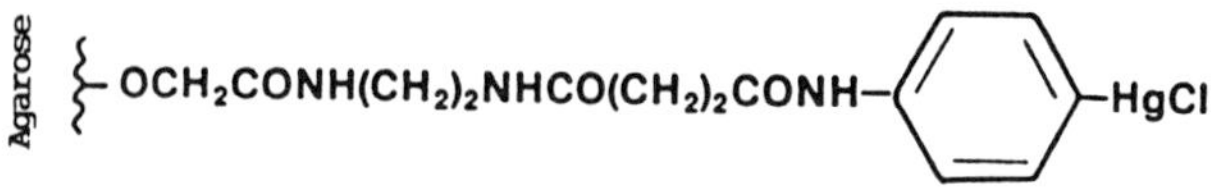

FIG. 1.   Chemical structure of the organomercurial matrix employed in the chromatographic isolation of transcriptionally active nucleosomes. The thiol-binding phenylmercury moiety is linked to agarose through an organic "spacer" that increases its accessibility to the reactive SH groups of the nucleosomes.

available with a phenylmercury concentration of $2-5 \ \mu M/\text{ml}$ (Affi-Gel 501, Bio-Rad Laboratories cat. no. 153–5101). The matrix is washed extensively in large volumes of deionized, distilled water to remove the isopropyl alcohol used to store it. The matrix is then equilibrated in buffer C [10 mM Tris-HCl, pH 7.4, 25 mM NaCl, 25 mM KCl, 2% (w/v) sucrose, 5 mM sodium butyrate, 5 mM ethylenediaminetetraacetic acid, disodium salt (EDTA), 0.1 mM PMSF, 0.1 mM EPNP]. The matrix must not be allowed to dry during any of the washing and equilibration steps. It is allowed to settle by gravity during the washing and equilibration procedures, and the supernatant fluid is removed by suction. A 10-ml Econo column (Bio-Rad Laboratories) is packed with 1 cm $\times$ 6 cm of equilibrated matrix, and washed with buffer C for 1 hour at 4°C prior to loading the nucleosomes.

The thiol capacity of the column can easily accommodate the total thiol content of the proteins present in the released nucleosome fraction (S1). An S1 fraction containing 1.0 mg of DNA contains about 2 mg of protein. Titration of the protein thiol groups under nondenaturing conditions, using 5,5'-dithiobis(2-nitrobenzoic acid) (Ellman, 1959; Soper et al., 1979), indicates a total of ~ 45 nM of reactive SH groups (~ 63 nM after denaturation in SDS), an amount far smaller than the $2-5 \ \mu M/\text{ml}$ Hg^{II} content of the organomercurial matrix.

b.   Nucleosome Binding to the Mercury Column. A Two-Step Elution Procedure for Preparation and Subfractionation of Transcriptionally Active Nucleosomes.   Chromatography of the nucleosomes is carried out at 4°C. The S1 fraction containing the released nucleosomes (~ 1.0 mg DNA) is adjusted to 5 mM EDTA and loaded onto the equilibrated mercury column at a flow rate of 20 ml/hour. Fractions of 1.5 ml are collected, and their absorbancy at 260 nm is measured. When almost all of the applied sample has entered the matrix, buffer C is added to wash any unbound nucleosomes from the column, monitoring the A_{260} until it returns to baseline.

One class of transcriptionally active nucleosomes is retained on the column through salt-labile associations with thiol-reactive nonhistone proteins, such as HMG-1 and HMG-2 (Walker et al., 1990). Those nucleosomes are released by adding 0.5 M NaCl to the eluting buffer.

When the A_{260} of the eluate returns again to baseline, the remaining nucleosomes, which are covalently linked to the organomercurial through the thiol groups of histone H3 (Walker *et al.*, 1990; Chen *et al.*, 1991), are displaced by adding 10 mM dithiothreitol (DTT) to the eluting buffer (buffer C containing 0.5 M NaCl). The release is monitored by measurements of the A_{260} (making appropriate correction for the absorbance due to DTT in the eluting buffer). Because the DTT-eluted nucleosomes are closely associated with nascent RNA chains, samples of every other fraction are also taken for DNA analysis by Hoechst assay (Cesarone *et al.*, 1979).

Following the elution of the DTT-released nucleosomes, the fractions of the eluates containing the unbound nucleosomes (peak 1), the 0.5 M NaCl-eluted nucleosomes (peak 2), and the DTT-eluted nucleosomes (peak 3) are pooled. The total amount of DNA in each nucleosome fraction is measured by A_{260} and by the Hoechst assay.

A typical distribution of DNA in the nucleosome fractions of 3T3 cells is shown in Fig. 2. Under these conditions of endonuclease digestion, about 80% of the released nucleosomes pass directly through the mercury column. The two Hg-bound fractions contain $18.5 \pm 7.2\%$ of the total DNA applied to the column, of which $62 \pm 4.8\%$ is eluted in 0.5 M NaCl and $38 \pm 4.6\%$ is eluted in 10 mM DTT. More extensive endonuclease digestions result in a disproportionate loss of the DTT-eluted nucleosomes (Walker *et al.*, 1990).

C. Characterization of the Nucleosome Fractions

1. DNA SIZING

DNA is prepared from samples of the three nucleosome peaks separated by mercury-affinity chromatography, using standard DNA purification procedures (Sambrook *et al.*, 1989).

To remove RNA, the samples are treated with 50 μg/ml RNase A (DNase-free RNase A, Boehringer Mannheim) for 1 hour at 37°C. Sodium dodecyl sulfate (SDS) is added to 0.1% (w/v), and the samples are incubated with 100 μg/ml proteinase K (DNase free, Boehringer Mannheim) for 2 hours at 37°C. The DNA is extracted twice with phenol/chloroform, precipitated in ethanol (Sambrook *et al.*, 1989), and collected by centrifugation at 75,000 g in a Beckman SW28 rotor for 30 minutes at 4°C. The DNA is dissolved in 10 mM Tris-HCl, pH 7.4, 1 mM EDTA, and the concentration of each sample is determined by its absorbancy at 260 nm and by the Hoechst assay.

The purified DNA is sized electrophoretically in either 2% agarose or 9% polyacrylamide gels as described by Maniatis *et al.* (1975). After staining with ethidium bromide, the gels are photographed with Polaroid 107C film (Fig. 2). The negative is scanned in a laser densitometer (LKB) and the nucleosmal

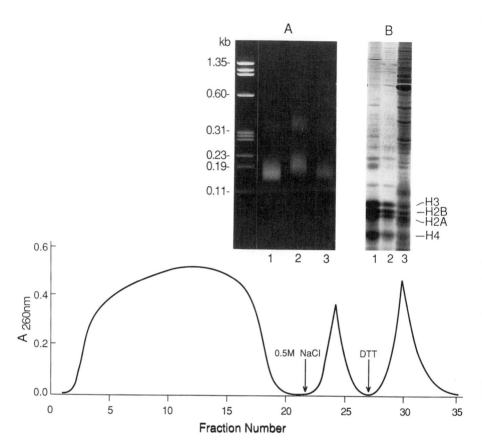

FIG. 2. Fractionation of nucleosomes by mercury-affinity chromatography. The mixture of active and inactive nucleosomes released by a limited micrococcal nuclease digestion of isolated 3T3 cell nuclei was applied to the organomercurial-agarose column. After elution of the unbound nucleosomes (peak 1), the Hg-bound nucleosomes were eluted in two steps. The first step used 0.5 M NaCl to release nucleosomes bound to the column through salt-labile associations with SH-reactive nonhistone proteins (peak 2). The second step used 10 mM dithiothreitol (DTT) to release the nucleosomes covalently linked to the column through the cysteinyl-thiol groups of histone H3 (peak 3). This step removed all nucleosomes from the column. Peak 1 originates from non-transcribed chromatin; peaks 2 and 3 contain actively transcribed DNA sequences (see Fig. 3). (A) DNA sizes of the different nucleosome fractions. Left lane, HaeIII restriction fragments of ϕ X174 DNA used as size markers; lane 1, DNA of unbound nucleosome fraction; lane 2, DNA of nucleosomes eluted in 0.5 M NaCl; lane 3, DNA of nucleosomes eluted in 10 mM DTT. Note that the DNA length of the monomeric nucleosomes eluted in 0.5 M NaCl (lane 2) exceeds the DNA lengths observed in the unbound (lane 1) or DTT-eluted nucleosome monomers (lane 3), indicating the presence of "linker" DNA sequences in the salt-eluted fraction. These linkers are probably shielded from endonuclease attack by associated thiol-reactive proteins, such as HMG-1 and HMG-2 (Walker et al., 1990). (B) Electrophoretic separations of the nucleosomal proteins of 3T3 cells. Lane 1, proteins of the unbound nucleosome fraction; lane 2, proteins of nucleosome fraction eluted in 0.5 M NaCl; lane 3, proteins of the nucleosome fraction eluted in 10 mM DTT. Note the presence of all four core histone (H2A, H2B, H3, and H4) in all fractions. The stoichiometry of the core histones in all fractions has been established (Chen et al., 1990; Walker et al., 1990).

peak positions are compared with those of DNA restriction fragments of know molecular weight.

The peak areas permit an estimate of the proportions of monomeric, dimeric, and oligomeric nucleosome fragments in each fraction (Walker et al., 1990; Chen et al., 1990). Such an analysis of DNA size in the three chromatographically separated nucleosome peaks shows that monomer DNA length in the 0.5 M NaCl-eluted nucleosomes is about 25 bp longer than that of the unbound nucleosomes (170 bp) or of the DTT-eluted nucleosome fraction (170 bp) (Chen et al., 1990). The longer DNA length indicates the persistence of "linker" DNA sequences in the nucleosomes that are retained on the column through salt-labile associations with DNA-binding proteins containing reactive thiol groups. The results suggest that those proteins are associated with the linker DNA in transcriptionally active chromatin and that they provide some shielding from endonuclease attack. The role of the high-mobility group proteins HMG-1 and HMG-2 and other DNA-binding non-histone proteins in the Hg-binding of transcriptionally active nucleosomes is considered elsewhere (Walker et al., 1990).

2. HISTONE ANALYSIS

To measure the proportions of the nucleosomal core histones in the chromatographically separated nucleosome fractions, samples are taken from each fraction for electrophoretic analysis. The samples are successively dialyzed against 5 mM sodium butyrate, 0.1 mM PMSF, 0.1 mM EPNP, and 2.5 mM sodium butyrate, 0.1 mM PMSF, 0.1 mM EPNP prior to dialysis against 0.1% (v/v) acetic acid, using dialysis membranes with a 2000 molecular weight limit cutoff. The protein content of each sample is determined by the Coomassie Blue dye-binding procedure (Bradford, 1976), and the samples are quickly frozen in an acetone/frozen CO_2 bath and lyophilized.

Electrophoretic separations are carried out in 15% polyacrylamide gels containing 0.1% SDS (Laemmli, 1970). The protein bands are stained with 0.1% Coomassie Blue-R250 salt in 40% methanol, 10% acetic acid for 2 hours. The gels are destained in 40% methanol, 10% acetic acid until the protein bands are clearly visible (Fig. 2).

Gels are photographed and the corresponding diapositives are scanned in a laser densitometer for quantitation of the proportions of the core histones. In 3T3 cell nucleosome fractions, and in all other cell types examined to date, histones H2A, H2B, H3, and H4 occur in stoichiometric proportions in all three nucleosome peaks (Sterner et al., 1987; Walker et al., 1990; Chen and Allfrey, 1987; Chen et al., 1990). It follows that there is no loss of histones in the transcriptionally active nucleosome fractions.

All of the histones in the nucleosome core are subject to enzymatically catalyzed modifications of structure that influence their interactions with the enveloping DNA strand. For histones H2B, H3, and H4, the most prevalent modification is the acetylation of one to four lysine residues in the NH_2-terminal regions of the polypeptide chains (for reviews, see Reeves, 1984; Matthews and Waterborg, 1985; Csordas, 1990). Because of the positive-charge neutralization resulting from acetylation of the lysine ε-amino groups, histone subfractions which differ in their degree of acetylation migrate at different rates in acid/urea/polyacrylamide gels. Histones H3 and H4, for example, each form five bands of decreasing mobility corresponding to the unacetylated form and its derivatives containing one, two, three, and four ε-N-acetyllysine residues, respectively.

To compare the levels of histone acetylation in the unbound and Hg-bound nucleosome fractions, the lyophilized proteins, prepared as described above, are suspended in 0.2 N H_2SO_4 to dissolve the histones. After centrifugation, the supernatant is added to 10 volumes of cold acetone to precipitate the histones. The histones are analyzed by electrophoresis in acid/urea/polyacrylamide gels containing 0.25% (v/v) Triton X-100 (Alfagame et al., 1974), as follows. The histones are dissolved in loading buffer (8 M urea, 5% acetic acid, 2.5% thioglycolic acid, 5% 2-mercaptoethanol, 0.25% Triton X-100, 0.01% pyronin Y) at a concentration of 1 mg/ml. Aliquots (20 μl) are applied to each lane of a stacking gel [4% mixture of 37.5/1 acrylamide/bisacryl-amide, 0.1% TEMED, 5% acetic acid, 4.4 M urea, 0.25% Triton X-100 0.1% $(NH_4)_2SO_4$] overlying the resolving gel [15% mixture of 37.5/1 acrylamide/bisacrylamide, 0.1% TEMED, 5% acetic acid, 8 M urea, 0.25% Triton X-100, 0.1% $(NH_4)_2SO_4$]. The running buffer is 5% acetic acid. The gels are stained with Coomassie Blue-R250, destained, and photographed. The diapositives are scanned with a laser densitometer, and the proportions of each acetylated form are calculated as a percentage of the total of unmodified and acetylated forms of that histone.

In 3T3 cells (and in all other cells types so far examined), the histones of the Hg-bound nucleosome fractions are hyperacetylated, while those of the unbound nucleosomes are deficient in this modification (Sterner et al., 1987; Allegra et al., 1987; Walker et al., 1990; Boffa et al., 1990). Given the evidence (see below) that the Hg-bound nucleosomes contain the transcriptionally active DNA sequences of the cell, this contrast in the distribution of the acetylated histones provides strong support for the proposal that histone acetylation facilitates transcription by releasing constraints on the DNA strand (Allfrey et al., 1964).

3. DNA Sequence Analysis

Purified DNA from each of the chromatographically separated nucleosome fractions is analyzed for its DNA sequence content by slot-blot hybridizations

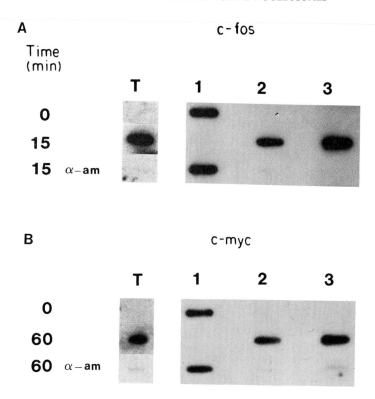

FIG. 3. Comparisons of the c-*fos* and c-*myc* DNA sequence contents of the unbound (peak 1), 0.5 M NaCl-eluted (peak 2), and 10 mM DTT-eluted (peak 3) nucleosomes of quiescent and serum-stimulated 3T3 cells. DNA was extracted from the fractions prepared at the indicated times after serum stimulation and hybridized to ^{32}P-labeled probes for c-*fos* and c-*myc*. The resulting slot-blot autoradiographs are compared for the three nucleosome fractions at each time point. (A) c-*fos* transcription (T) is compared with the distribution of c-*fos* DNA sequences in the unbound and Hg-bound nucleosomes of quiescent cells (time 0), after 15 minutes of serum stimulation (when c-*fos* transcription is maximal), and at 15 minutes followed by a 10-minute exposure of the permeabilized cells to 20 μg/ml α-amanitin. Note that retention by the mercury column of nucleosomes containing the c-*fos* DNA sequences corresponds with the timing of c-*fos* transcription, and that inhibition by α-amanitin results in the loss of Hg-binding of the c-*fos* nucleosomes within 10 minutes. (B) c-*myc* transcription (T) is compared with the distribution of c-*myc* DNA in the unbound and Hg-bound nucleosome fractions of quiescent cells (time 0), after 60 minutes of serum stimulation (when c-*myc* transcription is maximal), and at 60 minutes followed by a 10-minute exposure to α-amanitin. Note the correlation between transcription and Hg-binding by the nucleosomes of the c-*myc* gene, and the rapid change in nucleosome structure after α-amanitin treatment.

to ^{32}P-labeled DNA probes. Nucleosomal DNA (5–10 μg) from each fraction is blotted onto a prewetted nylon membrane (Zeta-Probe, Bio-Rad) in a slot-blot apparatus following the alkaline procedure (Reed and Mann, 1985). The membrane (DNA-side up) is placed over Whatman 3MM paper prewetted in 0.4 N NaOH for 10 minutes at 25°C. The membrane is then neutralized by immersion in 50 mM sodium phosphate buffer. When neutral, the membrane-bound DNAs are hybridized to the ^{32}P-labeled DNA probes as described by Sambrook et al. (1989). The filters are washed, dried, and exposed to Kodak X-Omat ARS film as described (Chen and Allfrey, 1987). After development, the autoradiograms are scanned with a laser densitometer and peak areas corresponding to each slot-blot are integrated.

Hybridizations using ^{32}P-labeled probes for c-*fos* and c-*myc* DNA sequences show striking differences in their nucleosomal distribution in quiescent and serum-stimulated 3T3 cells (Fig. 3). In the quiescent cells, both genes have very low levels of expression and their DNA sequences are found in the unbound, not in the Hg-bound nucleosome fractions. After 15 minutes of serum stimulation, when c-*fos* reaches the peak of its transcriptional activity, the c-*fos* DNA is recovered in both of the Hg-bound nucleosome fractions, but not in the unbound nucleosomes (Fig. 3).

At 15 minutes, c-*myc* transcription is minimal, and its DNA remains in the unbound nucleosome fraction, but, at 60 minutes, when c-*myc* is fully active, its sequences are not detected in the unbound nucleosomes and all of the hybridization signal appears in the Hg-bound nucleosomes.

These changing distributions of the oncogenic DNA sequences in the bound and unbound nucleosome fractions are in full agreement with the timing and extent of their transcription as determined by run-on transcription assays on the isolated 3T3 cell nuclei (Fig. 3).

III.   Advantages and Limitations of the Method

The major advantage of this procedure is that it permits the isolation of transcriptionally active DNA sequences together with the proteins involved in maintaining the active state. It has already provided new information on the differences in the postsynthetic modifications of histones in active and inactive chromatin. For example, histone H2A is phosphorylated in the unbound nucleosomes but not in the Hg-bound nucleosome fractions (Sterner et al., 1987). This difference suggests that the phosphorylation of histone H2A, unlike the acetylation of histones H2B, H3, and H4, may correlate with the inactivity of the associated DNA sequences in transcription. Moreover, the differences in the distributions of phosphorylated H2A and hyperacetylated

H2B, H3, and H4 provide convincing evidence that mercury-affinity chromatography provides a meaningful separation of nucleosomes with different properties and functions. There are also major differences in the nonhistone protein content of the Hg-bound and unbound nucleosome fractions. These include the presence of covalently linked DNA topoisomerase I in the nucleosomes eluted in 10 mM DTT, but not in the unbound nucleosome fractions. Such covalent linkages of the enzyme to the DNA of the Hg-bound nucleosomes are readily detectable even in the absence of the topoisomerase I inhibitor camptothecin, which is usually added to stabilize the covalent bond between the topoisomerase I and the cleaved DNA strand (Wortman *et al.,* 1991).

The presence in the DTT-eluated nucleosomes of the c-*fos* protein (a component of transcription factor AP-1) and its absence from the unbound or 0.5 M NaCl-eluted nucleosome fractions (Zong *et al.,* 1990) also illustrate the great potential of the method for identifying proteins involved in transcriptional control mechanisms.

A second advantage of the procedure is its capacity to monitor with accuracy the timing and extent of transcription of a given DNA sequence. The experiments on c-*fos* activation and repression in 3T3 cells show that the changes in nuclesome structure are reversible and rapid (within 15 minutes) (Fig. 3). The transient nature of the activated state of nucleosomes and the kinetics of nucleosome "refolding" along a specific DNA sequence can be studied by using inhibitors such as α-amanitin to block the progression of the transcription complex (Fig. 3).

Because of the relatively tight coupling of transcription of a given DNA sequence and the Hg-binding of its constituent nucleosomes, it becomes possible to compare nucleosome structures in transcribed and nontranscribed domains of the same gene. For example, in genes such as c-*myc* and c-*myb*, in which a block in RNA chain elongation is a major mechanism of transcriptional control, the separation of nucleosomes by mercury-affinity chromatography (combined with hybridizations to DNA probes for different domains of the gene) makes it possible to compare nucleosome structures on either side of the elongation block, and to monitor the timing and progression of nucleosomal changes downstream of the block when transcription proceeds.

The ability to isolate and analyze the active nucleosomes opens the way to the identification of the molecular mechanisms employed in "unfolding" the nucleosome core during gene activation, and, hopefully, to the ability to duplicate those changes in *in vitro* systems.

What are the limitations of the method? Probably the most serious is the inability to analyze *all* the nucleosomes of the nucleus. The extreme lability of transcriptionally active nucleosomes during prolonged digestions with micrococcal nuclease limits the analysis to about 10% of the total nuclear DNA. The nucleosomes of many nontranscribed DNA sequences, such as the α-globin gene of HeLa cells, are not released during such limited MNase digestions (Sterner *et al.,* 1987). Moreover, there is good evidence that other transcribed gene sequences, such as the active C_γ IgL immunoglobulin DNA sequences, are not readily released under these conditions, indicating that a substantial fraction of the active nucleosomes remains associated with the nuclear matrix (Rose and Garrard, 1984).

A second complication inherent in the preparation of the nucleosomes by endonuclease digestion is the possibility that proteins originally bound to the active nucleosomes may become detached under the conditions employed in isolation of the nuclei and treatment with micrococcal nuclease. This may be the reason for the very low content of histone H1, HMG-14 and HMG-17 observed in the Hg-bound nucleosome fractions (Allegra *et al.,* 1987).

Conversely, proteins not associated with the active nucleosomes *in situ* may appear in the Hg-bound nucleosome fractions. This is expected because any thiol-reactive proteins released during nuclear digestions would bind to the mercury column and be eluted along with the nucleosomes in the DTT eluate. Many of these adventitious contaminants can be eliminated by gel chromatography on Sephacryl S-200 before fractionation of the nucleosomes (Sterner *et al.,* 1987; Allegra *et al.,* 1987; Walker *et al.,* 1990; Boffa *et al.,* 1990); others are removed when the Hg-binding nucleosome fractions are further purified by density-gradient centrifugation (Zong *et al.,* 1990).

Other problems and potential artifacts will certainly arise, but the evidence that mercury-affinity chromatography can separate the nucleosomes of transcriptionally active and inactive DNA sequences is conclusive, and the procedure should find many new applications in the study of how chromatin structure is altered during gene activation.

REFERENCES

Alfagame, C. R., Zweidler, A., Mahowald, A., and Cohen, L. H. (1974). Histones of *Drosophila* embryos. Electrophooretic isolation and structural studies. *J. Biol. Chem.* **249,** 3729–3736.

Allegra, P., Sterner, R., Clayton, D. F., and Allfrey, V. G. (1987). Affinity chromatographic purification of nucleosomes containing transcriptionally active DNA sequences. *J. Mol. Biol.* **196,** 379–388.

Allfrey, V. G., Faulkner, R., and Mirsky, A. E. (1964). Acetylation and methylation of histones and their possible role in the regulation of RNA synthesis. *Proc. Natl. Acad. Sci. U.S.A.* **51,** 786–794.

Bellard, M., Gannon, F., and Chambon, P. (1977). Nucleosome structure. III. The structure and transcriptional activity of the chromatin containing the ovalbumin and globin genes in chick oviduct nuclei. *Cold Spring Harbor Symp. Quant. Biol.* **42,** 779–791.

Bellard, M., Kuo, M. T., Dretzen, G., and Chambon, P. (1980). Differential nuclease sensitivity of the ovalbumin and α-globin chromatin regions in erythrocytes and oviduct cells of laying hens. *Nucleic Acids Res.* **8,** 2737–2750.

Bloom,, K. S., and Anderson, J. N. (1978). Fractionation of hen oviduct chromatin into transcriptionally active and inactive regions after selective micrococcal nuclease digestion. *Cell (Cambridge, Mass.)* **15**, 141–150.

Bloom, K. S., and Anderson, J. N. (1979). Conformation of ovalbumin and globin genes in chromatin during differential gene expression. *J. Biol. Chem.* **254**, 10532–10539.

Bloom, K. S., and Anderson, J. N. (1982). Hormonal regulation of the conformation of the ovalbumin gene in chick oviduct chromatin. *J. Biol. Chem.* **257**, 13018–13027.

Bode, J., Henco, K., and Wingender, E. (1980). Modulation of the nucleosome structure by histone acetylation. *Eur. J. Biochem.* **110**, 143–152.

Boffa, L. C., Vidali, G., Mann, R. S., and Allfrey, V. G. (1978). Suppression of histone deacetylation *in vivo* and *in vitro* by sodium butyrate. *J. Biol. Chem.* **253**, 3364–3366.

Boffa, L. C., Walker, J., Chen, T. A., Sterner, R., Mariani, M. R., and Allfrey, V. G. (1990). Factors affecting nucleosome structure in transcriptionally active chromatin: Histone acetylation, nascent RNA, and inhibitors of RNA synthesis. *Eur. J. Biochem.* **194**, 811–823.

Bradford, M. (1976). A rapid and sensitive method for the quantitation of micrograms of protein utilizing the principle of protein-dye binding. *Anal. Biochem.* **72**, 248–254.

Cesarone, C. F., Bolognesi, C., and Santi, L. (1979). Improved microfluorometric DNA determination in biological material using 33258 Hoechst. *Anal. Biochem.* **100**, 188–197.

Chan, S., Attisano, L., and Lewis, P. N. (1988). Histone H3 thiol reactivity and acetyltransferases in chicken erythrocyte nuclei. *J. Biol. Chem.* **263**, 15643–15651.

Chen, T. A., and Allfrey, V. G. (1987). Rapid and reversible changes in nucleosome structure accompany the activation, repression, and super-induction of murine fibroblast proto-oncogenes c-*fos* and c-*myc*. *Proc. Natl. Acad. Sci. U.S.A.* **84**, 5252–5256.

Chen, T. A., Sterner, R., Cozzolino, A., and Allfrey, V. G. (1990). Reversible and irreversible changes in nucleosome structure along the c-*fos* and c-*myc* oncogenes following inhibition of transcription. *J. Mol. Biol.* **212**, 481–493.

Chen, T. A., Smith, M. M., Le, S., Sternglanz, R., and Allfrey, V. G. (1991). Nucleosome fractionation by mercury-affinity chromatography. Contrasting distribution of transcriptionally active DNA sequences and acetylated histones in nucleosome fractions of wild-type yeast cells and cells expressing a histone H3 gene altered to encode a cysteine-110 residue. *J. Biol. Chem.* **266**, 6489–6498.

Csordas, A. (1990). On the biological role of histone acetylation. *Biochem. J.* **265**, 23–38.

Dimitriadis, G. J., and Tata, J. R. (1980). Subnuclear fractionation by mild micrococcal nuclease treatment of nuclei of different transcriptional activities causes a partition of expressed and non-expressed genes. *Biochem. J.* **187**, 467–477.

Dulbecco, R., and Freeman, G. (1959). Plaque production by the polyoma virus. *Virology* **8**, 396–397.

Dulbecco, R., and Vogt, M. (1954). Plaque formation and isolation of pure lines with poliomyelitis viruses. *J. Exp. Med.* **99**, 167–182.

Ellman, G. (1959). Tissue sulfhydryl groups. *Arch. Biochem. Biophys.* **82**, 72–77.

Garel, A., Zolan, M., and Axel, R. (1977). Genes transcribed at diverse rates have a similar conformation in chromatin. *Proc. Natl. Acad. Sci. U.S.A.* **74**, 4867–4871.

Gottesfeld, J. M., and Butler, P. J. G. (1977). Structure of transcriptionally active chromatin subunits. *Nucleic Acids Res.* **4**, 3155–3175.

Gottesfeld, J. M., Garrard, W. T., Bagi, G., and Wilson, R. F. (1974). Partial purification of the template-active fraction of chromatin. *Proc. Natl. Acad. Sci. U.S.A.* **71**, 2193–2197.

Greenberg, M. E., and Ziff, E. B. (1984). Stimulation of 3T3 cells induces transcription of the c-*fos* proto-oncogene. *Nature (London)* **311**, 433–437.

Hebbes, T. R., Thorne, A. W., and Crane-Robinson, C. (1988). A direct link between core histone acetylation and transcriptionally active chromatin. *EMBO. J.* **7**, 1395–1402.

Johnson, E. M., Matthews, H. R., Littau, V. C., Lothstein, L., Bradbury, E. M., and Allfrey, V. G.

(1978). The structure of chromatin containing DNA complementary to 19S and 26S ribosomal RNA in active and inactive stages of *Physarum Polycephalum. Arch. Biochem. Biophys.* **191**, 537–550.

Johnson, E. M., Sterner, R., and Allfrey, V. G. (1987). Altered nucleosomes of active nucleolar chromatin contain accessible histone H3 in its hyperacetylated forms. *J. Biol. Chem.* **262**, 6943–6946.

Laemmli, U. (1970). Cleavage of structural proteins during the assembly of the head of bacteriophage T. *Nature (London)* **222**, 680–686.

Lawson, G. M., Knoll, B. J., March, C. J., Wu, S., Tsai, M.-J., and O'Malley, B. W. (1982). Definition of 5′ and 3′ structural boundaries of the chromatin domain containing the ovalbumin multigene family. *J. Biol. Chem.* **257**, 1501–1507.

Lochlear, L., Jr., Ridsdale, J. A., Bazett-Jones, D. P., and Davie, J. R. (1991). Ultrastructure of transcriptionally competent chromatin. *Nucleic Acids Res.* **18**, 7015–7024.

Maniatis, T., Jeffrey, A., and van deSande, H. (1975). Chain length determination of small double and single stranded DNA molecules by polyacrylamide gel electrophoresis. *Biochemistry* **14**, 3787–3794.

Matthews, H. R., and Waterborg, J. H. (1985). Reversible modifications of nuclear proteins and their significance. *In* "The Enzymology of Post-Translational Modification of Proteins" (R. B. Freedman and H. C. Hawkins, eds.), Vol. 2, pp. 125–185. Academic Press, New York.

Norton, V. G., Imai, B. S., Yau, P., and Bradbury, E. M. (1989). Histone acetylation reduces nucleosome core particle linking number change. *Cell (Cambridge, Mass.)* **57**, 449–457.

Oliva, R., Bazett-Jones, D. P., Lochchlear, L., and Dixon, G. H. (1990). Histone hyperacetylation can induce unfolding of the nucleosome core particle. *Nucleic Acids. Res.* **18**, 2739–2747.

Prior, C. P., Cantor, C. R., Johnson, E. M., Littau, V. C., and Allfrey, V. G. (1983). Reversible changes in nucleosome structure and histone H3 accessibility in transcriptionally active and inactive states of rDNA chromatin. *Cell (Cambridge, Mass.)* **34**, 1033–1042.

Reed, K. C., and Mann, D. A. (1985). Rapid transfer of DNA from agarose gels to nylon membranes. *Nucleic Acids Res.* **13**, 7207–7221.

Reeves, R. (1984). Transcriptionally active chromatin. *Biochim. Biophys. Acta* **782**, 343–393.

Riggs, M. G., Whittaker, R. G., Neumann, J. R., and Ingram, V. M. (1977). *n*-Butyrate causes histone modification in HeLa and Friend erythroleukemia cells. *Nature (London)* **268**, 462–464.

Rose, S. M., and Garrard, W. T. (1984). Differentiation-dependent chromatin alterations precede and accompany transcription of immunoglobulin light chain genes. *J. Biol. Chem.* **259**, 8534–8544.

Ruiz-Carrillo, A., and Allfrey, V. G. (1973). A method for the purification of histone H3 by affinity chromatography. *Arch. Biochem. Biophys.* **154**, 185–191.

Sambrook, J., Fritsch, E. F., and Maniatis, T. (1989). "Molecular Cloning. A Laboratory Manual." Cold spring Harbor Laboratory, Cold Spring Harbor, New York.

Sanders, M. M. (1978). Fractionation of nucleosomes by salt elution from micrococcal nuclease-digested nuclei. *J. Cell Biol.* **79**, 97–109.

Simpson, R. T. (1978). Structure of chromatin containing extensively acetylated H3 and H4. *Cell (Cambridge, Mass.)* **13**, 691–699.

Smith, J. B., Freeman, G., Vogt, M., and Dulbeecco, R. (1960). The nucleic acid of polyoma virus. *Virology* **12**, 185–196.

Smith, R. D., Yu, J., and Seale, R. L. (1984). Chromatin structure of the α-globin gene family in murine erythroleukemia cells. *Biochemistry* **23**, 785–790.

Soper, T. S., Jones, W. M., and Manning, J. M., (1979). Effects of substrates on the selective modification of the cysteinyl residues of D-amino acid transaminase. *J. Biol. Chem.* **254**, 10901–10905.

Sterner, R., Boffa, L. C., Chen, T. A., and Allfrey, V. G. (1987). Cell cycle-dependent changes in conformation and composition of nucleosomes containing human histone gene sequences. *Nucleic Acids Res.* **15,** 4375–4391.

Stoute, J. A., and Marzluff, W. R. (1982). HMG proteins I and II are required for transcription of chromatin by endogenous RNA polymerase. *Biochem. Biophys, Res, Commun.* **107,** 1279–1284.

Stratling, W. H., Dolle, A., and Sippel, A. E. (1986). Chromatin structure of the chicken lysozyme gene domain as determined by chromatin fractionation and micrococcal nuclease digestion. *Biochemistry* **25,** 495–502.

Tremethick, D. J., and Molloy, P. L. (1986). High mobility group proteins I and II stimulate transcription *in vitro* by RNA polymerases II and III. *J. Biol. Chem.* **261,** 6986–6992.

Walker, J., Chen, T. A., Sterner, R., Berger, M., Winston, F., and Allfrey, V. G. (1990). Affinity chromatography of mammalian and yeast nucleosomes. Two modes of binding of transcriptionally active mammalian nucleosomes to organomercurial-agarose columns, and contrasting behavior of the active nucleosomes of yeast. *J. Biol. Chem.* **265,** 5736–5746.

Weintraub, H., and Groudine, M. (1976). Chromosomal subunits in active genes have an altered conformation. *Science* **193,** 848–856.

Wortman, M., Chen, T. A., and Allfrey, V. G. (1991). Manuscript in preparation.

Wu, C., Wong, Y.-C., and Elgin, S. C. R. (1979). The chromatin structure of specific genes. II. Disruption of chromatin structure during gene activity. *Cell (Cambridge, Mass.)* **16,** 807–814.

Zong, C., Chen. T. A., and Allfrey, V. G. (1990). The c-*fos* oncogene protein is associated with transcriptionally active nucleosomes in murine fibroblasts. *J. Cell Biol.* **111,** 248a.

Chapter 13

The Nucleoprotein Hybridization Method for Isolating Active and Inactive Genes as Chromatin

CLAUDIUS VINCENZ, JAN FRONK,[1] GRAEME A. TANK, KAREN
FINDLING, SUSAN KLEIN, AND JOHN P. LANGMORE

Biophysics Research Division
University of Michigan
Ann Arbor, Michigan 48109

[1] Present address: Institute of Biochemistry, Warsaw University, 02 089 Warsaw, Poland.

337

I. Introduction

We describe a method for fractionating chromatin that is based on specific hybridization between soluble chromatin fragments and oligonucleotides. This method, termed nucloprotein hybridization, targets a specific gene solely on the basis of its DNA sequence. This technique is therefore able to isolate chromatin of defined DNA sequence irrespective of its functional state. Alternative protocols for chromatin fractionation have employed differential solubility of chromatin fragments after mild nuclease digestion (Bonner et al., 1975; Bloom and Anderson, 1978; Ridsdale and Davie, 1987; Strätling and Dölle, 1986; Strätling, 1987; Hanks and Riggs, 1986; Davis et al., 1983; Kumar and Leffak, 1986; Rose and Garrard, 1984; Razin et al., 1985), immunoprecipitation (Dorbic and Wittig, 1986, 1987; Hebbes et al., 1988), and mercury affinity columns (Allegra et al., 1987; Chen et al., 1990). These alternative techniques yield a fraction enriched in active or inactive chromatin representing an ensemble of different genes.

The seven steps of nucleoprotein hybridization are schematically shown in Fig. 1.

The feasibility of the technique was proven with the SV40 model system (Workman and Langmore, 1985a). SV40 could be mixed with an excess of exogenous sea urchin chromatin and then isolated to 88% purity corresponding to a 115-fold enrichment. Here, we describe the first purification of active and inactive cellular genes as chromatin using the nucleoprotein hybridization technique.

The Strongylocentrotus purpuratus sea urchin early histone gene repeat (SUEHGR) was chosen as a target for the first application of the technique for two reasons. First, the SUEHGR is a unit of 6.5 kb which is tandemly repeated head-to-tail about 500 times in the genome (Kedes, 1979). This reduces the amount of nuclei and enzymes needed to purify amounts large enough for biochemical analysis. The high copy number also reduces the enrichment necessary to obtain pure material. Second, the SUEHGR changes its transcriptional activity during early development in a well-established fashion (Weinberg et al., 1983). The genes are transcriptionally inactive in the early zygote, become activated at early morula stage, reach a maximum transcriptional rate round the 64 to 128-cell stage (corresponding to 8–10 hours of development under standard laboratory conditions), and then gradually become inactivated at the hatching blastula stage (18 hours).

The transcriptional rate is moderate (2–3 transcripts per gene per minute; Weinberg et al., 1983), which suggests that these genes are a suitable model. The majority of cellular genes are transcribed at such a moderate rate. Studies on specific chromatin fragments up to now were only possible on genes with

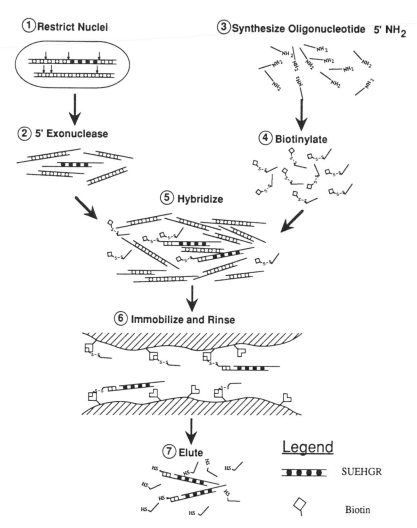

FIG. 1.   The nucleoprotein hybridization scheme for isolation of specific genes as chromatin. (1) Solubilization of gene size chromatin fragments by digestion of nuclei with restriction enzymes at low Mg^{2+} concentrations and EDTA lysis of the nuclei. (2) Generation of single-strand termini by T7 gene 6 exonuclease digestion. (3) Synthesis of an oligonucleotide complementary to the sequence adjacent to the restriction site on the targeted gene. (4) Biotinylation of the oligonucleotide. (5) Hybridization in solution between probe and chromatin fragments. (6) Immobilization of the oligonucleotide/chromatin hybrids on an avidin matrix. (7) Elution of the purified chromatin fragments by DTT cleavage of the disulfide bond in the linker of the biotin analog. The salt and temperature conditions throughout the procedure are compatible with intact chromatin.

unusual transcriptional rates (i.e., rRNA genes, Jones, 1978; Prior *et al.,* 1983; Amero *et al.,* 1988a,b; and chorion genes, Osheim *et al.,* 1985). Purification of SUEHGR chromatin should enable interesting comparisons between active and inactive genes, which should lead to a better understanding of the gene control mechanism *in vivo.*

We describe in this chapter independent assays for the critical steps of the procedure. The final enrichment of SUEHGR was > 700-fold. The overall yield was 2–15%. The purity measured by two independent methods was > 80%. The micrococcal nuclease digestion patterns for the isolated SUEHGR were identical to those for the nuclei prior to isolation. Electron microscopy (EM) of low-salt spreads confirmed that the isolated material was chromatin. The results show promise of being able to determine the histone variants and posttranslational modifications that are associated with the different functional states of multicopy genes such as the SUEHGR.

II.   Materials and Methods

Materials.   Vectrex avidin affinity matrix was from Vector Laboratories, Inc. Burlingame, CA. Sulfosuccinimidyl 2-(biotinamido)ethyl-1,3-dithiopropionate (NHS-SS-biotin) was purchased from Pierce. Glutathione (disodium salt, oxidized form, Grade IV, GSSG), leupeptin (hemisulfate salt), and *N*-lauroylsarcosine (Sarkosyl) were from Sigma. High-concentration *Sal*I (70 U/μl) was from Boehringer Mannheim. All other restriction enzymes, λ exonuclease, and T4 kinase were from Bethesda Research Laboratories. T7 gene 6 exonuclease was from United States Biochemical Corporation. Zeta-Probe nylon membrane was from Bio-Rad. [^{32}P]dCTP, [methyl-^{3}H]-thymidine, and L-[^{35}S]methionine were from Amersham.

The following solutions were used: SSPE, SSC, TAE, and TE were prepared as in Maniatis *et al.* (1982).

> Synthetic sea water: 1 liter contains 24.72 g of NaCl, 0.67 g of KCl, 1.36g of $CaCl_2 \cdot 2H_2O$, 4.66 g of $MgCl_2 \cdot 6H_2O$, 6.29 g of $MgSO_4 \cdot 7H_2O$, and 1.8 g of $NaHCO_3$
>
> Buffer A: 15 mM HEPES, pH 7.3, 60 mM KCl, 15 mM NaCl, 1 mM EDTA, 0.5 mM spermidine, and 0.15 mM spermine
>
> Chromatin digestion butter (buffer D): 50m M NaCl, 50 mM Tris, pH 8.0 3 mM $MgCl_2$, and 1 mM 2-mercaptoethanol
>
> Chromatin hybridization buffer (buffer H): 100 mM NaCl, 10 mM Tris, pH 8.0, 1 mM EDTA, 0.05% NP-40, and 0.02% azide

Affinity chromatography buffer (buffer C): buffer H containing 5 mM GSSG

Homogenization buffer (buffer Ho): buffer A containing 0.3 M sucrose, 5 mM iodoacetate, and 0.5 mM PMSF

Synthesis and Purification of the Biotinylated Oligonucleotides. Oligonucleotides were synthesized and 5′ amino-modified with Aminolink 2 according to the instructions of the manufacturer on an Applied Biosystems Model 391 DNA synthesizer at the University of Michigan DNA facility. The amino-modified oligonucleotide was cleaved from the support, HPLC-purified, and deprotected. Biotinylation was done at room temperature with 10 μg of 17^{mer} dissolved in 450 μl of 0.2 M HEPES, pH 7.7, to which 6 mg of solid NHS-SS-biotin was added and the reaction mixture was shaken for 1.5 hours. The insoluble material was removed by centrifugation and the supernatant was purified in two batches with a reverse-phase C$_4$ HPLC column with a pore size of 300 Å (VYDAC; The Separations Group, Hesperia, CA) using 0.1 M triethylammonium chloride, pH 7.0, as buffer and a 5–25% acetonitrile gradient in 20 minutes. Reanalysis of the isolated material on the same system gave a single peak eluting at 12% acetonitrile.

Figure 2 shows the pertinent restriction sites of the SUEHGR, the sequences of the synthetic oligonucleotides, and the hybridization probes used for filter hybridizations.

Sea Urchin Embryo Cultures and Nuclear Isolations. The details of the sea urchin culture and nuclear isolations are given in Vincenz *et al.* (1991). The important considerations are that milligram quantities of nuclei must be prepared without proteolytic or nucleolytic degradation. Only nonaggregated

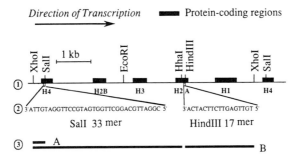

FIG. 2. Map of the *Strongylocentrotus purpuratus* early histone gene repeat. (1) Pertinent SUEHGR restriction sites and coding regions are indicated. (2) The sequences shown of the *Sal*I 33^{mer} and the *Hind*III 17^{mer} oligonucleotides are complementary to the 3′ strand upstream of the *Sal*I and *Hind*III site. (3) Hybridization probes. Fragment B does not include the *Hind*III 17^{mer} pCO2A contains one SUEHGR.

nuclei should be used in order to obtain reproducible solubilization efficiencies during restriction digestion.

Chromatin Solubilization. Reproducible solubilization is obtained if reduced levels of Mg^{2+} are used during restriction digestion and unnecessary centrifugation steps that promote aggregation of the nuclei are avoided. In a typical preparation, 2 mg of nuclei were thawed on ice and diluted to 0.5 mg/ml with buffer D containing 12 μM leupeptin. (All quantities of nuclei and chromatin are expressed as amount of DNA.) The solution was transferred into a dialysis bag (spectra/por #2, MW cut-off 12,000–14,000). *Hind*III (5 U/μg DNA) was added and the mixture dialyzed for 3 hours at 37°C against 1 liter of buffer D. Additional leupeptin was added to 6 μM and dialysis continued for 3 hours against 1 liter of buffer D containing 0.5 mM Mg^{2+}. The nuclei were then slowly lysed by dialysis against 1 mM EDTA for >4 hours at 4°C. Insoluble material was removed by 15 minutes centrifugation at 10,000 g. Solubilization efficiencies were determined by scintillation counting of aliquots. The nuclei were hydrolyzed for 3 hours in 10% trichloroacetic acid (TCA) at 90°C before counting to eliminate self-quenching. The same protocol was followed for solubilization with *Sal*I except that buffer D contained 100 mM NaCl during the first dialysis.

Exonuclease Digestion. An equal volume (~ 6 ml) of 20 mM Tris, pH 8.0, 2 mM $MgCl_2$, and 12 μM leupeptin was added to the solubilized chromatin followed by a 5-minute centrifugation at 2000 g. Four units of T7 gene 6 exonuclease per microgram of soluble chromatin was added to the supernatant followed by a 30-minute incubation at 37°C. An equal amount of exonuclease was added and the incubation continued for 30 minutes. Alternatively, λ exonuclease digestion was done in the same buffer with 5 U/μg chromatin for 3 hours. The reaction was stopped by adding EDTA to 1 mM. TCA solubility was determined by adding carrier salmon sperm DNA (100 μg/ml) and cold TCA to 6%. The supernatant was counted after 15 minutes centrifugation at 10,000 g.

Hybridization. The chromatin (typically 400 μg in 12 ml) was concentrated 3 to 5-fold by dialysis for 3 hours at 4°C against buffer H containing 60% sucrose. Leupeptin was added to 20 μM, the clamps on the dialysis tubing were tightened, and the sucrose was removed by dialysis overnight at 4°C against buffer H. In our hands, this method of concentration avoids losses due to aggregation of the chromatin. Next, 5 mM GSSG, 5 mM iodoacetamide (IAcNH$_2$), 6 μM leupeptin, and 3.6 nM biotinylated oligonucleotide were added from concentrated stock solutions. The standard hybridization time was 24 hours at 37°C, after which the mixture was spun at 2000 g for 5 minutes.

Avidin D Affinity Chromatography. For a standard preparative isolation, 110 mg of dry Vectrex avidin D was used (4 mg per nanogram of 17^{mer}). The resin was reconstituted for 30 minutes in H_2O and then loaded into Bio-Rad

2-ml Poly-Prep columns. The resin was equilibrated with 10 ml of buffer C. The flow rate was $\sim 400 \mu l$ per minute throughout the chromatography. The hybridization supernatant was loaded onto the column. The flow-through was reloaded two additional times. The column was rinsed with 17.5 ml of buffer C followed by 7.5 ml of buffer C containing 150 mM NaCl, and 5 ml of 0.75 × buffer C. Three fractions were eluted ($\sim 100, 300,$ and 250 μl) with 0.75 × buffer H containing 50 mM freshly dissolved DTT. A final fraction (300 μl) was collected with buffer H containing 150 mM NaCl and 50 mM DTT.

Sarkosyl Gel Assay. Hybridization conditions were optimized using a simple gel assay (Fig. 5). The unmodified oligonucleotide was 5′ end-labeled using kinase and purified with a Bio Spin 6 column (Bio-Rad), deproteinized, and ethanol precipitated. After hybridization of the oligonucleotide to chromatin or DNA, the EDTA concentration was adjusted to 10 mM, and 0.1 volume of 10 × loading buffer (10 × TAE, 1.25% Sarkosyl, 20% Ficoll, 0.1% bromphenol blue) was added. Up to 50 μl of the hybridization mixture was loaded in the 2 × 5 × 8 mm wells of a 0.6% agarose gel containing 0.125% Sarkosyl and TAE buffer. The gel was run at 1.25 V/cm for 16 hours without being submerged in the TAE buffer (Ackerman *et al.,* 1983). The gel was ethidium bromide-stained and then either fixed in 7% TCA for 30 minutes and dried, or electrotransferred to a Zeta-Probe membrane and exposed to Kodak X-OMAT AR film.

Dot Blots. Quantitation of dot blots is an important assay for purity of the isolated chromatin. Each dot-blot experiment included at least duplicates of each sample. Control dots containing the purified SUEHGR fragment from the plasmid pCO2A (Overton and Weinberg, 1978) were used for calibration and a dilution series of pCO2A was used to determine the linear range of hybridization and film response for each experiment. Samples were deproteinized, ethanol precipitated, and resuspended in 500 μl of 1 × SSPE. DNA concentrations were determined by scintillation counting. Carrier tRNA (5 μg) was added to each sample. The blotting was done using a Schleicher and Schuell dot-blot device and a Zeta-Probe nylon membrane. Hybridizations were performed using random primer-labeled plasmid fragments. The highest stringency wash after hybridization was with 0.1 × SSC, 0.1% SDS, 30 minutes at 50°C (for details see, Vincenz *et al.,* 1991).

Denaturing Acrylamide Gels and Probing with Strand-Specific Probes. An 8 M urea, 6% acrylamide gel (Maniatis *et al.,* 1982) was run in order to analyze the length of the 3′ and 5′ termini of the SUEHGR chromatin. The DNA was electrotransferred to a Zeta-Probe membrane, fixed, and prehybridized. Hybridization with the 5′ kinase end-labeled oligonucleotide was done at 55°C in 1 M NaCl, 1% SDS, 50 $\mu g/ml$ salmon sperm DNA, 25 mM NaPO$_4$, pH 7.0, and 10% dextran sulfate for 24 hours. The washes were as described for the dot blot except the last wash was at 40°C with 0.1 × SSC and 0.1%

SDS. Stripping of the probe was achieved by incubation of the filter for 30 minutes in 0.4 M NaOH at 37°C followed by two washes in 2 × SSPE and prehybridization.

Quantitation of Autoradiograms and Ethidium Bromide-Stained Gels. Autoradiograms were digitized using a cooled charge-coupled device (CCD) camera (Star 1,Photometrics). The images were converted to optical densities on a Silicon Graphics IRIS workstation. Integrated band intensities were calculated using the EMPRO image-processing package. Quantitation of the dot blots was done similarly. The linearity of response was checked using the pCO2A dilution series. Ethidium bromide-stained gels were destained in H_2O for several hours. Fluorescence was recovered with the Star 1 camera using a Mineralight UV 254-nm lamp as light source. The illumination was at a ∼45° angle. The CCD camera was equipped with a Wratten orange filter and shielded from direct UV light. Bands containing <1 ng DNA were not visible by eye, but could be quantitated easily by 10-minute exposures. Fluorescence intensities were analyzed with EMPRO after the dark current and local background were subtracted.

Micrococcal Nuclease Digestion of Nuclei and Purified SUEHGR.   Nuclei were resuspended to OD_{260} of ∼1 in buffer A containing 1 mM $CaCl_2$, preincubated at 37°C for 3 minutes, and incubated with different amounts of micrococcal nuclease for 5 minutes. The digestion was stopped by adding EDTA (10 mM) and SDS (0.5%). The total Ca^{2+} concentration of the DTT-released fractions (∼0.2 µg/ml) was adjusted to 2 mM, 3 µg of tRNA was added, and preincubation and digestion done as for nuclei. The DNA was purified, electrophoresed in 2% agarose, and electrotransferred to Zeta-Probe. Hybridizations were performed as described.

Low-Salt Chromatin Spreads.   Chromatin fractions were desalted on Sepharose 4B at 4°C. The elution buffer was 5 mM triethanolamine, pH 7.0, and 0.2 mM EDTA. The void volume peak was fixed with 0.4% glutaraldehyde at 4°C for 12–18 hours. Copper EM grids (400 mesh) were coated with a carbon film, put on top of a droplet containing ethidium bromide (30 µg/ml) for 15 minutes, and dried on filter paper, to facilitate adsorption of chromatin. Spreading (Thoma *et al.,* 1979) was done by adding benzyldimethylalkylammonium chloride (BAC) to the fixed chromatin (2.5 µg/ml). After 30 minutes on ice, 7.5-µl aliquots were placed on Parafilm, and the grids were put on top for 5 minutes. The grids were washed by floating them for 10 minutes on top of H_2O followed by a 3-second rinse in ethanol. The air-dried grids were rotary shadowed with 80:20 Pt:Pd.

Electron Microscopy and Nucleosome Counting.   A Zeiss EM 902 was used for microscopy. The grids with the purified SUEHGR were searched for molecules. Each molecule that came into the field was photographed at 30,000× magnification. The beads on each molecule were counted three times from prints at 100,000× magnification.

III. Results

A. Solubilization of Chromatin Fragments with Restriction Enzymes

The chromatin was solubilized by digestion with restriction enzymes at low Mg^{2+} concentrations (Workman and Langmore, 1985b). For the isolation of the whole SUEHGR repeat, the single cutter HindIII was used. Figure 3

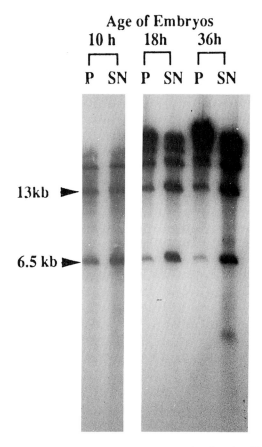

Age of Embryos

10 h	18h	36h
P SN	P SN	P SN

13kb ▶

6.5 kb ▶

FIG. 3. Southern blot of chromatin fractions after HindIII digestion. Nuclei isolated from embryos at the indicated time points were digested with HindIII, lysed, and centrifuged at 10,000 g for 5 minutes. The DNA was purified from the supernatants (SN) and the pellets (P), and electrophoresed in 0.6% agarose. The Southern blot was probed with nick-translated pCO2A. The weak band at 4 kb could be due to sequence polymorphism in the SUEHGR (Yager et al., 1987).

shows the result of such a digestion of sea urchin nuclei from different developmental stages. SUEHGR monomers (6.5 kb) and multimers are found in the insoluble (P) and soluble (SN) fractions. The use of dialysis instead of centrifugation (as used earlier, Workman and Langmore, 1985b) for the buffer changes reduced clumping of the fragile nuclei and increased the yield of solubilization (15–30%). Figure 3 shows that the yield was not limited solely by the efficiency of restriction. The chromatin in the pellet was cut with similar efficiency as the soluble chromatin. The active SUEHGR (10 hours) was cut more frequently than the inactive gene (18, 36 hours), but the solubilization efficiencies were similar for all stages of development (15–30%). Therefore, entrapment in the nuclear debris seems to limit chromatin solubility. The HindIII-solubilized chromatin was not enriched in SUEHGR sequences. On the other hand, SalI cut bulk chromatin less efficiently than the SUEHGR, enriching the SUEHGR 3- to 5-fold during solubilization (not shown).

B. Exonuclease Digestion

Chromatin fragments with 3′ single-stranded ends were produced by digestion with λ exonuclease or T7 gene 6 exonuclease. The alternative approach of using ExoIII was not attempted, because this enzyme was reported to have some double-strand nicking activity (Riley and Weintraub, 1978). Low monovalent salt conditions were chosen in order to avoid chromatin precipitation in the presence of 0.5 mM Mg^{2+}. The proteins associated with the chromatin have an inhibitory effect on the exonucleases. After extensive digestion of the soluble chromatin, only a small fraction of the [^{3}H]thymidine-labeled DNA was TCA soluble (1–3% acid solubility after 1–3 hours at 37°C).

Figure 4 shows that a 5′ exonuclease exposed intact 3′ termini adjacent to the SalI site of SUEHGR chromatin. After restriction, exonuclease treatment, and deproteinization, the DNA was cut with XhoI to obtain a small terminal fragment (363 bp) that could be separated with high resolution on a denaturing gel. The Southern blot of the gel was probed first with a single-stranded oligonucleotide complementary to the 3′ upstream strand of the SalI site (Fig. 4A) in order to measure the size of the nondigested strand. The sharp bands in lanes 2 and 3 indicate that the exposed ends were intact. The same blot was stripped and then probed with double-stranded probe A (Fig. 4B) in order to measure the length of both terminal strands. Quantitation of the autoradiograms revealed that >80% of SUEHGR terminal fragments had single-stranded ends and more than 50% of the molecules had 3′ termini larger than 50bp. The signal on top of the gel represents XhoI–XhoI fragments derived from SUEHGR multimers. T7 gene 6 exonuclease produced similar termini adjacent to the HindIII site (not shown).

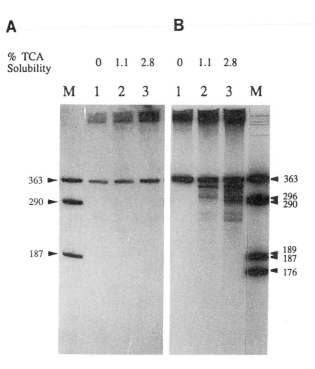

FIG. 4. Denaturing gel and strand-specific probing of λ exonuclease-treated chromatin. Chromatin was solubilized with *Sal*I and digested with 5 U of λ exonuclease per microgram of DNA. The DNA was purified, digested with *Xho*I, electrophoresed in 8 *M* urea, 6% acrylamide, and electrotransferred to Zeta Probe. (A) Initial probing with the single-stranded oligonucleotide *Sal*I 33^{mer}. (B) The same filter stripped and reprobed with double-stranded probe A (Fig. 2). M, partial digest of the SUEHGR as molecular weight markers. Lane 1 was incubated without exonuclease for 60 minutes. Lanes 2 and 3 were incubated with exonuclease for 5 and 60 minutes, respectively.

C. Chromatin Oligonucleotide Hybridization

High salt and temperature are known to maximize the rate and specificity of DNA hybridizations. Such conditions are not compatible with intact chromatin structure (Kinoshita, 1976; Ausio *et al.*, 1986). However, we have previously shown that 100 m*M* salt and 37°C were stringent enough for specific hybridization to SV40 minichromosomes (Workman and Langmore, 1985a). We developed a simple assay to determine the stringency and kinetics of hybridization of the oligonucleotide to SUEHGR chromatin.

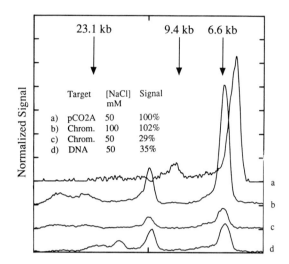

FIG. 5. Densitometer traces of Sarkosyl gel autoradiogram. ^{32}P-labeled oligonucleotide probe (*Hind*III 17[mer], 5 ng/ml) was hybridized in buffer H for 8 hours with the following targets: (a) plasmid DNA; (b) and (c) chromatin; and (d) DNA from solubilized and exonuclease-treated chromatin. All targets were digested with *Hind*III and T7 gene 6 exonuclease. The chromatin was from 18-hour nuclei. Procedures as described in Section II. The sodium chloride concentrations during hybridization are shown in the inset. All traces were normalized to a constant amount of loaded SUEHGR (determined from dot blots). The integrated signal intensities over all bands are shown. The plasmid fragment had a higher mobility due to more extensive digestion with T7 gene 6 exonuclease.

Figure 5 shows that the SUEHGR chromatin or DNA was specifically hybridized to an oligonucleotide at 100 mM NaCl and 37°C. Exonuclease-treated chromatin or DNA was hybridized with a 5′ ^{32}P-labeled oligonucleotide. The hybridization mixtures were electrophoresed on a gel containing enough Sarkosyl for the chromatin to comigrate with DNA of the same size (Ackerman *et al.,* 1983). Clearly, the ^{32}P-labeled oligonucleotide bound to DNA of the same size as the SUEHGR chromatin. The specificity was similar for pCO2A and for sea urchin chromatin (trace a versus b) and for chromatin and genomic DNA (trace c versus d). Therefore, the chromosomal proteins do not prevent hybridization. Hybridization was not completed after 8 hours in 50 mM NaCl (trace c versus b). The *Sal*I 33[mer] hybridized to *Sal*I-cut chromatin with similar efficiency and stringency (not shown).

Using this assay, the minimum amount of T7 gene 6 exonuclease enzyme required to achieve maximal hybridization efficiency was determined. A plateau value was reached at ~ 50 U/mg of DNA. This corresponds to an acid solubility of 2.8%. For standard preparative hybridizations 8 U of T7 gene 6 exonuclease/mg DNA was added. This corresponds to a hybridization efficiency of ~70% of the maximum achieved with a limit exonuclease digest.

λ Exonuclease-treated chromatin gives rise to a strong broad signal in the 4-kb region of the Sarcosyl gel and a specific hybridization signal that was only half that of T7 gene 6-treated chromatin. Additional studies revealed that the λ exonuclease bound nonspecifically to the oligonucleotide (not shown). This exonuclease is a highly processive enzyme and therefore binds strongly to oligonucleotides. The reduction in hybridization efficiency was probably due to the λ exonuclease remaining attached to the exposed single-stranded tails. We therefore used T7 gene 6 exonuclease for all preparative isolations.

The kinetics of the reaction were studied with the Sarcosyl assay and plotted according to a pseudo-first-order kinetics in Fig. 6. The $t_{1/2}$ at a probe concentration of 0.7 nM were 7 hours at 100 mM NaCl and 10 hours at 50 mM NaCl. These rates display the NaCl dependence expected for a hybridization reaction. These rates are about an order of magnitude higher then the rates measured for the hybridization of mercurated plasmid fragments to chromatin or DNA (Workman and Langmore, 1985a).

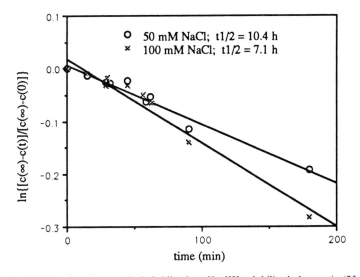

FIG. 6.   Kinetics of nucleoprotein hybridization. *Hind*III-solubilized chromatin (25 μg/ml) was hybridized to the 5' ^{32}P-labeled *Hind*III 17^{mer} (4 ng/ml) at the indicated NaCl concentrations, electrophoresed in a Sarcosyl gel, and electrotransferred to Zeta-Probe. Amounts of oligonucleotide hybridized to the SUEHGR monomer were quantitated by autoradiography and plotted according to pseudo-first-order kinetics. The signal intensity after 55 hours of hybridization time was assumed to be c (∞). Hybridization rate constants were $1.8 \times 10^4 \pm 0.09\ M^{-1}\mathrm{sec}^{-1}$ and $3.7 \times 10^4 \pm 0.23\ M^{-1}\mathrm{sec}^{-1}$ for 50 and 100 mM NaCl, respectively.

D. Affinity Chromatography

The chromatin oligonucleotide hybrids are separated from the bulk chromatin using an avidin affinity matrix. The bound hybrids are released by cleavage of the disulfide bond in the biotin linker (Fig. 7). The critical parameters determining the performance of the affinity chromatography are (1) efficiency of binding, (2) efficiency of release, and (3) extent of binding and release of non-SUEHGR chromatin.

All these parameters depend on the type of affinity system that is chosen. The biotin/avidin interaction should give a high binding efficiency ($K_{diss} = 10^{-15}$). The use of 5′ amino-modified oligonucleotides allows the synthesis and HPLC purification of a fully biotinylated probe. The distance between the disulfide bond and the biotin moiety is important to achieve efficient release. A longer distance is required for a streptavidin matrix (1.7 nm) than for an avidin matrix (0.7 nm, Herman *et al.*, 1986).

Other biotinylation schemes and hybridization probes were tested. The use of *Exo*III-treated plasmid fragments as hybridization probes led to high background binding. Such probes were easily biotinylated using cleavable linker photobiotin (Clonetech). This photoactivatable biotin analog reacts very inefficiently with oligonucleotides. Oligonucleotides can be enzymatically biotinylated with terminal transferase and Bio-19-SS-dUTP (gift from Tim Herman, Medical College of Wisconsin at Milwaukee). Lower yields are observed with such a combination, probably due to partial cleavage of the SS bond during the incubation in the terminal transferase buffer containing (0.2 mM) DTT.

The most difficult parameter to control was the amount of non-SUEHGR chromatin that bound to the affinity matrix and was released with DTT. The amount of this background chromatin limits the maximum enrichment

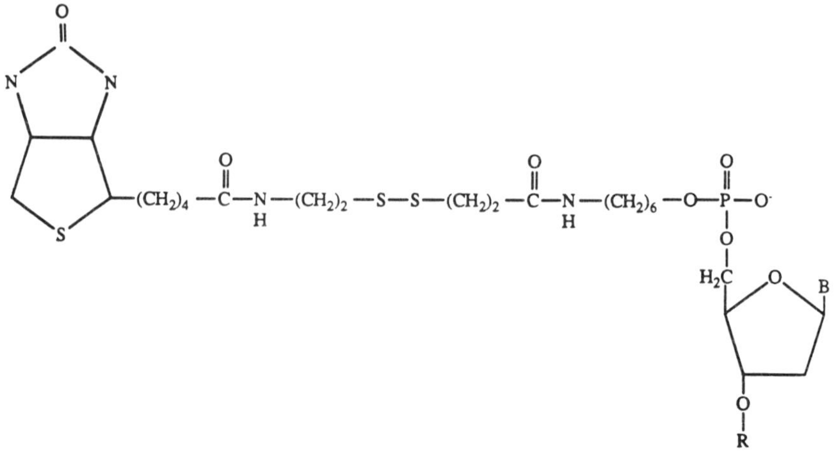

FIG. 7. Structure of the NHS-SS-biotin analog attached to the 5′ end of the oligonucleotide.

obtainable. The Vectrex avidin D matrix gave less background than strept-avidin agarose, perhaps because it does not have pores. Therefore the entire surface is available for binding to the oligonucleotide and the large oligonucleotide/chromatin hybrids. The binding capacity of Vectrex avidin D was 1 ng 17^{mer}/mg of dry matrix (not shown). Addition of oxidized glutathione (GSSG) and $IAcNH_2$ to the hybridization and rinsing buffers decreased the background 3-fold. A further reduction of the background chromatin was achieved by rinsing the column with 150 mM NaCl and eluting the chromatin in 75 mM NaCl.

Figure 8 shows a typical elution profile. The SUEHGR content of the different fractions was determined by dot-blot hybridization. The enrichment

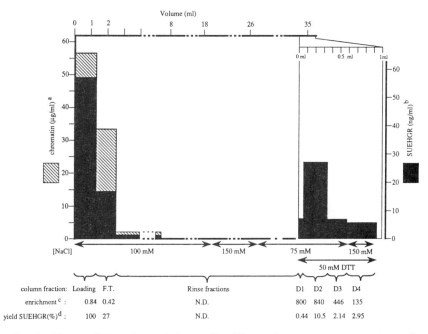

FIG. 8. Vectrex affinity column elution profile of *Strongylocentrotus purpuratus* chromatin after nucleoprotein hybridization. *Hind*III-solubilized chromatin from 18-hour nuclei was hybridized to the biotinylated *Hind*III 17^{mer} for 24 hours. Affinity chromatography was performed as described. F. T. is the flow-through fraction. NaCl concentrations of the column buffers are indicated. (a) Chromatin concentrations were determined by ^{3}H scintillation counting. (b) The SUEHGR concentration was calculated by multiplying the chromatin concentration by the purity. Conversion from enrichment (determined on dot blots) to purity was made using the results from the gel analysis of the D2 fraction, which showed that an enrichment of 840× corresponded to 100% purity (see Fig. 9). (c) The enrichment was determined by comparing the dot-blot hybridization signals of the column fractions to those of nuclear DNA with probe B. (d) The yield of the column procedure is the enrichment multiplied by the amount of chromatin in the fraction compared to the amount loaded onto the column.

was calculated by dividing the hybridization signal per nanogram of DNA for the enriched fraction by the signal for genomic DNA. The peak DTT fraction contained > 800-fold enriched SUEHGR sequences. The purity can by calculated by multiplying the enrichment factor by the SUEHGR content in the genome. We determined the genomic SUEHGR content by comparing the hybridization signal of the genomic DNA to that of a plasmid fragment containing the SUEHGR. The SUEHGR was found to be $0.15\% \pm 0.05\%$ of the sea urchin genome, which is substantially lower than the 0.25–0.45% previously reported (Weinberg et al., 1975). More accurately, the purity was calculated by comparing the dot-blot signal of the column fractions with the signal from similar amounts of the isolated pCO2A SUEHGR insert on the same filter. The purities determined in this fashion for the peak DTT fraction for four independent isolations were 83, 89, 105 and 132%.

Figure 9 shows the result of an independent assay of the purity and molecular weight distribution of the isolated SUEHGR. The chromatin

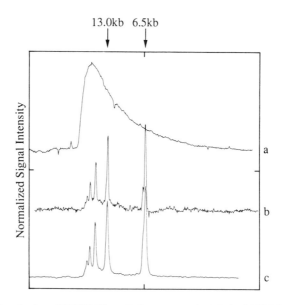

FIG. 9. Determination of SUEHGR purity by agarose gel analysis of DNA from DTT-eluted chromatin. Nucleoprotein hybridization was performed as described in the methods using HindIII-solubilized chromatin. The DNA from different column fractions was purified and electrophoresed in 0.6% agarose. One-dimensional projections of fluorescence intensities and autoradiographic densities are shown. The traces were scaled to give the same peak height for the strongest band. (a) Ethidium bromide fluorescence of the chromatin after hybridization (loading fraction. (b) Ethidium bromide fluorescence of the DTT-eluted chromatin (isolated SUEHGR). (c) Southern blot of lane b hybridized with probe B.

fractions were deproteinized and electrophoresed. The ethidium bromide fluorescence was recorded. Subsequently, a Southern blot of the same gel was probed with probe B to determine the distribution of the SUEHGR sequences. The DNA size distribution of the solubilized chromatin and the purified SUEHGR are shown in Fig. 9, traces a and b, respectively. The SUEHGR distribution is shown in the Southern blot (trace c). Traces b and c are superimposable. Thus, the fluorescence between the dimer to pentamer bands was not due to background but due to overlapping SUEHGR bands. Using this method, the purity of two preparative isolations was calculated to be 108 and 80%. Comparison of the molecular weight distribution on Southern blots of purified SUEHGR and unpurified soluble SUEHGR chromatin show that the SUEHGR chromatin has been purified as integral multiples of 6.5 kb without any degradation of the DNA (not shown).

To verify that the enrichment process is due to hybridization of chromatin to the oligonucleotide and not due to some other process, a negative control was performed. HindIII-solubilized chromatin without exonuclease treatment was used in the hybridization reaction with the 17^{mer} oligonucleotide. No hybridization was detectable by the Sarkosyl gel assay. Affinity chromatography of this mock hybridization mixture gave a small (< 50-fold) but reproducible enrichment (not shown). We interpret this result as a partial strand invasion that produces hybrids that are stable under the affinity column conditions. We conclude, however, that the much more significant enrichment seen by nucleoprotein hybridization was due to sequence-specific hybridization to single-stranded ends.

E. Characterization of the Isolated SUEHGR by Micrococcal Nuclease Digestion

Micrococcal nuclease is an established probe for the investigation of the SUEHGR chromatin structure (Wu and Simpson, 1985; Fronk et al., 1990). To assay if the main structural features were maintained during the purification, isolated SUEHGR chromatin and nuclei from different developmental stages were digested with micrococcal nuclease followed by Southern hybridization with SUEHGR sequences. Figure 10 shows the results of this comparison. The isolated SUEHGR (lanes 5, 6, 10) show clear nucleosomal ladders for the inactive chromatin (18 and 36 hours) that are virtually identical to those of intact nuclei (lanes 4, 9). The isolated 10 hour SUEHGR have a diffuse pattern at low levels of digestion (lane 2) and a shorter repeat length upon more extensive digestion (lane 3). These are characteristic features for the active SUEHGR (Wu and Simpson, 1985; Fronk et al., 1990). This pattern, indicative of nonregularly spaced nucleosomes, is also preserved during the isolation process. These results clearly establish that the purified

A

Age of
embryos

	10h			18h					36h	
	N.	i. S.		N.	i. S.		i. S.		N.	i. S.

Micr.Nucl.
(U / ml)

	5	**0.5**	**2**	**5**	**1**	**2**	**1**	**3**	**20**	**1**	
M	1	2	3	4	5	6	7	8	9	10	M

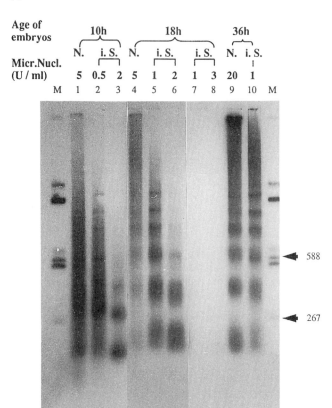

← 588

← 267

B

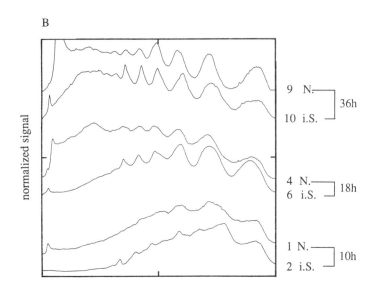

normalized signal

9 N. ⎤ 36h
10 i.S. ⎦

4 N. ⎤ 18h
6 i.S. ⎦

1 N. ⎤ 10h
2 i.S. ⎦

material is chromatin with the same main structural features as in the nuclei. Significant nucleosome loss did not occur during isolation. We conclude that the core histone proteins are still bound to their physiologically relevant DNA sequence.

F. Electron Microscopy of the Isolated Material

The isolated SUEHGR is estimated to have a molecular weight of $\sim 9 \times 10^6$/monomer. Electron microscopy is particularly suited to a give a low-resolution picture of molecules in this size range. Electron microscopy also allows the study of individual molecules, which enables an assessment of the homogeneity of the isolated SUEHGR. Such information is not obtainable by biochemical techniques. Figure 11A shows a low-salt BAC spread of the DTT-eluted material originating from 18-hour nuclei. The molecules have the expected beads-on-a-string structure typical of inactive chromatin. The number of beads on each of molecules was counted. The result of this analysis is given in Fig. 11B. In this highly pure preparation ($> 90\%$), the SUEHGR monomers appear homogeneous and contain 27 ± 3 beads. This is close to the expected 29.5 nucleosomes calculated using 220 bp as the average repeat length of SUEHGR (Wu and Simpson, 1985). The counting of beads on higher oligomers was less reliable because strands overlap due to incomplete spreading. These micrographs confirm the results of micrococcal nuclease digestion, and show the feasibility of using EM for a comparative study of active and inactive genes.

G. Histone Integrity and Histone Exchange

Nucleoprotein hybridization requires extended incubations at 37°C. Proteins of the bulk fractions were analyzed by SDS-PAGE as shown in Fig. 12. Proteolysis of the histone proteins was completely prevented by leupeptin.

FIG. 10. Micrococcal nuclease digestion of isolated SUEHGR. SUEHGR chromatin was purified from nuclei isolated at the indicated times of embryogenesis and digested with different amounts of micrococcal nuclease. Nuclei of the same batches were also digested. The DNA was purified and electrophoresed in 2% agarose. The nuclear lanes (lanes 1, 4, 9) contained ~ 5 μg and the isolated SUEGHR lanes (2, 3, 5, 6, 10) contained < 50 ng total DNA. Lanes 7 and 8 contained DTT-released material from negative control where the chromatin target was not exonucleased. These lanes showed a weak pattern similar to lanes 5 and 6 after overexposure of the filter. (A) Autoradiogram of the Southern blot probed with pCO2A containing one SUEHGR. (B) Scaled densitometer traces of the indicated lanes of the same autoradiogram. N., Nuclei; i.S., isolated SUEHGR.

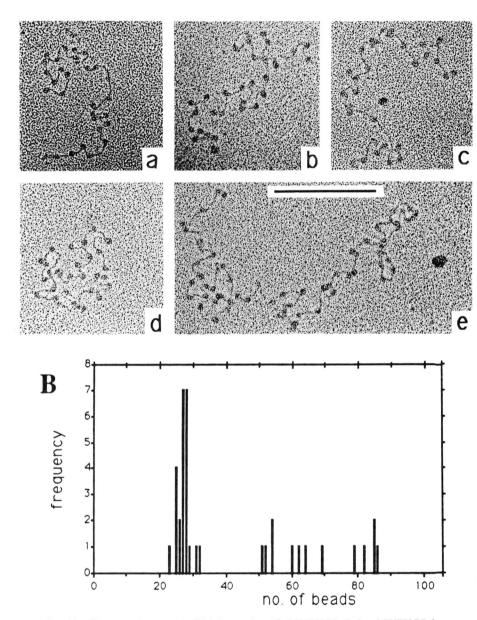

Fig. 11. Electron micrographs of BAC spread purified SUEHGR. Isolated SUEHGR from 18-hour nuclei were desalted on a Sepharose 4B column and BAC spread with the microdroplet method. The grids were rotary shadowed with Pt : Pd. Bar, 0.2 μm. (A) Micrographs a–d show SUEHGR monomers; micrograph e shows a dimer molecule. (B) Histogram of the number of beads on each molecule. Three molecules for which the standard deviation of three countings was >10% of the mean were not included. Six molecules had more than 100 beads.

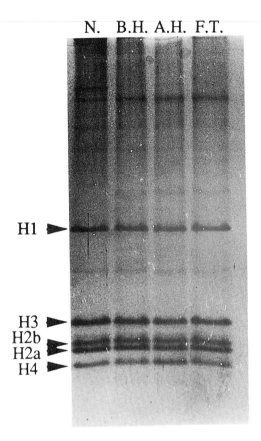

FIG. 12.   Coomassie Blue-stained SDS-PAGE of 2.5-μg aliquots of protein at different stages of nucleoprotein hybridization. N., Nuclei; B. H., before hybridization; A. H., after hybridization; F. T., flow-through fraction.

The four core histones as well as H1 are quantitatively preserved throughout the multiple dialysis, digestions, dilutions, and affinity chromatography steps. Unfortunately, the electrophoretic pattern of the purified SUEHGR chromatin could not be determined because of insufficient material for conventional visualization by Coomassie blue staining or autoradiography of ^{35}S-labeled proteins.

The conditions employed during nucleoprotein hybridization were designed to preserve the DNA protein interactions and minimize protein exchange. The results obtained with SV40 justified these conditions (Workman and Langmore, 1985a). In the following experiment, we addressed the question of histone exchange between bulk chromatin and SUEHGR. [3H]-Thymidine-labeled and exonuclease-treated chromatin was mixed with

[^{35}S]methionine-labeled, nonexonuclease-treated chromatin. After hybridization to the 17^{mer} oligonucleotide, the amount of ^{35}S-labeled protein in the DTT-eluted fraction was determined. Only a small percentage of the loaded radioactivity was eluted from the columns with DTT (0.025 ± 0.004% from the hybridizable sample and 0.026 ± 0.003% from a nonhybridizable control where the chromatin had not been exonuclease treated). Thus, most of the labeled protein eluted was not bound to the hybrids. The DTT-released chromatin was precipitated and the proteins were analyzed by gel electrophoresis. Each of the five histone bands was excised, H_2O_2 hydrolyzed, and counted by liquid scintillation (Tishler and Epstein, 1968). The ratio of ^{35}S counts recovered in the histone bands to the amount of [^{3}H]thymidine used for the ethanol precipitation of each sample was determined for the different chromatin fractions. Complete exchange would produce the same labeled histone to labeled DNA ratio in the DTT fraction as in the flow-through. Comparing these ^{35}S/^{3}H ratios of enriched and flow-through fractions of two independent experiments gave the following estimates of core histone exchange: Exp 1, 32 ± 24%; Exp 2, 17 ± 16%. The standard deviations refer only to the statistical uncertainty of the scintillation counting. It was not possible to measure H1 exchange using this procedure due to the small number of counts in the H1 band. The EM and micrococcal nuclease results argue also strongly for retention of the original chromatin organization (see Section IV).

IV. Discussion

A. The Technique

The results presented here clearly show that nucleoprotein hybridization can be used to purify the sea urchin early histone genes as chromatin. Independent assays were developed for the exonuclease and the hybridization reactions to facilitate their optimization. The applicability of the method to other genes would only require efficient solubilization of the gene with restriction enzyme and efficient production of single-stranded ends by exonuclease digestion. The fact that efficient nucleoprotein hybridization has now been achieved at all four sites tested (two in SUEHGR and two in SV40) makes us confident that the technique is of general use for the enrichment of eukaryotic genes as chromatin. The most significant limitation seems to be the limited number of genes that occur in sufficient abundance (i.e., >0.1% of the genome) to isolate biochemical quantities.

The digestion of nuclei with restriction enzymes produced soluble chromatin fragments with discrete lengths (Fig. 3) containing a full complement of histone proteins (Fig. 12). Earlier protocols for the restriction of chromatin have used high Mg^{2+} concentrations (e.g., Prior et al., 1983; Reynolds et al., 1983; McGhee et al., 1981), which cause chromatin to aggregate, making cleavage and solubilization difficult. Protocols using low Mg^{2+} concentrations (0.5 mM) produce higher accessibility and solubility (Workman and Langmore, 1985b), and have general applicability to the solubilization of intact, long chromatin. Only a fraction of the cleaved chromatin molecules became soluble (Fig. 3), which we attribute to entrapment of the genes on or within the nuclear debris. Recent progress in understanding the nuclear disassembly process might eventually lead to more efficient solubilization (Peter et al., 1990; Ward and Kirschner, 1990).

A solubilization efficiency of 15–30% is acceptable so long as subfractionation does not occur. Fortunately, the sea urchin early histone genes seem to be structurally homogeneous in the active as well as in the inactive state. The micrococcal nuclease digestion patterns of 18- and 36-hour SUEHGR (Fig. 10) show a very clear pattern, with very little background beneath the ladder. There is, for example, no band at ~ 290 bp which is a predominant feature of the active chromatin. In addition, it has been shown that DNaseI-hypersensitive sites disappear completely after gene inactivation (Fronk et al., 1990). The inactive, purified SUEHGR chromatin also seemed to be structurally homogeneous, as indicated by the narrow distribution of the number of beads per SUEHGR monomer (Fig. 11). The homogeneity of the active genes is also indicated by the micrococcal nuclease digests (Fig. 10). Clearly, the major fraction of 10-hour SUEHGR chromatin is different from the 18-hour chromatin (lanes 1 and 2 versus 4 and 5). However, there might be some fraction that behaves like inactive chromatin. The weak but sharp bands seen at higher molecular weight (lane 2) could be due to a minor fraction of SUEHGR chromatin resembling the inactive chromatin. Other features of the gel are ambiguous, e.g., the band observed at ~ 440 bp (best visible in lane 3) could represent a dimer with the 220-bp repeat of inactive chromatin or be a compact trimer with a ~ 146-bp repeat of active chromatin. Because only a minor fraction of the 10-hour genes seems to have an inactive structure, the biochemical and structural studies of the isolated SUEHGR chromatin should not be seriously hindered. Earlier studies of the SUEHGR of *Psammechinus miliaris* concluded that all of the copies of the genes (both at early blastula and hatching blastula) had a very homogeneous chromatin structure, as assayed by nuclease hypersensitivity and protection (Bryan et al., 1983).

We conclude that the isolated genes are homogeneous and are representative of the entire population of SUEHGR within the cells. Thus, SUEHGR are more homogeneous than rRNA genes, which seem to coexist in populations with different structures and functions (Conconi et al., 1989).

The extent of exonuclease digestion of soluble chromatin was limited by the chromatin structure. A limiting value of $\sim 3\%$ TCA solubility was reached under standard incubation conditions, similar to earlier results with SV40 (Workman and Langmore, 1985a), chicken, and sea urchin chromatin (not shown). The single-strand gel assay showed that both the *Sal*I and *Hind*III 5' termini are efficiently digested, leaving $> 80\%$ of the SUEHGR with intact 3' overhangs of an average size of ~ 50 bp (Fig. 4).

The use of the nonporous avidin D matrix resulted in the expected decrease of background chromatin. The enrichments obtained using identical column conditions were similar for chromatin from different developmental stages. Adding GSSG and $IAcNH_2$ to the hybridization buffer and GSSG and 150 mM NaCl to the column rinsing buffer reduced the background further so that chromatin from 18-hour embryos could be enriched > 700-fold in a single affinity step. The eluted material was $> 80\%$ pure, contained chromatin pieces of discrete length (Fig. 9), and had properties of the native nuclear chromatin (Fig. 10).

Control experiments have shown that the exchange of core histones is minimal. Preservation of the original core histone organization is also indicated by the micrococcal nuclease digestion patterns (Fig. 10) and the EM analysis (Fig. 11). The EM analysis revealed that the SUEHGR monomers had 27 ± 3 nucleosomes. Because the average repeat length of SUEHGR in nuclei is 220 bp (Fronk *et al.*, 1990), we expected 29.5 nucleosomes per SUEHGR monomer. This slight discrepancy could be due to gaps in the nucleosomal repeat, which would not affect the digestion ladder, or to the inherent difficulties in counting nucleosomes by EM. The experiments presented measure only exchange that occurs during the prolonged incubations after the exonuclease steps, and were limited by the low specific activity of the proteins. Earlier experiments using SV40 (Workman and Langmore, 1985a) assayed exchange throughout the isolation procedure and found less than 10% loss or exchange of the core histones.

We were not surprised by the lack of significant loss or exchange because previous studies had repeatedly found that core histones do not exchange at ~ 100 mM monovalent salt. Germond *et al.* (1976) reported negligible loss or exchange of histones from SV40 minichromosomes during 30-minute incubations at $37°$C with an equal amount of 100 mg/ml SV40 form I DNA, as long as the salt concentration was less than 700 mM NaCl. Crémisi *et al.* (1976) were unable to detect protein exchange after 2-hour incubations of minichromosomes with DNA. Ilyn *et al.* (1971) found little exchange of histone and nonhistone proteins after 24-hour incubations of sheared chromatin with exonogenous tRNA, DNA, and other chromatin at 120 mM NaCl. Even nucleosome sliding was not detectable at 100–150 mM NaCl (Crémisi *et al.*, 1976; Glotov *et al.*, 1982).

The presence of H1 on the purified SUEHGR is indicated by the preservation of the native micrococcal nuclease repeat length (Fig. 10), because H1 stripping of chromatin and incubation at 37°C under physiological salt have led invariably to the formation of compact dimers (Spadafora et al., 1979; Watkins and Smerdon, 1985). On the other hand, H1 exchange on bulk chromatin is expected during isolation of the nuclei, enzymatic digestions, and nucleoprotein hybridization (Caron and Thomas, 1981). We were unable to determine whether histone H1 exchange onto the SUEHGR took place.

B. Comparison with Alternative Methods

We have devised a general method to purify a chromatin fragment with a specific sequence. This has enabled us to purify an RNA polymerase II gene as chromatin. Nucleoprotein hybridization can be used to isolate the same chromatin in different functional states. This makes possible direct comparative studies between active and inactive chromatin fragments of the same DNA sequence. The technique is complementary to other techniques that have been designed to isolate specific genes as chromatin or subfractionate the genome according to transcriptional activity or specific protein binding.

Simpson and co-workers have devised a general method for isolation of specific genes as chromatin ([11], this volume). The targeted gene is first cloned into a multicopy yeast plasmid and later isolated as a minichromosome. The strength of this system is that any gene can be cloned and regulated by a chosen promoter and upstream regulatory sequences, which allows experimental manipulation of the genes in order to dissect the molecular mechanism of gene regulation. The only potential limitations to this approach are the difficulties in purifying minichromosomes from cytoplasmic contaminants, and the uncertainties in expressing multiple copies of the genes in a nonnative environment, which might be limiting in critical factors.

Many techniques have been devised to enrich for chromatin that has properties that seem to be specific for active genes. A particular advantage of these techniques is that they can be used to study single-copy genes. Among these approaches are immunoprecipitation (e.g., Dorbic and Wittig, 1986, 1987; Hebbes et al., 1988), mercury-Sepharose affinity chromatography (e.g., Allegra et al., 1987; Chen et al., 1990), differential solubilization (e.g., Bonner et al., 1975; Bloom and Anderson, 1978; Ridsdale and Davie, 1987; Strätling and Dölle, 1986; Strätling, 1987; Hanks and Riggs, 1986; Davis et al., 1983; Kumar and Leffak, 1986; Rose and Garrard, 1984; Razin et al., 1985), or DNase hypersensitivity (Dawson et al., 1989). These techniques cannot directly target a specific gene, because enrichment depends on physical properties or protein composition rather than DNA sequence. Chapter 12 of this volume details the method of affinity chromatography on mercury-Sepharose, which specifically isolates nucleosomes with exposed SH groups associated with active genes. This technique is complementary to the nucleoprotein technique for study of active nucleosomes; however, it is not suited for study of the differences between the active and inactive states of particular genetic loci. Chapter 14, of this volume describes the use of UV corss-linking and immunoprecipitation, which is particularly suited to study binding of known proteins, including topoisomerase I, RNA polymerase II, and TATA factors, which are bound to specific sites in the genome. This method has the potential to be used as a control for loss of nonhistone proteins during nucleoprotein hybridization, and for correlating changes in histone composition, as determined by nucleoprotein hybridization, with changes in nonhistone composition.

At present, it is reasonable to use nucleoprotein hybridization to purify eukaryotic genes representing >0.1% of the genome. Less abundant genes cannot be purified to homogeneity by single-step biotin-affinity chromatography. Indications are that the limiting factor is the binding of extraneous chromatin to the avidin D columns rather than low stringency of hybridization. For higher eukaryotes, the current enrichment limits applications to multicopy genes. Practical limitations in the number of cells and the amount of enzymes will restrict the yield to a few nanograms of specific genes, requiring microanalytical analysis of the material by radiography of proteins radioiodinated to high specific activity and by EM of the isolated molecules.

C. Future Directions

The availability of a purified gene as native chromatin enables one to carry out a variety of studies. The first studies will use two-dimensional gel electrophoresis to investigate changes in protein composition and covalent modifications of the SUEHGR during development. The method as presented can be used for the study of SUEHGR-specific core histone proteins. Of particular interest is the identification of covalent modifications and their relation to gene activity. Because histones are the most abundant proteins, they should be easily detectable on autoradiograms of gels containing radioiodinated SUEHGR proteins. Tyrosine labeling of histone proteins should give 10^4 times higher specific activity than was achieved *in vivo* with $[^{35}S]$methionine.

Better measurements of H1 exchange have to be performed before its association can be correlated with SUEHGR gene activity. This applies also for nonhistone proteins. An additional purification step such as size-exclusion chromatography might be required to separate SUEHGR-associated higher-molecular-weight proteins from background proteins. The sensitivity of *in vitro* iodination and Western blotting should allow the detection of proteins occurring in a single copy per SUEHGR. Isolation of proteins with a low binding constant to their native chromatin fiber might require the use of lower salt concentrations or reversible fixation (Solomon and Warshawski, 1985) to avoid loss or exchange. We have shown that efficient hybridization is also observed at 50 mM NaCl (Fig. 6).

Isolated SUEHGR is also an ideal target for functional studies investigating the effect of nucleosomes and chromatin on transcriptional processes. Workman et al. (1988) and Workman and Roeder (1987) have shown that interactions of nucleosomes with the RNA polymerase II transcriptional machinery play a role in reconstituted systems. Purification of chromatin fragments with specific sequences would overcome the difficulties of these model systems that rely on faithful reconstitution.

The demonstration that nucleoprotein complexes can be isolated to homogeneity by the cleavable biotin/Vectrex avidin D system suggests other approaches to the isolation of specific nucleoprotein complexes. Transient transfection systems are commonly used for transcription studies of mutated genes (e.g., Hesse et al., 1986). Direct information about the protein composition could be obtained upon reisolation of the plasmids as chromatin. Components of the mRNA processing machinery have also been fractionated using biotin affinity systems (Grabowski and Sharp, 1986). Our improvements in the specificity of the affinity chromatography will benefit these other studies.

We have demostrated the feasibility of investigating the isolated material by EM (Fig. 11). Inter-SUEHGR homogeneity and intra-SUEHGR heterogeneity of nucleosome positioning will be studied by EM. Immunostaining techniques could also be used to map proteins along single SUEGHR molecules (Harper et al., 1984; Bustin et al., 1976).

Nucleoprotein hybridization could be combined with EM to study the structure of single-copy genes. If a heavy atom label can be put onto the hybridization oligonucleotide, specific genes could be identified from micrographs of partially purified genes. Tagged oligonucleotides have been identified previously in the EM (Brantley and Beer, 1989). Therefore, partial purification by nucleoprotein hybridization, combined with gene-specific tagging would make EM studies of single-copy genes feasible. For example, a 5-kb single-copy gene from Drosophila would comprise $\sim 0.004\%$ of the genome. Assuming a 1000-fold enrichment and 10% yield during nucleoprotein hybridization, 1 mg of nuclei should yield ~ 100 ng of chromatin containing ~ 4 ng ($\sim 10^8$ molecules) of the specific genes. Assuming a 200:1 ratio for gene-specific to nonspecific tagging with a heavy atom label, $>80\%$ of the tagged molecules would be chromatin fragments from the specific gene. It should be practical to achieve $>200:1$ specificity of tagging by EM, because copies of the specific gene could be identified both by the presence of a heavy atom label at one terminus and by the observed molecular length. Thus, EM tagging combined with enrichment by nucleoprotein hybridization would make it feasible to study single-copy genes.

ACKNOWLEDGMENTS

We are grateful to Dr. Shawn P. Williams and Michel F. Smith for setting up the image-processing system. This work was supported by grants NIH GM 27937, NSF PCM-8317039, and NSF DCB-9007295 and the University of Michigan Office of the Vice President for Research. CV was supported by a Rackham Predoctoral fellowship.

REFERENCES

Ackerman. S., Bunick, D., Zandomeni, R., and Weinmann, R. (1983). RNA polymerase II ternary transcription complexes generated *in vitro. Nucleic Acids Res.* **11,** 6041–6064.

Allegra, P., Sterner, R., Clayton, D. F., and Allfrey, V. G. (1987). Affinity chromatographic purification of nucleosomes containing transcriptionally active DNA sequences. *J. Mol. Biol.* **196,** 379–388.

Amero, S. A., Ogle, R. C., Keating, J. L., Montoya, V. L., Murdoch, W. L., and Grainger, R. M. (1988a). The purification of ribosomal RNA gene chromatin from *Physarum polycephalum. J. Biol. Chem.* **263,** 10725–10733.

Amero, S. A., Montoya, V. L., Murdoch, W. L., Ogle, R. C., Keating, J. L., and Grainger, R. M. (1988b). The characterization of ribosomal RNA gene chromatin from *Physarum polycephalum. J. Biol. Chem.* **263,** 10734–10744.

Ausio, J., Sasi, R., and Fasman, G. D. (1986). Biochemical and physiological characterization of chromatin fractions with different degrees of solubility isolated from chicken erythrocyte nuclei. *Biochemistry* **25,** 1981–1988.

Bloom, K. S., and Anderson, J. N. (1978). Fractionation of hen oviduct chromatin into transriptionally active and inactive regions after selective micrococcal nuclease digestion. *Cell (Cambridge, Mass.)* **15,** 141–150.

Bonner, J., Gottesfeld, J., Garrard, W., Billing, R., and Uphouse, L. (1975). Isolation of template active and inactive regions of chromatin. *In* "Methods in Enzymology" (B. D. O'Malley and J. G., Hardman, eds.), Vol. 40, pp. 97–102. Academic Press, New York.

Brantley, J. D., and Beer, M. (1989). Gene-specific labeling of chromatin for electron microscopy. *Gene Anal. Tech.* **6,** 75–78.

Bryan, P. N., Olah, J., and Birnstiel, M. L. (1983). Major changes in the 5' and 3' chromatin structure of sea urchin histone genes accompany their activation and inactivation in development. *Cell (Cambridge, Mass.)* **33,** 843–848.

Bustin, M., Goldblatt, D., and Sperling, R. (1976). Chromatin structure visualization by immunoelectron microscopy. *Cell (Cambridge, Mass.)* **7,** 297–304.

Caron, F., and Thomas, J. O. (1981). Exchange of histone H1 between segments of chromatin. *J. Mol. Biol.* **146,** 513–537.

Chen, T. A., Sterner, R., Cozzolino, A., and Allfrey, V. G. (1990). Reversible and irreversible changes in nucleosome structure along the c-fos and c-myc oncogenes following inhibition of transcription. *J. Mol. Biol.* **212,** 481–493.

Conconi, A., Widmer, R. M., Koller, T., and Sogo, J. M. (1989). Two different chromatin structures coexist in ribosomal RNA genes throughout the cell cycle. *Cell (Cambridge, Mass.)* **57,** 753–761.

Crémisi, C., Pignatti, P. F., and Yaniv, M. (1976). Random location and absence of movement of the nucleosomes on SV40 nucleoprotein complex isolated from infected cells. *Biochem. Biophys. Res. Commun.* **73,** 548–554.

Davis, A. H., Reudelhuber, T. L., and Garrard, W. T. (1983). Variegated chromatin structures of mouse ribosomal RNA genes. *J. Mol. Biol.* **167**, 133–155.

Dawson, B. A., Herman, T., and Lough, J. (1989). Affinity isolation of transcriptionally active murine erythroleukemia cell DNA using a cleavable biotinylated nucleotide analog. *J. Biol. Chem.* **264**, 12830–12837.

Dorbic, T., and Wittig, B. (1986). Isolation of oligonucleosomes from active chromatin using HGM17 specific monoclonal antibodies. *Nucleic Acids Res.* **14**, 3363–3376.

Dorbic, T., and Wittig, B. (1987). Chromatin from transcribed genes contains HMG17 only downstream from the starting point of transcription. *EMBO J.* **6**, 2393–2399.

Fronk, J., Tank, G. A., and Langmore, J. P. (1990). Chromatin structure of the developmentally regulated early histone genes of the sea urchin *Strongylocentrotus purpuratus*. *Nucleic Acids Res.* **18**, 5255–5263.

Germond, J. E., Bellard, M., Oudet, P., and Chambon, P. (1976). Stability of nucleosomes in native and reconstituted chromatin. *Nucleic Acids Res.* **3**, 3173–3193.

Glotov, B. O., Rudin, A. V., and Severin, E. S. (1982). Conditions for sliding of nucleosomes along DNA: SV40 minichromosomes. *Biochim. Biophys. Acta* **696**, 275–284.

Grabowski, P. J., and Sharp, P. A. (1986). Affinity chromatography of splicing compolexes: U2, U5, and U4 + U6 small nuclear ribonucleoprotein particles in the spliceosomes. *Science* **233**, 1294–1299.

Hanks, S. K., and Riggs, M. G. (1986). Selective insolubility of active hsp70 chromatin. *Biochem. Biophys. Acta* **867**, 124–134.

Harper, F., Florentin, Y., and Puvion, E. (1984). Localization of T-antigen on simian virus 40 minichromosomes by immunoelectron microscopy. *EMBO J.* **3**, 1235–1241.

Hebbes, T. R., Thorne, A. W., and Crane-Robinson, C. (1988). A direct link between core histone acetylation and transcriptionally active chromatin. *EMBO J.* **7**, 1395–1402.

Herman T. H., Lefever, and Shimkus, M. (1986). Affinity chromatography of DNA labeled with chemically cleavable biotinylated nucleotide analogs. *Anal. Biochem.* **156**, 48–55.

Hesse, J. E., Nickol, J. M., Lieber, M. R., and Felsenfeld, G. (1986). Regulated gene expression in transfected primary chicken erythrocytes. *Proc. Natl. Acad. Sci. U.S.A.* **83**, 4312–4316.

Ilyn, Y. V., Varshavsky, A. Y., Mickelsaar, U. N., and Georgiev, G. P. (1971). Studies on deoxyribonucleoprotein structure. Redistribution of proteins in mixtures of deoxyribonucleoproteins, DNA and RNA. *Eur. J. Biochem.* **22**, 235–245.

Jones, R. W. (1978). Histone composition of a chromatin fraction containing ribosomal deoxyribonucleic acid isolated from the macronucleus of *Tetrahymena pyriformis*. *Biochem. J.* **173**, 155–164.

Kedes, L. H. (1979). Histone genes and histone messengers. *Annu. Rev. Biochem.* **48**, 837–870.

Kinoshita, S. (1976). Properties of sea urchin chromatin as revealed by means of thermal denaturation. *Exp. Cell Res.* **102**, 153–161.

Kumar, S., and Leffak, M. (1986). Assembly of active chromatin. *Biochemistry* **25**, 2055–2060.

McGhee, D. J., Wood, W. I., Dolan, M., Engel, J. D., and Felsenfeld, G. (1981). A 200 base pair region at the 5' end of the chicken adult β-globin gene is accessible to nuclease digestion. *Cell (Cambridge, Mass.)* **27**, 45–55.

Maniatis, T., Fritsch, E. F., and Sambrook, J. (1982). "Molecular Cloning: A Laboratory Manual." Cold Spring Harbor Laboratory Cold Spring Harbor, New York.

Osheim, Y. N., Miller, O. L., Jr., and Beyer, A. L., (1985). RNP particles at splice junction sequences on *Drosophila* chorion transcripts. *Cell (Cambridge, Mass.)* **43**, 143–151.

Overton, G. C., and Weinberg, E. S. (1978). Length and sequence heterogeneity of the histone gene repeat unit of the sea urchin, *S. purpuratus. Cell (Cambridge, Mass.)* **14**, 247–257.

Peter, M., Nakagawa, J., Dorée, Labbé, J. C., and Nigg. E. A. (1990). *In vitro* disassembly of the nuclear lamina and M phase-specific phosphorylation of lamins by cdc2 kinase. *Cell (Cambridge, Mass.)* **61**, 591–602.

Prior, C. P., Cantor, C. R., Johnson, E. M., Littau, V. C., and Allfrey, V. G. (1983). Reversible changes in nucleosome structure and histone H3 accessibility in transcriptionally active and inactive states of rDNA chromatin. *Cell (Cambridge, Mass.)* **34,** 1033–1042.

Razin, S. V., Yarovaya, O. V., and Georgiev, G. P. (1985). Low ionic strength extraction of nuclease-treated nuclei destroys the attachment of transcriptionally active DNA to the nuclear skeleton. *Nucleic Acids Res.* **13,** 7427–7444.

Reynolds, W. F., Bloomer, L. S., and Gottesfeld, J. M. (1983). Control of 5S RNA transcription in *Xenopus* somatic cell chromatin: Activation with an oocyte extract. *Nucleic Acids Res.* **11,** 57–75.

Ridsdale, J. A., and Davie, J. R. (1987). Chicken erythrocyte polynucleosomes which are soluble at physiological ionic strength and contain linker histones are highly enriched in β-globin gene sequences. *Nucleic Acids. Res.* **15,** 1081–1096.

Riley, D., and Weintraub, H. (1978). Nucleosome DNA is digested to repeats of 10 bases by exonuclease III. *Cell (Cambridge, Mass.)* **13,** 281–293.

Rose, S. M., and Garrard, W. T. (1984). Differentiation-dependent chromatin alterations precede and accompany transcription of immunoglobulin light chain genes. *J. Biol. Chem.* **259,** 8534–8544.

Solomon, M. J., and Warshawski, A. (1985). Formaldehyde-mediated DNA-protein crosslinking: A probe for *in vivo* chromatin structures. *Proc. Natl. Acad. Sci. U.S.A.* **82,** 6470–6474.

Spadafora, C., Oudet, P., and Chambon, P. (1979). Rearrangement of chromatin structure induced by increasing ionic strength and temperature. *Eur. J. Biochem.* 100, 225–235.

Strätling, W. H. (1987). Gene-specific differences in the supranucleosomal organization of rat liver chromatin. *Biochemistry* **26,** 7893–7899.

Strätling, W. H., and Dölle, A. (1986). Chromatin structure of the chicken lysozyme gene domain as determined by chromatin fractionation and micrococcal nuclease digestion. *Biochemistry* **25,** 495–502.

Thoma, F., Koller, T., and Klug, A. (1979). Involvement of histone H1 in the organization of the nucleosome and of the salt-dependent superstructures of chromatin. *Cell Biol.* **83,** 403–427.

Tishler, P. V., and Epstein, C. J. (1968). A convenient method of preparing polyacrylamide gels for liquid scintillation spectrometry. *Anal. Biochem.* **22,** 89–98.

Vincenz, C., Fronk, J., Tank, G. A., and Langmore, J. P. (1991). Nucleoprotein hybridization: A method for isolating active and inactive genes as chromatin. *Nucleic Acids Res.* **19,** 1325–1326.

Ward, G. E., and Kirschner, M. W. (1990). Identification of cell cycle-regulated phosphorylation sites on nuclear lamin C. *Cell (Cambridge, Mass.)* **61,** 561–567.

Watkins, J. F., and Smerdon, M. J. (1985). Nucleosome rearrangement *in vitro*. 1. Two phases of salt-induced nucleosome migration in nuclei. *Biochemistry* **24,** 7279–7287.

Weinberg, E. S., Overton, G. C., Shutt, R. H., and Reeder, R. H. (1975). Histone gene arrangement in the sea urchin, *Strongylocentratus pupuratus*. *Proc. Natl. Acad. Sci. U.S.A.* **72,** 4815–4819.

Weinberg, E. S., Hendricks, M. B., Hemminki, K., Kuwabara, P. E., and Farrelly, L. A. (1983). Timing and rates of synthesis of early histone mRNA in the embryo of *Strongylocentrotus purpuratus*. *Dev. Biol.* **98,** 117–129.

Workman, J. L., and Langmore, J. P. (1985a). Nucleoprotein hybridization: A method for isolating specific genes as high molecular weight chromatin. *Biochemistry* **24,** 7486–7497.

Workman, J. L., and Langmore, J. P. (1985b). Efficient solubilization and partial purification of sea urchin histone genes as chromatin. *Biochemistry* **24,** 4731–4738.

Workman, J. L., and Roeder, R. G. (1987). Binding of transcription factor TFIID to the major late promoter during *in vitro* nucleosome assembly potentiates subsequent initiation by RNA polymerase II. *Cell (Cambridge, Mass.)* **51,** 613–622.

Workman, J. L., Abmayr, S. M., Cromlish, W. A., and Roeder, R. G. (1988). Transcriptional regulation by the immediate early protein of pseudorabies virus during *in vitro* nucleosome assembly. *Cell (Cambridge, Mass.)* **55,** 211–219.

Wu, T.-C., and Simpson, R. T. (1985). Transient alterations of the chromatin structure of sea urchin early histone genes during embryogenesis. *Nucleic Acids Res.* **13,** 6185–6203.

Yager, L. N., Kaumeyer, J. F., Lee, I., and Weinberg, E. S. (1987). Insertion of an intermediate repetitive sequence into a sea urchin histone-gene spacer. *J. Mol. Evol.* **24,** 346–356.

Chapter 14

Protein–DNA Cross-Linking as a Means to Determine the Distribution of Proteins on DNA in Vivo

DAVID S. GILMOUR

Department of Molecular and Cell Biology
Pennsylvania State University
University Park, Pennsylvania 16802

ANN E. ROUGVIE

Department of Cell and Developmental Biology
Harvard University
Cambridge, Massachusetts 02138

JOHN T. LIS

Section of Biochemistry, Cell and Molecular Biology
Cornell University
Ithaca, New York 14853

METHODS IN CELL BIOLOGY, VOL. 35

I. Introduction

The temporal and spatial distributions of particular proteins on specific DNA sequences can provide key insights into the regulation and mechanics of transcription and chromosome replication, as well as basic information on chromatin structure. Often, these distributions are inferred from a variety of biochemical and genetic analyses. Biochemical approaches can provide detailed and quantitative information on the binding of proteins to DNA *in vitro*, but they do not permit the examination of these interactions in living cells. In contrast, genetic approaches identify functional cis-acting DNA sequence elements and trans-acting factors, but do not permit the direct and quantative analysis of interactions between these components. More recently, *in vivo* footprinting has provided a means of examining biochemical interactions of proteins on DNA in intact cells and nuclei; however, the method does not identify the protein responsible for a particular footprint. Here, we describe a method for directly mapping the distribution of a protein on DNA in living cells by ultraviolet (UV)-induced cross-linking (Gilmour and Lis, 1984, 1985). UV irradiation of cells cross-links protein to DNA at points of contact. The protein–DNA adducts are purified from cells, and then the protein of interest is immunoprecipitated with an appropriate antibody. The distribution of the protein on DNA in the cell is revealed by identifying specific restriction fragments that coprecipitate with the protein via their UV-induced covalent attachment to the protein.

The UV-cross-linking method has provided a unique opportunity to evaluate mechanistic questions *in vivo*. First, measurements of the density of RNA polymerase along the length of the *Salmonella typhimurium* leucine operon in wild-type cells and in an attenuator mutant strain by UV cross-linking tested predictions of the attenuation model of transcriptional regulation (Gilmour and Lis, 1984). Second, the method was used to demonstrate that *Drosophila* topoisomerase I is recruited during heat shock to activated transcription units and not to sequences flanking these transcribed sequences (Gilmour *et al.*, 1986). Third, the method provided a direct measure of the relative *in vivo* density of RNA polymerase II on a variety of *Drosophila* genes (Gilmour and Lis, 1985). Even the density of RNA polymerase on different closely related members of a gene family could be assessed by determining the amount of cross-linking of RNA polymerase to the different restriction fragments that contain different family members (Gilmour and Lis, 1985, 1987). Finally, the analysis of the distribution of RNA polymerase within the uninduced *Drosophila hsp70* gene identified a higher concentration of RNA polymerase at its 5′ end than on the body of the gene (Gilmour and Lis, 1986). More recently, this distribution has also been observed on a variety of other *Drosophila* genes (Rougvie and Lis, 1990). This suggested that RNA polymerase II has ready access to certain promoters, but appears to be impeded in its ability to progress beyond the 5′ region of these genes. Although nuclear run-on assays provided an independent means of showing the high concentration of RNA polymerase on the 5′ end of these genes in isolated nuclei (Rougvie and Lis, 1988, 1990), the ability to cross-link these polymerases to DNA in intact cells using UV light indicates that the association of polymerase with the 5′ ends of genes is not occurring artifactually during the purification of nuclei.

Two conditions must be met to use the *in vivo* cross-linking technique. First, it is essential to have an antibody that binds with sufficient affinity so that the immune complexes can survive stringent washing of the immunoprecipitate. These washes are required to reduced the amount of nonspecifically associated DNA. Second, the DNA-binding protein must cross-link to DNA during the UV irradiation. The efficiency with which a particular protein cross-links to DNA *in vivo* is difficult to predict, but we expect that simple *in vitro* tests of the cross-linking of the protein of interest and to its radioactive DNA target may provide a good indication of the *in vivo* efficiency. We have recently described one such *in vitro* analysis of a TATA-dependent complex that forms on the hsp70 gene promoter (Gilmour *et al.,* 1990). The efficiency of cross-linking protein to DNA could be enhanced by incorporating modified bases, such as 5-bromodeoxyuridine, into DNA (Lin and Riggs, 1974), although one must consider that the introduction of these analogs into cellular DNA may alter the protein–DNA complexes being examined.

Our *in vivo* cross-linking studies to date indicate that the technique has broad applicability. We have successfully used it to examine RNA polymerases in bacteria (Gilmour and Lis, 1984), *Drosophila* cells in culture (Gilmour and Lis, 1985, 1986), *Drosophila* embryos (Rougvie, 1989), and mammalian cells (D. Gilmour, unpublished results). We have also used the technique to localize topoisomerase I (Gilmour *et al.,* 1986), histone H1 (Gilmour and Lis, 1987), and a protein recruited to heat-shock puffs, B52 (Champlin *et al.,* 1991), on genes in *Drosophila* cells. It is important to note, however, that the efficiency of cross-linking is quite low. We estimate that approximately 0.1% of the RNA polymerase II 215-kDa subunit that is associated with the actively transcribed *hsp70* genes becomes covalently attached to the DNA following a 10-minute irradiation. Nevertheless, the method can easily detect the presence of a single molecule of RNA polymerase on a single-copy gene of *Drosophila*.

The following cross-linking protocol is for *Drosophila* cells. By taking into consideration cell numbers and genome sizes, the method should be easily adaptable to other systems. When testing a new application, consider using RNA polymerase II as a test case, since, in most cases, its distribution on DNA can be predicted from transcription measurements.

II.   Ultraviolet Cross-Linking Protocol for *Drosophila* Cells

A.   *Drosophila* Schneider II Cell Line Culture

1. We culture the *Drosophila* Schneider II cell line at 22°C in Sheilds and Sang media (Sang, 1981) containing the following (in grams/liter):

1.5	L-a-Alanine	0.25	Oxalacetic acid
0.25	L-b-Alanine	0.25	L-Phenylalanine
0.6	L-Arginine-HCl	0.4	L-Proline
0.3	L-Asparginine	0.36	L-Serine
0.3	L-Aspartic acid	0.5	L-Threonine
0.2	L-Cysteine-HCl	0.25	L-Tyrosine
0.6	L-Glutamine	0.1	L-Tryptophan
7.2	Potassium glutamate	0.4	L-Valine
6.6	Sodium glutamate	10	Glucose
0.5	L-Glycine	2.5	Bactopeptone
0.55	L-Histidine	2.0	Yeastolate
0.25	L-Isoleucine	0.76	$CaCl_2$
0.4	L-Leucine	2.16	$MgSO_4$
0.86	L-Lysine-HCl	0.78	$NaH_2PO_4 \cdot H_2O$
0.25	L-Methionine	1.06	Bis-Tris

Add the above mix to 760 ml of glass-distilled H_2O and stir for 30 minutes. Add 50 mg of choline and 0.5 g of $KHCO_3$. Adjust the pH of the solution to 6.8 with NaOH, and bring the volume to 1 liter. Remove undissolved particles by passing the medium through Schleichter and Schuell 24-cm prepleated filters. The media is filter-sterilized and stored at 4°C in 450-ml aliquots. Just before use, supplement with 10 mM HEPES, pH 6.8, and with 10% fetal calf serum, which has been heat-inactivated by incubation at 65°C for 30 minutes. We expect that other cell lines and culture media can also be used (see Section III).

The cells are maintained in T-flasks and diluted 1/7 (to an approximate density of 2×10^6 cells/ml) once every 5 to 7 days. For growth in spinner flasks, the cells are diluted 1/7 and grown to density of 7×10^6 to 1×10^7 cells/ml. RNA polymerase II can easily be detected on the transcriptionally active *hsp70* gene in as few as 10^6 heat-shocked cells.

2. Cells are heat shocked by circulating water from a 36.5°C water-bath through the jacket of a spinner flask for 20 minutes.

B.   Irradiation

1. Cells in culture medium are placed in a dish of a size that the depth of the solution does not exceed 0.5 cm. This dish is chilled by placing it in a second dish containing ice. Since most of our studies have been directed to the heat-shock genes, we have chosen to irradiate the cells in the medium in which they are growing so as to avoid inadvertently inducing the heat-shock genes. There is significant absorbance of the UV light by the medium, since at least 10-fold more damage can be inflicted on the DNA when the cells are irradiated in phosphate-buffered saline than when they are irradiated in culture medium (G. Thomas, personal communication). Consequently, a particular dosage of UV irradiation will probably produce more cross-links in cells suspended in phosphate-buffered saline than in culture medium.

2. The filter cover of a short-wavelength UV transilluminator (Chromato-Vu transilluminator, Model C-61, UV Products) is removed to expose the bulbs. These bulbs should be clear in appearance (UV products # 340008 01); bulbs producing long-wavelength UV light appear white and do not efficiently produce protein–DNA cross-links. As an alternative to the transilluminator, we have also used a shop-made apparatus consisting simply of a bank of six 15-W mercury bulbs mounted in an open-ended box. We recommended that the bulb box be of stainless steel or wood lined with aluminum foil to reflect light toward the open end. The transilluminator or bulb box is inverted and supported so that the bulbs are approximately 10 cm from the surface of the cells. The cells are irradiated for 10 minutes. Since UV light of this intensity is dangerous, one should perform the irradiation in a shielded area or darkroom and use eye protection.

Alternatively, cells can be irradiated with a Xenon Model 457 Micropulsar, equipped with a Xenon Corp 815D Xenon flash lamp (Gilmour and Lis, 1986). We typically use a setting of 12,000 V. A trigger wire is attached to the cathode end of the lamp and then extended along the length of the lamp. A 60-mH inductor is placed in series with the lamp, providing sufficient inductance so the lamp does not explode at these power levels. This setup allows 40-μsec pulses of UV light to be delivered to the sample, making kinetic measurements of protein recruitment to specific segments of DNA feasible.

3. Following the irradiation, the cells are transferred to a chilled centrifuge tube. Cells adhering to the bottom of the dish can be easily removed by vigorously pipetting medium on them.

4. Collect cells by centrifugation at 2000 g in a clinical centrifuge for 5 minutes at 4°C.

C. Isolation of DNA and Protein–DNA Adducts

1. Resuspend cells at no more than 10^8 cells/ml in cold 100 mM KCl, 50 mM NaCl, 5 mM MgCl$_2$, 10 mM Tris-Cl, pH 7.4, 0.5 mM PMSF (phenyl-methylsulfonyl fluoride, freshly added from a 100 mM stock dissolved in isopropanol and stored at 4°C).

2. Lyse cells by the addition of 0.1 volume of 10% NP40. Incubate on ice for 5 minutes with intermittent vortexing.

3. Collect nuclei by centrifugation at 2000 g for 10 minutes at 4°C. Discard the supernatant.

4. Resuspend the nuclei from the *Drosophila* cells in a volume of nuclear suspension buffer [100 mM NaCl, 10 mM Tris-Cl, pH 8.0, 1 mM EDTA, 0.5 mM PMSF (add fresh), 0.1% NP40] as indicated in Table I, and then lyse nuclei with the addition of 0.1 volume of 20% Sarkosyl.

5. Shear the DNA with five passages through a 20-gauge needle followed by five passages through a 26-gauge needle. Alternatively, the DNA may be sheared using a Brinkman Polytron or by very vigorous vortexing. Remove insoluble nuclear debris by centrifugation at 2000 g for 10 minutes.

6. Overlay the nuclear lysate on CsCl step gradients prepared as described in Table I. The CsCl solutions also contain 0.5% Sarkosyl and 1 mM EDTA. Centrifuge the preparations for the times indicated in Table I. Since only a small portion of protein cross-links to the DNA, protein–DNA adducts copurify with the chromosomal DNA in the CsCl gradient, whereas free protein remains at the top of the gradient.

7. After ultracentrifugation, collect fractions by puncturing the bottom of the tube with an 18-gauge needle or by carefully inserting a glass capillary tube from the top down to the bottom of the tube. Locate DNA-containing fractions by running ~ 3-μl aliquots on a 0.7% agarose/Tris acetate–EDTA gel containing 0.5 μg/ml ethidium bromide. The CsCl will disrupt the dye front, but DNA can easily be located after running the gel for 30–60 minutes. This also allows an assessment of potential degradation of DNA as well as indicating the location of RNA.

8. Pool the DNA-containing fractions and dialyze exhaustively at 4°C (24 hours) against three changes of 0.2% Sarkosyl, 50 mM Tris-Cl, pH 8.0, 2 mM EDTA. Sarkosyl is included in this buffer to ensure efficient recovery of protein–DNA adducts.

TABLE I

GUIDE FOR PROCESSING CROSS-LINKED *Drosophila* CELLS

Number of cells	$<3 \times 10^8$	$3–5 \times 10^9$	$>5 \times 10^9$
Nuclear lysis			
Nuclear suspension buffer	0.9 ml	2.7 ml	9.0 ml
20% Sarkosyl	0.1 ml	0.3 ml	1.0 ml
Ultracentrifugation			
Rotor size (Beckman)	SW60	SW41	SW28
CsCl steps			
1.75 g/cm^3	1.3 ml	4.5 ml	18.5 ml
1.50 g/cm^3	0.9 ml	2.3 ml	6.0 ml
1.30 g/cm^3	0.7 ml	1.5 ml	3.5 ml
Speed (rpm)/time (hr at 20°C)	30K/~20	37K/~24	26K/30
	(93,000 g)	(167,000 g)	(89,000 g)

CsCl preparation

Since CsCl is very hydroscopic, it is necessary to check the density by refractive index. To prepare 100 ml of the CsCl stock solutions:

Density (g/cm^3)	CsCl (g)	SE^a (ml)	Refractive index
1.75	100	75	1.404
1.50	66.7	83.3	1.381
1.30	40.0	90	1.363

^{a} Adjust SE solution so that the final concentrations of the 100 ml of CsCl stock contains 0.5% Sarkosyl, 1 mM EDTA, pH 8.0.

9. Protein–DNA adducts can be stored frozen at $-70°C$ until use. The adducts are stable for at least 6 months.

D. Restriction Enzyme Digestion of DNA

The presence of Sarkosyl ensures an efficient recovery of protein–DNA adducts but also inhibits DNA cutting by restriction enzymes. Many restriction enzymes have been found to function in 0.2% Sarkosyl upon addition of Triton X-100 (or NP40) to 1%. Presumably, the Triton X-100 reduces the concentration of monomeric Sarkosyl by forming mixed micelles. We have also successfully performed the restriction digests and immunoprecipitations with the level of Sarkosyl reduced to 0.1% and the level of NP40 at 0.5%. A systematic analysis of whether or not these lower levels of detergents improve the efficiency of the restriction enzymes has not been carried out.

1. DNA samples should be tested for their ability to be digested with selected restriction enzymes by performing small-scale pilot digestions. Typically, the concentration of DNA in the digestion mixture is approximately

25 μg/ml. For *Drosophila* cells, this corresponds to cutting DNA from 5×10^7 cells in a volume of 2 ml. Use buffers as recommended by the supplier, but supplemented to 0.2% Sarkosyl and 1% Triton-X 100. Dilution of DNA ~ 2-fold may be necessary to achieve efficient cutting.

2. Incubate samples at 37°C for 10 minutes prior to enzyme addition. Multiple additions of enzyme result in more complete digestion. The completeness of digestion can be checked on minigels. As a general guide for the amount of enzyme to use, 1 μg of DNA is digested with enough enzyme so that the units of enzyme multiplied by the hours of digestion at 37°C is equal to at least five.

3. Stop reactions by adding 0.1 volume of 0.2 M EDTA, pH 8.0.

4. Centrifuge reactions for 10 minutes in a microfuge to remove any particulates.

5. Save 5–10% of the DNA sample to represent the input DNA (Totals) for use in quantifying the DNA recovered from the immunoprecipitate.

E.   Immunoprecipitation

1. Add antibody to the above restriction enzyme-digested DNA (minus the Totals). Polyclonal antibodies may result in better recovery; however, we have successfully used monoclonal antibodies in this procedure. Each antibody must be titered individually, but for a rough guideline, we use 0.5 μl of a polyclonal rabbit RNA polymerase II antiserum for preparations from 10^7 *Drosophila* cells. Keep in mind that most of the protein has been removed by the CsCl gradient, so that only small quantities of antibody are likely to be required to achieve maximum precipitation. Incubate the antibody with the sample for at least 3 hours on ice.

2. Collect the needed volume of formalin-fixed Staph A (Calbiochem; stored at 4°C or in aliquots at -70°C) by centrifugation for 60 seconds in an Eppendorf microfuge. Resuspend the Staph A in 5 volumes of 0.2% Sarkosyl, 50 mM Tris-Cl, pH 8.0, 2 mM EDTA. Add 15 μl of this Staph A suspension (this is an excess) per 0.5 μl of polyclonal antibody. Incubate at 4°C for at least 1 hour on a rocker platform. Some antibodies, for instance certain mouse monoclonals, are not efficiently recognized by Staph A. Thus, incubation with a second antibody (i.e., rabbit anti-mouse heavy- and light-chain serum for mouse monoclonals) before addition of Staph A may be necessary. This is done by incubating the sample with the primary antibody for ~ 3 hours and then adding the secondary antibody (~ 2-fold excess over the first) and incubating for an additional hour.

3. Collect immune complexes with a 60-second centrifugation in an Eppendorf microfuge (if the sample is too large, the centrifugation can be done in a Sorval and the sample transferred to an Eppendorf tube for washes).

Wash pellet 3 times at 4°C by resuspending completely in 500 μl of cold 0.2% Sarkosyl, 50 mM Tris-Cl, pH 8.0, 2mM EDTA, mixing well, and pelleting for 60 seconds. Wash 5–7 additional times in 100 mM Tris-Cl, pH 9.0, 0.5 M LiCl, 1% NP40, 1% deoxycholate. Binding of antibodies is pH dependent, so it may be necessary to lower the pH of the wash buffers to 8.0 for some monoclonal antibodies.

4. Elute complexes from Staph A by resuspending the pellet in 100–200 μl 50 mM Tris·Cl pH 8.5, 2 mM EDTA, 1% SDS, 1.5 μg/ml sonicated carrier DNA (or 20 μg/ml glycogen, see Section IV, below) and by shaking for 10 minutes on a vortex shaker at room temperature. Spin for 60 seconds in a microfuge and transfer the supernatant to a new tube. Repeat the elution 3 additional times and pool the supernatants.

5. Precipitate DNA from the eluates by adding 0.5 volume of 7.5 M ammonium acetate and 2 volumes of ethanol. At this point, also precipitate DNA from the small portion of material (the Totals) that was set aside prior to the addition of antibody; process this in parallel with the immunoprecipitated samples for the remaining steps. Precipitate samples at −20°C for at least 2 hours. Collect DNA by spinning samples in a microfuge for 15 minutes at 4°C. Dry the DNA and resuspend in 200–400 μl of 0.5% SDS, 10 mM Tris-Cl, pH 8.0, 10 mM EDTA.

6. Add RNase A to 100 μg/ml and incubate at 37°C for 1 hour.

7. Add proteinase K to 50 μg/ml and incubate at 60°C for at least 5 hours. Precipitate the DNA with ethanol and ammonium acetate as described above.

8. Resuspend dried DNA in DNA loading buffer.

F. Gel Electrophoresis of DNA Samples

1. Prepare 1% agarose gel with 40 mM Tris-acetate, 1mM EDTA buffer (Maniatis *et al.*, 1982). Tris–borate–EDTA buffer (Maniatis *et al.*, 1982) is not recommended because the DNA in the immunoprecipitated samples (presumably containing residual cross-linked peptide) migrates more slowly than the DNA in the Total fractions under these conditions; this variation is not observed with Tris–acetate–EDTA buffer.

2. Load a dilution series (2–4 lanes) in the range of 0.2 to 2% of the DNA removed prior to the addition of antibody (Totals) along side the immunoprecipitated samples.

3. Following electrophoresis, stain the gel with 1 μg/ml of ethidium bromide and photograph.

G. Blotting and Hybridization

1. Soak the gel in 0.06 M HCl for 15 minutes, soak in 0.4 N NaOH until dye turns blue again, and then blot to Gene Screen Plus (Du Pont) in 0.4 N NaOH.

After transfer is complete (~4–6 hours), neutralize the filter by soaking in 0.2 *M* Tris (pH 7.5)–2 × SSC until the tracking dyes return to blue.

2. Soak DNA blot for at least 15 minutes at 42°C in hybridization mix (6 × SSC, 50% formamide, 1% SDS, 10% dextran sulfate, 100 μg/ml carrier DNA). Dilute denatured probe into a small amount of hybridization mix and add directly to the remaining hybridization mix that surrounds the DNA blot. Hybridization probes, produced either by nick-translation or by random priming (Feinberg and Vogelstein, 1984), generally must have specific activities of between 10^8 and 10^9 cpm/μg. Continue hybridization at 42°C overnight. Wash three times for 10 minutes each in 2 × SSC, 0.5% SDS at room temperature, followed by three times 15 minutes each in 0.1 × SSC, 0.1% SDS at 55°C. Expose the blot at −70°C to preflashed Kodak XAR-5 film using an intensifying screen.

H.   Data Analysis

The autoradiograph resulting from a typical cross-linking experiment is shown in Fig. 1. To determine the distribution of a protein on DNA, the

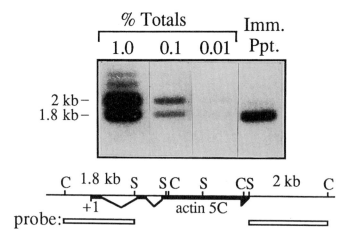

FIG. 1.   Analysis of the distribution of RNA polymerase II on the *actin 5C* gene. DNA and protein–DNA complexes purified from 1 × 10^8 *Drosophila* Schneider line II cells were restriction digested with *Sal*I and *Cla*I, and a portion removed to serve as Totals (lanes 1.0, 0.1, and 0.01%). RNA polymerase II–DNA complexes were immunoprecipitated from the remainder (Imm. Ppt.) and analyzed by Southern analysis along with the Totals. The autoradiogram of the filter probed with a mixture of two *actin 5C* gene fragments is shown, with a restriction map of the *actin 5C* gene (indicated by long black line) below. There is likely an additional *Sal*I site very close to the *Cla*I site at the 3′ end of the gene. Appropriate exposures of the filter can be quantified by densitometry comparing the amount of DNA present in the immunoprecipitate track to the amount of the same fragment present in the Totals.

hybridization signal of a given band in the immunoprecipitate lane is compared relative to the signal of the same band in the input DNA present in the Total lanes. Thus, the relative density of a protein across many DNA fragments may be assessed. Such a filter may be reprobed multiple times to examine the distribution of the protein on additional DNA segments.

III.   Alternative Applications

A.   Application of the Cross-Linking Technique to Cultured Mouse Fibroblasts

Grow cells in monolayer in several petri dishes or in one 12-inch square disposable tissue culture dish until dense but subconfluent. Remove media, rinse cells once in cold phosphate-bufferred saline (PBS), and cover with ~ 0.5 cm of cold PBS. Irradiate the cells for 5 minutes. Harvest the cells by scraping them with a rubber policeman into a Dounce homogenizer. Lyse cells by adding NP40 to 1% and homogenizing with several strokes of the pestle. Spin out nuclei and continue as described for the *Drosophila* cells.

B.   Application to *Drosophila* Embryos

Collect embryos and suspend in cold PBS on ice. Irradiate embryos for 10 minutes in a glass petri plates on ice such that the final depth of PBS does not exceed 0.5 cm. Use no more than a monolayer of settled embryos, and agitate embryos once during the irradiation. Collect embryos by filtration onto Nitex (nylon monofilament bolting cloth, 100–μm mesh, Tetko, Inc.). Dechorionate embryos in 50% Chlorox, 0.5% Triton X-100, 0.7% NaCl for 2 minutes. Collect embryos on Nitex and wash with cold PBS. Transfer embryos to Teflon/glass homogenizer in 10 ml of solution A (0.5 M sucrose, 10 mM Tris-Cl, pH 8.0, 0.5 mM EGTA, 0.1 mM EDTA, 0.5 mM DTT, 2 mM magnesium acetate, 0.5 mM fresh PMSF) and homogenize by 10 strokes with a tight-fitting pestle. Do not allow the embryos to bunch up in front of the pestle: glass homogenizers have been observed to shatter under high pressure. Filter once through Miracloth (CalBiochem), add an equal volume of solution B (1.5 M sucrose, 10 mM Tris-Cl, pH 8.0, 5 mM magnesium acetate, 0.5 mM DTT, 0.5 mM PMSF), mix, and spin through a 10-ml cushion of solution B in a swinging bucket SW28 rotor at 11,000 rpm (16,000 g) for 20 minutes. Resuspend pellet in nuclear lysis buffer and proceed as described for cultured cells.

IV. Additional Notes

1. Not all enzymes seem to work well in the presence of Sarkosyl and Triton-X 100. Those enzymes that have consistently worked well are *Ava*I, *Ava*II, *Bam*HI, *Bgl*I, *Bst*NI, *Cla*I, *Dde*I, *Eco*RI, *Eco*RV, *Pvu*II, *Xho*I, *Stu*I, *Sal*I, *Hin*dIII, *Hin*fI, *Hae*III, and *Sau*3AI. Digestion by *Pst*I and *Xba*I has been somewhat variable, while *Hha*I, *Hin*cII, and *Hae*II have not worked. Most companies sell a variety of enzymes at high concentration that are convenient for use in cross-linking experiments.

2. Glycogen (as described by Boehringer Mannheim in BMBiochemica Vol. 3, #2 April 1986) has been successfully used instead of sonicated DNA as the carrier in the immunoprecipitation step (see Section II,E):

    2a. Add LiCl to 0.4 *M*.

    2b. Add glycogen (cat. no. 901 393) to 20–40 μg/ml.

    2c. Add 2.5 volumes of cold ethanol and incubate at $-20°C$ for 2–3 hours.

    2d. Collect pellet by centrifugation at 12,000 rpm for 15 minutes in the microfuge. Remove the supernatant and rinse the pellet in 70% ethanol.

    2e. The glycogen dissolves readily in 10 m*M* Tris-Cl, pH 8.0, 1m*M* EDTA and does not seem to interfere with restriction digests or with gel electrophoresis.

3. Protein A Sepharose does not efficiently recover large IgG–protein–DNA complexes, and, if possible, formalin-fixed Staph A cells should be used to recover protein–DNA adducts that have been fragmented with restriction enzymes. Protein A Sepharose may be used, if the DNA has been sonicated to a small size [see Methods of Gilmour and Lis (1985) for discussion].

REFERENCES

Champlin, D. T., Frasch, M., Saumweber, H., and Lis, J. T. (1991). Characterization of a *Drosophila* protein associated with boundaries of transcriptionally active chromatin. *Genes Dev.* (in press).

Feinberg, A. P., and Vogelstein, B. (1984). A technique for radiolabeling DNA restriction endonuclease fragments to high specific activity. *Anal. Biochem.* **137,** 266–267.

Gilmour, D. S., and Lis, J. T. (1984). Detecting protein–DNA interactions *in vivo*: Distribution of RNA polymerase on specific bacterial genes. *Proc. Natl. Acad. Sci. U.S.A.* **81,** 4275–4279.

Gilmour, D. S., and Lis, J. T. (1985). *In vivo* interactions of RNA polymerase II with genes of *Drosophila melanogaster*. *Mol. Cell. Biol.* **5,** 2009–2018.

Gilmour, D. S., and Lis, J. T. (1986). RNA polymerase II interacts with the promoter region of the noninduced hsp70 gene in *Drosophila melanogaster* cells. *Mol. Cell. Biol.* **6,** 3984–3989.

Gilmour, D. S., and Lis, J. T. (1987). Protein-DNA crosslinking reveals dramatic variation in RNA polymerase II density on different histone repeats of *Drosophila*. *Mol. Cell. Biol.* **7,** 3341–3344.

Gilmour, D. S., Pflugfelder, G., Wang, J. C., and Lis, J. T. (1986). Topoisomerase I interacts with transcribed regions in *Drosophila* cells. *Cell (Cambridge, Mass.)* **44,** 401–407.

Gilmour, D. S., Dietz, T. J., and Elgin, S. C. R. (1990). UV-Crosslinking identifies four polypeptides that require the TATA box to bind to the hsp70 promoter. *Mol. Cell. Biol.* **10,** 4233–4238.

Lin, S.-Y., and Riggs, A. D. (1974). Photochemical attachment of *lac* repressor to bromodeoxyuridine-substituted *lac* operator by ultraviolet radiation. *Proc. Natl. Acad. Sci. U.S.A.* **71,** 947–951.

Maniatis, T., Fritsch, E. F., and Sambrook, J. (1982). "Molecular Cloning: A Laboratory Manual." Cold Spring Harbor Laboratory, Cold Spring Harbor, New York.

Rougvie, A. E. (1989). Ph.D. Thesis, Cornell University, Ithaca, New York.

Rougvie, A. E., and Lis, J. T. (1988). The RNA polymerase II molecule at the 5' end of the uninduced hsp70 gene of *D. melanogaster* is transcriptionally engaged. *Cell (Cambridge, Mass.)* **54,** 795–804.

Rougvie, A. E., and Lis, J. T. (1990). Post-initiation transcriptional control in *Drosophila melanogaster*. *Mol. Cell. Biol.* **10,** 6041–6045.

Sang, J. H. (1981). *Drosophila* cells and cell lines. *In* "Advances in Cell Culture" (K. Maramorosch, ed.), Vol. 1, pp. 125–182. Academic Press, New York.

Chapter 15

Protein–DNA Interactions in Vivo—Examining Genes in Saccharomyces cerevisiae and Drosophila melanogaster by Chromatin Footprinting

MELISSA W. HULL

Department of Biological Chemistry
University of Michigan
Ann Arbor, Michigan 48109

GRAHAM THOMAS

Department of Cellular and Developmental Biology
Harvard University
Cambridge, Massachusetts 02138

JON M. HUIBREGTSE

Laboratory of Tumor Virus Biology
National Cancer Institute
National Institutes of Health
Bethesda, Maryland 20892

DAVID R. ENGELKE

Department of Biological Chemistry
University of Michigan
Ann Arbor, Michigan 48109

METHODS IN CELL BIOLOGY, VOL. 35

I. Introduction

Interactions of proteins with chromosomal DNA control a variety of cellular processes including gene transcription, DNA packing, replication, recombination, and DNA repair. Early work in elucidating these processes depended on classical genetics to identify the protein factors and regulatory DNA sequences, and on relatively low resolution physical methods to characterize the nucleoprotein structures after release from the cell. In recent years, however, there has been a proliferation of tools with which to study DNA–protein interactions at high resolution *in vitro*. Specific cloned DNA sequences can be assembled with proteins isolated from cellular extracts to provide details on the binding of each factor as well as its effect on the overall structure and activity of the complex. Mutations engineered into both the DNA target and the protein factors provide further means by which to test initial hypotheses.

One high-resolution method of mapping DNA–protein contacts that can be used with both reconstituted (Tulluis, 1989) and cellular assemblies (Becker and Schutz, 1988) is DNA footprinting. This technique was originally designed to locate the binding sites of purified proteins on end-labeled DNA fragments by determining which DNA base pairs become selectively resistant to DNA cleavage reagents. To perform *in vitro* DNase footprinting, DNA with and without protein bound is partially digested at less than 1 nick per molecule, and then electrophoresed on a denaturing polyacrylamide gel to generate a ladder of bands that will be missing rungs where protein protects specific DNA sequences. It was quickly realized that this method could also be used to examine multicomponent assemblies even in situations where the individual protein components were not highly purified.

The strength of *in vitro* footprinting is that it provides a high-resolution contact map with protein components that can be defined biochemically and assembled stepwise. However, the significance of these complexes must be independently established. Inaccuracies in *in vitro* models can occur because it is difficult to reproduce *in vivo* DNA–protein interactions correctly. Direct *in vivo* examination of the cellular DNA–protein assemblies provides an evaluation of the fidelity of *in vitro* complexes. In addition, it is often possible to take DNA target or protein factor mutations and compare their effects on the relevant process *in vivo* with their effects on protein binding both *in vitro* and *in vivo*.

The utility of *in vivo* footprinting alone is also limited, primarily because of the investigator's inability to manipulate and characterize what is bound into the DNA complex at any given time. The inability to add components one at a time often means that chromosomal footprinting patterns are very complex. Even in cases where it is possible to control the level and activity of trans-acting regulatory proteins *in vivo*, it is not obvious whether any correlated changes in chromatin structure are the result of direct binding of the proteins or indirect influences on complex formation. Thus, chromatin footprinting is most effective when used in parallel with biochemical studies as a predictive or confirmatory tool.

A number of different approaches to chromosomal footprinting have been developed with a common overall strategy. The cellular chromatin is first attacked with DNA cleavage or modification reagents, followed by deproteinization and DNA isolation. Only then is a detection label, usually radioactive, placed at a single site on the genome near the region of interest and the distance between the cleavage/modification sites and the labeled site determined. Identical methods can also be used to characterize *in vitro* complexes in parallel experiments. Although generally more difficult, the postlabeling techniques allow *in vitro* experiments not possible using end-labeled liner DNA (e.g., the effects of topology on protein–DNA complexes) and allow direct visual comparisons between chromosomal and reconstituted complexes. The best choice among the available methods for any particular application will depend on a number of factors, particularly the size of the genome in the organism under study. The next section briefly surveys several viable alternatives and discusses their strengths and weaknesses. The last two sections give detailed protocols for approaches used to map DNase I sensitivity in *Saccharomyces cerevisiae* and methidium propyl-EDTA.Fe (II) (MPE.Fe [II]) sensitivity in *Drosophila melanogaster*.

II.   Technical Considerations

A.   Alternatives to DNA Footprinting

1.   REGION-SPECIFIC MICROCOCCAL NUCLEASE OR DNASE SENSITIVITY

Examination of the nuclease sensitivity of chromatin has taken many forms. Some methods ask what the general sensitivity of a region is to DNases by digesting nuclei and examining the DNA released into the soluble fraction versus the DNA fraction retained in the nuclei (Weintraub and Groudine, 1976). A similar approach that gives more information about the general architecture of a region is to digest partially with micrococcal nuclease, which hits preferentially between, rather than within, nucleosomes (Gottesfeld and Bloomer, 1980; Nedospasov and Georgiev, 1980; Wu *et al.,* 1979b). Electrophoretic separation of the released DNA on agarose followed by blotting and probing with a labeled probe to an entire region will reveal the presence and spacing of nucleosomes by the appearance of a "ladder" of bands corresponding to mono-, di-, tri-, and higher order nucleosome arrays. This approach has the advantage that it finds chromatin structural features that are not dependent on the precise position of the nucleosome, but rather are region-specific and might be missed using position-specific techniques like footprinting alone.

2.   ENDONUCLEASE HYPERSENSITIVITY

A disadvantage of footprinting is that the results are negative in character; the vast majority of DNA molecules must have the protein bound at a specific site in order to see an absence of bands (i.e., a footprint). Low-resolution methods used to detect position-specific hypersensitivity have also been available for some time (Weintraub, 1985). These techniques have the advantage that they detect a positive signal. Sites or small regions that are unusually sensitive to the single-strand-specific nuclease S1 (Larsen and Weintraub, 1982), DNase I (Wu *et al.,* 1979a; Elgin, 1988), or a restriction endonuclease (McGhee *et al.,* 1981) have often been correlated to physiological change in the activity of the DNA, such as tissue-specific induction of a transcription promoter. In nuclei these can be localized within a region by several methods that measure the distance from the cleavage site to a fixed point on the nearby DNA, where a detection label is specifically annealed after the DNA is isolated. The advantages of these methods derive from the fact that they detect the appearance of a signal corresponding to increased cleavage at one or a cluster of sites. This both makes signal detection at single sites on large, complex genomes more sensitive and allows the detection of a "positive" signal in cases where the change is present only a small proportion of the time or in a minority of the cell population, on average.

3. EXONUCLEASE III FOOTPRINTING

Another technique that generates a positive signal is exonuclease footprinting (Wu, 1984). Chromatin is digested to one side of the region of interest with a restriction endonuclease and then a processive exonuclease is added. The exonuclease digests toward the region of interest until it is blocked, presumably by a stably bound protein. Using an indirect end-labeled probe (see below) from the other side of the region of interest, the blocked position appears as a new product not seen after similar treatment of deproteinized DNA. This method is more sensitive than the footprinting described below and is able to detect protein occupation on a low percentage of templates. However, it detects only the outer borders of protein complexes where exonuclease progression is first strongly stopped and requires a convenient nearby restriction endonuclease site that is itself fully exposed in the chromatin structure.

B. DNA Footprinting

To provide both single nucleotide resolution and improved interpretation of the nucleoprotein structure, several methods of DNA footprinting have been developed that can be used either in whole cells or nuclei, and can be directly compared to DNA footprinting results using cloned DNA and defined protein fractions. Alternative footprinting protocols can be sorted by two main variables: the reagent used to attack the exposed DNA and the method used to detect the cleavage sites when the resulting DNA is isolated. When choosing a modification or cleavage reagent, there are three criteria to be kept in mind: the degree to which the attacking methods perturb the physiological structure originally present in the cell, the degree of resolution and type of information desired from the footprint, and the technical difficulty and reproducibility of the methods.

1. CLEAVAGE REAGENTS

DNase I was the first reagent used in footprinting *in vitro* and is commonly used as a reagent for chromatin. We typically use DNase I because the combined steric radii of the DNase and the chromatin proteins give large, obvious footprints where access to the DNA is limited by tightly bound proteins. In cases where it is not clear in advance precisely what interactions are being examined, obvious signals are a distinct advantage. Alternatively, any cleavage reagent used for footprinting of reconstituted complexes can also be used on chromatin after cell lysis. Smaller, chemical reagents that penetrate closer to the actual DNA–protein contacts will give considerably greater resolution and should be considered for elucidating finer points. For example, dimethyl sulfate (DMS) (Giniger et al., 1985; Palmieri, and Tavey, 1990) probes protein–guanine interactions in the major groove, while hydroxyl radical cleavage (Gross et al., 1990) detects minor groove DNA–protein interactions as well as the periodicity of the helical twist.

Permeabilization of cells to DNase and many other agents requires lysis into an artificial buffer. This is clearly of concern in its potential for perturbing the pre-existing structures, especially if the nuclei are to be subjected to lengthy isolation procedures before the attack reagent is added. Even if it is shown that RNA transcripts can be made de novo from the isolated chromatin template, it is not clear that all of the genes are active or that all of the original structures are still present. The best way around the perturbation concern is probably to use several different types of attack reagents (i.e., nucleases in lysates and chemicals in intact cells) where practical, since they are subject to different potential artifacts. There are also ways to lessen the chances of artifacts in lysate footprinting alone. The lysis should be performed as gently and quickly as possible and the time between lysis and stopping the attack reaction should be mimimized. For DNase I, we typically use hypotonic lysis assisted by Dounce homogenization or a nonionic detergent, although several different methods were initially compared. DNase I is added immediately after lysis for as short a digestion time as is experimentally practical. To monitor the possibility that nucleoprotein structure rearranges during the period of digestion, a time course can be run in which DNase is added at various times after lysis. For the tRNA genes (see Fig. 2), we found that the footprint patterns were constant for approximately 30 minutes. Additional precautions might include harvesting and preparing the cells in a medium or buffer that induces or represses the desired physiological response.

An alternative to lysing the cell is to use chemical reagents that enter the cell. This approach is referred to as in vivo footprinting, as opposed to chromatin footprinting, and might be less disturbing to nucleoprotein structure than cell lysis. All such reagents might perturb the system, however. Chemicals such as DMS enter the cell to attack DNA, but they also attack other cellular constituents. This efficiently kills the cell and has the potential to inactivate chromatin protein components directly or to disrupt intracellular equilibria.

One method used to circumvent this is to deliver a relatively benign reagent specifically to the DNA by intercalation, with cleavage occurring in response to light activation (Becker and Wang, 1984; Becker *et al.,* 1989). This is subject to different potential problems, since intercalators can distort nucleoprotein complexes by locally extending and unwinding the DNA helix. Becker and Wang (1984) also showed that ultraviolet (UV) light could be used in the absence of intercalators to sensitize the DNA. The DNA can be damaged directly with UV light and the photoproducts visualized as primer extension stops (see below) (Selleck and Majors, 1987a,b). This procedure is the ideal in that it is not disruptive to the native structures. However, the results are usually small but quantitative differences in band intensity because bound proteins are at best translucent to UV light. There are also possibilities for using structure-specific reagents in intact cells, as has been demonstrated using permanganate to detect open promoter complexes in *Escherichia coli* (Sasse-Dwight, and Gralla, 1990).

2.   DETECTION

The exact positions of DNA cleavages across a region can be determined by variations on two basic methods. The first, called "indirect end labeling" (Nedospasov and Georgiev, 1980; Wu, 1980), was also used by Church and Gilbert (1984) for genomic DNA sequencing. Briefly, it involves purifying the DNA after partial chromatin digestion, cleaving to completion on one side of the region of interest with a restriction endonuclease, separating the unlabeled DNA by size on a gel, blotting the DNA onto a membrane support, and annealing a unique detection probe to a site immediately adjacent to the common endpoint created by the restriction endonuclease. Thus, each single-stranded piece of DNA from the desired region (and only the desired region) is "indirectly" end-labeled at the restriction site and the data have the same form as a footprint in which the end of a unique DNA fragment is labeled in advance. Although this technique is technically demanding, it has the advantage that high-specific-radioactivity nucleic acid probes of various types can be used that give stronger signals than the primer extension methods described in the next section and that the same blot can be reused to examine a different location.

For organisms with genomes the size of yeast, we have found that primer extension (Fig. 1) gives satisfactory signals and has the advantage that it can be used with reagents that modify but do not cleave exposed positions of the bases that are involved in hydrogen bonding. Detailed methods using three different primer extension reactions are given in the next section. This scheme,

DEPROTEINIZED DNA   SITE-SPECIFIC PROTEIN BOUND

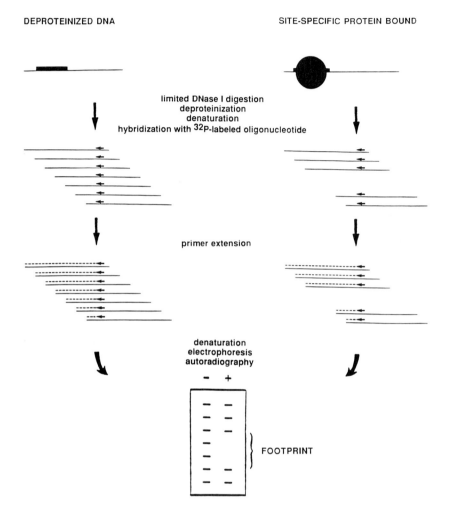

FIG. 1.   Schematic representation of primer extension footprinting. Unlabeled, deproteinized DNA and chromatin are partially digested or modified in parallel. DNA is then purified and denatured. A radioactively labeled oligonucleotide primer (arrow) is annealed to a unique site and extended to the cleavage site, followed by separation on a DNA sequencing gel and detection of only the new extension products. Protection of the DNA cleavage by a stably bound protein will result in a "footprinting," or absence of bands corresponding to the region of DNA bound by the protein as compared to the deproteinized DNA lanes.

which can also be used to footprint on closed circular and supercoiled templates *in vitro* (Huibregste *et al.,* 1987; Sasse-Dwight and Gralla, 1988), introduces the labeled, fixed end in a manner analogous to chain termination DNA sequencing. A labeled oligonucleotide is annealed to the purified DNA and extended with a DNA polymerase to the modification or cleavage sites created by the attack reagent. The oligonucleotide primer, which becomes the labeled end of the DNA in the footprint, can be chosen to hybridize to virtually any unique position near the region of interest, obviating the need to have convenient restriction endonuclease sites in the vicinity.

After testing a number of different DNA polymerases for both sequencing and footprinting of yeast DNA, we have found the most useful to be avian myeloblastosis virus (AMV) reverse transcriptase, modified T7 DNA polymerase, and *Thermus aquaticus* (*Taq*) DNA polymerase. We are presenting protocols for the use of all three enzymes because we find that the degree to which each enzyme pauses or terminates over a given DNA sequence varies in an unpredictable fashion, and we often find it necessary to determine empirically which will give the cleanest primer extension results. The use of *Taq* DNA polymerase is potentially the most flexible because this enzyme operates at higher temperatures, and longer oligonucleotides can be used with higher annealing and extension temperatures. This increases the chances of unique priming on more complex templates. The reactions can also be performed in an automated thermal cycler, with multiple rounds of primer annealing and extension giving linear increases in signal. Further, new techniques are coming into use that can increase the signal by using anchored polymerase chain reactions to amplify the initial primer extension products geometrically (Mueller and Wold, 1989). We have not found the amplification necessary for yeast, and therefore avoid the extra steps and the possibility of selective signal amplification. However, for larger genomes the extra sensitivity would be required and has proved effective in detecting cleavage sites and intrinsic DNA methylation (Pfeifer *et al.,* 1989, 1990).

Detailed protocols are given below for footprinting chromatin from freshly lysed *Saccharomyces cerevisiae* using modified T7 DNA polymerase, AMV reverse transcriptase, or *Taq* DNA polymerase for primer extension. Data generated using the three enzymes are presented in Figs. 2, 3, and 4, respectively. This is followed by a protocol for footprinting by indirect end-labeling applied to a promoter in *Drosophila melanogster*, with the chemical cleavage reagent methidium propyl-EDTA.FeII (Fig. 5). These particular experiments have been chosen because they not only give examples of the data generated by the protocols, but also illustrate the types of information that can be obtained when DNA control sequences and transacting factors are manipulated *in vivo*. Results and interpretations for the various experiments are discussed in the figure legends.

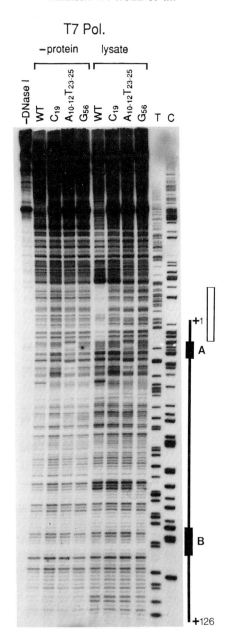

III. Protocols for DNase I Footprinting of *Saccharomyces cerevisiae* Chromatin by Primer Extension Techniques

A. Digestion of Chromatin in Permeablized Cells with DNase I

1. Inoculate 250 μl of a freshly saturated overnight culture into 250 ml of synthetic complete medium. (Cells grown in rich media my be harder to spheroplast.) Grow to OD_{600} 0.5–1.0, about 16 hours.

2. Measure the volume and the OD_{600} of each culture. Adjust the volumes of cells used so that the number of "Units" of each culture used is the same (no. of units = OD_{600} × no. of ml).

FIG. 2.   Footprinting of yeast, *SUP53* tRNA gene alleles with modified T7 DNA polymerase extension. Yeast tRNA genes contain internal promoter elements, the A and B boxes, with little dependence on specific sequences upstream of the transcription start site. Transcription factor TFIIIC binds to both internal promoters *in vitro* as an initial step in transcription activation (Kassavetis *et al.*, 1990). We have footprinted the yeast *SUP53* tRNA gene and several variants containing internal promoter mutations, both as single-copy chromosomal genes and when present on multiple-copy plasmids, with comparable results (Huibregste and Engelke, 1989, 1990). Multiple-copy footprinting with modified T7 DNA polymerase extension is shown here for the gene with wild-type promoters (WT lanes) or internal promoter mutations (C_{19}, others) that reduce transcription at least 10-fold. The labeled 16mer primer annealed to the sense strand 50 bp downstream of the gene. One round of annealing and primer extension was performed with a 94/42°C cycle. The tRNA coding sequences, starting with + 1, are indicated by lines to the right of the panel and the positions of the A box and B box are shown as solid rectangles. C_{19} has a single point mutation in the A box; $A_{10-12}T_{23-25}$ has multiple point mutations in the A box; G_{56} has a single point mutation in the B box. All of these mutations reduce transcription at least 10-fold (Newman *et al.*, 1983). Digestions of chromatin from freshly lysed cells ("lysate" lanes) are compared with the corresponding deproteinized DNA lanes to locate protection and enhanced cleavages, presumably due to protein binding. The only clear footprint on the wild-type gene, marked by an open bar, occurs at postions from − 40 to + 15 with respect to the start point of transcription. Little or no protection is seen over the A and B box internal promoter, with the only indication of low-level TFIIIC binding to the gene being some intensified cleavages between the boxes that are characteristic of TFIIIC footprints *in vitro* (Huibregste and Engelke, 1989). This and concurrent assembly experiments *in vitro* (Kassavetis *et al.*, 1989, 1990) led to a model of transcriptional activation where TFIIIC binds to the internal promoters as an "assembly factor," facilitating the entry of RNA polymerase III and other factor(s) at the initiation site. Subsequently, the initiation site complex is more stably bound than is TFIIIC. Confirmation that TFIIIC is required for initiation site complex assembly *in vivo* is provided by the fact that all of the internal promoter mutations substantially reduce or eliminate the initiation site footprint. However, only the point mutation in the B box primary binding site for TFIIIC eliminates the enhanced cleavages characteristic of TFIIIC association. (Reproduced with permission from Huibregtse and Engelke, 1989.)

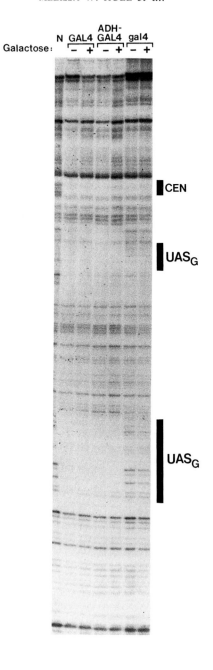

3. Harvest cells by room temperature centrifugation at 2000 *g* for 3 minutes.

4. Resuspend cells in 15 ml of 40 m*M* ethylenediaminetetraacetic acid, sodium salt (NaEDTA) (pH 8). Add 100 μl of 2-mercaptoethanol. Incubate for 3–5 minutes at room temperature. Transfer cells to 30-ml screw-cap tubes.

5. Centrifuge cells at 2000 *g* for 3 minutes at room temperature.

6. Resuspend each tube of cells in 10 ml of 1.1 *M* sorbitol, 20 μl of 0.5 *M* EDTA, pH 8, 30 μl of 1 *M* dithiothreitol (DTT), 20 mg of Zymolyase 20,000 (from Seikagaku Kogyo).

7. Incubate cells at 30°C for 10–25 minutes, or until spheroplasted. (To determine what fraction of the cells have been spheroplasted after a given length of time, dilute two 100-μl aliquots of cells into 900 μl of H_2O or 900 μl of 1.1 *M* sorbitol. Votex for 30 seconds to break open cells in H_2O. When the cells have been spheroplasted to completion, the OD_{600} of the cells diluted into H_2O should be half that of the cells in sorbitol.) The incubation can be performed either by shaking cells at approximately 100 rpm in a 30°C incubator, or putting the cells in a waterbath and gently mixing them once or twice during the incubation. It is important not to lyse the spheroplast by vigorous mixing.

8. Centrifuge cells as in step 4.

9. Resuspend cells in 0.1 ml of hypotonic buffer (100 m*M* NaCl, 6 m*M* Tris-HCl, pH 7.4, 6 m*M* $MgCl_2$) per 10 units. Homogenize in a loose-fitting Dounce homogenizer several times to break apart clumps of cells. Aliquot

---

FIG. 3.   Footprinting of the yeast *GAL2* upstream activator sequences (UAS_G) with extension by AMV reverse transcriptase. Transcription of the single-copy, chromosomal *GAL2* gene by RNA polymerase II is regulated by galactose via two UAS_G elements upstream of the start of transcription that are known to interact with the GAL4 transcription factor. The figure shows data obtained from a 16mer to the sense strand annealed 60 bp downstream of the most proximal UAS_G. One round of annealing and primer extension was performed with a 94/42°C cycle. Parallel DNase I footprinting was performed on three strains, one that was wild-type for the GAL4 transactivator (GAL4), one in which the GAL4 gene had been deleted (gal4), and one in which the GAL4 protein was overproduced (ADH-GAL4). Primer extensions on undigested DNA or deproteinized DNA digested with DNase I are shown for comparison. Experiments were also performed under growth conditions that were either inducing (+) or repressing (−) for *GAL2* transcription. No strong footprint was observed at the start point of transcription or the presumptive TATA promoter elements (data not shown), but either wild-type or overproduced levels of GAL4 protein give obvious footprints at the proximal UAS_G and weaker protection at the distal UAS_G. The constant footprint more distal from the primer over a region marked CEN coincides with a previously noted centromere subregion-like sequence (Bram and Kornberg, 1987). Note that the presumptive GAL4 protein binding to the UAS_G elements takes place when the transcription of *GAL2* is either induced by GAL4 protein or repressed, suggesting that carbon source transcriptional control of this gene acts at a level other than regulating GAL4 binding to the DNA.

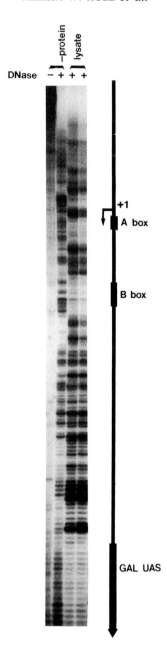

300 μl of cells at 10-second intervals into 1.5-ml microcentrifuge tubes containing 30 μl of hypotonic buffer or hypotonic buffer with diluted DNase I. (50 mg/ml DNase I stocks are stored at $-20°C$ in H_2O, thawed on ice, and diluted 1:100, 1:200 into cold hypotonic buffer immediately before use.)

10. Incubate for 5 minutes at room temperature.

11. Stop reaction by addition of 330 μl of 1 M NaCl, 50 mM Tris-Cl, pH 7.9, 2% sodium dodecyl sulfate (SDS), 50 mM EDTA.

12. Extract with an equal volume of phenol/chloroform until the interface disappears, usually 2 or 3 times.

13. Add an equal volume of isopropanol (it is not necessary to add any additional salt), invert several times to mix, incubate at room temperature for 5 minutes, centrifuge for 15 minutes, remove supernatant, wash pellet with 75% ethanol, and dry the pellet.

14. Resuspend in 400 μl of $H_{10}E_{0.1}$ (10 mM HEPES, 0.1 mM EDTA, pH 7.9) with 12.5 μl of 1 mg/ml RNase.

15. Incubate for 3 hours to overnight (as convenient) at 37°C.

16. Add an equal volume of isopropanol and 0.1 volume of 1 M NaCl, invert several times to mix, incubate at room temperature for 5 minutes, centrifuge in a microfuge for 15 minutes, remove supernatant, wash pellet with 75% ethanol, dry, and resuspend in 50 μl of $H_{10}E_{0.1}$.

17. Run 3 μl of each sample on a 1% agarose gel to verify (a) that the DNA which has not been treated with DNase runs as a single broad high-molecular-weight band (about 20 kb) and (b) that the DNase-treated DNA is smeared

---

FIG. 4. *Taq* DNA polymerase was used to footprint across a tRNA gene and UAS_G juxtaposed on a single-copy plasmid in yeast. Phenotypic data shows that wild-type yeast cells carrying this plasmid transcribe the tRNA gene but that the UAS_G is not able to stimulate transcription from an adjacent polI II promoter (M. W. Hull and D. R. Engelke, unpublished data). The footprinting analysis was performed to determine whether or not the tRNA gene interfered with pol II transcription at the level of GAL4 binding to the UAS_G. In the example shown here, the tRNA gene is oriented such that the GAL promoter and UAS_G are approximately 80 bp downstream from the tRNA transcription terminal site. The primer is a 28mer which hybridizes 120 bp away from the nearest end of the tRNA coding region. Fifteen cycles of 94/72°C synthesis were performed in a thermocycler. The tRNA gene consistently displays strong footprints across the upstream region (noted in Fig. 2) and also across the internal promoter regions. (This is typical of tRNA gene internal promoters with a high affinity for TFIIIC.) In addition, there is a footprint covering the UAS_G sequence, and enhanced cleavages immediately upstream of the footprint. This suggests that the GAL4 protein is bound even though GAL4-induced transcription is inhibited. Two caveats make this conclusion questionable with the data shown, however. Poor processivity by *Taq* polymerase near the primer gives a ladder of stops at the bottom of the gel, in the control lane, and interferes with clear footprinting at the UAS_G. In addition, the UAS_G is immediately flanked by a *Bam*HI site, which is a known target for an abundant, but poorly characterized DNA-binding protein in yeast (Marschalek and Dingermann, 1988). Further work, involving footprinting of strains which are gal4$^-$ or have mutated GAL4-binding sites, is necessary to clarify such an issue.

from about 200 bp to about 20 kb. (Since different regions of DNA vary in their overall sensitivity to DNase, the optimal amount of DNase digestion will depend on the region under examination.)

18. In some cases, we have found that restriction enzyme digestion of the DNA decreases the intensity of artifactual primer extension stops seen near the primer when using *Taq* polymerase. In these cases, restriction digestion is performed overnight with 3 units of a restriction enzyme per 50 μl of DNA (resuspended as specified above), and the DNA is used directly from the restriction endonuclease reaction, without phenol extraction or ethanol precipitation. We have not found it necessary to digest the DNA before prime rextension reactions with modified T7 DNA polymerase or reverse transcriptase.

B. Digestion of Purified DNA with DNase I

1. Dilute 50 mg/ml DNase I stock 1 : 1000 with hypotonic buffer. (Further dilutions, i.e., 1 : 2 or 1 : 4, may be necessary.) Add 4 μl of diluted DNase to 14 μl of DNA premixed with 22 μl of hypotonic buffer.
2. Let the reaction proceed for 30 seconds at room temperature.
3. Stop the reaction with addition of 2 μl of 0.5 M EDTA.
4. Add 2.5 volumes of 95% ethanol and 0.1 volume of 1 M NaCl, invert to mix, incubate on dry ice for 10 minutes, centrifuge for 15 minutes, wash pellet with 1 volume 95% ethanol, dry, and resuspend DNA in 14 μl of $H_{10}E_{0.1}$.
5. Run an aliquot of the DNA on a 1% agarose gel to check the extent of DNase digestion.

C. Primer Extensions

The DNase I-treated DNA is analyzed by primer extention with dNTP mix as described below. The primer extension will stop at DNase I cleavage sites, resulting in discrete-sized bands on the sequencing gel. To determine the position of the DNase I cleavages, sequencing reactions are carried out with the same primer and the ddNTP mixes described below.

1. MODIFIED T7 DNA POLYMERASE

Buffers and Nucleotide Mixes
    10 × buffer: 0.2 M HEPES, pH 7.5, 0.1 M MgCl$_2$, 0.5 M NaCl
    10 × dNTP mix: 3 mM dATP, dTTP, dCTP, and dGTP

10 × ddNTP mixes: 0.3 m*M* appropriate ddNTP, 3.0 m*M* each of dATP, dTTP, dCTP, and dGTP

Procedure. For each primer extension reaction, combine the following in a microcentrifuge tube: 3 µl of 10 × buffer, 1.5 µl of 0.1 *M* DTT, 3 µl of 10 × dNTP or ddNTP mix, 10–15 µg yeast DNA, 1–2 µl of ^{32}P-end-labeled oligonucleotide (200,000–500,000 cpm, with a specific radioactivity of 3,000–6,000 Ci/mmol) prepared as described in Sambrook *et al.* (1989), and water to give a total volume of 29 µl.

1. Heat at 95–100°C for 5 minutes.
2. Centrifuge for a few seconds to pellet ᵗ   at has condensed on the sides of the tube.
3. Place at 37–44°C for 15 minutᵉ
4. Add 2 U of modified T7 DN   ᵈ States Biochemicals Sequenase) diluted into 1X buffer.
5. Continue incubation at 37–44°C for 3 minutes.
6. Additional rounds of primer extension may be done by repeating steps 1–5.
7. Stop the reaction by addition of 4 µl of solution containing 1% SDS, 0.1 *M* NaEDTA, pH 8, and 1 mg/ml proteinase K. Incubate for 10 minutes at 50°C.
8. Add 2.5 volumes of ethanol to precipitate DNA. Freeze on dry ice, then microfuge for 15 minutes, remove the supernatant, wash the pellet with 75% ethanol, and dry the pellet. Resuspend in 4 µl of loading buffer (98% formamide, 10 m*M* NaOH, 1 m*M* EDTA, 0.1% bromophenol blue, and 0.1% xylene cyanol). Heat at 95–100°C for 2 minutes prior to loading on a sequencing gel prepared as described in Sambrook *et al.* (1989).

2. AVIAN MYOBLASTOSIS VIRUS REVERSE TRANSCRIPTASE

Buffer and Nucleotide Mixes

10 × buffer: 0.5 *M* Tris, pH 8.3, 0.4 *M* KCl, 0.01 *M* DTT, 0.06 *M* MgCl$_2$

5 × dNTP mix: 0.625 m*M* dATP, dCTP, dGTP, dTTP.

5 × ddATP mix: 0.5 m*M* dCTP, dGTP, dTTP, 0.05 m*M* dATP, 8 µ*M* ddATP

5 × ddCTP mix: 0.5 m*M* dATP, dGTP, dTTP, 0.05 m*M* dCTP, 4 µ*M* ddATP

5 × ddGTP mix: 0.5 m*M* dATP, dCTP, dTTP, 0.05 m*M* dGTP, 12 µ*M* ddGTP

5 × ddTTP mix: 0.5 m*M* dATP, dCTP, dGTP, 0.05 m*M* dTTP, 4 µ*M* ddTTP

Procedure.   For each primer extension reaction, combine the following in a microcentrifuge tube: 3 μl of 10 $\times$ buffer, 6 μl of 5 $\times$ dNTP mix or 5 $\times$ ddNTP mix, 10–15 μg of yeast DNA, 1–2 μl of ^{32}P-end-labeled oligonucleotide, and water to give a total volume of 29 μl.

1. Heat at 95–100°C for 5 minutes.
2. Centrifuge for a few seconds to bring down liquid that has condensed on the sides of the tube.
3. Place at 37–50°C for 15 minutes. (Temperature optimum depends on primer and is usually determined empirically.)
4. Add 11 U of AMV reverse transcriptase (Life Sciences, Inc.).
5. Continue incubation at 37–50°C for 20 minutes.
6. Additional rounds of primer extension may be done by repeating steps 1–5.
7. Stop the reaction by addition of 4 μl of solution containing 1% SDS, 0.1 M EDTA, pH 8.0, and 1 mg/ml proteinase K. Incubate for 10 minutes at 50°C.
8. Add 2.5 volumes of ethanol to precipitate DNA. Freeze on dry ice, then microfuge for 15 minutes, remove the supernatant, wash the pellet with 75% ethanol, and dry the pellet. Resuspend in 4 μl of loading buffer. Heat at 95–100°C for 2 minutes prior to loading on a sequencing gel prepared as described in Sambrook *et al.* (1989).

3.   *Taq* DNA POLYMERASE

Buffer and Nucleotide Mixes
    10 $\times$ *Taq* buffer: 500 mM KCl, 100 mM HEPES, pH 8.4, 25 mM MgCl$_2$
    10 $\times$ dNTP: 1 mM each dATP, dCTP, dGTP, dTTP
    10 $\times$ ddATP: 1 mM dCTP, dGTP, dTTP, 25 μM dATP, 1 mM ddATP
    10 $\times$ ddCTP: 1 mM dATP, dGTP, dTTP, 100 μM dCTP, 1 mM ddCTP
    10 $\times$ ddGTP: 1 mM dATP, dCTP, dTTP, 100 μM dGTP, 500 μM ddGTP
    10 $\times$ ddTTP: 1 mM dATP, dCTP, dGTP, 100 μM dTTP, 1 mM ddTTP
Procedure.   For each primer extension reaction combine: 10–20 μg of DNA, 2 U of *Taq* DNA polymerase (Cetus), 5 μl of 10 $\times$ buffer, 5 μl of 10 $\times$ nucleotides, 1–2 μl of ^{32}P-labeled oligonucleotide, H$_2$O to 50 μl. Carry out 15 rounds of primer extension, cycling between 94°C for 1.5 minutes, the annealing temperature for 4 minutes, and the extension temperature (72°C) for 3 minutes. (If the oligonucleotide is annealed at 72°C, the total incubation time at 72°C can be shortened to 6 minutes.) Add 2.5 volumes of ethanol to precipitate. Freeze on dry ice, then spin for 15 minutes, wash pellet with 75% ethanol and dry the pellet. Add 4 μl of loading buffer. Heat at 95–100°C for 2 minutes prior to loading onto a sequencing gel prepared as described in Sambrook *et al.* (1989).

D. Additional Considerations

1. PRIMER EXTENSION CONTROLS

Since DNase I has some sequence specificity, it is necessary to run a control lane of DNA cleaved after purification to compare with the DNA digested in cell lysates. It is also a good idea to run a lane with undigested DNA to identify the intrinsic polymerase stops. The priming specificity of any new oligonucleotide should be tested by sequencing reactions with yeast DNA; this also provides markers for the footprinting reactions. Most oligonucleotide primers will be specific when used on plasmid DNA because of the low sequence complexity.

2. DETERMINATION OF OLIGONUCLEOTIDE ANNEALING TEMPERATURE

Sequencing reactions should initially be performed at a variety of temperatures to determine the temperature which maximizes the specificity of and signal from the oligonucleotide primer. Oligonucleotides which are 50% AT and 50% GC and lack stretches of four or more Gs or Cs in a row usually work well.

IV. Protocols for MPE.Fe (II) Footprinting of *Drosophila melanogaster* Chromatin by Indirect End-Labeling Techniques

Achieving an acceptable signal-to-noise ratio with high-resolution indirect end-labeling is quite challenging, and the advent of the ligation-mediated polymerase chain reaction (Mueller and Wold, 1989) might eventually make this technique redundant. However, there is lingering doubt as to what biases in data recovery may be introduced by the ligation/amplification combination. Thus, we have included a fairly explicit protocol for the indirect end-labeling approach and illustrated the technique in Fig. 5. Additional information is available in a number of the other reviews on this method (Becker and Schutz, 1988; Saluz and Jost, 1987, 1989).

The basic requirements for detection by indirect end-labeling are (1) a restriction site close to the region of interest (e.g., < 250 bp) for an enzyme (the reference enzyme) that does not cut again in this region, and (2) a subcloned fragment to make the probe which is ideally bounded at one end by the reference enzyme's cleavage site and is 100–150 bp in length. The probes used should be single stranded and short (e.g., < 150 bp), so that the data obtained from the blot do not arise from within the target sequence of the probe. It should also be of the highest specific activity and concentration possible to achieve high sensitivity and efficient hybridization. Procedures for the use of RNA probes generated by phage T7 and T3 RNA polymerases and ssDNA probes synthesized on phage M13 templates are described below.

Indirect end-labeling, like many other methods, can give rise to artifacts. A prominent cleavage site within the probe sequence will create an additional reference end, resulting in a superposition of formation on the sequencing gel. Similar problems arise as the distance between the probe sequence and the reference cleavage site increases. To prevent misinterpretation, data should be generated from more than one reference end. In addition, the target sequence for the probe must be unique in the DNA mixture being analyzed. Sequencing reactions visualized with the same probe, and commonly run as position markers, will confirm the specificity of the probe. Any cross-reactivity of the probe will result in uninterpretable sequence information.

A. Sample Preparation

The preparation of nuclei and the partial digestion of the chromatin with methidium propyl-EDTA.Fe (II) [MPE.Fe (II)], as described below, have been detailed in several references (Wu *et al.*, 1979a; Cartwright and Elgin,

---

FIG. 5.   Footprinting of the proximal regulatory region of the *Drosophila hsp26* promoter using methidium propyl-EDTA. Fe (II) in isolated nuclei. Nuclei from 6- to 18- hour embryos were prepared and digested with MPE with or without a prior heat shock as previously described (Wu *et al.*, 1979b; Cartwright and Elgin, 1986; Thomas and Elgin, 1988). The purified DNA was cut to completion with *Dra*I and analyzed as described in the test using a probe stretching from −170 to −269 relative to the start of transcription (Thomas and Elgin, 1988). MPE is an intercalator that carries an Fe (II) atom in a tethered EDTA group to the DNA. This reagent, while not sequence neutral, is far less specific than DNase I can consequently give quite precise limits for a bound protein. In this case, a prominent footprint is visible resulting from the binding of heat-shock factor (HSF) on induction. Less prominent with this reagent is the TATA box-associated footprint (see also Wu, 1984; Thomas and Elgin, 1988) and the CT factor (Gilmour *et al.*, 1989), which binds *in vitro* to the −85 to −130 polypurine/polypyrimidine tract. The failure to obtain clear footprints for these proteins may reflect partial occupancy due to displacement by the intercalator or due to the preparation method for obtaining nuclei. This heat-shock element is one of two that are required for *hsp26* expression on heat shock. DNase I footprinting of this region has shown that the second element (∼250 bp 5′ of the first) is also occupied upon heat shock and that intervening DNA is bound to a nucleosome (Thomas and Elgin, 1988).

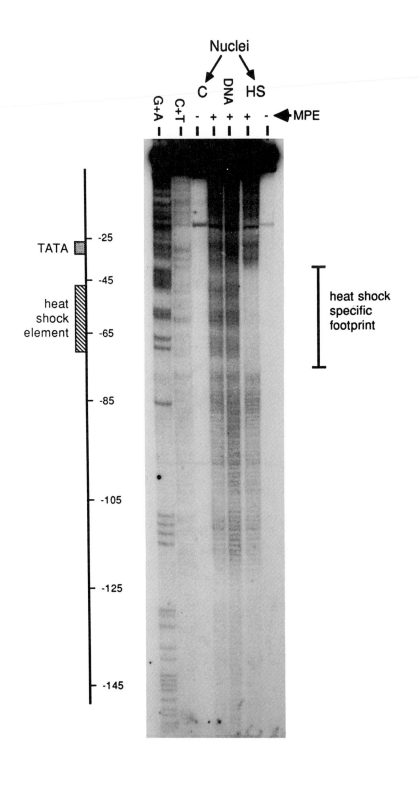

1986, 1989; Thomas and Elgin, 1988). The basic protocol is as follows:

1. Gather 20–30 g of fresh 6- to 18-hour *Drosophila* embryos as described by Elgin and Miller (1978) and flash-freeze in liquid nitrogen.
2. Add the frozen embryos to 300 ml of freshly made 1% NaCl, 1% Triton, 50% commercial bleach to dechorionate. Stir for 5 minutes to ensure that all lumps of embryos are dispersed.
3. Filter through 116-mesh Nitex screen and wash with several liters of distilled water.
4. Fold over the Nitex, press the embryos dry between paper towels, then place on ice.
5. Add one-fourth of the embryos to 30 ml of buffer A1 (60 mM KCl, 15 mM NaCl, 0.5 mM spermidine, 0.15 mM spermine, 0.5 mM DTT, 0.1 mM PMSF, 15 mM Tris-Cl, pH 7.4, 1 M sucrose, 1 mM EDTA, 0.1 mM EGTA) and homogenize with 10 strokes in a stainless steel homogenizer.
6. Pour the homogenate through two layers of miracloth placed in a filter funnel over a 250-ml flask on ice. (Miracloth is now available under the product name "Dottie Dustcloths" from Habco Products, P. O. Box 696; Bound Brook, NJ 08805. Use only the green, untreated, miracloths.)
7. Repeat steps 5 and 6 three times for remaining embryos.
8. Squeeze the miracloth dry into the flask, place the debris into the homogenizer, and add 30 ml of buffer A1. Homogenize with 10 strokes.
9. Filter homogenate as in step 6.
10. Split the filtrate into 50-ml centrifuge tubes and spin at 500 g for 9 minutes at 4°C.
11. Decant the supernatant into 50-ml screw-cap tubes, add 0.01 volume of 20% Triton X-100 to each tube, vortex for 25–30 seconds, and pour into fresh 50-ml centrifuge tubes.
12. Spin at 4400 g for 10 minutes.
13. Aspirate off the supernatant, making sure to remove the top fatty layer.
14. Add 5 ml/tube of buffer A1 and vortex to resuspend nuclei.
15. To two 30-ml Corex tubes add 20 ml of buffer A1.7 (same as A1 buffer, but with 1.7 M sucrose), and layer 10 ml of the nuclei on top of each. Gently distrub the interface with a stream of bubbles blown from a Pasteur pipette.
16. Spin at 12000 g for 25 minutes at 4°C in a swinging bucket rotor.
17. Aspirate off the supernatant and resuspend in 15 mM Tris-Cl, pH 7.4, 15 mM NaCl, 60 mM KCl, 1 mM EDTA, 0.25 M sucrose at a concentration of 10^9 nuclei per milliliter. Immediately before adding the MPE.F (II), warm the suspension to 25°C.

18. From a stock solution of MPE in H_2O ($E_{488} = 5994$; stored at $-80°C$), make a $10 \times$ MPE.Fe (II) solution by mixing MPE and freshly dissolved ferrous ammonium sulfate to equal concentrations (typically $100–250$ mM) in 15 mM Tris-Cl, pH 7.4, 15 mM NaCl, 60 mM KCl, 0.25 M sucrose and allow the complex to form for 1–2 minutes. (MPE was obtained from Professor Peter Dervan, Division of Chemistry and Chemical Engineering, California Institute ⌐   ⌐hnology, Pasedena, CA 91125.)

19. Activate the complex by adding DTT ⌐   immediately dilute into the nuclei.

20. Remove 500-μl aliquots at 5-minute intervals and add to tubes containing 1/9th volume of 50 mM bathophenathroline disulfonate to stop the reaction.

21. Add EDTA to 12.5 mM, SDS to 0.5%, and treat overnight with 50 mg/ml proteinase K at 37°C.

22. Extract once with phenol/$CHCl_3$/isoamyl alcohol (25 : 24 : 1) and once with ether.

23. Add RNase A to 50 mg/ml and incubate at 37°C for 2 hours.

24. Add SDS to 0.5% and proteinase K to 50 mg/ml and incubate at 37°C for 3 hours and repeat step 22.

25. Add sodium acetate to 0.2 M and precipitate the DNA with 2.5 volumes of ethanol.

26. Resuspend the DNA in an appropriate volume of 10 mM Tris, pH 7.5, 1 mM NaEDTA (pH 8). Cut the purified DNA to completion with the appropriate reference enzyme, phenol/chloroform extract, ethanol precipitate, and resuspend in 3 μl of loading buffer. Ten to 12.5 μg of DNA per lane is sufficient to analyze a single-copy gene in *Drosophila* with an exposure time of 48–96 hours with an intensifying screen. If a repeated sequence is the target (e.g., the histone genes; Gilmour *et al.* 1989), an 8- to 9-hour exposure is all that is needed.

B. Marker Preparation

Markers can be made either by cleavage with different combinations of restriction enzymes to generate discrete size fragments or by sequencing reactions. The former are useful for development work as they are easy to prepare, while the latter are essential for accurate mapping of footprints.

1. RESTRICTION MARKERS

Digest 20 μg of genomic DNA to completion with the reference enzyme. Split this into several aliquots and digest each one with a different second enzyme, the combination of which spans the region of interest. Phenol/chloroform extract each digest, mix the aqueous layers, ethanol precipitate the DNA, and resuspend in formamide loading buffer. Since the restriction cuts are limited to a few locations, the signal will be stronger than for a sequencing reaction and so 20 µg of genomic DNA is sufficient for about 10 gels.

2. SEQUENCING MARKERS

Twenty micrograms of genomic DNA is prepared according to the protocol described by Maxam and Gilbert (1980) for a single sequencing reaction. Note that the precise alignment of the chemical-sequencing markers with the samples depends on which side of the cleavage is being labeled, since chemical sequencing eliminates a nucleoside (Maxam and Gilbert, 1980). Thus, the equivalent oligonucleotide in the sequencing ladder runs one base lower than its counterpart in DNase I-digested samples when mapping from the 3' end. However, when mapping from the 5' end, the equivalent oligonucleotide in the sequencing ladder carries a 3' phosphate group, causing it to run a further half a nucleotide ahead, thus displacing the markers 1.5 nucleotides from the DNase I-cut samples (Sollner-Webb and Reeder, 1979).

In our hands, the standard G + A reaction (Maxam and Gilbert, 1980) has worked most consistently on genomic DNA:

1. Digest 20 µg of genomic DNA to completion with the reference enzyme, phenol/chloroform extract twice, ether extract twice, ethanol precipitate, and reprecipitate.
2. Resuspend the DNA in 35 µl of dH_2O and equilibrate at 20°C. Add 35 µl of formic acid at 20°C, mix, and incubate for 5 minutes at this temperature.
3. Stop the reaction with 200 µl of 0.3 M sodium acetate, pH 7.0, at 0°C and precipitate with 750 µl of ethanol. Place on dry ice for 10 minutes and centrifuge in a microfuge in the cold for 20 minutes.
4. Dry the DNA and resuspend it in 200 µl of 0.3 M sodium acetate, pH 7.0, on ice for 15–20 minutes with intermittent vortexing. Add 600 µl of ethanol, place on dry ice for 10 minutes, and spin in a microfuge in the cold for 20 minutes. Rinse once with 70% ethanol.
5. Dry the pellet and resuspend in 150 µl of freshly diluted 1 M piperidine and heat at 90°C for 1 hour.
6. Freeze and lypophilize 4-6 times, resuspending in 50 µl of dH_2O each time.
7. Finally, redissolve in 3 µl of loading buffer and heat in a boiling waterbath for 5 minutes before loading.

C.   Blot Preparation

1.   GEL ELECTROPHORESIS

Sequencing gels [8 M urea, 6% acrylamide: bis (19:1), 135 mM Tris base, 45 mM boric acid, 2.5 mM EDTA, pH 8.9; dimensions 0.4 mm × 19cm × 40 or 80 cm] are run at either 35 W (40-cm gels) or 60 W (80-cm gels). This particular Tris–borate buffer (Anderson, 1981) was adopted to maintain buffer capacity when running the 80-cm gels for up to 8 hours. The appropriate electrophoresis time is determined by running labeled DNA fragments of known size on a test run. Run the experimental gel for the appropriate length of time, loading more running dye in the two wells bracketing the samples every 30 minutes and again 2 minutes before stopping the gel to improve the visibility of the sample region of the gel in subsequent steps.

2.   ELECTROPHORETIC TRANSFER OF DNA TO A MEMBRANE

Electroblotters are available commercially (e.g., Idea Scientific) or can be made according to the design of Church and Gilbert (1984) (Fig. 6). An alternative method is hydrokinetic transfer (Gross *et al.,* 1988). It is important to have a filter which will retain fragments down to about 60–100 bp, such as 0.45-μm Nytran membrane (Schleicher & Schuell) or Gene Screen (DuPont).

1. Cut the filter and allow it to thoroughly wet in transfer buffer (50 mM Tris, 50 mM boric acid, 1.5 mM EDTA).
2. Separate the gel plates and float the gel off the plate (gel side down) into 2 liters of transfer buffer above a sheet of Blot Block (Schleicher & Schuell) or 3MM (Whatman) paper. Submerge the gel and gently move it to lie above the filter paper. Soak the gel for 5 minutes to reduce background.
3. Lay the nylon membrane over the gel, using the dye samples as a guide, and gently press it down against the gel to remove air bubbles.
4. Place another piece of filter paper over the nylon membrane, and again gently press down and smooth out, removing bubbles.
5. Place this "sandwich" in the electroblotter, membrane side down, to transfer downward on to it, as less gas comes off the anode.
6. Transfer the DNA for 30 minutes at 5.5 V/cm (cm = distance between electrodes). This will take up to 2 A.
7. After transfer is complete, remove the filter from the sandwich, blot off the excess moisture, and dry it thoroughly.
8. Cross-link the DNA to the filter by exposing the DNA side for 10 seconds to a four-tube short-wave transilluminator (without the lid) at a distance of 15 cm.

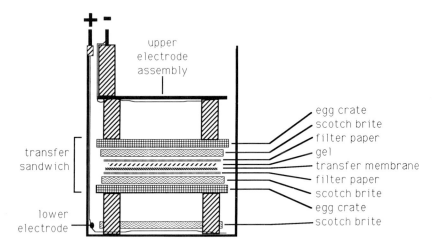

FIG. 6. Electroblotter. This schematic drawing shows all the essential elements required in a custom-made sequencing gel elctroblotter as described by Church and Gilbert (1984). All the unlabeled items are plexiglass or, in the case of the electrodes, platinum wire. It should be noted that this design holds a considerable volume of buffer and so the outer box walls should be at least 1 cm thick. Only two exotic components are required in order to build a blotter. These are Scotch Brite cleaning and finishing roll (grade A VFN, available from 3M Co. via local distributors) and egg-crate louvre panels, the thick plastic grid which provides rigid support for the gel while allowing current to pass through. This latter component is designed for use in fluorescent light assemblies and may be available locally (it can also be obtained from AIN Plastics Inc, 249 E Sanford Boulevard, Mt. Vernon, NY 10550). The electrodes should zig-zag across the width of the box and the upper electrode assembly. In our case, the electrode had a height of 20 cm and a "wavelength" of 7–8 cm and was affixed with blobs of silicone glue. The transfer sandwich is held together with three loops of plastic tubing tied around it and rests in cross-cuts made in the posts below. Similar cross-cuts in the posts of the upper electrode assembly locate it on the top of the transfer sandwich. The posts are 3.5 cm long and serve as spacers to ensure that the field lines are reasonably parallel as they pass through the gel. Transfer is from the top down as the lower electrode will then produce less gas. The gas which is produced is held below the lowest layer of Scotch Brite and thus prevented from accumulating in the egg crate and inhibiting transfer.

D.   Probe Preparation

1.   RIBOPROBE SYNTHESIS

Templates cloned into the Stratagene vector "Bluescribe," which has phage T7 and T3 RNA polymerase promoters on either side of an M13 mp19 polylinker, work well; however, any vector of similar composition should be suitable. The protocol that follows is basically according to the manufacturer's instructions, except that the labeling ribonucleotide was not supplemented with unlabeled material in order to maximize the specific activity of the probes. Of course, the labeling ribonucleotide must be kept at a sufficiently high concentration to achieve full-length runoff products. This requires a somewhat higher concentration for T3 (8 μM) than T7 (2–4 μM) when incorporating [α-^{32}P]UTP into a 100-bp template. It may be neccessary to lyophilize the labeled ribonucleotide to increase its concentration. Greater than 60% incorporation is usually achieved in 2 hours.

2. T7 RNA Polymerase Probe Synthesis

1. Lyophilize 250 μCi of >3000 Ci/mM [α-^{32}P]UTP (e.g., ICN Radiochemicals) and resuspend in 10 μl of dH$_2$O containing 2 μg of template (previously linearized to generate the appropriate runoff template).
2. Add 10 μl of a 2 × mix [80 mM Tris-Cl, pH 8.0, 16 mM MgCl$_2$, 100 mM NaCl, 800 μM each of ATP, GTP, and CTP, 60 mM DTT, 3.2 U/μl RNasin (Promega), 4 mM spermidine, 5 U/μl T7 RNA polymerase] and incubate at 37°C for 2 hours.
3. Ethanol precipitate the probe twice and resuspend it in hybridization solution.

3. T3 RNA Polymerase Probe Synthesis

1. Dry down the 250 μCi of [$\alpha-^{32}$P]UTP with the 2 μg of template in a siliconized tube.
2. Dilute 9 μl of the 2 × mix described above (made without T7 RNA polymerase) with 6 μl of dH$_2$O and thoroughly resuspend the template/isotope in 5 μl of the resulting solution.
3. Add 1 μl (50 U) of T3 RNA polymerase. Incubate for 2 hours at 37°C in a low-evaporation container.
4. Ethanol precipitate the probe twice and resuspend in hybridization mix.

4. M13 Probe Synthesis

Starting with the probe sequence subcloned into the appropriate M13 vector (e.g., mp18/19), synthesis is carried out on single-stranded DNA prepared according to standard procedures (Sambrook *et al.,* 1989). This protocol should allow greater than 50% of the incorporated radionucleotide to be in the first 100 bases synthesized. Initial synthesis is followed by a cold chase and the probe is released by cutting with a restriction enzyme on the distal side of the insert from the primer. The short, hot probe is then purified on a standard sequencing gel to remove all other DNA sequences. Two clones in vectors of opposite polylinker orientation are required to do both strands.

1. Combine 13 μg of M13 template and 5 pmol of the appropriate primer in a final volume of 28 μl of annealing buffer (10 mM Tris-Cl, pH 7.4, 400 mM NaCl, 2mM EDTA). Heat this mixture to 50°C for 1 hour and place at 37°C for several hours or overnight as convenient.
2. Add 10 μl (100 μCi) of each [α-^{32}P]dNTP ($>$3000 Ci/mM), 4 μl of 100 mM DTT, 8 μl of Klenow buffer (100 mM Tris-Cl, pH 7.4, 100 mM MgCl$_2$), 11 μl of dH$_2$O, and 8 μl (40 U) of Klenow DNA polymerase. Incubate at 37°C for 1 hour.
3. Chase by adding 4 μl of a 10 mM dNTP mix and 2 μl more of Klenow for 15 minutes at 37°C.
4. Stop the reaction by adding 5 μl of 0.5 M EDTA.
5. Remove the unincorporated nucleotides in a spin column and ethanol precipitate.
6. Resuspend and digest to completion with the appropriate restriction enzyme to release the probe as a uniform length piece of DNA. The inclusion of nonlabeled plasmid in the digestion allow this to be monitored by running a small aliquot on a gel.
7. Phenol/chloroform extract the digestion mixture and ethanol precipitate the DNA.
8. Resuspend the dry pellet in loading buffer, boil for 5 minutes, and load onto a standard 6% sequencing gel.
9. Autoradiograph the wet gel covered in plastic wrap for approximately 30 seconds to find the desired band.
10. Cut out the segment of the gel containing the band and put it in 1 ml of 10 mM Tris-Cl, pH 7.4, 1 mM EDTA, 100 mM NaCl at 37°C for several hours, change the solution, and repeat.
11. Pool the two aliquots, concentrate them by butanol extraction and add them directly to the hybridization mix.

E.   Characterization of Probes

ESTIMATION OF T_m

The probes used in this procedure are all relatively short and their sequence composition can significantly alter their T_m. It is useful to measure the T_m under experimental conditions, as described below, in order to determine the optimal temperature for the washes.

1. Prepare approximately 3 μg of your template DNA from the clone in one fragment, cutting at sites as close as possible to the probe sequence. Phenol/chloroform extract, ether extract, and remove a 250-ng aliquot

to make the probe. Ethanol precipitate this and the remaining 2.75 μg separately.

2. Resuspend the larger aliquot in 50 μl of loading buffer.

3. Resuspend the smaller aliquot in water, boil, and end label with polynucleotide kinase by the exchange reaction using [γ-^{32}P]ATP in standard protocols (e.g., Sambrook *et al.,* 1989). Remove the unincorporated nucleotides, precipitate, and resuspend in 5 μl of loading buffer.

4. Cast a sequencing gel that has three wells with widths of 5 mm, 140 mm, and 5 mm, in that order.

5. Boil both DNA samples and load 2.5 μl of the labeled DNA into the outer lanes (as markers) and the unlabeled DNA into the central lane. Run the gel until the insert is well separated from the other DNA, then electroblot onto nylon and UV cross-link as described above.

6. Make an autoradiograph of the filter to find out where the insert is and cut it out in a 2-cm-wide strip.

7. Synthesize the labeled probe from the same template as described above and hybridize it to the strip overnight at 37°C (see below).

8. Rinse the filter in wash solution A (see below) for three changes at room temperature.

9. Put 1 liter of final wash solution in a beaker on a heating stirrer, and suspend a thermometer and the hybridized filter in the solution so that it can be easily removed (a hemostat is convenient). Switch on the heater and the stirrer so that the temperature rises at a moderate pace.

10. Starting at 35°C, cut off a short length of the filter with scissors every 5°C up to 90°C.

11. Make an autoradiograph of the dried filter fragments in order and you have a melting curve. Wash at 10 to 20°C below the T_m, depending on the background observed. The same temperature is also used for hybridization.

F. Hybridization and Washing Conditions

The hybridization and washing conditions are the same as those described by Church and Gilbert (1984) with the modified hybridization solution of Marchionni and Gilbert (1986). The temperature for hybridization and washing will vary from probe to probe and should be optimized as described in the previous section.

Large stock solutions should be prepared of 1 M Na(HPO$_4$) (71 g of anhydrous Na$_2$HPO$_4$, 4 ml of 85% H$_3$PO$_4$ per liter), cheap 20% SDS (w/v; Sigma) for wash solutions, good 20% SDS (w/v; Boehringer Mannheim, Ultra Pure) for the hybridization mix, and 0.4 M NaEDTA, pH 7.5.

1. HYBRIDIZATION CONDITIONS

To make the hybridization mix [500 mM Na(HPO$_4$), 250 mM NaCl, 1% bovine serum albumin (BSA), 7% SDS, 1 mM EDTA], add the following in order: 500 ml of 1 M Na(HPO$_4$), 10 g of BSA (fraction V), 2.5 ml of 0.4 M EDTA, 350 ml of good 20% SDS. Make up to 950 ml with dH$_2$O, then add 14.6 g of NaCl dissolved in a final volume of 50 ml of dH$_2$O. The solution will immediately become viscous at room temperature after this last step. Warm the solution before working with it, as it loses its viscosity even at 37°C and any precipitated SDS will redissolve. We routinely filtered each batch of the (warm) solution through our transfer membrane in a pressure filtration device.

1. Thoroughly wet the filter in H$_2$O. Seal it in a bag (Seal-a-Meal, Sears) leaving approximately 1 cm between the filter and the seals all the way round, and so that there is a narrow, ~2.5 × 10 cm spout at one corner for the introduction and expulsion of solutions.

2. Depending on filter size, use a funnel to pour in approximately 50–150 ml of the hybridization mixture, exclude all bubbles, and seal the spout. Prehybridize for at least 30 minutes. Occasionally, move any new bubbles into the spout.

3. Cut the tip of the spout and push out all the solution. Introduce the probe with a transfer pipette (10 ml for a narrow blot, 15 ml for a full 20 × 40 cm filter), exclude as many bubbles as possible, and reseal.

4. Move the hybridization solution around the blot and back into the sealed spout to move any remaining bubbles into the latter, where they can remain for the remainder of the hybridization, and redistribute the solution evenly over the filter (check with a monitor).

5. By now the solution will have returned to room temperature and become viscous, so place the filter at the hybridization temperature (determined as above) until warm and repeat the previous step.

6. Seal the bag inside another bag along with 100–200 ml of H$_2$O if you are using an oven (other the hybridization solution will evaporate overnight). This is because Seal-a-Meal is slightly permeable to water and there is a very high surface area-to-volume ratio with this bag.

7. The solution should be moved around at least two, but preferably more, times during the incubation to ensure even hybridization and so that any new bubbles that have formed can be moved in to the spout. Hybridize for 18 hours.

2. WASHING CONDITIONS

How the filter is washed depends on what equipment is available. Since the filters are large, washing in a waterbath requires a large waterbath. Washes can also be done in bags in an oven, but our best results were achieved by doing a series of short washes in a large plexiglass tray on an orbital shaker as originally described by Church and Gilbert (1984) using washing solutions that were brought to temperature overnight in the hybridization oven.

1. Preheat 2 liters of washing solution A [40 mM Na(PO$_4$), 5% cheap SDS, 0.5% BSA, 1 mM EDTA] and 8 liters of washing solution B [40 mM Na(PO$_4$), 1% cheap SDS, 1 mM EDTA].

2. Pour a shallow layer of wash solution A into the tray to keep the filter moist until the first wash is added.

3. Push the hybridization solution into the spout and cut off the opposite end of the bag. Pull out the filter and place in the tray; pour on the first wash.

4. Wash twice, for 5 minutes each, in 1 liter of washing solution A and seven times, for 5 minutes each, in 1 liter of washing solution B.

5. Seal the filter in a fresh bag with the last liter of washing solution B and place at the washing temperature for 30 minutes. Agitate every 10 minutes.

6. Remove the filter from the bag and rinse three times, for 5 minutes each, in 1 liter of 0.1 M Na(HPO$_4$) at room temperature. Retain the last rinse in the tray.

7. Seal the filter in a new bag with 25 ml of RNase A solution (10 mM Tris-Cl, pH 8.0, 300 mM NaCl, 10 μg/ml RNase A) and place at 37°C for 2 hours. Move the solution around occasionally.

8. Remove the filter from the bag and rinse for 5–10 minutes in the 0.1 M Na(HPO$_4$).

9. Dry the blot thoroughly, tape it to filter paper, and expose it directly against film.

ACKNOWLEDGMENTS

This work was supported by NSF grant DMB-8901559 to D. R. Engelke, University of Michigan, and by PHS grant GM31532 to S. C. R. Elgin, Washington University.

REFERENCES

Anderson, S. (1981). Shotgun DNA sequencing using cloned DNase I-generated fragments. *Nucleic Acids Res.* **9**, 3015–3027.

Becker, P. B., and Schutz, G. (1988). Genomic footprinting. *In* "Genetic Engineering" (J. K. Setlow, ed.), Vol. 10, pp. 1–19. Plenum, New York.

Becker, M. M., and Wang, J. C. (1984). Use of light for footprinting DNA *in vivo*. *Nature (London)* **309**, 682–687.

Becker, M. M., Wang, Z., Grossmann, G., and Becherer, K. A. (1989). Genomic footprinting in mammalian cells with ultraviolet light. *Proc. Natl. Acad. Sci. U.S.A.* **86**, 5315–5319.

Bram, R. J., and Kornberg, R. D. (1987). Isolation of a *Saccharomyces cerevisiae* centromere DNA-binding protein, its human homolog, and its possible role as a transcription factor. *Mol. Cell. Biol.* **7**, 403–409.

Cartwright, I. L., and Elgin, S. C. R. (1986). Nucleosomal instability and induction of new upstream protein-DNA associations accompany activation of four small heat shock protein genes in *Drosophila melanogaster. Mol. Cell. Biol.* **6,** 779–791.

Cartwright, I. L., and Elgin, S. C. R. (1989). Nonenzymatic cleavage of chromatin. *In* "Methods in Enzymology" (P. M. Wasserman and R. D. Kornberg, eds.), Vol. 170, pp. 359–369. Academic Press, San Diego, California.

Church , G. M., and Gilbert, W. (1984). Genomic sequencing *Proc. Natl, Acad. Sci. U.S.A.* **81,** 1991–1995.

Elgin, S. C. (1988). The formation and function of DNase I hypersensitive sites in the process of gene activation. *J. Biol. Chem.* **263,** 19259–19262.

Elgin, S. C., and Miller, D. W. (1978). *In* "Mass Rearing of Flies and Mass Production and Harvesting of Embryos in Genetics and Biology of Drosophila" (M. Ashburner and T. R. F. Wright, eds.), Vol. 2A. Academic Press, London.

Ewel, A., Jackson, J. R., and Benyajati, C. (1990). Alternative DNA-protein interactions in variable-length internucleosomal regions associated with the *Drosophila ADH* distal promoter expression. *Nucleic Acids Res.* **18,** 1771–1781.

Gilmour, D. S., Thomas, G. H., and Elgin, S. C. R. (1989). *Drosophila* nuclear proteins bind to regions of alternating C and T residues in gene promoters. *Science* **245,** 1487–1490.

Giniger, E., Varnum, S. M., and Ptashne, M. (1985). Specific DNA binding of GAL4, a positive regulatory protein of yeast. *Cell (Cambridge, Mass.)* **40,** 767–774.

Gottesfeld, J. M., and Bloomer, L. S. (1980). Nonrandom alignment of nucleosomes on 5S RNA genes of *X. laevis. Cell (Cambridge, Mass.)* **21,** 751–760.

Gross, D. S., Collins, K. W., Hernandez, E. M., and Garrard, W. T. (1988). Vacuum blotting: A simple method for transferring DNA from sequencing gels to nylon membranes. *Gene* **74,** 347–356.

Gross, D. S., English, K. E., Collins, K. W., and Lee, S. (1990). Genomic footprinting of the yeast *HSP82* promoter reveals marked distortion of the DNA helix and constitutive occupancy of heat shock and TATA elements. *J. Mol. Biol.* **216,** 611–631.

Huibregste, J. M., and Engelke, D. R. (1989). Genomic footprinting of a yeast tRNA gene reveals stable complexes over the 5′-flanking region. *Mol. Cell. Biol.* **9,** 3244–3252.

Huibregste, J. M., and Engelke, D. R. (1990). Direct sequence and footprint analysis of yeast DNA by primer extension. *In* "Methods in Enzymology" (C. Guthrie and G. R. Fink, eds.), Vol. 194, pp. 550–562. Academic Press, San Diego, California.

Huibregste, J. M., Evans, C. F., and Engelke, D. R. (1987). Comparison of tRNA gene transcription complexes formed *in vitro* and *in nuclei. Mol. Cell. Biol.* **7,** 3212–3220.

Kassavetis, G. A., Riggs, D. L., Negri, R., Nguyen, L. H., and Geiduschek, P. (1989). Transcription factor IIIB generates extended DNA interactions in RNA polymerase III transcription complexes on tRNA genes. *Mol. Cell. Biol.* **9,** 2551–2566.

Kassavetis, G. A., Braun, B. R. Nguyen, L. H., and Geiduschek, E. P. (1990). *S. cerevisiae* TFIIIB is the transcription initiation factor proper of RNA polymerase III, while TFIIIA and TFIIIC are asssembly factors. *Cell (Cambridge, Mass.)* **60,** 235–245.

Larsen, A., and Weintraub, H. (1982). An altered DNA conformation detected by S1 nuclease occurs at specific regions in active chick globin chromatin. *Cell (Cambridge, Mass.)* **29,** 609–622.

McGhee, J. D., Wood, W., Donal, M., Engel, J. D., and Felsenfeld, G. (1981). A 200 base pair region at the 5′ end of the chicken adult β-globin gene is accessible to nuclease digestion. *Cell (Cambridge, Mass.)* **27,** 45–55.

Marchionni, M., and Gilbert, W. (1986). The triosephosphate isomerase gene from maize: Introns antedate the plant-animal divergence. *Cell (Cambridge, Mass.)* **46,** 133–141.

Marschalek, R., and Dingermann, T. (1988). Identification of a protein factor binding to the 5′-flanking region of a tRNA gene and being involved in modulation of tRNA gene transcription *in vivo* in *Saccharomyces cerevisiae. Nucleic Acids Res.* **16,** 6737–6752.

Maxam, A. M., and Gilbert, W. (1980). Sequencing end-labelled DNA with base-specific chemical cleavages. *In* "Methods in Enzymology" (L. Grossman and K. Moldave, eds.), Vol. 65, pp. 499–560. Academic Press, New York.

Mueller, P. R., and Wold, B. (1989). *In vivo* footprinting of a specific enhancer by ligation mediated PCR. *Science* **246,** 780–786.

Nedospasov, S. A., and Georgiev, G. P. (1980). Non-random cleavage of SV40 DNA in compact minichromosomes and free in solution by M. nuclease. *Biochem. Biophys. Res. Commun.* **92,** 532–539.

Newman, A. J., Ogden, R. C., and Abelson. (1983). tRNA gene transcription in yeast: Effects of specified base substitutions in the intragenic promoter. *Cell (Cambridge, Mass.)* **35,** 117–125.

Palmieri, M., and Tavey, M. G. (1990). Genomic footprinting: Detection of putative regulatory proteins in the promoter region of the interferon alpha-1 gene in normal human tissues. *Mol. Cell. Biol.* **10,** 2554–2561.

Pfeifer, G. P., Steigerwald, S. D., Mueller, P. R., Wold, B., and Riggs, A. D. (1989). Genomic sequencing and methylation analysis by ligation mediated PCR. *Science* **246,** 810–813.

Pfeifer, G. P., Tanquay, R. L., Steigerwald, S. D., and Riggs, A. D. (1990). *In vivo* footprint and methylation analysis by PCR-aided genomic sequencing: Comparison of active and inactive X chromosomal DNA at the CpG island and promoter of human PGK-1. *Genes. Dev.* **4,** 1277–1287.

Saluz, H. P., and Jost, J. P. (1987). A laboratory guide to genomic sequencing. "BioMethods," Vol. 1. Birkhauser, Boston.

Saluz, H. P., and Jost, J. P. (1989). Genomic sequencing and *in vivo* footprinting. Anal. Biochem. **176,** 201–208.

Sambrook, J., Fritsch, E. F., and Maniatis, T. (1989). "Molecular Cloning: A Laboratory Manual." Cold Spring Harbor Laboratory, Cold Spring Harbor, New York.

Sasse-Dwight, S., and Gralla, J. D. (1988).

Sasse-Dwight, S., and Gralla, J. D. (1990). Role of eukaryotic-type functional domains found in the prokaryotic enhancer receptor factor s[54]. *Cell (Cambridge, Mass.)* **62,** 945–954.

Selleck, S. B., and Majors, J. (1987a). *In vivo* DNA-binding properties of a yeast transcription activator protein. *Mol. Cell. Biol.* **7,** 3260–3267.

Selleck, S. B., and Majors, J. (1987b). Photofootprinting *in vivo* detects transcription-dependent changes in yeast TATA boxes. *Nature (London)* **325,** 173–177.

Sollner-Webb, B., and Reeder, R. H. (1979). The nucleotide sequence of the initiation and termination sites for ribosomal RNA transcription in *X. laevis. Cell (Cambridge, Mass.)* **18,** 485–499.

Thomas, G. H., and Elgin, S. C. R. (1988). Protein/DNA architecture of the DNase I hypersensitive region of the *Drosophila hsp26* promoter. *EMBO J.* **7,** 2191–2201.

Tullius, T. D. (1989). Physical studies of protein-DNA complexes by footprinting. *Annu. Rev. Biophys. Biophys. Chem.* **18,** 213–237.

Weintraub, H. (1985). High-resolution Mapping of S1- and DNase I-hypersensitive sites in chromatin. *Mol. Cell. Biol.* **5,** 1538–1539.

Weintraub, H., and Groudine, M. (1976). Chromosomal subunits in active genes have altered conformation. *Science* **193,** 848–856.

Wu, C. (1980). The 5′ ends of *Drosophila* heat shock genes in chromatin are hypersensitive to DNase I. *Nature (London)* **286,** 854–860.

Wu, C. (1984). Two protein-binding sites in chromatin implicated in the activation of heat-shock genes. *Nature (London)* **309,** 229–234.

Wu, C., Bingham, P. M., Livak, K. J., Holmgren, R., and Elgin, S. C. R. (1979a). The chromatin structure of specific genes: I. Evidence for higher order domains of defined DNA sequence. *Cell (Cambridge, Mass.)* **16,** 797–806.

Wu, C., Holmgren, R., Livak, K., Wong, Y.-C., Elgin, S. C. R. (1979b). The chromatin structure of specific genes: II. Disruption of chromatin structure during gene activity. *Cell (Cambridge, Mass.)* **16,** 807–814.

Part IV. Reconstitution of Functional Complexes

Chapter 16

Control of Class II Gene Transcription during in Vitro Nucleosome Assembly

JERRY L. WORKMAN, IAN C. A. TAYLOR, AND
ROBERT E. KINGSTON

Department of Molecular Biology
Massachusetts General Hospital
Boston, Massachusetts 02114
and
Department of Genetics
Harvard Medical School
Boston, Massachusetts 02115

ROBERT G. ROEDER

Laboratory of Biochemistry and Molecular Biology
The Rockefeller University
New York, New York 10021

419

I.   Introduction

Over the last several years extensive structural studies have indicated that the 5′ ends of transcriptionally active genes in cellular chromatin are usually contained in nuclease-hypersensitive (nucleosome-free) regions which are often occupied by sequence-specific transcription factors (reviewed in Elgin, 1988; Gross and Garrard, 1988). These studies suggest that nucleosomes and some transcription factors play opposing roles in the regulation of transcription. Thus, by excluding nucleosomes, the binding of certain transcription factors, alone or in combination, may generate accessible chromatin conformations over class II promoters. Support for this theory has come from genetic studies in yeast. Interruption of histone gene expression in yeast, resulting in nucleosome depletion, allows permissive transcription from several promoters in the absence of either upstream elements or the corresponding induced factors that are normally required for high-level transcription (Kim *et al.,* 1988; Han *et al.,* 1988; Han and Grunstein, 1988). These studies indicate that nucleosomes play an important role in the regulation of gene transcription by suppressing the basal levels of promoter activity (mediated by general factors) and, consequently, by enhancing the dependence of transcription on the action of regulatory factors.

Several laboratories have used *in vitro* approaches to investigate the involvement of nucleosomes in class II gene transcription. The assembly of the adenovirus major late promoter into nucleosomes (either by the components in a *Xenopus* oocyte extract or by self-assembly of histones into nucleosomes by dilution from high ionic strength) renders the promoter refractory to subsequent transcription initiation by RNA polymerase II and accessory factors (Knezetic and Luse, 1986; Lorch *et al.,* 1987, Workman and Roeder, 1987). However, if a stable preinitiation transcription complex is formed on the promoter prior to nucleosome assembly in *Xenopus* egg or oocyte extracts, it remains largely active in subsequent transcription (Matsui, 1987; Workman and Roeder, 1987; Knezetic *et al.,* 1988). These *in vitro* studies of a class II gene, as well as earlier studies on the class III 5S RNA genes (reviewed in Brown, 1984; Wolffe and Brown, 1988), begin to recapitulate functionally the natural events suggested by the *in vivo* studies mentioned above.

These approaches allow *in vitro* studies of the role that nucleosomes play in the accurate regulation of transcription. In addition, they can be used to address the role of regulatory factors in establishing the transcriptional potential of class II genes during or subsequent to chromatin assembly. In this chapter, we describe the preparation of reagents and *in vitro* protocols for nucleosome assembly and transcription from class II promoters at physiological ionic strength. Additional protocols for nucleosome assembly can be found in Wassarman and Kornberg (1989). The protocols described here are useful for addressing the ability of transcription factors to prevent or reverse nucleosome-mediated repression of promoter activity and the ability of such factors to bind nucleosomal DNA. In addition, we describe assays that involve concurrent nucleosome and transcription complex assembly to ask whether the competitive nucleosome assembly pathway creates conditions under which the regulatory activity of particular factors is more apparent.

II.   *In Vitro* Transcription of Class II Genes

A.   Accurate Transcription in Nuclear Extracts from Mammalian Cells

Accurate initiation of transcription from purified DNA templates is achieved in whole-cell or nuclear extracts but not by purified RNA polymerase II alone. This indicated that additional proteins (general transcription factors, see below) are required for recognition of eukaryotic promoters (Weil *et al.*, 1979; Manley *et al.*, 1980; Dignam *et al.*, 1983a,b). While there are several protocols for preparing extracts or concentrated fractions from such extracts which support accurate transcription initiation, a simple and commonly used protocol is that of Dignam *et al.* (1983a,b). This procedure involves the preparation of extracts from isolated mammalian cell nuclei, thus taking advantage of the nuclear localization of transcription components. A detailed protocol for the preparation of these extracts has recently been described (Abmayr and Workman, 1990).

Nuclear extracts are dialyzed into buffer D (100 mM KCl, 20% glycerol, 20 mM HEPES, pH 7.9, 0.2 mM EDTA, 0.5 mM DTT, and 0.2 mM PMSF) at the final stages of preparation and are diluted and supplemented for transcription reactions. A standard transcription reaction of 25 μl can include up to 15 μl of nuclear extract with final concentrations of 60 mM KCl, 12% glycerol, 7 mM MgCl$_2$, 0.12 mM EDTA, 0.3 mM DTT, 12 mM HEPES, pH 7.9, 5–25 μg/ml template DNA, and nucleoside triphosphates. The exact concentration of each nucleoside triphosphate depends on the assay chosen for RNA synthesis. If RNA is analyzed by primer extension or S1 nuclease digestion of complementary probes (for protocols see, Ausubel *et al.*, 1990; Sambrook *et al.*, 1989), 0.6 mM of each nucleotide is used. Transcripts can also be assayed by run-off transcription from linear templates. In this case the concentration of GTP is reduced to 25 μM and includes 5 μCi of [α-^{32}P]GTP/reaction. Alternatively, transcription can be assayed from circular templates through the G-less cassette (which contains no G residues on the transcribed strand), if placed immediately downstream of the cap site such that a correctly initiated RNA contains no G residues (Sawadogo and Roeder, 1985). In this assay GTP is omitted, UTP is the labeled nucleotide at 25 μM concentration, and both 0.1 mM 3'-O-methyl-GTP (a chain terminator) and 15 units of RNase T1 are included. These last two components limit the appearance of nonspecifically initiated transcripts (which would include G residues) from the transcription analysis. However, because of contaminating GTP in extracts, it is important to determine the background level of transcription signal (using a promoter-less construct) to ensure that it is low relative to the basal level of transcription initiated at the promoter (Taylor and Kingston, 1990).

Transcription reactions are incubated for 60 minutes at 30°C. The reactions are then quenched by the addition of 75 μl of stop mix (67 mM sodium acetate, pH 5.5, 6.7 mM EDTA, 0.33% SDS, and 0.66 mg/ml carrier yeast tRNA) followed by phenol extraction and then chloroform extraction. RNA is precipitated on dry ice after the addition of 250 μl of 95% ethanol with 0.1 M sodium acetate (pH 5.5) and subsequently pelleted in a microfuge. This RNA pellet can be resuspended and analyzed by primer extension or S1 nuclease protection. For labeled RNA it is often useful to reduce the concentration of unincorporated nucleotides further by two additional ethanol precipitations in order to increase the RNA signal over background. This is done by resuspending the wet pellet in 100 μl of 67 mM sodium acetate (pH 5.5) and reprecipitating as above. The final pellets are briefly vacuum dried and resuspended in 20 μl of deionized formamide with 0.002% xylene cyanol and bromophenol blue and loaded onto a 7 M urea, 4% acrylamide gel with 1 × TBE running buffer. The gel is run until the xylene cyanol reaches the bottom, washed with water for 15 minutes with shaking (to remove the urea), dried, and exposed to film.

B.   Isolation of Transcription Factors

There are two well-characterized classes of transcription factors whose activities are readily demonstrated *in vitro*. These include regulatory factors which usually act via gene-specific upstream sequence elements and general initiation factors which act via core promoter elements common to most class II genes (i.e., TATA elements). Because the action of regulatory factors must ultimately (directly or indirectly) affect the activity of the general transcription factors, investigations of the control of transcription during nucleosome assembly must involve independent or combined manipulation of these two classes of factors.

The functions of independently isolated factors can readily be assayed *in vitro* in appropriate reconstituted systems. However, when the regulatory factor of interest is present in nuclear extracts (the most common source of general factors), a transcription system depleted of the regulatory factor is required. Immobilized oligonucleotide affinity depletion methods have been successfully used to separate specific regulatory factors from general factors (Scheidereit et al., 1987), as have immunodepletion methods with specific antibodies coupled to protein A–Sepharose (Pognonec and Roeder, 1991). However, standard chromatographic procedures can separate regulatory factors from general factors, as well as individual general factors from each other. This allows the reconstitution of more purified systems in which the effects of specific regulatory factors can be tested (during nucleosome assembly) in conjunction with all the specific subsets of the general factors.

Gene-specific regulatory factors, either ubiquitous or cell specific, can be purified directly from cellular extracts by affinity chromatography techniques or following expression of corresponding cDNAs in various prokaryotic or eukaryotic expression vectors (for protocols, see Ausubel et al., 1990; Sambrook et al., 1989). In contrast most of the general factors must still be isolated by conventional biochemical fractionation techniques from whole-cell or nuclear extracts. At present, such studies have resulted in the identification of at least six general factors (designated TFIIA, TFIIB, TFIID, TFIIE, TFIIF, and TFIIG) that are implicated in accurate transcription initiation by RNA polymerase II on cloned DNA templates (reviewed in Saltzman and Weinmann, 1989; Sawadogo and Sentenac, 1990; see also Sumimoto et al., 1990). Of these factors, TFIID appears to play a preeminent role in preinitiation complex assembly, since it binds to the TATA box and facilitates subsequent binding of RNA polymerase II and the remaining general factors (Nakajima et al., 1988; Van Dyke et al., 1988; Buratowski et al., 1989). Although there is considerable information regarding these factors, a better understanding of their exact number and mechanism of action awaits their complete purification and cloning. TFIID provides an example of a general factor which has been cloned, expressed, and successfully employed for functional studies in coupled nucleosome assembly–transcription systems (Meisterernst et al., 1990; Workman et al., 1991).

A current fractionation scheme (Sumimoto et al., 1990) based on earlier procedures (Dignam et al., 1983a,b; Sawadogo and Roeder, 1985; Reinberg and Roeder 1987) is shown in Fig. 1. The isolated general factors are usually dialyzed into buffer D (see above) and are analyzed functionally as described above. RNA polymerase II is usually purified separately by an independent procedure (reviewed in Sawadogo and Sentenac, 1990), most conveniently from the nuclear pellets which are a by-product of the nuclear extract preparations (Reinberg and Roeder, 1987). Also potentially relevant are additional cofactors (reviewed in Lewin, 1990; Ptashne and Gann, 1990) that have been suggested from indirect evidence to be required for certain DNA-

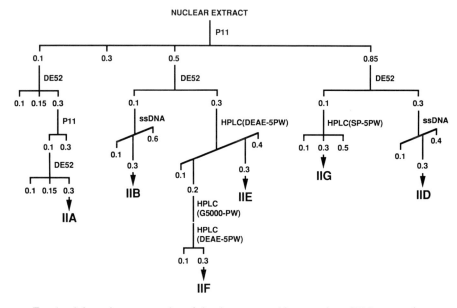

FIG. 1.   Schematic representation of the chromatographic separation of HeLa general transcription factors. The samples in buffers containing 0.1 *M* KCl were loaded onto the indicated
columns and eluted either stepwise (horizontal lines) or with gradients (oblique lines). The numbers represent the molar concentrations of KCl. Adapted from Fig. 1 of Sumimoto *et al.* (1990).
Proc. Natl. Acad. Sci. U.S.A. **87**, 9158–9162: Copyright National Academy of Science USA.

binding proteins to stimulate transcription by the general transcription
machinery. Direct evidence for the presence and function of such factors in
reconstituted systems containing highly purified general factors and upstream
activators has been obtained recently (Meisterernst *et al.,* 1991, and references
therein). Such factors will likely prove important for studies with reconstituted
nucleosome assembly–transcription systems, as these systems become more
defined.

III.   Nucleosome Assembly under Physiological Conditions

The histones which make up the nucleosome core particle (H2A, H2B, H3,
and H4) are capable of self-assembly into nucleosomes when mixing with
purified DNA at monovalent salt concentrations greater than 1 *M* is followed

by slow dialysis or dilution to lower ionic strengths. If mixed at physiological ionic strengths, rapid binding of the basic histones to nucleic acids results in precipitation instead of ordered nucleosome assembly (reviewed in Laskey and Earnshaw, 1980). Nucleosome assembly at physiological salt concentrations is achieved by the inclusion of acidic components (i.e., polyglutamic acid; see Stein, 1989), which serve as a "histone buffer" and allow an orderly assembly of nucleosomes (reviewed in Rhodes and Laskey, 1989). In extracts from *Xenopus* eggs or oocytes this function is carried out by acidic nucleosome assembly factors which complex with histones. These include nucleoplasmin (apparent MW 29,000–33,000) which exist as a pentamer (Laskey *et al.,* 1978; Dingwall *et al.,* 1982) and N1/N2 (apparent MW 105,000–110,000) (Kleinschmidt and Franke, 1982). The observation that purified nucleoplasmin alone functions in nucleosome assembly *in vitro* suggests that it can bind to all the core histones (Laskey *et al.,* 1978; Sealy *et al.,* 1986). However, in *Xenopus* egg extracts nucleoplasmin has been found to be complexed with H2A and H2B, while N1/N2 is complexed with H3 and H4 (Kleinschmidt *et al.,* 1985; Dilworth *et al.,* 1987; Kleinschmidt *et al.,* 1990).

A.   Preparation of Nucleosome Assembly Extracts

A simple protocol for preparing nucleosome assembly extracts from *Xenopus* oocytes (the oocyte S-150) has been detailed (Shimamura *et al.,* 1989a). Below is an alternative protocol slightly modified from that of Laskey *et al.,* (1978). Although it describes the preparation of nucleosome assembly-competent extracts from *Xenopus* eggs, the same procedure can also be used to prepare extracts from oocytes (Sealy *et al.,* 1986; Workman and Roeder, 1987).

Preparation of High-Speed Xenopus Egg Extracts

1. Prime a large well-fed female *X. laevis* with human chorionic gonadotropin by injecting 250 units into the dorsal lymph sac in the morning. In the evening, inject the animal with an additional 500–1000 units and place in a breeding tank containing collection buffer (120 mM NaCl, 2 mM KCl, 1 mM MgSO$_4$, 5 mM HEPES, pH 7.4, and 0.1 mM EDTA). Collect eggs overnight.

2. Dejelly eggs by swirling gently in a 600 ml beaker with 200 ml of 2% cysteine (pH 7.8) with two changes of solution for a total of 15 min.

3. Wash dejellied eggs by gently pouring in collection buffer and decanting 10 times. Remove damaged eggs during decanting and any remaining damaged eggs with a pipette upon inspection in a petri dish.

4. Rinse eggs quickly twice in ice-cold distilled water and pour into a glass Dounce homogenizer. Remove excess water down to the surface of the packed eggs.

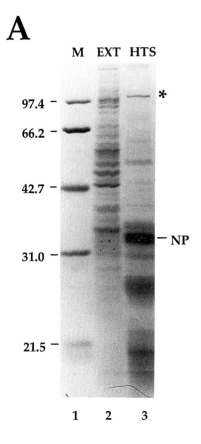

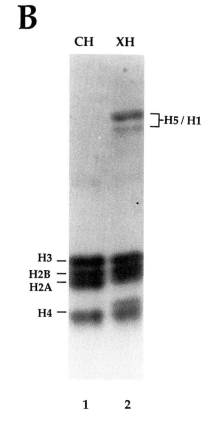

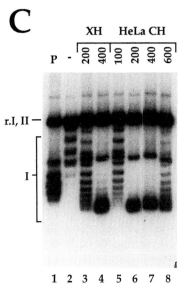

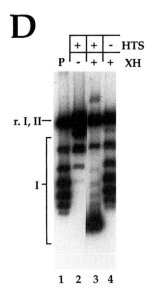

5. Homogenize one stroke with a size B glass pestle. Pour the homogenate into a test tube and add 0.66 volume of HA buffer (120 mM KCl, 2 mM MgCl$_2$, 20 mM Tris-HCl, pH 7.5).

6. Centrifuge at 1300 g_{av} (2500 rpm with a Beckman JS 4.2 rotor in a J6 centrifuge) for 25 minutes at 4°C. Remove the cloudy-gray center supernatant, avoiding the pelleted material and the plug at the top. Place supernatants into swinging bucket ultracentrifuge tubes (i.e., Beckman SW55).

7. Centrifuge at 150,000 g_{av} (40,000 rpm, SW55 rotor) for 30 minutes at 4°C. Remove center clear (yellow tinted) supernatant and recentrifuge for 30 minutes at 150,000 g. Remove second supernatant.

8. Gently extract any remaining lipids with an equal volume of 1,1,2-trichlorotrifluoroethane. Separate the phases by spinning for 2 minutes at 1000 g_{av} (2200 rpm, JS 4.2 rotor) and then repeat the extraction. To the final extract add phenylmethylsulfonyl fluoride (PMSF) to a concentration of 1 mM.

These extracts are competent in assembling nucleosomes onto exogenously added DNA. Heat treatment of these extracts and pelleting the precipitated proteins reduces the protein concentration to 2–5% of that in the original extract. The heat-treated supernatant is depleted in transcriptional activity, topoisomerase activity, and histones but is very highly enriched for nucleoplasmin.

FIG. 2. Reconstitution of nucleosome assembly *in vitro*. (A) The proteins from a high-speed *Xenopus* egg extract (EXT) (lane 2) and the heat-treated supernatant (HTS) prepared from that extract (lane 3) were separated on a 12% acrylamide SDS gel. Twenty micrograms of protein was loaded in each lane. NP marks the nucleoplasmin band and the asterisk illustrates a protein band of the correct molecular weight to be N1/N2. The sizes in kilodaltons of molecular weight markers run in lane 1 are indicated. (B) Ten micrograms of HeLa core histones, purified by hydroxylapatite chromatography (lane 1), and *Xenopus* erythrocyte histones, purified by acid extraction (lane 2), were run on a 15% acrylamide SDS gel. The core histones (H2A, H2B, H3, and H4) are indicated, as well as the H1 and H5 present in the erythrocyte histones. (C) Reconstitution of nucleosome assembly with the purified histones. One hundred nanograms of plasmid DNA (P, lane 1), internally labeled according to Razvi *et al.* (1983), was incubated with the HTS in the absence of histones (lane 2) or with the nanogram quantities of *Xenopus* erythrocyte (XH) or HeLa core histones (HeLa CH) indicated. The deproteinized DNA was run on a 1% agarose gel. The location of closed circular (I), relaxed closed circular (r.I), and nicked (II) plasmid bands are indicated. The center band in lanes 4, 6 and 7 indicates linear plasmid. (D) Both the heat-treated supernatant (HTS) and histones (XH) are required for nucleosome assembly. Assembly reactions included HTS and/or *Xenopus* erythrocyte histones (XH) as indicated. Reactions omitting either component contained the appropriate buffer. Other designations are as in C.

Preparation of Heat-Treated Egg Supernatants

1. Divide the egg extracts into 1-ml aliquots in 1.5-ml tubes.
2. Place into an 80°C water bath for exactly 10 minutes and then immediately into a ice/water bath for 5 minutes.
3. Centrifuge tubes in a microfuge for 2 minutes at 14,000 rpm (approximately 12,000 g) and 4°C.
4. Take off the heat-treated supernatants, aliquot, freeze in liquid N_2, and store at −80°C.

The protein composition of the egg extracts and the heat-treated supernatants is shown in Fig. 2A. Note that the heat-treated supernatant contains only a fraction of the proteins present in the egg extracts and is highly enriched in nucleoplasmin. Also present in the heat-treated supernatant is a protein (marked by an asterisk) whose apparent molecular weight is similar to that of N1/N2 (approximately 110 kDa). However, this protein is present in much lower quantities than nucleoplasmin, which suggests that nucleosome assembly with this supernatant is facilitated primarily by nucleoplasmin. Nucleoplasmin can be purified to homogeneity from these supernatants in only two columns (Sealy *et al.*, 1986, 1989). However, we have found that nucleosome assembly with the heat-treated supernatant functions over a wider range of conditions than assembly with purified nucleoplasmin. In particular, the heat-treated supernatant works efficiently when diluted with transcription factors and at different histone concentrations. The nucleosome assembly capacity of the nucleoplasmin in the heat-treated supernatant may be assisted by the presence of lesser amounts of N1/N2 or other histone-binding components in the supernatant (Sealy *et al.*, 1986).

B.   Purification of Histones

Nucleosome assembly with the heat-treated supernatant requires supplementation with histones. Thus, both nucleosome assembly and any transcriptional effects can be shown to be due to the addition of purified histone proteins. Histones from any cell type or fractionation scheme could be used to reconstitute nucleosome assembly. For a review on histone isolation, see von Holt *et al.* (1989). Described below are two simple protocols for the preparation of histones that are functional in nucleosome assembly. The first, more simple, protocol involves acid extraction of histones from *Xenopus* erythrocyte nuclei, which contain lower concentrations of other proteins than most cell types. The second procedure involves the purification of core histones by hydroxylapatite chromatography and is modified from the protocol of Stein and Mitchell (1988). The second protocol is preferable since the purified core histones are separated from linker histones (i.e., H1 or H5) and are more active in nucleosome assembly. This latter method is applicable to purifying histones from different cell types.

Acid Extraction of Histones from Xenopus Erythrocytes

1. Pith and dissect a frog to expose the heart. Clip the tip of the ventricle with a small scissors so the heart pumps blood through the wound.

2. Invert the animal over a large petri dish placed on ice, containing 20 ml of WB with 15 units/ml heparin (WB: 120 mM NaCl, 5 mM KCl, 3 mM $MgCl_2$, and 10 mM HEPES, pH 7.4).

3. Pellet cells for 5 minutes at 1000 g (2200 rpm, JS 4.2 rotor) and 4°C. Remove the supernatant and the very top slightly white layer of cells. Resuspend the pellet in 10 ml of WB and filter through four layers of cheesecloth to remove any clots. Pellet the cells and wash 2–3 times in 10 ml of WB or until the supernatant is clear.

4. Lyse the cells by resuspending in 10 ml of MLB (60 mM KCl, 15 mM NaCl, 3 mM $MgCl_2$, 15 mM PIPES, pH 6.5, 0.1% NP40, and freshly added 0.5 mM PMSF and 1 mM sodium tetrathionate). Pellet nuclei for 5 minute at 1500 g (2700 rpm, JS 4.2 rotor) and wash at least two additional times until the pelleted nuclei are white.

5. Wash nuclei, as in step 4, twice in MB (same as MLB but omitting NP40).

6. Resuspend the nuclear pellet with a type B pestle in a glass homogenizer in 20 pellet volumes of 0.4 N H_2SO_4. Mix continuously by rotating or gently stirring at 4°C for 1 to 2 hours.

7. Centrifuge for 15 minutes at 12,000 g_{av} (12,500 rpm, Sorvall SS34 rotor) and 4°C. Remove and save the supernatant. Reextract the pellet with 20 volumes of 0.4 N H_2SO_4 for 1 hour and repellet.

8. Combine first and second supernatants, add 5 volumes of ethanol, and precipitate at -20°C overnight.

9. Pellet precipitate at 12,000 g_{av} (12,500 rpm, SS34 rotor) for 15 minutes at 4°C, remove the supernatant, wash the pellet with acetone, and vacuum dry.

10. Resuspend in 50 mM Tris, pH 8.0, 5 mM EDTA to 0.5 ml, or a sufficient volume to dissolve the pellet within a few hours and store frozen.

Purification of Core Histones by Hydroxylapatite Chromatography

1. Collect and concentrate HeLa cells from 3 liters of spinner cultures and wash with phosphate-buffered saline (137 mM NaCl, 2.7 mM KCl, 4.3 mM $Na_2HPO_4 \cdot 7H_2O$, 1.4 mM KH_2PO_4, pH 7.3).

2. Lyse cells by resuspending in 20 pellet volumes of lysis buffer (10 mM Tris-HCl, pH 8.0, 3 mM $MgCl_2$, 0.25 M sucrose, 1% NP40, 0.5 mM PMSF) and homogenize several strokes with a type B pestle. Alternatively, the nuclear pellets remaining after nuclear extract preparation can be used (Dignam *et al.,* 1983a) and are resuspended by homogenization in the same buffer.

3. Pellet nuclei for 20 minutes at 3600 g_{av} (4200 rpm, JS 4.2 rotor). Wash nuclei twice in lysis buffer and twice in rinse buffer (same as lysis buffer but omitting NP40). Homogenize gently if necessary to resuspend the pellets.

4. Determine the concentration of resuspended nuclei by diluting an aliquot in 2 M NaCl and reading the A_{260}. Take an amount of nuclei containing approximately 12 mg of DNA (240 A_{260} units) and resuspend in 50 ml of LSB (0.4 M NaCl, 10 mM Tris-HCl, pH 8.0, 1 mM EDTA, 0.5 mM PMSF). Stir gently for 15 minutes at 4°C. Pellet for 15 minutes at 3600 g_{av} (4200 rpm, JS 4.2 rotor), and wash once more in LSB.

5. Resuspend nuclei in 50 ml of MSB (0.6 M NaCl, 50 mM NaPO4, pH 6.8, 0.5 mM PMSF). Stir the resuspension gently for 10 minutes to lyse the nuclei.

6. While stirring slowly, add 20 g of dry Bio-gel HTP hydroxylapatite (Bio-rad) and allow the resin to swell to a paste.

7. Suspend the paste in the minimal volume of MSB necessary to pour the resin into a column. Pour the column and collect the eluant, which contains partially pure histone H1 (Stein and Mitchell, 1988).

8. Wash the column with 10 column volumes of MSB.

9. Elute core histones from the column with a step of HSB (2.5 M NaCl, 50 mM NaPO4, pH 6.8, 0.5 mM PMSF). Monitor the protein concentration of the fractions, and pool the peak fractions.

10. Concentrate the core histones to 2–4 mg/ml using low-molecular-weight cut-off centriprep concentrators (Amicon) and store frozen.

C. Nucleosome Assembly of Plasmid DNA

The components prepared in the previous section can assemble nucleosomes *in vitro* at physiological salt concentrations. The protein composition of the heat-treated egg supernatant (HTS) and the purified histones are shown in Fig. 2A and B, respectively. Nucleosome assembly can be assayed by micrococcal nuclease or DNase I digestion, gradient sedimentation, electron microscopy, and plasmid supercoiling (for the various protocols, see Wassarman and Kornberg, 1989). Nucleosome assembly is most easily assayed by the introduction of negative supercoils into closed circular plasmid DNA. In this instance, it is necessary to supplement nucleosome assembly reactions with topoisomerase I. The topoisomerase continuously relaxes the template such that the stable supercoils resulting from nucleosome assembly are readily apparent when the subsequently deproteinized DNA is run on agarose gels. Large amounts of topoisomerase I are required in these reactions [30 units (Promega)/μg DNA]. A standard reaction is described below.

Protocol for Reconstituting Nucleosome Assembly

1. Prepare 10 μl of HTS : histone mix for each reaction. This mix is 80%

HTS and 20% of the appropriate histone dilution in 10 mM Tris-HCl (pH 8.0), 1 mM EDTA. Final histone concentrations in this mixture should be approximately 10–40 μg/ml. Incubate this mixture for 15 minutes at 20°C.

2. Start individual assembly reactions by mixing 10 μl of HTS : histone mix, 5 μl of buffer D, nuclear extract, or isolated factors, and 1 μl of 100 μg/ml plasmid DNA. Either unlabeled or internally labeled closed circular plasmid DNA (Razvi *et al.*, 1983) can be used. Incubate at 30°C for the desired time (we routinely use 1 hr).

3. Quench the reactions with 5 μl of DSB (3% SDS, 0.1 M EDTA, 50 mM Tris-HCl, pH 8.0, 25% glycerol, and 0.002% bromophenol blue and xylene cyanol). Add 1 μl of 10 mg/ml proteinase K and incubate for 30 minutes at 37°C.

4. Load the samples directly onto a 1% agarose TBE gel. It is best to run gels slowly (overnight is best) such that linear DNA does not migrate near the most supercoiled forms. Note that the inclusion of chloroquine in the gel and running buffer can be used to shift relative mobility of the most supercoiled forms, allowing different ranges of resolution (Shure and Vinograd, 1976).

5. DNA bands can be visualized by staining the gel with ethidium bromide, which should be added after the gel has run. If residual RNA in the extracts obscures the DNA bands, it can be removed by the addition of 5 μg of DNase-free RNase A followed by a 30-minute incubation at 37°C, prior to the proteinase K digestion in step 3. When labeled DNA is used the agarose gels are dried and exposed to film.

Figure 2C illustrates the products of assembly reactions. When histones were omitted, the moderate supercoils (form I) present in the input labeled plasmid (P) were relaxed by the topoisomerase. When either *Xenopus* erythrocyte histones (XH) or HeLa core histones (CH) were included, the closed circular plasmid was resupercoiled in a histone concentration-dependent manner. While nucleosome assembly of the nicked plasmid (form II) is not revealed in this assay, glycerol gradient sedimentation and micrococcal nuclease digestions demonstrate that this plasmid form is also assembled into nucleosomes in these reactions. At an excess of histones relative to the capacity of the HTS (600 ng HeLa CH), supercoiling decreased. Note that the HeLa core histones supercoiled the plasmid more efficiently at lower histone concentrations than the acid-extracted *Xenopus* erythrocyte histones. This difference is probably due to a combination of the method of preparation, the source of the histones, and the presence of H1 and H5. Figure 2D illustrates that both the HTS and purified histones are required for nucleosome assembly. In the absence of the HTS, instead of assembling nucleosomes, the free histones precipitated plasmid DNA, inhibiting relaxation by topoisomerase I.

D. Mononucleosome Assembly with Short DNA Fragments

The interactions of transcription factors and nucleosomes can be addressed both through the functional studies described in the following sections and through DNA-binding studies (band shift and footprinting) in which nucleosomes are used as the substrate for factor binding. Mononucleosomes can be assembled with short DNA fragments (150–200 bp) containing binding elements of interest using either salt dilution protocols (Lorch *et al.*, 1987), the reconstituted assembly system described above, or with purified nucleoplasmin (Rhodes and Laskey, 1989). These complexes in turn can be used as substrates to address the ability of particular factors to bind to nucleosomal DNA. For example, reconstituted nucleosomes have been used to illustrate the binding of steroid receptors to nucleosomes (Perlmann and Wrange, 1988; Pina *et al.*, 1990; Archer *et al.*, 1991) and to demonstrate the differential binding activities of GAL4 derivatives versus human heat-shock factor to nucleosomal DNA (Taylor *et al.*, 1991). With reconstituted systems, nucleosome assembly is performed at physiological salt concentrations; thus, it is possible to prebind factors and address the stability of their binding to subsequent nucleosome assembly.

The same reactions conditions described above will assemble mononucleosomes on short DNA fragments. It is convenient to include the same amount of total DNA as used in the assembly reactions above (i.e., 100 ng/reaction). This can include only specific fragment or can also include nonspecific fragments, such as extracted mononucleosome DNA. Alternatively, plasmid DNA or poly(dIdC) can be used. As with the plasmid assembly reaction described above, changing the histone concentrations modulates the extent of mononucleosome assembly.

Figure 3 illustrates the assembly of mononucleosomes with a 170-bp end-labeled fragment. Addition of this fragment to assembly reactions (which also contain 100 ng plasmid DNA) in the absence of histones did not alter the mobility of the probe on a nondenaturing acrylamide gel. When 200 ng of HeLa core histones was also included, a large fraction of the fragment was shifted into a nucleosome complex. When identical reactions were sedimented on 5–20% sucrose gradients (Fig. 3B), the fragment sedimented as free DNA when histones were omitted and as a nucleosome core complex (11S) when histones were included. Figure 3C illustrates the nucleosome complex as detected on a 1.5% agarose minigel and illustrates that the gradient-purified 11S complexes migrate at the same position.

While the extent of nucleosome assembly on short fragments is less efficient than on plasmid DNA, increasing concentrations of histones can drive a fragment almost completely into the nucleosome complex (for example, at

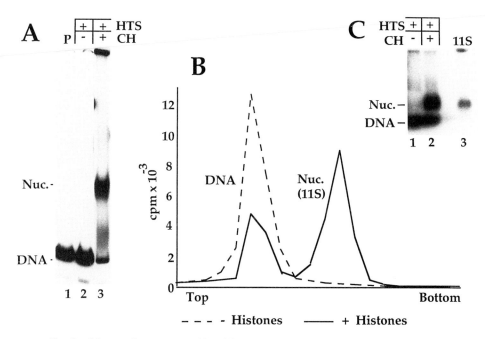

FIG. 3.   Mononucleosome assembly with short DNA fragments. Assembly reactions contained 100 ng of unlabeled plasmid DNA and 5 ng of a 170-bp end-labeled probe. Reactions contained heat-treated supernatant (HTS) and HeLa core histones (CH) as indicated. (A) Complexes resulting from the assembly reactions were separated on a 4% acrylamide nondenaturing TAE gel. Lane 1 illustrates the migration of the labeled probe. Lanes 2 and 3 illustrate the migration of the probe after assembly reactions omitting histones or including 200 ng of core histones, respectively. The free DNA and nucleosome core complexes are indicated. (B) Sedimentation of assembly products from reactions as in A through 5–20% sucrose gradients. The sedimentation profile from the reaction in which histones were omitted is indicated by the dashed line and that from reactions including core histones is illustrated by the solid line. The naked DNA and 11S nucleosome peaks are indicated. Twenty fractions were collected from each gradient and the top of the gradient is on the left. (C) Complexes from assembly reactions as in A were run on a 1.5% agarose minigel in 10 mM Tris, pH 8.0, 1 mM EDTA. Core histones were included or omitted as indicated. Lanes 3 illustrates the migration of the 11S complexes isolated from the sucrose gradient shown in B.

approximately 400 ng of HeLa core histones). Thus, factor binding can be assayed immediately after nucleosome assembly. As with plasmid supercoiling, however, an excess of histones over the capacity of the heat-treated supernatant will inhibit mononucleosome assembly. Thus, histone concentrations should be carefully titrated. Alternatively, the 11S nucleosome complexes can be gradient purified prior to the analysis of factor binding.

IV. Nucleosome Assembly and Transcription

The effect of assembling the major late promoter into nucleosomes on transcription has been addressed *in vivo* by assembling template DNA into nucleosomes in *Xenopus* oocyte or egg extracts or by salt dilution reconstitution of a single nucleosome over the promoter (Knezetic and Luse, 1986; Lorch *et al.*, 1987; Matsui, 1987; Workman and Roeder, 1987; Knezetic *et al.*, 1988). These studies have illustrated that, while nucleosomes repress promoter activity in transcription, previously formed preinitiation transcription complexes (containing minimally the TATA factor TFIID) confer a degree of resistance to subsequent nucleosome-mediated repression of transcription. Concurrent nucleosome assembly, using the reconstituted system described here, has been used to limit preinitiation complex formation, creating an early rate-limiting step in transcription. Proteins which stimulate preinitiation complex formation (i.e., the immediate-early protein of Pseudorabies virus, Abmayr *et al.*, 1988; the upstream stimulatory factor USF, Carcamo *et al.*, 1989; and GAL4 derivatives, Horikoshi *et al.*, 1988) display a greater fold of transcriptional stimulation during nucleosome assembly than without nucleosome assembly (Workman *et al.*, 1988, 1990, 1991). Thus, transcription is more dramatically regulated by these proteins *in vitro* under conditions of nucleosome assembly.

A. Template Assembly in Extracts versus the Reconstituted System

As stated above, complete (i.e., not heat-treated) *Xenopus* egg or oocyte extracts are able to assemble exogenous DNA into nucleosomes using an endogenous pool of histones, assembly factors, and topoisomerases (Laskey *et al.*, 1978; Glikin *et al.*, 1984). Such extracts have been used for nucleosome assembly in the analysis of class II promoter function (Knezetic and Luse, 1986; Matsui, 1987; Workman and Roeder, 1987; Knezetic *et al.*, 1988; Corthesy *et al.*, 1990). Nucleosome assembly in complete extracts is dependent on N1 as well as nucleoplasmin (Dilworth *et al.*, 1987) and thus presumably proceeds by a more natural mechanism (Kleinschmidt *et al.*, 1990) than assembly facilitated primarily by nucleoplasmin alone. In addition, nucleosome assembly of plasmid DNA in whole extracts can generate physiological spacing of nucleosomes which has not been accomplished with reconstituted systems (reviewed in Rhodes and Laskey, 1989).

The disadvantage of using complete *Xenopus* egg or oocyte extracts is that the extracts contain endogenous transcription activity, some class II gene regulatory factors, and histones. A primary concern (especially in oocyte extracts) is the appearance of transcripts which are insensitive to 1 μg/ml α-amanitin (which inhibits RNA polymerase II when included during the transcription assay) and results from endogenous RNA polymerase III activity (Glikin and Blangy, 1986; Knezetic and Luse, 1986; Workman and Roeder, 1987). These transcripts can show some dependence on the presence of class II promoters in the template DNA (J. L. Workman and R. G. Roeder, unpublished observation, 1987). In addition, a recent study has demonstrated the function of an endogenous upstream factor (NF1) during nucleosome assembly in a complete extract (Corthesy et al., 1990). Thus, studies involving the addition of isolated upstream regulatory factors to nucleosome assembly/transcription reactions may be complicated in complete extracts by the possibility that related DNA-binding proteins may already be present. Finally, nucleosome assembly with complete extracts does not allow an investigation of the function of histones from different sources (i.e., cell types or fractionation schemes) or simple modulation of the rate and extent of nucleosome assembly by histone titration, as in the reconstituted system (Section III,C and IV,D).

Thus, complete extracts are less amenable to manipulation and supplementation than reconstituted systems. However, due to the more native conformation of the assembled template (i.e., physiologically spaced nucleosomes), the use of complete extracts for nucleosome assembly and transcription studies remains valuable. Below is a general protocol for nucleosome assembly in these extracts and subsequent transcriptional analysis. To measure RNA polymerase II transcription in the presence of egg or oocyte extracts, primer extension or nuclease protection analysis of RNA should be performed and carefully controlled with α-amanitin (Matsui, 1987; Corthesy et al., 1990). Alternatively, assembled templates can be isolated by gel filtration or glycerol gradient sedimentation and subsequently analyzed in transcription (Workman and Roeder, 1987). We emphasize that the exact protocol used should result from a series of carefully controlled experiments monitoring the extent of nucleosome assembly achieved, the α-amanitin sensitivity of observed transcription, and the dependence of transcription on added transcription factors for which specific functions are invoked.

Nucleosome Assembly and Transcription with Complete Egg or Oocyte Extracts

1. Preincubate template DNA at 5–25 μg/ml in 10 μl of a mixture containing 6 μl of HeLa nuclear extract, reconstituted transcription factors or buffer D (see Section II,A) with final conditions of 12% glycerol, 60 mM KCl, 30 mM HEPES, pH 8.0, 1–7 mM MgCl$_2$, 0.6 mM EDTA, and 0.6 mM ATP.

2. Preincubate for 1 hour at 30°C.

3. Dilute the preincubation reactions with 30–40 μl of *Xenopus* egg or

oocyte extract (nucleosome assembly may be facilitated by the addition of approximately 10 units of topoisomerase I; Matsui, 1987).

4. Incubate at 20°C for 3–4 hours for nucleosome assembly to proceed. Nucleosome assembly can be assayed following this incubation as described in Section III,C.

5. Supplement the assembly reactions with any required general transcription factors not added during the preincubation step (usually in 10–20 μl) and bring the whole reaction back into transcription conditions (12% glycerol, 60 mM KCl, 7 mM MgCl$_2$, and HEPES to 20 mM, final pH 8.0, and NTPs as described in Section II,A).

6. Transcribe for 1 hour at 30°C.

7. Quench the reactions with 3 volumes of transcription stop mix; extract and analyze the RNA by primer extension or nuclease protection (Section II,A).

Alternatively, transcription can be assayed after isolation of the assembled templates if labeled template DNA were included. Following the assembly step (step 4 above) follow steps 5A to 9A.

5A. Apply each assembly reaction to a 0.5-ml Sepharose CL-4B gel filtration column (Pharmacia), poured in a 1-ml syringe, at 4°C. Columns can be run in 10 mM Tris, pH 8.0, 1 mM EDTA, 5% glycerol, or in 60 mM KCl, 12% glycerol, 20 mM HEPES, pH 8.0, 1 mM EDTA.

6A. Pool peak fractions in the excluded volume, count an aliquot of each, and make the concentration of templates from all reactions equal by dilution with running buffer.

7A. Take 10–20 μl of the isolated templates, supplement with transcription factors, and bring the total volume back into transcription conditions as in step 5.

8A. Transcribe for 1 hour at 30°C.

9A. Quench reactions and analyze RNA as described above (step 7). In addition, labeled RNA may be analyzed as in Section II,A by including three ethanol precipitations.

Background transcripts resulting from α-amanitin resistant elongation complexes (i.e., polIII) in the isolated templates may obscure specific transcripts. This can be alleviated by omitting the labeled nucleotide and additional required factors at step 7 and allowing these complexes to elongate for 1 hour at 30°C. Additional factors and labeled nucleotide are then added and newly initiated transcription is assayed (Workman and Roeder, 1987).

The following sections describe nucleosome assembly and transcription studies utilizing the reconstituted system described in Section III,C.

B. Nucleosome-Mediated Repression of Promoter Activity

The assembly of template DNA into nucleosomes prior to transcriptional analysis allows a functional assay for the repression of transcription and for the function of factors capable of binding to nucleosomal DNA and rescuing promoter function. A protocol for the assembly of template DNA into nucleosomes immediately prior to *in vitro* transcription assays is described below.

Assembly of Nucleosomes prior to Transcription Analysis

1. Prepare HTS:histone mixes as described above (80% HTS and 20% histone dilution) and incubate at 20°C for 15 minutes.

2. To 10-μl aliquots of these mixes, add 3 units topoisomerase I (Promega) and 100 ng of template DNA (total volume = 11–12 μl).

3. Allow nucleosome assembly to proceed for 1 hour at 30°C.

4. Add 5–10 μl of either nuclear extract or a reconstitution of general and regulatory transcription factors.

5. Add 10 μl of transcription additive (TA), 20% glycerol, 20 mM HEPES, pH 8.4, 15.5 mM MgCl$_2$, and 3 × the final concentration of NTPs, dependent on the RNA assay as indicated in Section II,A. A second template, which yields transcripts distinguishable from that of the nucleosome assembled template, can be added here as an internal DNA control (see Fig. 4A).

6. Transcribe at 30°C for 1 hour.

7. Quench each reaction with 75 μl of transcription stop mix and extract and precipitate the RNA as described in Section II,A. To increase the signal-to-noise ratio for labeled RNA, it is best to precipitate the RNA three times.

8. Analyze unlabeled RNA by primer extension or S1 nuclease protection. Analyze labeled RNA directly on a urea/acrylamide gel.

RNA products generated from such an experiment are shown in Fig. 4A. One of two templates (each containing the major late promoter but generating different length transcripts) was present during the assembly step and the second was added as free DNA at the start of the transcription assay. In the absence of histones (i.e., after a mock nucleosome assembly reaction), both templates were transcribed and moderately stimulated by the addition of the upstream factor USF. In contrast, when histones were included, no transcripts were detected from the assembled template while transcription from the subsequently added DNA template was unimpaired. This result shows that the repression of transcription by nucleosome assembly was dependent on added histones, was mediated in cis, and was not reversed by the presence of USF (Workman *et al.*, 1990).

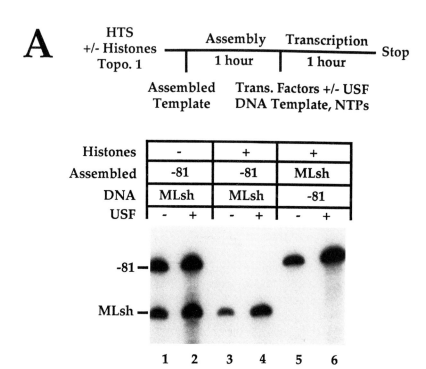

A

HTS +/- Histones Topo. 1	Assembly 1 hour	Transcription 1 hour	Stop
	Assembled Template	Trans. Factors +/- USF DNA Template, NTPs	

Histones	-		+		+	
Assembled	-81		-81		MLsh	
DNA	MLsh		MLsh		-81	
USF	-	+	-	+	-	+

-81 —

MLsh —

1 2 3 4 5 6

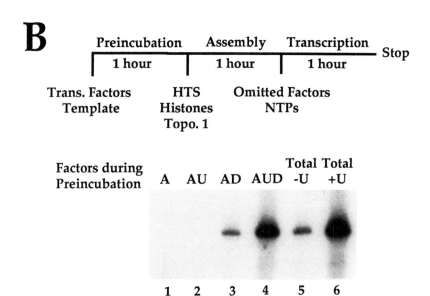

B

Preincubation 1 hour	Assembly 1 hour	Transcription 1 hour	Stop
Trans. Factors Template	HTS Histones Topo. 1	Omitted Factors NTPs	

Factors during Preincubation	A	AU	AD	AUD	Total -U	Total +U

1 2 3 4 5 6

C. Alleviation of Nucleosome-Mediated Repression

The formation of preinitiation complexes on different promoters and their stability to nucleosome assembly can be assayed in the following protocol. In addition, the effects of regulatory proteins on the number or stability of preinitiation complexes can be addressed.

Formation of Preinitiation Complexes prior to Nucleosome Assembly

1. Prepare a mixture containing 4–8 μl of buffer D, nuclear extract, or reconstituted general transcription factors and any regulatory factors. Supplement with 0.25 volume of $MgCl_2$ dilutions to bring the final concentration to between 1 and 5 mM. The $MgCl_2$ concentrations should be sufficient to favor the binding of regulatory factors; however, at higher $MgCl_2$ concentrations, larger concentrations of topoisomerase I are required during the subsequent nucleosome assembly reaction.

2. Add 100 ng of template DNA and preincubate at 30°C for 1 hour to allow preinitiation complex formation.

3. After 45 minutes, prepare NTS: histone mixes as above and incubate for 15 minutes at 20°C.

4. Start nucleosome assembly by adding 10 μl of the HTS: histone mixture and 3 units of topoisomerase I to each factor: template preincubation reaction.

5. Incubate at 30°C for 1 hour.

6. Start transcription by adding 10 μl of TA (as in Section IV,B, step 5) to each reaction and any additional factors required for transcription. (Note that

---

FIG. 4.   Order of addition experiments addressing nucleosome-mediated repression of transcription and the effect of preformed preinitiation transcription complexes. (A) Transcriptional repression of nucleosome-assembled templates relative to naked DNA. Assembly/transcription reactions were as described in the text and utilized two templates containing the major late promoter −81 (which produces a 390-base transcript) and MLsh (which produces a 350-base transcript). The RNA products from these templates are shown. As indicated, 100 ng of one template (Assembled) was included in the assembly step while 100 ng of the other was added as purified DNA (DNA) along with general transcription factors ±USF at the start of the transcription assay. Assembly reactions included or omitted 400 ng of *Xenopus* erythrocyte histones as indicated and transcription reactions included or omitted USF as indicated. Adapted from Fig. 3 of Workman *et al.* (1990) *EMBO J.* **9,** 1299–1308: By permission of Oxford University Press. (B) Factor requirements for the formation of preinitiation complexes stable to subsequent nucleosome assembly. Shown are RNA products from 100 ng of -81, which was preincubated with factors indicated as described in the text and included TFIIA and TFIID. Total reconstitutions during preincubation included both of these as well as TFIIB/TFIIE/TFIIF/TFIIG and RNA polymerase II. USF was included during preincubation where indicated (+U). Following nucleosome assembly with 400 ng of *Xenopus* erythrocyte histones, factors omitted during the preincubation step were added. The template DNA was −81. Adapted from Fig. 4 of Workman *et al.* (1990) *EMBO J.* **9,** 1299–1308: By permission of Oxford University Press.

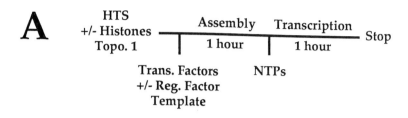

A

HTS
+/- Histones
Topo. 1 ├─── Assembly ──┬── Transcription ──┤ Stop
              1 hour   1 hour

Trans. Factors   NTPs
+/- Reg. Factor
Template

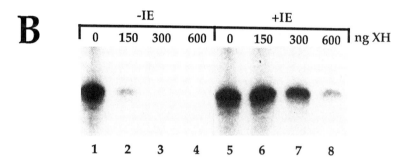

B

	-IE				+IE			
0	150	300	600	0	150	300	600	ng XH

1   2   3   4   5   6   7   8

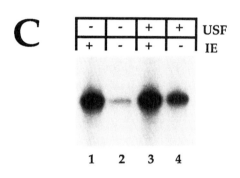

C

-	-	+	+	USF
+	-	+	-	IE

1   2   3   4

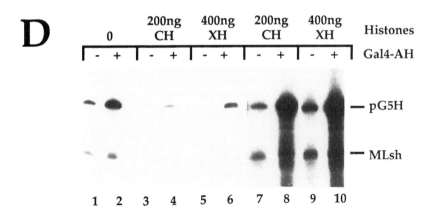

D

	0	200ng CH	400ng XH	200ng CH	400ng XH	Histones
	- +	- +	- +	- +	- +	Gal4-AH

—— pG5H

—— MLsh

1   2   3   4   5   6   7   8   9   10

the total volume of added transcription factors/reaction should not exceed 10 μl. This includes those added during the preincubation and during the transcription step.)

7. Incubate for 1 hour at 30°C.
8. Quench reactions and analyze RNA as described above.

An analysis of RNA products generated from a preincubation/nucleosome assembly experiment is shown in Fig. 4B. Preincubation of the template, containing the major late promoter, with TFIIA or USF did not prevent nucleosome-mediated repression of the promoter. When TFIID or all the general factors were included during the preincubation the promoter retained some transcriptional activity. If TFIID or all the general factors were present during preincubation, USF increased the number of preinitiation complexes formed (Workmen *et al.*, 1990) and thus increased the subsequent levels of transcription observed.

D. Transcriptional Regulation during Nucleosome Assembly

The order of addition experiments described above allows an assessment of the ability of nucleosomes to prevent preinitiation complex formation and the

---

FIG. 5. Transcriptional regulation during *in vitro* nucleosome assembly. (A) The protocol used as described in the text. (B) The effect of nucleosome assembly on transcriptional stimulation by the Pseudorabies virus immediate-early protein (IE). Shown are the run-off RNA products generated from the 100 ng of linear plasmid DNA bearing the major late promoter following assembly reactions which included the concentrations of *Xenopus* erythrocyte histones (XH) indicated. Assembly reactions also included partially purified IE protein (+ IE) or the corresponding fraction from mock infected cells (− IE). General transcription factors were provided by a HeLa nuclear extract. Adapted from Fig. 1B of Workman *et al.* (1988) *Cell (Cambridge, Mass.)* **55**, 211–219: Copyright Cell Press. (C) The combined effect of USF and IE during nucleosome assembly. Shown are the run-off RNA products generated from 100 ng of linear plasmid DNA bearing the major late promoter after assembly reactions which contained 300 ng of *Xenopus* erythrocyte histones. In addition to a total reconstitution of general transcription factors (as described in the legend to Fig. 4), assembly reactions contained IE and/or USF as indicated. When either was omitted, the corresponding buffer or mock fraction was included. (D) Transcriptional regulation by a derivative of GAL4 during nucleosome assembly. Each assembly reaction contained 50 ng of pG5H (an HSP70 promoter bearing five GAL4 sites and generating a 385-base transcript) and 50 ng of MLsh (see legend to Fig. 4). RNA products generated from each template are shown. Assembly reactions contained or omitted GAL4-AH as indicated. In addition, assembly reactions contained the quantities of either HeLa core histones (CH) or *Xenopus* erythrocyte histones (XH) indicated. Lanes 7–10 are from a longer exposure of lanes 3–6, respectively, in order to illustrate the increase in the ratio of induced (+ GAL4-AH) to basal (− GAL4-AH) levels of transcription under conditions of nucleosome assembly. General transcription factors were provided by a HeLa nuclear extract.

resistance of preinitiation complexes to subsequent nucleosome assembly. Concurrent nucleosome and transcription complex assembly protocols can be used to ask whether particular regulatory factors increase the formation of preinitiation complexes such that complex formation more effectively competes with nucleosome assembly for occupancy of promoter sequences.

Concurrent Nucleosome and Transcription Complex Assembly

1. Prepare HTS : histone mixtures with increasing histone concentrations. Incubate for 15 minutes at 20°C.

2. For each reaction take 10 μl of HTS : histone mixtures and add 3 units of topoisomerase I (10 units/μl) and 5–10 μl of either nuclear extract or isolated general transcription factors. Include or omit (for controls) the regulatory factor of interest. This mixture is now competent in both nucleosome and transcription complex assembly.

3. Add 100 ng of template DNA and incubate for 1 hour at 30°C.

4. Following the assembly reaction, promoter activity in transcription is assayed by adding 10 μl of TA (as above) and incubation is continued for 1 hour at 30°C.

5. Extract and analyze RNA as described above.

Increasing concentrations of histones lead to an increase in the rate and extent of nucleosome assembly and a decrease in the subsequent levels of transcription observed. The relevance of nucleosome assembly to the regulation of transcription is revealed by the relative increase in the ratio of induced to basal levels of transcription (i.e., in the presence or the absence of the regulatory factor). As shown in Fig. 5B, the addition of histones to the assembly reactions dramatically increases the extent of transcriptional enhancement by the immediate-early protein of Pseudorabies virus (IE). In the absence of histones there is little or no stimulation of transcription resulting from the presence of IE. However, under the suppressive conditions of nucleosome assembly, the presence of IE results in a dramatic increase in the number of transcription complexes formed and thus in the levels of transcription observed (Workman *et al.*, 1988).

Similar studies have indicated that transcriptional stimulation by the upstream factor USF also increase several-fold under conditions of nucleosome assembly (Workman *et al.*, 1990). Figure 5C illustrates the combined effect of both IE and USF during nucleosome assembly. The combined effect of both of these factors in the nucleosome assembly/transcription assay is neither synergistic nor additive. Transcription from templates assembled in the presence of both proteins approximately equals that of the more potent IE protein alone. This observation suggests that IE and USF effect a common step (in preinitiation complex formation) that becomes limiting under conditions of nucleosome assembly.

Transcriptional stimulation by a bacterially produced derivative of the yeast GAL4 protein also increases several-fold under conditions of nucleo-some assembly (Workman *et al.*, 1991). In the absence of histones, GAL4-AH (a fusion protein of the DNA-binding and dimerization domains of GAL4 fused to an artificial amphipathic helix; Giniger and Ptashne, 1987; Lin *et al.*, 1988) moderately stimulated transcription from a human HSP70 promoter bearing five GAL4-binding sites (pG5H). When 200 ng of HeLa core histones or 400 ng of *Xenopus* erythrocyte histones was included to pro-mote nucleosome assembly, basal transcription from pG5H and the major late promoter was repressed. However, transcripts were still detectable from pG5H when GAL4-AH was also present under these conditions, suggesting that the fold stimulation was increased. This is clearly evident in the longer exposure of the lanes from reactions containing histones. The stimulation (GAL4-AH induced/basal levels) of transcription is much greater under the suppressed conditions of nucleosome assembly than during the permissive transcription of naked DNA (i.e., in the absence of histones).

V. Conclusions

In vitro functional and structural studies of nucleosome and chromatin reconstitution should allow an investigation into the individual roles of multi-ple factors regulating a single transcription unit. Functional studies can ad-dress the role of a factor in establishing the transcriptional potential of a promoter in chromatin as well as in the initiation and elongation of tran-scription. Binding studies can be used to address the ability of a factor to bind to nucleosomal DNA and the fate of the nucleosome upon such binding. Such functions of regulatory factors may correlate with particular DNA-binding domains (those capable of binding to nucleosomal DNA) or activa-tion domains (those stimulating preinitiation complex formation).

These *in vitro* approaches provide assays for the function of chromosomal structural proteins in transcription. Studies on the class III RNA 5S genes indicate that histone H1 augments nucleosome-mediated repression in cellu-lar chromatin (reviewed in Wolffe and Brown, 1988) and in minichromo-somes assembled in an oocyte extract at long nucleosome repeat lengths (Shimamura *et al.*, 1989b). While the reconstituted system described here does not homogeneously space nucleosomes in the short 1-hour reactions (Workman *et al.*, 1990), prolonged exposure to H1 may generate nucleosome spacing as shown in other systems (Stein and Mitchell, 1988). This could allow an assessment of the role of H1 in the repression of class II genes (i.e., by stabilizing nucleosome position, blocking factor accessibility to linker DNA, and/or condensing nucleosomes into thick fibers). Furthermore, it is now possible to perform biochemical assays to assess directly the effect of histone modifications (i.e., histone acetylation; reviewed in Csordes, 1990) or that of additional chromosomal proteins (i.e., HMG 14 and 17; reviewed in Weisbrod, 1982) on the repression of preinitiation complex formation and the passage of RNA polymerase II through nucleosomes.

ACKNOWLEDGMENTS

This work was supported by a grant from Hoechst A. G. to REK, a grant from the National Cancer Institute, CA 42567, to RGR and general support from the Pew Trusts to The Rockefeller University. JLW is a Special Fellow of The Leukemia Society of America, Inc.

REFERENCES

Abmayr, S. M., and Workman, J. L. (1990). Preparation of nuclear and cytoplasmic extracts from mammalian cells. *In* "Current Protocols in Molecular Biology" (F. Ausubel, R. Brent, R. Kingston, D. Moore, J. Seidman, J. Smith, and K. Struhl, eds.), Vol. 2, pp. 12.1.1–12.1.19. Greene Publishing Associates and Wiley-Interscience, New York.

Abmayr, S. M., Workman, J. L., and Roeder, R. G. (1988). The pseudorabies immediate early protein stimulates *in vitro* transcription by facilitating TFIID: Promoter interactions. *Genes Dev.* **2**, 542–553.

Archer, T. K., Cordingly, M. G., Wolford, R. G., and Hager, G. L. (1991). Transcription factor access is mediated by accurately positioned nucleosomes on the mouse mammary tumor virus promoter. *Mol. Cell. Biol.* **11**, 688–698.

Ausubel, F. M., Brent, R., Kingston, R. E., Moore, D. D., Seidman, J. G., Smith, J. A., and Struhl, K. (1990). "Current Protocols in Molecular Biology." Greene Publishing Associates and Wiley-Interscience, New York.

Brown, D. D. (1984). The role of stable complexes that repress and activate eucaryotic genes. *Cell (Cambridge, Mass.)* **37**, 359–365.

Buratowski, S., Hahn, S., Guarente, L., and Sharp, P. A. (1989). Five intermediate complexes in transcription initiation by RNA polymerase II. *Cell (Cambridge, Mass.)* **56**, 549–561.

Carcamo, J., Lobos, S., Merino, A., Buckbinder, L., Weinmann, R., Natarajan, V., and Reinberg, D. (1989). Factors involved in specific transcription by mammalian RNA polymerase II: Role of factors IID and MLTF in transcription from the adenovirus major late and IVa2 promoters. *J. Biol. Chem.* **264**, 7704–7714.

Corthesy, B., Leonnard, P., and Wahli, W. (1990). Transcriptional potentiation of the vitellogenin B1 promoter by a combination of both nucleosome assembly and transcription factors: An *in vitro* dissection. *Mol. Cell. Biol.* **10**, 3926–3933.

Csordas, A. (1990). On the biological role of histone acetylation. *Biochem. J.* **265**, 23–38.

Dignam, J. D., Lebovitz, R. M., and Roeder, R. G. (1983a). Accurate transcription initiation by RNA polymerase II in a soluble extract from isolated mammalian nuclei. *Nucleic Acids Res.* **11**, 1475–1489.

Dignam, J. D., Martin, P. L., Shastry, B. S., and Roeder, R. G. (1983b). Eukaryotic gene transcription with purified components. *In* "Methods in Enzymology" (R. Wu, L. Grossman and K. Moldave, eds.), Vol. 101, pp. 582–598. Academic Press, New York.

Dilworth, S. M., Black, S. J., and Laskey, R. A. (1987). Two complexes that contain histones are required for nucleosome assembly *in vitro*: Role of nucleoplasmin and N1 in *Xenopus* egg extracts. *Cell (Cambridge, Mass.)* **51**, 1009–1018.

Dingwall, C., Sharnick, S. V., and Laskey, R. A. (1982). A polypeptide domain that specifies migration of nucleoplasmin into the nucleus. *Cell (Cambridge, Mass.)* **30**, 449–458.

Elgin, S. C. R. (1988). The formation and function of DNase I hypersensitive sites in the process of gene activation. *J. Biol. Chem.* **263**, 19259–19262.

Giniger, E., and Ptashne, M. (1987). Transcription in yeast activated by a putative amphipathic helix linked to a DNA binding unit. *Nature (London)* **330**, 670–672.

Glikin, G. C., and Blangy, D. (1986). *In vitro* transcription by *Xenopus* oocyte RNA polymerase III requires a DNA topoisomerase II activity. *EMBO J.* **5**, 151–155.

Glikin, G. C., Ruberti, I., and Worcel, A. (1984). Chromatin assembly in *Xenopus* oocytes: *In vitro* studies. *Cell (Cambridge, Mass.)* **37**, 33–41.

Gross, D. S., and Garrard, W. T. (1988). Nuclease hypersensitive sites in chromatin. *Annu. Rev. Biochem.* **57**, 159–197.

Han, M., and Grunstein, M. (1988). Nucleosome loss activates downstream promoters *in vivo*. *Cell (Cambridge, Mass.)* **55**, 899–906.

Han, M., Kim, U. J., Kayne, P., and Grunstein, M. (1988). Depletion of histone H4 and nucleosomes activates the PHO5 gene in *Saccharomyces cerevisiae*. *EMBO J.* **7**, 2221–2228.

Horikoshi, M., Carey, M. F., Kakidani, H., and Roeder, R. G. (1988). Mechanism of action of a yeast activator: Direct effect of GAL4 derivatives on mammalian TFIID-promoter interactions. *Cell (Cambridge, Mass.)* **54**, 665–669.

Kim, U. J., Han, M., Kayne, P., and Grunstein, M. (1988). Effects of histone H4 depletion on the cell cycle and transcription in *Saccharomyces cerevisiae*. *EMBO J.* **7**, 2211–2219.

Kleinschmidt, J. A., and Franke, W. W. (1982). Soluble acidic complexes containing histones H3 and H4 in nuclei of *Xenopus laevis* oocytes. *Cell (Cambridge, Mass.)* **29**, 799–809.

Kleinschmidt, J. A., Fortklamp, E., Krohne, G., Zentgraf, H., and Franke, W. W. (1985). Coexistence of two different types of soluble histone complexes in nuclei of *Xenopus laevis* oocytes. *J. Biol. Chem.* **260**, 1166–1176.

Kleinschmidt, J. A., Seiter, A., and Zentgraf, H. (1990). Nucleosome assembly *in vitro*: Separate histone transfer and synergistic interaction of native histone complexes purified from nuclei of *Xenopus laevis* oocytes. *EMBO J.* **9**, 1309–1318.

Knezetic, J. A., and Luse, D. S. (1986). The presence of nucleosomes on a DNA template prevents initiation by RNA polymerase II *in vitro*. *Cell (Cambridge, Mass.)* **45**, 95–104.

Knezetic, J. A., Jacob, G. A., and Luse, D. S. (1988). Assembly of RNA polymerase II preinitiation complexes before assembly of nucleosomes allows efficient initiation of transcription on nucleosomal templates. *Mol. Cell. Biol.* **8**, 3114–3121.

Laskey, R. A., and Earnshaw, W. C. (1980). Nucleosome assembly. *Nature (London)* **286**, 763–767.

Laskey, R. A., Honda, B. M., Mills, A. D., and Finch, J. T. (1978). Nucleosomes are assembled by by an acidic protein which binds histones and transfers them to DNA. *Nature (London)* **275**, 416–420.

Lewin, B. (1990). Commitment and activation at Pol II promoters: A tail of protein–protein interactions. *Cell (Cambridge, Mass.)* **61**, 1161–1164.

Lin, Y. S., Carey, M. F., Ptashne, M., and Green, M. R. (1988). GAL4 derivatives function alone and synergistically with mammalian activators *in vitro*. *Cell (Cambridge, Mass.)* **54**, 659–664.

Lorch, Y., LaPointe, J. W., and Kornberg, R. D. (1987). Nucleosomes inhibit the initiation of transcription but allow chain elongation with the displacement of histones. *Cell (Cambridge, Mass.)*. **49**, 203–210.

Manley, J. L., Fire, A., Cano, A., Sharp, P. A., and Gefter, M. L. (1980). DNA-dependent transcription of adenovirus genes in a soluble whole-cell extract. *Proc. Natl. Acad. Sci. U.S.A.* **77**, 3855–3859.

Matsui, T. (1987). Transcription of adenovirus 2 major late and peptide IX genes under conditions of *in vitro* nucleosome assembly. *Mol. Cell. Biol.* **7**, 1401–1408.

Meisterernst, M., Horikoshi, M., and Roeder, R. G. (1990). Recombinant yeast TFIID, a general transcription factor, mediates activation by the gene-specific factor USF in a chromatin assembly assay. *Proc. Natl. Acad. Sci. U.S.A.* **87**, 9158–9162.

Meisterernst, M., Roy, A. L., Lieu, H. M., and Roeder, R. G. (1991). Activation of class II gene transcription by regulatory factors is potentiated by a novel mechanism. *Cell* **61**, 1–20.

Nakajima, N., Horikoshi, M., and Roeder, R. G. (1988). Factors involved in specific transcription by mammalian RNA polymerase II: Purification, genetic specificity, and TATA box-promoter interactions of TFIID. *Mol. Cell. Biol.* **8**, 4028–4040.

Perlmann, T., and Wrange, O. (1988). Specific glucocorticoid receptor binding to DNA reconstituted in a nucleosome. *EMBO J.* **7**, 3073–3079.

Pina, B., Bruggemeier, U., and Beato, M. (1990). Nucleosome positioning modulates accessibility of regulatory proteins to the mouse mammary tumor virus promoter. *Cell (Cambridge, Mass.)* **60**, 719–731.

Pognonec, P., and Roeder, R. G. (1991). Recombinant 43-kilodalton USF binds to DNA and activates transcription in a manner indistinguishable from that of natural 43/44-kDa USF. *Mol. Cell. Biol.*, in press.

Ptashne, M., and Gann, A. A. F. (1990). Activators and targets. *Nature (London)* **346**, 329–331.

Razvi, F., Gargiulo, G., and Worcel, A. (1983). A simple procedure for parallel sequence analysis of both strands of 5′-labeled DNA. *Gene* **23**, 175–183.

Reinberg, D., and Roeder, R. G. (1987). Factors involved in specific transcription by mammalian RNA polymerase II: Purification and functional analysis of initiation factors IIB and IIE. *J. Biol. Chem.* **262**, 3310–3321.

Rhodes, D., and Laskey, R. A. (1989). Assembly of nucleosomes and chromatin *in vitro In* "Methods in Enzymology" (P. M. Wassarman and R. D. Kornberg, eds.), Vol. 170, pp. 575–585. Academic Press, San Diego, California.

Saltzman, A. G., and Weinmann, R. (1989). Promoter specificity and modulation of RNA polymerase II transcription. *FASEB J.* **3**, 1723–1733.

Sambrook, J., Fritsch, E. F., and Maniatis, T. (1989). "Molecular Cloning: A Laboratory Manual." Cold Spring Harbor Laboratory, Cold Spring Harbor, New York.

Sawadogo, M., and Roeder, R. G. (1985). Factors involved in specific transcription by human RNA polymerase II: Analysis by a rapid and quantitative *in vitro* assay. *Proc. Natl. Acad. Sci. U.S.A.* **82**, 4349–4398.

Sawadogo, M., and Sentenac, A. (1990). RNA polymerase B (II) and general transcription factors. *Annu. Rev. Biochem.* **59**, 711–754.

Scheidereit, C., Heguy, A., and Roeder, R. G. (1987). Identification and purification of a human lymphoid-specific octamer-binding protein (OTF-2) that activates transcription of an immunoglobin promoter *in vitro. Cell (Cambridge, Mass.)* **51**, 783–793.

Sealy, L., Cotton, M., and Chalkley, R. (1986). *Xenopus* nucleoplasmin: Egg vs. oocyte. *Biochemistry* **25**, 3064–3072.

Sealy, L., Burgess, R. R., Cotton, M., and Chalkley, R. (1989). Purification of *Xenopus* egg nucleoplasmin and its use in chromatin assembly *in vitro. In* "Methods in Enzymology" (P. M. Wassarman and R. D. Kornberg, eds), Vol. 170, pp. 612–630. Academic Press, San Diego, California.

Shimamura, A., Jessee, B., and Worcel, A. (1989a). Assembly of chromatin with oocyte extracts. *In* "Methods in Enzymology" (P. M. Wassarman and R. D. Kornberg, eds.), Vol. 170, pp. 603–612. Academic Press, San Diego, California.

Shimamura, A., Sapp. M., Rodriguez-Campos, A., and Worcel, A. (1989b). Histone H1 represses transcription from minichromosomes assembled *in vitro*. *Mol. Cell. Biol.* **9**, 5573–5584.

Shure, M., and Vinograd, J. (1976). The number of superhelical turns in native SV40 virion DNA and minicol DNA determined by the band counting method. *Cell (Cambridge, Mass.)* **8**, 215–226.

Stein, A. (1989). Reconstitution of chromatin from purified components. *In* "Methods in Enzymology" (P. M. Wassarman and R. D. Kornberg, eds.), Vol. 170, pp. 585–603. Academic Press, San Diego, California.

Stein, A., and Mitchell, M. (1988). Generation of different nucleosome spacing periodicities *in vitro*: Possible origin of cell type specificity. *J. Mol. Biol.* **203**, 1029–1043.

Sumimoto, H., Ohkuma, Y., Yamamoto, T., Horikoshi, M., and Roeder, R. G. (1990). Factors involved in specific transcription by mammalian RNA polymerase II: Identification of general transcription factor TFIIG. *Proc. Natl. Acad. Sci. U.S.A.* 87, 9158–9162.

Taylor, I. C. A., and Kingston, R. E. (1990). Factor substitution in a human HSP70 promoter: TATA dependent and TATA independent interactions. *Mol. Cell. Biol.* **10**, 165–175.

Taylor, I. C. A., Workman, J. L., Schuetz, T. J., and Kingston, R. E. (1991). Facilitated binding of GAL4 and heat shock factor to nucleosomal templates: Differential function of DNA binding domains. *Genes Dev.* **5**, 1285–1298.

Van Dyke, M., Roeder, R. G., and Sawadogo, M. (1988). Physical analysis of transcription preinitiation complex assembly on a class II gene promoter. *Science* **241**, 1335–1338.

von Holt, C., Brandt, W. F., Greyling, H. J., Lindsey, G. G., Retief, J. D., Rodrigues, J. A., Schwager, S., and Sewell, B. T. (1989). Isolation and characterization of histones. *In* "Methods in Enzymology" (P. M. Wasserman and R. D. Kornberg, eds.), Vol. 170. pp. 431–542. Academic Press, San Diego, California.

Wassarman, P. M., and Kornberg, R. D., eds (1989). Nucleosomes. "Methods in Enzymology" Vol. 170. Academic Press, San Diego, California.

Weil, P. A., Luse, D. S., Segall, J., and Roeder, R. G. (1979). Selective and accurate initiation of transcription at the Ad2 major late promoter in a soluble system dependent on purified RNA polymerase II and DNA. *Cell (Cambridge, Mass.)* **18**, 469–484.

Weisbrod, S. (1982). Active chromatin. *Nature (London)* **297**, 289–295.

Wolffe, A. P., and Brown, D. D. (1988). Developmental regulation of two 5S ribosomal RNA genes. *Science* **241**, 1626–1632.

Workman, J. L., and Roeder, R. G. (1987). Binding of transcription factor TFIID to the major late promoter during *in vitro* nucleosome assembly potentiates subsequent initiation by RNA polymerase II. *Cell (Cambridge, Mass).* **51**, 613–622.

Workman, J. L., Abmayr, S. M., Cromlish, W. A., and Roeder, R. G. (1988). Transcriptional regulation by the immediate early protein of pseudorabies virus during *in vitro* nucleosome assembly. *Cell (Cambridge, Mass.)* **55**, 211–219.

Workman, J. L., Roeder, R. G., and Kingston, R. E. (1990). An upstream transcription factor, USF (MLTF), facilitates the formation of preinitiation complexes during *in vitro* nucleosome assembly. EMBO J. **9**, 1299–1308.

Workman, J. L., Taylor, I. C. A., and Kingston, R. E. (1991). Activation domains of stably bound GAL4 derivatives alleviate repression of promoters by nucleosomes. *Cell (Cambridge, Mass)* **64**, 533–544.

Chapter 17

Systems for the Study of Nuclear Assembly, DNA Replication, and Nuclear Breakdown in Xenopus laevis Egg Extracts

CARL SMYTHE AND JOHN W. NEWPORT

Department of Biology
University of California, San Diego
La Jolla, California 92093

METHODS IN CELL BIOLOGY, VOL. 35

I.   Introduction

A definitive characteristic of the eukaryotic cell is the existence of a compartment containing the genomic DNA of that cell. The eukaryotic nucleus is bounded by a nuclear envelope, consisting of an inner and outer membrane. The nuclear envelope contains many nuclear pores, which control the transport of macromolecules into and out of the nucleus, thus regulating nuclear content and communication with the cytoplasmic environment. The nuclear lamina located between the inner nuclear membrane and chromatin serves to support nuclear shape and creates sites for attachment and organization of the chromatin. In eukaryotic cells which undergo "open" mitoses, all of these nuclear structures must be disassembled and reassembled once during the cell cycle. Although much is known about the physical composition and organization of the intact nucleus (Newport and Forbes, 1987; Gerace and Burke, 1988), much less is understood about the temporal regulation of nuclear assembly, DNA replication, and subsequent nuclear disassembly during mitosis.

In the amphibian *Xenopus laevis*, egg fertilization is followed by a period of rapid and synchronous cell division, which occurs in the complete absence of transcription or of any increase in total mass of the egg. Thus, during oogenesis, the egg must synthesize and store a large reservoir of nuclear components for use during the rapid proliferation which occurs during early development. The existence of such a reservoir of nuclear components has permitted the development of systems for the study of nuclear dynamics *in vitro*. Described below are conditions for preparing cell-free extracts from *Xenopus* eggs which allow the study of these nuclear dynamics and which have proved fruitful in elucidating structural and functional requirements for nuclear reassembly, DNA replication, and nuclear diassembly.

II.   Obtaining *Xenopus* Eggs and Isolation of Sperm Chromatin

All reagents may be obtained from Sigma Chemical Co.

A.   Induction of Oocyte Maturation and Ovulation

Female *Xenopus* frogs are maintained in water tanks (45 cm × 28 cm × 28 cm) containing 40 liters of 0.1 *M* NaCl at an approximate density of one animal per liter. In order to induce the maturation of oocytes, each frog is primed by injection with 100 units (0.5 ml) of pregnant mare serum gonadotropin 3–10 days before eggs are required. Gonadotropin is introduced into the dorsal lymph sac by subcutaneous injection using a 25-gauge needle. If necessary, frogs may be restrained by immersion in ice-cold water for 5 minutes prior to injection. Mature eggs may be obtained by injecting each frog with 500 units (0.5 ml) of human chorionic gonadotropin 12–16 hours before use. The route of injection is the same as that used for priming.

B. Egg Collection

After injection to induce ovulation, each frog is transferred to an individual tank containing 2.5 liters of either 100 mM NaCl (for S and M extracts) or modified amphibian Ringer's solution (MMR; 100 mM NaCl, 2 mM KCl, 1 mM MgSO$_4$, 2 mM CaCl$_2$, 0.1 mM EDTA, 5 mM HEPES, pH 7.8, for oscillating extracts). Frogs generally begin to lay eggs 8–10 hours after injection and may continue to do so for a period of up to 15 hours. Maintaining frogs in individual tanks ensures that poor-quality eggs laid by a particular frog may be discarded. Eggs are recovered from each tank by pouring off the excess solution and transferring the eggs to a glass beaker.

C. Preparation of Demembranated Sperm Chromatin

Xenopus sperm provide an abundant source of chromatin, which after removal of plasma and nuclear membranes may be easily purified. The following is a modification of the method of Lohka and Masui (1983).

Solutions

Buffer T: 15 mM PIPES, 15 mM NaCl, 5 mM EDTA, 7 mM MgCl$_2$, 80 mM KCl, 0.2 M sucrose, pH 7.4 (4 ml)

Buffer R: Buffer T + 3% bovine serum albumin (3 ml)

Buffer S: Buffer T + 20 mM maltose + 0.05% lysolecithin (0.3 ml)

Apparatus

Clinical centrifuge

Procedures

1. Anesthetize a male frog by immersion in ice-containing water for 5 minutes. Sacrifice the animal by cranial dislocation. Open peritoneal cavity via a mid-line incision through the abdominal wall using dissection scissors and move the organs of the intestinal tract to one side. Remove the pair of testes, off-white in colour, and each 5–8 mm in length, which are located in the mid body, on either side of the mid line, and ventral to the kidneys.

2. Rinse testes in buffer T and transfer testes to a microcentrifuge tube containing 1 ml of buffer T. Mince testes with forceps to release sperm into buffer T.

3. Centrifuge the suspension in a clinical centrifuge at 170 g (1000 rpm) for 10 seconds. Retain the supernatant containing the sperm and reextract the pellet with 1 ml of buffer T. Recentrifuge as before and combine supernatants.

4. Centrifuge suspension at 1350 g (2800 rpm) for 2 minutes. The resultant pellet consists of a whitish upper region consisting predominantly of the sperm, and a lower pink layer containing red blood cells and somatic cells. Resuspend only the upper portion of the pellet in 0.5 ml of buffer T by trituration with a Pipetman p200, and repeat centrifugation and resuspension steps twice.

5. Resuspend pellet in 100 μl of buffer T and add 300 μl of buffer S and incubate at 23°C for 5 minutes.

6. Add 1.2 ml of buffer R and centrifuge at 300 g (1300 rpm) for 10 minutes.

7. Resuspend pellet in 200 μl of buffer R and dilute with 1000 μl of buffer R and recentrifuge at 300 g (1300 rpm) for 10 minutes.

8. Finally, resuspend sperm chromatin in 50 μl of buffer T. Dilute an aliquot of the sperm chromatin preparation 100-fold in buffer T and estimate the sperm chromatin concentration using a hemocytometer. Typically, concentrations of 40,000 per microliter are obtained. Sperm may be stored in 5-μl aliquots at -70°C indefinitely.

III.   Preparation of *Xenopus* Extracts and Assays Associated with Nuclear Assembly/Disassembly

All of the radiochemicals described below may be obtained from ICN. Histone H1, pepstatin A, chymostatin, aprotinin, and leupeptin are from Boehringer. Bacteriophage λ DNA is obtained from New England Biolabs. Versilube F-50 oil is obtained from Andpak-EMA (San Jose, CA). All other reagents may be obtained from Sigma Chemical Co.

A.   The S Phase Extract

Mature *Xenopus* eggs are physiologically arrested in metaphase of the second meiotic division and therefore contain high levels of maturation (or mitosis) promoting factor (MPF). MPF consists of at least two components, the protein kinase p34^{cdc2} (Dunphy *et al.*, 1988) and cyclin (Draetta *et al.*, 1989), and a complex of these polypeptides is essential for the mitotic activity of p34^{cdc2}. Egg lysis in the absence of phosphatase inhibitors or Ca^{2+} chelators results in a transient release of Ca^{2+} from intracellular stores. This brings about the specific proteolysis of cyclin, which causes the distruction of MPF, allowing the extract to exit from M phase and enter S phase.

Eggs require no homogenization but rather are lysed by centrifugation; the yolk cortex and much of the pigment are sedimented in the pellet, while cytoplasmic constituents are released into the supernatant. Lysis is performed in the presence of cycloheximide to prevent any further protein synthesis which might allow the renewed accumulation of cyclin. Addition of either demembranated sperm chromatin (Lohka and Masui, 1984) or protein-free bacteriophage λ DNA (Newport, 1987) to the cytoplasmic S extract allows the assembly of functional nuclei around these DNA templates. Such nuclei formed *in vitro* are morphologically indistinguishable from normal eukaryotic nuclei, and undergo semiconservative DNA replication (Blow *et al.,* 1989).

Stock Solutions. Storage temperatures are indicated in parentheses.

Cytochalasin B, 5 mg/ml in DMSO ($-20°C$).
Protease inhibitor cocktail: Aprotinin, leupeptin, each 5 mg/ml in water ($-70°C$).

Solutions. Volumes required are indicated in parentheses.

2% cysteine, pH 7.7 (100 ml)
MMR: 100 mM NaCl, 2 mM KCl, 1 mM MgSO$_4$, 2 mM CaCl$_2$, 0.1 mM EDTA, 5 mM HEPES, pH 7.8 (200 ml)
Lysis buffer: 250 mM sucrose, 2.5 mM MgCl$_2$, 50 mM KCl, 50 μg/ml cycloheximide, 1 mM dithiothreitol, 10 mM HEPES, pH 7.7 (50 ml)

Apparatus
Clinical centrifuge
Sorvall RC refrigerated centrifuge (or equivalent), HB-4 swinging bucket rotor
Light microscope

Procedures
1. Dejelly about 25 ml of eggs in 100 ml of 2% cysteine, pH 7.7 (22°C). Allow the eggs to stand for up to 5 minutes, gently swirling them at intervals to facilitate removal of the jelly coat. Dejellying takes 3–5 minutes and is complete when eggs are tightly packed. The total volume of packed eggs will be about 5 ml.

2. Decant cysteine and wash eggs three times with 50 ml of MMR (22°C) to remove residual cysteine and jelly coats. Remove all but 20 ml of final MMR wash and transfer eggs to a petri dish. It is essential that eggs be suspended gently by swirling prior to transfer between containers as packed eggs tend to stick to glass and subsequently lyse.

3. Examine eggs under a light microscope and discard abnormal eggs using a Pasteur pipette. Normal eggs are 1.2–1.5 mm in diameter and consist of discrete animal (dark pigmented) and vegetal (pale yellow) regions of roughly equal size. Abnormal eggs may be grossly enlarged and/or display nonuniform pigmentation.

4. Rinse eggs three times with lysis buffer (22°C) and transfer eggs in a minimal volume to a 15-ml polypropylene centrifuge tube. Pack the eggs by centrifugation at 170 g (1000 rpm) for 30 seconds in a clinical centrifuge and remove excess buffer initially with a Pasteur pipette and finally with absorbent tissue paper. Add leupeptin and aprotinin to a final concentration of 10 μg/ml each and cytochalasin B (which prevents unwanted actin gelation in the extract) to 5 μg/ml.

5. Lyse the eggs by centrifugation at 12,000 g (Sorvall HB-4 rotor, 10,000 rpm or equivalent, 4°C, 10 minutes). Withdraw the cytoplasmic extract (2.5 ml) by puncturing the side of the tube (Fig. 1A) using a 21-gauge needle and a 5-ml syringe. The cytoplamic extract varies in color depending on precise egg pigmentation, and may either be gray or cream. The extract may be recentrifuged (10,000 rpm, 4°C, 10 minutes) to remove remaining contaminating yolk and particulate matter, and is then stored on ice until required. The nuclear reconstitution extract prepared in this manner consists of 75% cytoplasm and 25% lysis buffer and thus, 1 egg equivalent is approximately 0.625 μl of extract. The extract is stable for nuclear reconstitution experiments for up to 3 hours and may be diluted by up to 50% with lysis buffer with no loss in nuclear reconstitution efficiency.

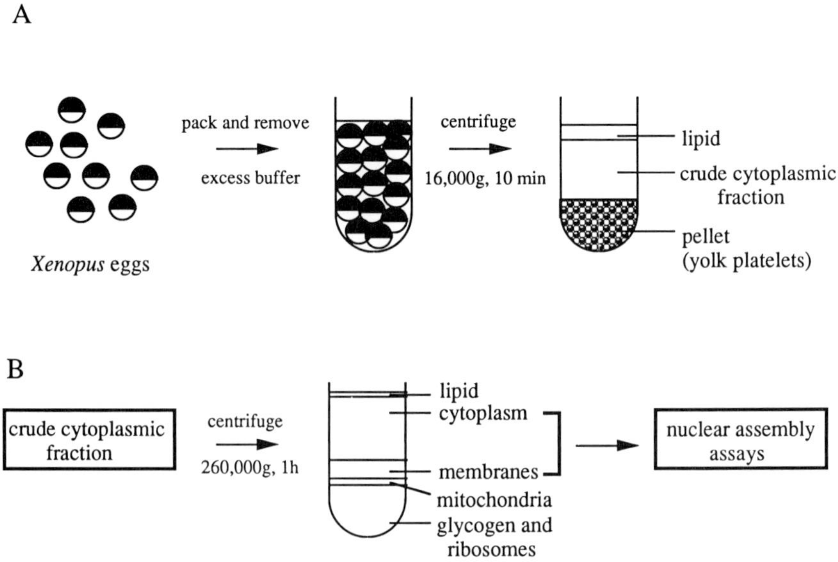

A

B

FIG. 1.   Flow diagram for the preparation of (A) the S phase extract and (B) membrane and cytosol fractions for nuclear assembly.

B.   The Fractionated S Phase Extract

The crude extract described above contains all of the molecular components necessary to assemble nuclei around metaphase chromosomes, sperm
chromatin, or protein-free DNA (Newport, 1987). The membranes and
soluble components required for this process may be separated from each
other and from other cellular components by further fractionation and this
allows long-term storage of the nuclear reassembly components (Wilson and
Newport, 1988).

Apparatus

    Beckman TL-100 table-top ultracentrifuge (or equivalent), TLS-55
    swinging bucket rotor

Procedures

1. Centrifuge the crude cytoplasmic fraction (step 5 above) at 260,000 *g*
(TLS-55 rotor or equivalent, 55,000 rpm, 1 hour, 4°C) This procedure generates a series of layers containing different subcellular components as indicated in Fig. 1B.

2. Remove the soluble fraction using a Pipetman p1000, taking care to
minimize the removal of any of the underlying membrane components. The
soluble fraction may be recentrifuged at 260,000 *g* (25 minutes, 4°C) to remove
any residual membrane. The resulting membrane-free supernatant (1 ml) may
either be used for nuclear reassembly assays directly or frozen in aliquots at
− 70°C. Frozen aliquots should be no larger than 25 µl, as the slower rate of
thawing obtained with larger volumes reduces the efficiency of subsequent
nuclear assembly.

3. Remove the uppermost golden, translucent membrane fraction (Fig. 1B,
~ 100 µl) using a Pipetman p200 with a sawn-off tip, taking care not to remove
any of the tan-colored layer directly below. Resuspend the membrane fraction
in lysis buffer (1 ml), containing protease inhibitor cocktail, and layer over lysis
buffer +0.5 *M* sucrose (200 µl) and recentrifuge at 34,000 *g* (TLS-55 rotor,
20,000 rpm, 20 minutes, 4°C). The membrane pellet is resuspended in lysis
buffer +0.5 *M* sucrose to 0.1 volume of the original cytoplasmic extract and
frozen in aliquots not larger than 5 µl.

C.   Nuclear Assembly Assays

Stock Solutions.   Storage temperatures are indicated in parentheses.

    ATP regenerating system components (− 20°C)
    0.2 *M* phosphocreatine, in 10 m*M* potassium phosphate buffer, pH 7.0
    0.2 *M* ATP, pH 7.0
    0.5 mg/ml creatine phosphokinase in 10 m*M* HEPES, pH 7.5, 50%
    glycerol

Procedures.   Rapidly thaw both soluble and membrane fractions. The soluble fraction is supplemented with an ATP regenerating system (20 mM phosphocreatine, 2 mM ATP, pH 7.0, 5 μg/ml creatine kinase). A typical reaction consists of 20 μl of soluble fraction, 2–3 μl of membrane fraction, and 0–1500 sperm chromatin/μl of extract. For reassembly around protein-free DNA, bacteriophage λ DNA (5–15 ng/μl of extract) may be used. Reactions may be incubated for up to 5 hours at room temperature (22°C) and nuclear assembly monitored as described in Section III,H.

Timing of Events in the Assembly of Nuclei.   The formation of nuclei is an ordered process involving discernible intermediate steps. The rate of complete nuclear reconstitution may vary slightly from extract to extract. If protein-free DNA is used for nuclear reassembly, then the DNA is first assembled into nucleosomes (0–40 minutes), and further organized to form a distinctive condensed sphere (40–80 minutes), involving the formation of a nuclear scaffold. Only after the chromatin is fully condensed do the nuclear lamina and membrane vesicles begin to assemble around the DNA and subsequently form an intact nuclear envelope (80–150 minutes). In contrast, membrane vesicles and the lamina begin binding to sperm chromatin as soon as the latter is added to the extract and thus, using this template, mature nuclei are formed more quickly ($\sim$60 minutes). DNA replication (Newport, 1987; and Fig. 2) occurs over a 30- to 60-minute period after the formation of an intact nuclear envelope.

D.   The M Phase Extract

MPF is known to be responsible for maintaining mature unfertilized *Xenopus* eggs in meiotic metaphase (Masui and Markert, 1971). The activity of MPF (consisting of cyclin and p34^{cdc2}) is regulated by mechanisms of phosphorylation/dephosphorylation (Dunphy and Newport, 1989; Morla *et al.*, 1989; Gautier *et al.*, 1989), and its extraction in stable form requires conditions which prevent the proteolytic destruction of the cyclin component, and which maintain its active phosphorylation state. This is achieved by lysing eggs in the presence of the Ca^{2+} chelator EGTA, together with the phosphatase inhibitor β-glycerophosphate and ATPγS. After egg lysis, MPF is recovered in a high-speed supernatant fraction. Addition of this M phase extract to reconstituted nuclei formed previously in an S phase extract brings about the mitotic disassembly of these nuclei, as judged by the loss of the nuclear envelope, lamin depolymerization, and chromosome condensation (Newport and Spann, 1987; Lohka and Maller, 1985; Miake-Lye and Kirschner, 1985).

Stock Solutions.   Storage temperatures are indicated in parentheses.

Cytochalasin B, 5 mg/ml in DMSO (-20°C)
Protease inhibitor cocktail: aprotinin, leupeptin, each 5 mg/ml in water (-70°C)
ATPγS, 100 mM in 50 mM Hepes, pH 7.8, 0.1 mM DTT (-20°C)

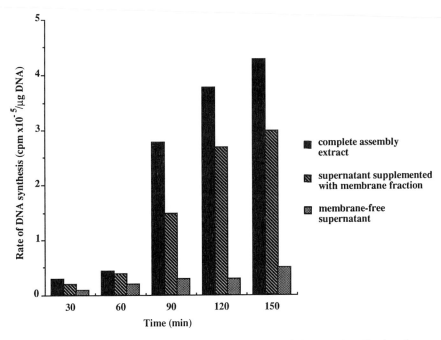

FIG. 2. Nuclear envelope formation is a prerequisite for efficient DNA replication. Sperm chromatin was added to each extract and the extracts were incubated at 22°C to initiate nuclear assembly (where appropriate). DNA replication was determined at the times indicated by measuring the incorporation of [^{32}P]dCTP into DNA during a 30-minute pulse. The incorporation of label into DNA was measured by scintillation counting of the DNA after agarose gel electrophoresis. DNA replication was measured in a complete S phase extract, a high-speed membrane-free supernatant obtained by centrifugation of the S phase extract at 200,000 g for 1 hr, and the high-speed supernatant supplemented with the membrane fraction. Membranes were added to a level equivalent to that present in the complete S phase extract. No DNA replication was observed with the membrane fraction alone.

Solutions

2% cysteine, pH 7.7 (100 ml)

MMR: 100 mM NaCl, 2 mM KCl, 1 mM MgSO$_4$, 2 mM CaCl$_2$, 0.1 mM EDTA, 5 mM HEPES, pH 7.8 (200 ml)

Buffer M: 240 mM sodium β-glycerophosphate, 60 mM EGTA, 45 mM MgCl$_2$, 1 mM dithiothreitol, pH 7.3 (50 ml)

Apparatus

Clinical centrifuge

Sorvall RC centrifuge (or equivalent), HB-4 rotor

Beckman TL-100 table-top ultracentrifuge (or equivalent), TLS-55 swinging bucket rotor

Procedure

1. Dejelly, wash, and sort eggs as described in steps 1–3 of Section III,A.

2. Rinse eggs three times with buffer M (22°C) and transfer eggs in a minimal volume to a 15-ml centrifuge tube. Pack the eggs by centrifugation at 170 g (1000 rpm) for 30 seconds in a clinical centrifuge and remove excess buffer initially with a Pasteur pipette and finally with absorbent tissue paper. Add leupeptin and aprotinin to a final concentration of 10 μg/ml each, cytochalasin B (which prevents unwanted actin gelation in the extract) to 5 μg/ml, and ATPγS to final concentration of 0.5 mM.

3. Lyse the eggs by centrifugation at 16,000 g (Sorvall HB-4 rotor or equivalent, 10,000 rpm, 10 minutes 4°C). Withdraw the cytoplasmic extract (2.5 ml) by puncturing the side of the tube (Fig. 1A) using a 21-gauge needle and a 5-ml syringe.

4. Dilute the cytoplasmic extract with an equal volume of ice-cold 0.33 × buffer M containing 0.5 mM ATPγS and 10 μg/ml aprotinin/leupeptin. Centrifuge the diluted extract at 260,000 g (TLS-55 rotor or equivalent, 55,000 rpm, 1 hour, 4°C).

5. Remove the soluble fraction using a Pipetmen p1000. The soluble fraction may be recentrifuged at 260,000 g (25 minutes, 4°C) to remove any residual membrane. The resulting membrane-free supernatant (1 ml) may either be used for nuclear disassembly assays directly or frozen in 100-μl aliquots at −70°C.

E.   Preparation of a Crude MPF Fraction

The soluble supernatant described above contains active MPF, capable of bringing about nuclear disassembly *in vitro*. A partially purified and concentrated preparation of MPF can be obtained as follows.

1. Add 0.43 volume of 3.6 M ammonium sulfate dissolved in 0.33 × buffer M to the soluble fraction obtained in Section III,C, step 7.

2. Centrifuge the precipitated proteins at 20,000 g (TLS-55 rotor, 15,000 rpm, 10 minutes, 4°C).

3. Discard the supernatant and redissolve the pellet in 1 volume of 0.33 × buffer M (containing 0.1 mM ATPγS) equivalent to one fifth of the volume of the original soluble fraction (Section III,D, step 5).

4. Dialyze this fraction against 30 volumes of 0.33 × buffer M containing 0.1 mM ATPγS for 6 hours. The MPF preparation may be frozen in 100-μl aliquot at −70°C and is stable for at least 6 months.

F.   Nuclear Disassembly Assays

M phase extracts are capable of bringing about the mitotic disassembly of purified rat liver or thymus nuclei (Newport and Spann, 1987), in addition to *Xenopus* nuclei reconstituted as described in Section III,A and C (Dunphy and Newport, 1988). The time required for complete nuclear membrane breakdown and chromosome condensation depends on the type of exogenous nuclei added. The most rapid disassembly occurs with reconstituted nuclei (20–25 minutes). In contrast, disassembly of rat liver nuclei, prepared by the method of Blobel and Potter (1966), may take up to 3 hours (Newport and Spann, 1987). For routine nuclear disassembly experiments using reconstituted nuclei, it is useful to prepare nuclei which may then be frozen at $-70°C$ and subsequently used when required.

Procedures

1. Prepare reconstituted nuclei as described in Section III,A and C using sperm chromatin (final concentration of $1000/\mu l$ of extract) as a template. Incubate nuclear assembly components (up to 500 μl as required) for 1 hour at 22°C, to allow DNA decondensation and nuclear envelope formation.

2. Divide the incubation mixture into covenient aliquots (not greater than 50 μl), freeze in liquid nitrogen, and store at $-70°C$.

3. Rapidly thaw frozen reconstituted nuclei when required, and mix 5 μl with an equal volume of M phase extract or partially purified MPF fraction. Incubate the mixture for up to 1 hour at 22°C and monitor the release of the nuclear envelope from the nuclei as described in Section III,H.

G. The Oscillating Extract

In *Xenopus* eggs, the first 12 cell divisions following fertilization are independent of new transcription and are biphasic, consisting of consecutive S and M phases with no measurable G_1 or G_2 phases. The oscillation between S phase and mitosis is regulated by the periodic activation and inactivation of MPF, which consists of at least two components, $p34^{cdc2}$ and cyclin. Unlike $p34^{cdc2}$, which is present in constant amounts during the cell cycle, levels of cyclin vary during the cycle; the protein accumulates during S phase and is abruptly degraded at the metaphase/anaphase transition. Progression into M phase has long been known to have a protein synthesis requirement, and this extract preserves both the protein synthetic capability of the egg and the ability to destroy cyclin specifically at the metaphase/anaphase transition.

Eggs are arrested in second meiotic metaphase and contain high levels of MPF, stabilized by another cytoplasmic effector (Masui and Markert, 1971), termed cytostatic factor (CSF). During normal fertilization, sperm not only contributes the paternal genome, but also provides the stimulus for embryonic development (know as egg activation) by bringing about the destruction of CSF and allowing normal cell-cycle oscillations to commence. The metaphase arrest may be overcome by parthenogenetic activation of an egg and, in the procedure described below, synchronous activation of eggs is achieved by

electrical stimulation. The resulting extract faithfully mimics the cell cycle of
the intact fertilized egg with respect to nuclear breakdown and reformation,
DNA replication, and MPF activation and inactivation (Hutchinson *et al.,*
1987; Murray and Kirschner, 1989).

Stock Solutions.   Storage temperatures are indicated in parentheses.

Cytochalasin B, 5 mg/ml in DMSO ($-20°$C)

Protease inhibitor cocktail: pepstatin A, chymostatin, leupeptin, each
5 mg/ml in water ($-70°$C)

ATP regenerating system ($-20°$C): 150 mM creatine phosphate, 20 mM
ATP, 2 mM EGTA, 20 mM MgCl$_2$, pH 7.7

Solutions

2% cysteine (100 ml), pH 7.7

0.2 × MMR: 20 mM NaCl, 0.4 mM KCl, 0.2 mM MgSO$_4$, 0.4 mM
CaCl$_2$, 0.02 mM EDTA, 1 mM HEPES, pH 7.8 (1000 ml)

Extract buffer (XB): 50 mM sucrose, 100 mM KCl, 0.1 mM CaCl$_2$, 1 mM
MgCl$_2$, 10 mM K-HEPES, pH 7.7 (200 ml)

Apparatus

Activation chamber: The chamber consists of a plexiglass box (11 cm ×
11 cm × 5 cm). The bottom of the interior and the underside of the lid
contain plate electrodes made of stainless steel. The bottom electrode
is covered to a depth of 1 cm with 2% agarose in 0.2 × MMR (Fig. 3).
The eggs rest on the agarose and are covered with 600 ml of 0.2 ×
MMR to allow electrical contact with the upper electrode. Activation
is achieved using a 12 V (AC) power supply

Timer/clock

Ice/water bath

Wide-bore Pasteur pipette: Cut the tip of a Pasteur pippette with a glass
saw and remove the lower part of the tip to make a pipette with a bore
of 4–5 mm

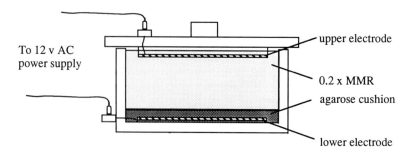

FIG. 3.   Electrical activation chamber for the parthenogenetic activation of *Xenopus* eggs.

Narrow-bore Pasteur pipette: Using a Bunsen flame, draw out a Pasteur
pipette to generate a pipette with a bore of ~0.5 mm

Clinical centrifuge

Beckman L-8 preparative ultracentrifuge (or equivalent), SW 50 swinging
bucket rotor, or equivalent which accepts 5-ml tubes

Procedures

1. Rinse about 25 ml of eggs, previously laid into MMR, with deionized
water (22°C) and allow eggs to stand in water for 10 minutes. This ensures that
the subsequent activation step is quantitative.

2. Add 1 ml of Versilube F-50 oil to each of two 5-ml ultracentrifuge tubes
(Beckman SW50). Layer 2 ml of XB (22°C) containing the protease inhibitor
cocktail (final concentration of 5 μg/ml per inhibitor) and cytochalasin B (final
concentration of 5 μg/ml) on top of the oil.

3. Dejelly and sort eggs as described in Section III,A,4 steps 1–3,

4. Activate eggs either by electrical stimulation or with the Ca ionophore
A23187. For electrical activation, transfer eggs to activation chamber filled
with 0.2 × MMR (22°C). Activation is achieved by applying two 3-second
pulses (12 V AC) separated by a 5-second pause. Successful activation of eggs
is judged by the contraction of the pigment in the animal hemisphere. At the
time of activation, set the timer running and ensure that all subsequent procedures up to step 6 are completed within 15 minutes of egg activation. Following activation, remove eggs from the activation chamber, place in a petri
dish, and rinse eggs with four changes of XB (22°C).

5. Using a wide-bore Pasteur pipette, transfer the eggs into the centrifuge
tubes (step 2, above) with a minimal volume of suspension buffer, to ensure
that cytochalasin B and protease inhibitors are kept as concentrated as possible. Eggs should fall through the XB and come to rest above the oil layer.

6. Centrifuge the eggs in a clinical centrifuge at 300 g (1300 rpm) for 60 seconds followed by 1350 g (280 rpm) for 20 seconds. This packs the eggs at the
bottom of the tube and allows the Versilube oil to rise above the eggs, thus
separating them from excess buffer which must be removed using a Pasteur
pipette. Small pockets of buffer which remain trapped between the eggs and
the wall of the tube should be removed using a narrow-bore Pasteur pipette.
Incubate the packed eggs at 22°C until 15 minutes has elapsed since activation.

7. Transfer the centrifuge tubes containing the eggs to an ice/water bath
and incubate at 0°C for a further 15 minutes.

8. Lyse the eggs by centrifugation at 12,000 g (Beckman SW50 rotor or
equivalent, 10,000 rpm, 10 minutes, 2°C). Withdraw the cytoplasmic extract
(volume 1.5–2.0 ml), which may be found either above or below the oil layer,
by puncturing the side of the tube (Fig. 4) using a 21-gauge needle and a 5-ml
syringe. The cytoplasmic extract varies in color depending on precise egg
pigmentation, and may be either gray or cream.

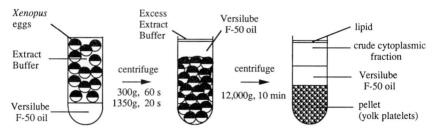

FIG. 4.   Flow diagram for the preparation of an oscillating extract from *Xenopus* eggs.

9. Add 0.05 volume of ATP regenerating system, and recentrifuge at 12,000 g (Beckman SW50 rotor or equivalent, 10,000 rpm, 10 minutes, 2°C) to remove remaining contaminating yolk and particulate matter. The extract may then be stored on ice for up to 2 hours until required.

10. Following the addition of demembranated sperm chromatin to the required concentration (100–500/μl), the cell-cycle oscillations may be initiated by warming the extract to 22°C.

Timing of Events in an Oscillating Extract.   When demembranated sperm chromatin is added to such extracts, while the extract is in interphase, the chromatin acts as a template for the formation of intact nuclei and the DNA undergoes replication. After 50–90 minutes in S phase, the extract initiates mitosis. At this time, the previously intact nuclei undergo nuclear envelope breakdown and chromosome condensation (Fig. 5). After approximately 20 minutes in mitosis, the extract returns to S phase as indicated by the reassembly of nuclear structure around the condensed chromatin. The chromatin decondenses and assumes an interphase morphology and an intact nuclear envelope is again observed. Cell-free oscillating extracts carry out two to three such cycles with uniform periods of 60–100 minutes, depending on the extract and the concentration of DNA present (Dasso and Newport, 1990).

H.   Monitoring Progress of Nuclear Assembly and Disassembly

Nuclear assembly and disassembly is readily followed either by phase-contrast microscopy or fluorescence microscopy using a light microscope (e.g., Zeiss Photomicroscope III) with exciter–barrier reflector combinations suitable for 3,3′-dihexyloxacarbocyanine (DHCC; fluorescein channel), and bisbenzamide (Hoechst 33258). Samples (1–2 μl) are removed at appropriate times and diluted on a microscope slide with an equal volume of fix/ stain buffer (200 mM sucrose, 10 mM HEPES, pH 7.5, 7.4% formaldehyde)

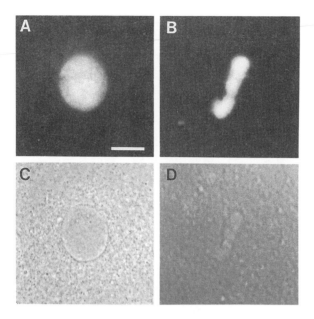

FIG. 5. Assembly and disassembly of nuclei in an oscillating extract. An oscillating extract was made from *Xenopus* eggs as described in Section III,G. Sperm chromatin was added to the extract at a concentration of 500/μl of extract. The sperm chromatin decondensed and nuclear assembly occurred, with the extract remaining in S phase for 70 minutes, as judged by the decondensed appearance of the DNA (A) and the presence of a nuclear envelope (C). The extract then entered mitosis for a 20-minute period as indicated by the condensation of the DNA (B) and the loss of the nuclear envelope (D). Following the mitotic period, the extract returned to S phase as judged by the reformation of nuclear envelope and decondensed chromatin (not shown). The nuclear envelope was observed by phase-contrast microscopy, while the DNA was visualized by fluorescence microscopy after staining with bisbenzamide. Bar, 5 μm.

containing either bisbenzamide (1 μg/ml) for staining DNA, and/or the lipophilic dye DHCC (1 μg/ml) for visualizing membranes.

I. Histone H1 Kinase Assays

The p34^{cdc2} component of MPF is a protein kinase, and it has been established that the growth-associated histone H1 kinase (Arion *et al.*, 1988) is identical to MPF. Thus, changes in the activity of MPF during the course of the cell cycle can be conveniently monitored by measuring histone H1 kinase activity (Fig. 6). As cAMP-dependent protein kinase is also a histone H1 kinase, specificity for p34^{cdc2} is maintained by the addition, to the assay mixture, of an inhibitor peptide of the former enzyme.

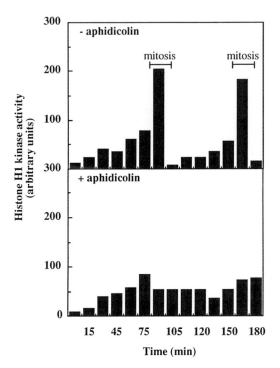

Fig. 6. Cell-cycle-dependent variation of histone H1 kinase activity. Histone H1 kinase activity was measured at the indicated times in an oscillating extract containing sperm chromatin $(400/\mu l)$ in the absence (upper panel) and presence (lower panel) of the DNA replication inhibitor aphidicolin (40 μg/ml). Kinase activity was quantitated by laser densitometry of an autoradiogram following gel electrophoresis of histone H1-containing samples. In the upper panel, mitosis refers to the time intervals in which the extracts were judged to be in mitosis by visual observation of the nuclei. The aphidicolin-treated extract failed to enter mitosis.

Stock Solutions.   Storage temperatures are given in parentheses. The letters A–F refer to the assay components in Table I.

    A. 200 mM *HEPES*, 50 mM EGTA, 100 mM $MgCl_2$, pH 7.3 (4°C)

    B. Histone H1, 2 mg/ml in H_2O (−20°C)

    C. Protein kinase inhibitor peptide (PKI), 100 μM in H_2O, (−20°C)

    D. ATP, pH 7.0, 2 mM H_2O (−20°C)

    E. [γ-^{32}P]ATP, which, when added to solution D, is sufficient to make the final specific activity $\approx$ 500,000–1,000,000 cpm/nmol (−20°C)

    F. H_2O

    EB buffer: 80 mM sodium glycerophosphate, 10 mM $MgCl_2$, 5 mM EGTA, pH 7.5 (4°C)

TABLE I

TABLE I

ASSAY MIXTURE FOR MEASURING HISTONE H1
KINASE ACTIVITY OF p34^{cdc2}

Assay components		Number of assays required	
		1 assay[a]	10 assays[a]
Buffer	(A)	2.0	20
Histone H1	(B)	2.0	20
PKI	(C)	2.0	20
ATP	(D + E)	2.0	20
H$_2$O	(F)	2.0	20

[a] All volumes given are in microliters.

Procedure

1. In order to assay histone H1 kinase activity in any of the above extracts, samples (2 μl) are removed from the extract as appropriate, diluted into an equal volume of EB buffer, and immediately frozen in liquid nitrogen for subsequent analysis.

2. Mix the reaction components according to Table I, such that there are 10 μl of reaction mixture per assay.

3. Maintain samples frozen in a dry ice/ethanol bath until immediately prior to assay. Initiate reactions at 1-minute intervals, by diluting each sample 60-fold with EB buffer and immediately adding 10 μl of the diluted enzyme to 10 μl of assay mixture. The incubation is continued for 10 minutes at 22°C. Stop reaction either by step 4 or 5.

4. Add 20 μl of 2 × SDS sample load buffer and electrophorese on 10% SDS-polyacrylamide gel. After electrophoresis, discard that portion of the gel containing the dye front (and free [^{32}P]ATP), fix the gel in 7% acetic acid, 10% methanol, and, after drying the gel, expose to preflashed X-ray film for direct autoradiography.

Alternatively:

5. Remove 15 μl of reaction mixture and pipet onto 1.5 cm × 1.5 cm phosphocellulose paper (Whatman, P81). Terminate reaction by immersion in 1% phosphoric acid. Wash filters for two 15-minute periods in 1% phosphoric acid, then 95% ethanol, dry, and count in scintillation fluid.

J. DNA Replication Assays

DNA replication may be assayed in any of the above extracts as follows.
Stock Reagents

$[\alpha\text{-}^{32}P]dATP\,(-20°C)$

Replication sample buffer: 8 mM EDTA, 0.13% phosphoric acid, 10% Ficoll, 5% SDS, 0.2% bromophenol blue, 80 mM Tris-Cl, pH 8.0 (22°C)

Proteinase K, 10 mg/ml in H_2O (−20°C)

Procedures

1. Remove 10-μl aliquots of extract incubation at 15-minute intervals and add each to 1 μCi of $[\alpha\text{-}^{32}P]dATP$.

2. Allow incubation to continue for a further 15 minutes. Terminate the reaction by the addition of 10 μl of replication sample buffer.

3. Digest each reaction mixture with proteinase K (final concentration 1 mg/ml) for 2 hours at room temperature.

4. Electrophorese on 0.8% agarose gel an after drying the gel, expose to preflashed X-ray film for direct autoradiography. The extent of DNA replication may be quantitated by laser densitometry of the autoradiogram, or by excising and counting relevant segments of the gel.

IV. Conclusions and Perspectives

Using the extracts described above, synthetic nuclei may be reconstituted around protein-free DNA or sperm chromatin. The *in vitro* assembled nuclei are morphologically indistiguishable from normal eukaryotic nuclei; they are surrounded by a double membrane, containing functional nuclear pores, and are lined with a peripheral nuclear lamina (Newport, 1987). Fractionation studies have shown that nuclear envelope assembly is initiated by the binding to chromatin of a specific set of membrane vesicles, which are distinct from the majority of endoplasmic reticulum-derived vesicles. The association of these vesicles with chromatin is mediated by a trypsin-sensitive integral membrane receptor (Wilson and Newport, 1988), and this association is regulated by reversible phosphorylation in a cell-cycle-dependent manner (Pfaller *et al.*, 1991).

Inhibitors may be added to, and particular proteins may be depleted from, these extracts in order to test theories of nuclear assembly, disassembly, and the cell-cycle control of these processes. Thus, the addition of the DNA topoisomerase inhibitor VM26 prevents DNA condensation and subsequent nuclear envelope assembly (Newport, 1987). Following initial envelope formation and insertion of nuclear pores, inhibition of transport through the nuclear pores by addition of wheat germ agglutinin (Finlay *et al.*, 1987) blocks the import of the embryonic lamin L_{III}, which prevents subsequent enlargement of the nuclear envelope. Immunodepletion of lamin L_{III} from the extract results in the formation of nuclei which are unable to replicate their

DNA (Newport *et al.*, 1990). Inhibition of DNA replication by aphidicolin prevents the subsequent activation of MPF and entry into mitosis, indicating the presence of a control system that monitors the replication state of DNA and regulates the activation of MPF (Dasso and Newport, 1990). Such approaches yield the conclusion that the structural organization within the nucleus is built up sequentially, such that the completion of one layer of organization serves as the foundation for the next, and that checkpoints exist to ensure that key stages are completed before the next stage is initiated.

Future applications of these systems will undoubtedly involve the development of assays to measure discrete steps in the pathways of nuclear assembly, genome replication, and nuclear disassembly. The progressive biochemical dissection of these processes, coupled with genetic and immunological approaches, will facilitate the identification of, and assignment of function to, proteins involved in processes as diverse as nuclear vesicle binding to chromatin, nuclear vesicle fusion and envelope growth, nuclear transport, as well as signal transduction pathways which regulate cell-cycle progression.

ACKNOWLEDGMENTS

The authors would like to thank their colleagues, Eva Meier, Colin Macaulay, Mary Dasso, Sally Kornbluth, Philippe Hartl, and Fang Fang for much helpful advice on the compilation of these methodologies. Work on this subject in J. W. N's laboratory was supported by a National Institutes of Health grant (GM334234).

REFERENCES

Arion, D., Meijer, L., Brizuela, L., and Beach, D. (1988). cdc2 is a component of the M-phase specific H1 kinase: Evidence for identity with MPF. *Cell (Cambridge, Mass.)* **55**, 371–378.

Blobel, G., and Potter, V. (1966). Nuclei from rat liver: Isolation method that combines purity with high yield. *Science* **154**, 1662–1665.

Blow, J. J., Sheehan, M. A., Watson, J. V., and Laskey, R. A. (1989). Nuclear structure and the control of DNA replication in the *Xenopus* embryo. *J. Cell Sci.* Suppl. 12, 183–185.

Dasso, M., and Newport, J. W. (1990). Completion of DNA replication is monitored by a feedback system that controls the initiation of mitosis *in vitro*: Studies in *Xenopus*. *Cell (Cambridge, Mass.)* **61**, 811–823.

Draetta, G., Luca, F., Westendorf, J., Ruderman, J., and Beach, D. (1989). cdc2 protein kinase is complexed with cyclin A and B: Evidence for proteolytic inactivation of MPF. *Cell (Cambridge, Mass.)* **56**, 829–838.

Dunphy, W. G., and Newport, J. W. (1988). Mitosis-inducing factors are present in a latent form during interphase in the *Xenopus* embryo. *J. Cell Biol.* **106**, 2047–2056.

Dunphy, W. G., and Newport, J. W. (1989). Fission yeast p13 blocks mitotic activation and tyrosine dephosphorylation of the *Xenopus* cdc2 protein kinase. *Cell (Cambridge, Mass.)* **58**, 181–191.

Dunphy, W. G., Brizuela, L., Beach, D., and Newport, J. W. (1988). The *Xenopus* cdc2 protein is a component of MPF, a cytoplasmic regulator of mitosis *Cell (Cambridge, Mass.)* **55**, 423–431.

Finlay, D. R., Newmeyer, D. D., Price, T. M., and Forbes, D. J. (1987). Inhibition of *in vitro* nuclear transport by a lectin that binds to nuclear pores. *J. Cell Biol.* **104**, 189–200.

Gautier, J., Matsukawa, T., Nurse, P., and Maller, J. (1989). Dephosphorylation and activation of *Xenopus* p34 protein kinase during the cell cycle. *Nature (London)* **339**, 626–629.

Gerace, L., and Burke, B. (1988). Functional organisation of the nuclear envelope. *Annu. Rev. Cell Biol.* **4**, 335–374.

Hutchinson, C., Cox, R., Drepaul, R., Gomperts, M., and Ford, C. (1987). Periodic DNA synthesis in cell-free extracts in *Xenopus* eggs. *EMBO J.* **6**, 2003–2010.

Lohka, M., and Maller, J. (1985). Induction of nuclear envelope breakdown, chromosome condensation and spindle formation in cell-free extracts. *J. Cell Biol.* **101**, 518–523.

Lohka, M., and Masui, Y. (1983). Formation *in vitro* of sperm pronuclei and mitotic chromosomes induced by amphibian ooplasmic components. *Science* **220**, 719–721.

Lohka, M., and Masui, Y. (1984). Roles of cytosol and cytoplasmic particles in nuclear envelope assembly and sperm pronuclear formation in cell-free preparations from amphibian eggs. *J. Cell Biol.* **98**, 1222–1230.

Masui, Y., and Markert, C. L. (1971). Cytoplasmic control of nuclear behavior during meiotic maturation of frog oocytes. *J. Exp. Zool.* **177**, 129–146.

Miake-Lye, R., and Kirschner, M. (1985). Induction of early mitotic events in a cell-free system. *Cell (Cambridge, Mass.)* **41**, 165–175.

Morla, A. O., Draetta, G., Beach, D., and Wang, J. Y. J. (1989). Reversible tyrosine phosphorylation of cdc2: Dephosphorylation accompanies activation during entry into mitosis. *Cell (Cambridge, Mass.)* **58**, 193–203.

Murray, A. W., and Kirschner, M. W. (1989). Cyclin synthesis drives the early embryonic cell cycle. *Nature (London)* **339**, 287–292.

Newport, J. W. (1987). Nuclear reconstitution *in vitro*: Stages of assembly around protein-free DNA. *Cell (Cambridge, Mass.)* **48**, 205–217.

Newport, J. W., and Forbes, D. J. (1987). The nucleus: Structure, function and dynamics. *Annu. Rev. Biochem.* **56**, 535–565.

Newport, J. W., and Spann, T. (1987). Disassembly of the nucleus in mitotic extracts: Membrane vesicularization, lamin disassembly, and chromosome condensation are independent processes. *Cell (Cambridge, Mass.)* **48**, 219–230.

Newport, J. W., Wilson, K. L., and Dunphy, W. G. (1990). A lamin-independent pathway for nuclear envelope assembly. *J. Cell Biol.* **111**, 2247–2259.

Pfaller, R., Smythe, C., and Newport, J. W. (1991). Assembly/disassembly of the nuclear envelope membrane: Cell-cycle dependent binding of nuclear membrane vesicles to chromatin *in vitro*. *Cell (Cambridge, Mass.)* **65**, 209–217.

Wilson, K. L., and Newport, J. W. (1988). A trypsin sensitive receptor on membrane vesicles is required for nuclear envelope formation *in vitro*. *J. Cell Biol.* **107**, 57–68.

Chapter 18

In Vitro Nuclear Protein Import Using Permeabilized Mammalian Cells

STEPHEN A. ADAM, RACHEL STERNE-MARR,[1] AND LARRY GERACE

Department of Molecular Biology
Research Institute of Scripps Clinic
La Jolla, California 92037

I. Introduction

Molecular traffic between the cytoplasm and nucleus occurs through the nuclear pore complex, an elaborate supramolecular structure with a mass of approximately 125×10^6 D that spans the nuclear envelope (Reichelt *et al.,* 1990; for reviews, see Gerace and Burke, 1988; Goldfarb, 1989). The pore complex contains an aqueous channel with a diameter of about 10 nm, which allows rapid nonselective diffusion of ions, metabolites, and other small molecules between the cytoplasm and nucleus (Paine *et al.,* 1975). Most

---

[1] Present address: Merck, Sharpe and Dohme, West Point, Pennsylvania 19486.

METHODS IN CELL BIOLOGY, VOL. 35

proteins and RNAs are too large to diffuse across this aqueous channel at physiologically significant rates, and instead utilize mediated mechanisms to cross the pore complex. The use of ligand-coated gold probes has allowed the direct visualization of protein and RNA transport across the pore complex, and has demonstrated that pore complexes are involved in concurrent protein import and RNA export (Feldherr *et al.*, 1984; Dworetzky *et al.*, 1988). Thus, macromolecular transport across the pore complex appears to be an intricate, highly coordinated process.

Mediated import of proteins into the nucleus is specified by short amino acid sequences intrinsic to the proteins called nuclear location sequences (NLS; Gerace and Burke, 1988). The best characterized NLS is that of the simian virus 40 (SV40) large T antigen, and consists of a stretch of 7 residues (P-K^{128}K-K-R-K-V) that is necessary and sufficient for nuclear import of the T antigen (Kalderon *et al.*, 1984; Lanford and Butel, 1984). Most well-characterized NLSs are similar to this prototype, in that they consist of a stretch of 3–5 basic residues, often flanking a proline or a glycine residue (Chelsky *et al.*, 1989). The peptide context of NLSs within proteins is relatively unimportant, since they can function in many different internal regions of a single protein (Roberts *et al.*, 1987). An important advance that has facilitated functional analysis of the import process is the finding that synthetic peptides containing a NLS are capable of directing the nuclear import of a protein to which they have been chemically coupled (Goldfarb, 1988; Lanford *et al.*, 1986).

In vivo and *in vitro* studies have revealed several major features of mediated protein import into the nucleus, including a dependence on temperature and a requirement for a functional NLS, ATP, and an intact nuclear envelope (Newmeyer *et al.*, 1986; Newmeyer and Forbes, 1988; Richardson *et al.*, 1988). Under conditions of reduced temperature and ATP depletion, gold particles coated with NLS-containing proteins become associated with the cytoplasmic surface of the pore complex, representing putative arrested intermediates in nuclear import (Newmeyer and Forbes, 1988; Richardson *et al.*, 1988). Subsequent to "docking" at the pore complex, particles are translocated through the central channel of the pore complex (the site of primary diffusional restriction) by a process that involves channel "gating" (Feldherr *et al.*, 1984; Akey and Goldfarb, 1989). This allows protein-coated gold particles with a diameter of at least 25 nm to be transported across the pore complex. However, almost no information is known about the biochemical basis for these transport steps.

Cell-free systems promise to be important tools to study the biochemistry of protein import into the nucleus. For an *in vitro* system to be considered physiologically relevant, it must reproduce key features of mediated transport across the pore complex seen *in vivo*. Hence, the system should result in the concentration of proteins in nuclei from the surrounding medium in a fashion dependent on NLS, ATP, and physiological temperature. Moreover, it is essential that proteins enter the nucleus through the nuclear pore complex. This requires that the nuclear envelopes of transporting nuclei be intact and retain the same diffusional restrictions as seen *in vivo*. That is, nonuclear proteins which are too large to diffuse into nuclei *in vivo* (i.e., globular proteins larger than approx. 40 kDa; Paine *et al.,* 1975) should also be excluded *in vitro*. This requirement is essential, since nuclei frequently lose their "intactness" during cell lysis, particularly with lysis procedures involving mechanical homogenization. If nuclei are not intact, transport probes could passively diffuse into nuclei through ruptures in the nuclear envelope, and could accumulate in the broken nuclei by binding to certain NLS-binding proteins that have been found to occur in the nuclear interior (e.g., Adam *et al.,* 1989; Meier and Blobel, 1990). Furthermore, this binding process could be stimulated by ATP and physiological temperatures. Nuclear transport *in vivo* is inhibited with anti-nuclear pore complex antibodies (Featherstone *et al.,* 1988) and wheat germ agglutinin, a lectin that binds to specific glycoproteins of the pore complex (Yoneda *et al.,* 1987; Dabauvalle *et al.,* 1988), and this also can be used as a criterion for *in vitro* nuclear entry via a pore complex route (Finlay *et al.,* 1987).

Two cell-free nuclear import systems characterized up to now adequately satisfy the criteria discussed above (Adam *et al.,* 1990; Newmeyer *et al.,* 1986). A number of other systems show ATP-dependent association of proteins with isolated nuclei, but the integrity of the nuclear envelope has not been rigorously demonstrated in these cases, and the significance of these systems is not yet clear (Markland *et al.,* 1987; Imamoto-Sonobe *et al.,* 1988; Kalinich and Douglas, 1989; Silver *et al.,* 1989; Parniak and Kennady, 1990).

One of the systems that yields authentic nuclear import is based on *Xenopus* egg extracts. The egg extracts have strong nuclear assembly activity and can "reseal" the ruptured nuclear envelopes of isolated rat liver nuclei and also can assemble intact nuclei around exogenous chromatin or chromosomes to produce transport-competent nuclei (Newmeyer *et al.,* 1986). This system is powerful for studying *de novo* assembly of the nuclear envelope (Finlay and Forbes, 1990), but is limited because a source of *Xenopus* eggs must be maintained and the user is restricted to use of *Xenopus* egg cytosol to support import. In addition, while cytosol is required in this nuclear transport system, it is difficult to distinguish cytosolic factors essential for nuclear resealing or assembly from those directly involved in transport across the pore complex (Newmeyer and Forbes, 1990).

We have developed a second *in vitro* system for physiological nuclear import that utilizes permeabilized cultured cells grown on glass coverslips (Adam *et al.,* 1990). The advantages of this system include the ease with which it can be set up, and the wide variety of cells that can be used as sources of permeabilized cells and cytosol. The plasma membrane of most cells can be selectively and efficiently perforated with the cardiac glycoside digitonin, due to the plasma membrane's proportionally higher cholesterol content compared to most other intracellular membranes. Under appropriate detergent conditions that permeabilize the plasma membrane, intracellular membranes, including the nuclear envelope, remain intact and much of the cytoplasmic cytoskeleton is preserved. At the same time, soluble contents of the cell diffuse through the digitonin-induced holes in the plasma membrane. When permeabilized cells are supplemented with exogenous cytosol and ATP, the nuclei of these cells rapidly accumulate a fluorescent protein containing a nuclear location sequence. This accumulation meets the criteria established for authentic nuclear import in that it is specific for a functional NLS, is dependent on ATP, temperature, and an intact nuclear envelope, and can be inhibited by the addition of wheat germ agglutinin.

Transport in permeabilized cells absolutely requires the addition of exogenous cytosol. The cytosol is inactivated by treatment with the sulfhydryl alkylating reagent N-ethylmaleimide (NEM; Adam *et al.,* 1990). Detailed analysis of NEM inactivation of the system suggests that at least three different cytosolic components are required for protein import, include one NEM-insensitive component and two NEM-sensitive components. We have fractionated erythrocyte cytosol and have isolated one of the NEM-sensitive transport factors, which is a receptor for NLS (Adam *et al.,* 1989). This system will prove very useful for isolating additional transport factors and for characterizing the role of cytoplasmic components in nuclear import.

II.   Methods

A.   Preparation of Cytosol Fractions

1.   CHOICE OF CELL TYPE

A valuable feature of this system is the ability to use cytosol prepared from a wide variety of vertebrate cells. Among other things, this could be useful for analyzing regulation of transport during cell growth and development (e.g., Feldherr and Akin, 1990). Cell types from which we have successfully obtained import-competent cytosol are indicated in Table I. When cultured cells are used to prepare cytosol extracts, it is necessary for the cells to be in log phase growth for optimal activity. Stationary cultures, as well as cell types that do not rapidly divide, do not yield highly active cytosol. In addition, we

TABLE I

Sources of Cytosol and Permeable Cells

Cytosol	Permeable cells
Rabbit reticulocyte	HeLa (human)
HTC (rat hepatoma)	NRK (normal rat kidney)
Rat erythrocyte	BRL (buffalo rat liver)
Bovine erythrocyte	
Xenopus laevis oocyte	
Rat liver	

find that cells propogated in suspension culture using spinner flasks yield cytosol that is more active than monolayer cells. This may be due to the fact that it is easier to thoroughly wash large quantities of suspension cells and to obtain compact cell pellets prior to homogenization.

For most of our routine work, we use rabbit reticulocyte lysate obtained from commercial distributors (Promega Biotech, Milwaukee, WI) for cytosol. These extracts have consistently higher import activity than cultured cell extracts. We also have found that cytosol prepared by hypotonic lysis of red blood cells (RBCs) is equivalent in activity to reticulocyte extracts. RBCs can be easily obtained in large quantities from appropriate sources of fresh blood and provide a convenient source of material for biochemical fractionation of soluble import components. While mammalian RBCs are nondividing and lack nuclei, it appears that the cytosolic components required for nuclear import stably persist in RBCs from earlier developmental stages.

We also have found that extracts prepared from *Xenopus laevis* oocytes can support import in this system. In contrast to the previously described *in vitro* system that requires membrane components present in egg extracts to reseal rat liver nuclei (Newmeyer *et al.*, 1986), in our system high-speed supernatants of oocyte lysates that are devoid of membranes are capable of supporting import. It is likely that oocytes from other organisms also would be able to provide import-competent cytosol. Although we have not extensively investigated the possibility of using yeast cytosol for transport in permeabilized mammalian cells, our preliminary attempts to prepare active cytosol from *Saccharomyces cerevisiae* have been unsuccessful. We feel that this may be due to a nonspecific inhibitory factor, and that with appropriate conditions it may be possible to obtain transport-competent cytosol from yeast.

Finally, we have found that high-speed supernatants prepared from concentrated homogenates of rat liver also support *in vitro* nuclear import. Like lysates of RBCs, cytosol prepared from this and other tissue sources may prove useful for isolating factors involved in protein import.

2.   HOMOGENIZATION OF CELLS

For preparation of cytosol to support *in vitro* nuclear import, we have extensively used two lines of cultured mammalian cells grown in suspension: HeLa cells and a rat hepatoma cell line (HTC). The homogenization conditions described below also can be used for other cultured cell lines. Suspension cultures of HeLa cells or HTC cells are grown in Joklik's modified minimum essential medium with 5% fetal bovine serum (FBS), 20 mM HEPES, pH 7.2, and penicillin/streptomycin. Cultures are maintained at 37°C in 1-liter microcarrier flasks (Bellco Glass Co.) mixing at 30–50 revolutions/minute. Exponentially growing cultures are collected by centrifugation at 250 g for 10 minutes at 4°C in a Beckman J6B refrigerated centrifuge equipped with a JS5.2 4-liter swinging bucket rotor and washed at least two times with cold phosphate-buffered saline (PBS), pH 7.4, by resuspension and centrifugation. The cells are then washed with 10 mM HEPES, pH 7.3, 110 mM potassium acetate, 2 mM magnesium acetate, and 2 mM dithiothreitol (DTT) and pelleted. The cell pellet is gently resuspended in 1.5 volumes of lysis buffer [5 mM HEPES, pH 7.3, 10 mM potassium acetate, 2 mM magnesium acetate, 2 mM DTT, 20 μM cytochalasin B, 1 mM phenylmethylsulfonyl fluoride (PMSF), and 1 μg/ml each aprotinin, leupeptin, and pepstatin (Boehringer Mannheim Biochemicals, Indianapolis, IN)] and swelled for 10 minutes on ice. Because the cytosol is very sensitive to inactivation by oxidation, it is important to keep reducing agents present at all times. The cells are lysed by five strokes in a tight-fitting stainless steel or glass Dounce homogenizer. The resulting homogenate is centrifuged at 1500 g for 15 minutes to remove nuclei and cell debris. The supernatant is then sequentially centrifuged at 15,000 g for 20 minutes in a Beckman JA 20 fixed-angle rotor followed by 100,000 g for 30 minutes in a Beckman 70.1 Ti fixed-angle ultracentrifuge rotor. The final supertant is dialyzed extensively with a collodion membrane apparatus (molecular weight cut-off 25,000 D; Schleicher & Schuell Inc., Keene, NH) against import buffer (20 mM HEPES, pH 7.3, 110 mM potassium acetate, 5 mM sodium acetate, 2 mM magnesium acetate, 0.5 mM EGTA, 2mM DTT, and 1 μg/ml each aprotinin, leupeptin, and pepstatin) and frozen in aliquots in liquid nitrogen prior to storage at −80°C.

RBCs from many different mammalian species can be used to prepare active cytosol for transport, according to the following procedure that we use for rat erythrocytes. Blood is collected from several rats by cardiac puncture after CO_2 asphyxiation into 1 ml of acid–citrate–dextrose (ACD; 75 mM trisodium citrate, 38 mM citric acid, and 136 mM dextrose, pH 5.0)/7 ml blood. A 250-g rat typically yields from 3 to 8 ml blood. Prior to lysis, the blood is washed by several steps to separate erythrocytes from platelets and leucocytes. Immediately after bleeding, the blood is centrifuged at 500 g for 20 minutes at room temperature in a Beckman J6B refrigerated centrifuge equipped with a JS5.2 swinging bucket rotor. The plasma layer and buffy coat are aspirated and an equal volume of 2% gelatin (250 bloom; Sigma Chem. Co., St. Louis, MO) in PBS is added and gently mixed. The red cells are allowed to sediment undisturbed at 37°C for 30 minutes. If the red cells do not sediment, more gelatin may be added. The cloudy upper layer containing platelets and white blood cells is aspirated and the cells are washed four times in PBS by centrifugation at 1600 g in the same rotor as above. The cells are lysed by adding 1–2 volumes of ice-cold magnesium lysis buffer (5 mM HEPES, pH 7.3, 2 mM magnesium acetate, 1 mM EGTA, 2 mM DTT, 0.1 mM PMSF, 1 μg/ml each aprotinin, leupeptin, and pepstatin) and vortexing on a high setting for 30 seconds. The lysate is allowed to set on ice for 5 minutes before the cell debris is removed by centrifugation at 6000 g for 20 minutes at 4°C in a Beckman JS5.2 rotor. The clear supernatant is removed, avoiding the cloudy layer of red cell ghosts filling the lower half of the tube. This supernatant is centrifuged at 100,000 g for 30 minutes in a Beckman Ti 70.1 rotor and dialyzed into transport buffer as described above for reticulocyte lysate. After dialysis, it may be necessary to recentrifuge the extract to remove material that may precipitate during dialysis. Occasionally, some red cell ghosts may remain and can be removed by filtration of the extract through a 0.45-μm syringe filter. To prepare commercially obtained rabbit reticulocyte lysate for transport studies, the lysate is dialyzed into transport buffer as described above and is centrifuged at 100,000 g prior to dialysis and freezing.

Oocytes are prepared from *X. laevis* by dissection of the ovary into PBS. The ovary is cleaned of extraneous tissue, blotted to remove excess PBS, and placed in a glass Dounce homogenizer with 1 volume of 20 mM HEPES, pH 7.3, 110 mM potassium acetate, 2 mM magnesium acetate, 1 mM DTT, and 1 μg/ml each aprotinin, leupeptin, and pepstatin. The oocytes are crushed by two strokes of a very loose fitting pestle. The crude lysate is centrifuged at 2000 g for 10 minutes and the supernatant is removed with a Pasteur pipette, taking care to avoid the lipid layer floating on top and the diffuse cloudy layer near the bottom of the tube. This supernatant is then centrifuged in a Beckman 70.1 Ti rotor at 100,000 g for 45 minutes. Again, the clear supernatant is removed and saved, avoiding the upper and lower layers. This supernatant is then dialyzed against import buffer as described above.

B.   Preparation of Fluorescent Conjugates

In choosing a reporter protein to study mediated nuclear import, it is important to select a protein that is too large to diffuse passively through the pore complex (i.e., a globular protein larger than 40–60 KDa). The reporter protein that we commonly use to detect nuclear import is the 104-kDa naturally fluorescent protein allophycocyanin (APC; Calbiochem). This protein is chemically coupled to synthetic peptides containing the SV40 large T antigen NLS through an N-terminal cysteine on the peptides. Synthetic peptides containing the SV40 large T antigen wild type (CGGGPK[128]KKRKVED) or a mutant import-deficient (CGGGPK[128]NKRKVED) nuclear location signal are obtained from Multiple Peptide Systems (San Diego, CA) and, as appropriate, are purified by chromatography on an HPLC system. Prior to conjugation, the peptide should be reduced to maximize its ability to couple through the cysteine sulfhydryl. To accomplish this, 1 mg of peptide is resuspended in 50 mM HEPES, pH 7.0, and incubated with 5 mg of DTT for 1 hour at room temperature. A 1% (v/v) concentration of glacial acetic acid is added to the reduced peptides and the mixture is separated by chromatography on a 20 × 1 cm Sephadex G-10 column. The column is eluted with 1% acetic acid and 0.5-ml fractions are collected. The peptide elution profile is monitored by the use of the Ellman reaction. This involves addition of 10-μl aliquot of each fraction to a test tube containing 1 ml of 0.1 M sodium phosphate buffer, pH 7.4, and 5 mM EDTA. To each tube is then added 100 μl of 1 mM dithiobisnitrobenzoic acid (Ellman's reagent) in methanol. Fractions containing free sulfhydryl residues will yield a bright yellow color. Fractions containing peptide (eluting first off the column) are pooled, and the concentration of free cysteine-containing peptide is determined by Ellman's reaction as described above, using glutathione to prepare a standard curve.

To activate the APC for peptide conjugation, the protein is first dissolved in 0.1 M sodium phosphate, pH 8.0, at 2 mg/ml and dialyzed extensively against the same buffer. Sulfo-SMCC (Pierce Chem. Co.) is dissolved in dimethyl sulfoxide at 1 mg/ml and a 20-fold molar excess is immediately added to the APC and incubated for 30 minutes at room temperature. The activated protein is immediately separated from the unreacted sulfo-SMCC by desalting on a 10 × 1 cm Sephadex G-25 column equilibrated in 0.1 M sodium phosphate, pH 7.0. The peptides containing reduced N-terminal cysteine residues are mixed at a 50-fold molar excess with the activated APC. After overnight incubation in the dark at 4°C, the APC–peptide conjugates are separated from free peptides by desalting on Sephadex G-25. The blue color of the APC makes selection of appropriate fractions very easy. The number of peptides conjugated to the protein is estimated by mobility shift on sodium dodecyl sulfate-polyacrylamide gels and is usually 4–8 peptides per APC molecule for the conditions described above. The conjugates are then dialyzed against 10 mM HEPES, pH 7.3, 110 mM potassium acetate and concentrated to 1 mg/ml in a collodion membrane apparatus before storage in aliquots at −80°C. The concentration of the allophycocyanin conjugates is determined by a commercial dye binding assay (Pierce BCA, Pierce Chem. Co., or BioRad Protein Assay, BioRad).

In addition to APC, we also have successfully used peptide conjugates of rhodamine isothiocyanate-labeled bovine serum albumin (BSA) for transport studies. Highly purified crystallized BSA (e.g., Sigma Chemical Co.) should be used for this work so that the labeled BSA is not contaminated with other minor labeled proteins. Purified IgG conjugated to NLS peptides also has been used for transport studies (Lanford *et al.,* 1986), and IgG can be readily isolated by adsorption to immobilized protein A or protein G. Fluorescent conjugates of BSA or IgG are prepared as described below. Alternatively, fluorescent conjugates of BSA or IgG are commercially available (Sigma Chem. Co.). Whatever carrier protein is used as a reporter, the protein must be labeled with a fluorochrome prior to conjugation with peptides to avoid inactivation of the NLS by modification of essential amino acid residues.

For labeling a reporter protein with fluorochrome, a protein solution of 1–2 mg/ml must first be dialyzed extensively in 0.1 *M* sodium carbonate, pH 9.0. The fluorescein isothiocyanate (FITC) or tetramethylrhodamine isothiocyanate (TRITC) is dissolved in dimethyl sulfoxide at 1 mg/ml immediately prior to use. For each milligram of protein to be labeled, 25 μl of the fluorochrome is added slowly while mixing. The mixture is incubated overnight in the dark at 4°C. To stop the reaction, Tris-HCl, pH 8.0, is added to 50 m*M* and incubated for 1 hour in the dark. The conjugated protein is separated from free fluorochrome by chromatography through Sephadex G-25 or G-50 equilibrated in 0.1 *M* sodium phosphate, pH 8.0. The size of the column should be approximately 20 times the volume of the reaction mix. The labeled protein should then be concentrated to approximately 2 mg/ml and coupled to the peptide as described above for APC.

C. Permeabilization of Cells and Import Assay

Human (HeLa) cells and normal rat kidney (NRK) cells are grown on plastic petri dishes in Dulbecco's modified Eagle's medium containing 10% FBS and penicillin/streptomycin. Cultures are maintained in a humidified incubator at 37°C with a 5% CO_2 atmosphere. Twenty four to 48 hours before use in a transport assay, cells are removed from the plastic dishes by trypsinization and replated on 18 × 18 mm glass coverslips in 6-well multiwell plates. To permeabilize the cells, coverslips are rinsed in ice-cold import buffer and immersed in ice-cold import buffer containing 40 μg/ml digitonin (Calbiochem; diluted from a 40 mg/ml stock solution in dimethyl sulfoxide) in a small Coplin jar. The cells are allowed to permeabilize for 5 minutes, after which the coverslips are transferred to a Coplin jar containing cold import buffer. The coverslips are then drained and blotted to remove excess buffer, and inverted over a 50-μl drop of complete import mixture on a sheet of Parafilm in a humidified plastic box. The complete import mixture contains the following components: 50% cytosol, 100 n*M* APC–peptide conjugate,

20 mM HEPES, pH 7.3, 110 mM potassium acetate, 5 mM sodium acetate, 2 mM DTT, 0.5 mM EGTA, 1 mM ATP, 5 mM creatine phosphate (Calbiochem), 20 units/ml creatine phosphokinase (Calbiochem), and 1 μg/ml each aprotinin, leupeptin, and pepstatin. The optimum cytosol concentration should be determined for each cytosol preparation since some batches of cytosol are slightly inhibitory at higher concentrations or may cause cells to be lost from the coverslip. The entire box is then floated in a waterbath at 30°C. At the end of the assay, each coverslip is rinsed in import buffer and mounted on a glass microscope slide in a small amount of import buffer and the coverslip edges are sealed with nail polish.

If many samples are to be observed or if they cannot be examined immediately, the rinsed cells may be fixed by immersion in import buffer containing 4% formaldehyde. Samples are observed by phase-contrast and epifluorescence microscopy with a Zeiss Axiophot microscope equipped with 40 × and 63 × planapochromat objectives. The results of a typical import reaction are shown in Fig. 1. With an import system consisting of rabbit reticulocyte lysate and NRK cells, we obtain an average of 10- to 20-fold accumulation of the reporter protein in the nucleus with respect to the surrounding medium after 30 minutes. While approximately 95% of the nuclei on the coverslip show significant nuclear accumulation of the transport probe, there is extensive heterogeneity from nucleus to nucleus in the actual level of nuclear accumulation.

It must be stressed that, when the system is initially set up, appropriate controls must be done to authenticate import. First, it should be demonstrated that the nuclei of permeabilized cells are intact and exclude large proteins lacking NLSs. This feature constrains the import of large proteins to a mediated pathway. To determine nuclear integrity, we incubate permeabilized cells with anti-DNA antibodies in 0.2% gelatin in PBS for 15 minutes at room temperature, rinse the coverslips, and then incubate the coverslips with a secondary antibody (diluted in gelatin–PBS) to detect the anti-DNA antibodies. If the permeabilized cell nuclei are not intact, the nuclei become strongly labeled because of the large number of binding sites for anti-DNA antibodies in the nucleus (Adam et al., 1990). Another control involves demonstrating that reporter proteins coupled with peptides containing the nontransported mutant T antigen NLS sequence (see above) are not accumulated in the nucleus. A related experiment involves analysis of the import capability of the carrier protein alone that has been modified with the cross-linking reagent, since some carrier proteins may exhibit high nonspecific binding to the permeabilized cells and/or nuclei. Removal of ATP by hexokinase/glucose treatment of the cytosol should demonstrate strict ATP dependence (Newmeyer et al., 1986; Adam et al., 1990), and incubation at 0°C should indicate temperature dependence. Finally, nuclear import in ver-

Phase   **Import**

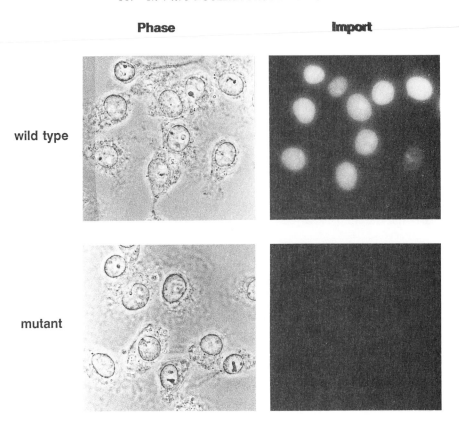

FIG. 1. Results of a typical import reaction. NRK cells were permeabilized with digitonin and incubated in complete transport mix as described in the text. The APC–peptide conjugate in each case was prepared with peptides containing either the wild-type SV40 T antigen NLS or a mutant, nonfunctional NLS.

tebrate cells should be inhibited by addition of wheat germ agglutinin to the reaction at 50 μg/ml (Finlay *et al.,* 1987; Adam *et al.,* 1990), which would argue that proteins enter the nuclei of permeabilized cells through the pore complexes.

D. Quantifying Import

The amount of import into an individual nucleus can be quantified by performing densitometry on photographic negatives. A field of cells is photographed with Kodak TMax film (ASA 400) and the film processed. A standard curve of fluorescence intensity values is obtained by preparing a dilution series of the fluorescent APC from about 1.5-fold to 50-fold dilution. Eight microliters of each dilution is pipetted onto a slide and mounted under an 18 × 18 mm glass coverslip. This dilution series is then photographed at the same exposure time as the cells. The negatives are scanned with a scanning laser densitometer (LKB Instruments, Inc., Gaithersburg, MD). Comparison to values generated from the standard curve can be used to determine the APC concentration in each nucleus. Alternatively, fluorescent images can be directly recorded from the microscope using a video camera, and image-processing software can be used to analyze the data for quantitation (Newmeyer and Forbes, 1990). Many image analysis systems are available to carry out the simple analysis required here.

III. Discussion

The nuclear import system described here is simple, efficient, and can be tailored to the needs of the individual laboratory. A wide range of different cell types can be used as the source of permeabilized cells or cytosol. We utilize a fluorescent reporter protein to measure nuclear import, because it is possible to visualize directly its nuclear accumulation and determine the range of nuclear accumulation given by a population of nuclei. We routinely use allophycocyanin for this assay because of its large size (104kDa), its naturally high fluorescence output, and its resistance to photobleaching, although we also have been successful with rhodamine-labeled BSA coupled to the synthetic peptides. Another widely used protein for analysis of nuclear import is *Xenopus* nucleoplasmin, a nuclear protein that is easily purified to homogeneity in sufficient quantities for these studies (Dingwall *et al.*, 1982). Unfortunately, in our hands, rhodamine-labeled nucleoplasmin exhibits high nonspecific binding to the permeabilized cells and is imported weakly. A nonfluorescent reporter protein also may be used to measure import if an antibody to the protein is available to measure its nuclear accumulation by immunofluorescence microscopy, although the protein must not be present in the cells used for the assay and the analysis is lengthened by the requisite immunofluorescence labeling.

In principle, it would be possible to use ^{125}I-labeled reporter proteins and gamma counting to quantitate this assay. However, in the latter case, it would be necessary to perform additional controls to determine what percentage of radioactivity becomes associated with permeabilized cells during the import assay due to adsorption to extranuclear components, as opposed to actual nuclear import. Inhibition of nuclear import by wheat germ agglutinin could be useful for this purpose.

While this permeabilized cell system has been developed to analyze nuclear protein import, we believe that it will also be applicable to analysis of RNA export from the nucleus. Virally infected cells may provide an especially useful permeabilized cell model for such studies, since large amounts of a relatively small number of transcripts are produced after infection of mammalian cells by certain viruses.

REFERENCES

Adam, S. A., Lobl, T. J., Mitchell, M. A., and Gerace, L. (1989). Identification of specific binding proteins for a nuclear location sequence. *Nature (London)* **337**, 276–279.

Adam, S. A., Sterne-Marr, R. E., and Gerace, L. (1990). Protein import in permeabilized mammalian cells requires soluble cytoplasmic factors. *J. Cell Biol.* **111**, 807–816.

Akey, C. W., and Goldfarb, D. S. (1989). Protein import through the nuclear pore complex is a multistep process. *J. Cell Biol.* **109**, 971–982.

Chelsky, D., Ralph, R., and Jonak, G. (1989). Sequence requirement for synthetic peptide mediated translocation to the nucleus. *Mol. Cell. Biol.* **9**, 2487–2492.

Dabauvalle, M.-C., Schulz, B., Scheer, U., and Peters, R. (1988). Inhibition of nuclear accumulation of Karyophilic proteins in living cells by microinjection of the lectin wheat germ agglutinin. *Exp. Cell. Res.* **174**, 291–296.

Dingwall, C., Sharnick, S., and Laskey R. (1982). A polypeptide domain that specifies migration of nucleoplasmin in the nucleus. *Cell (Cambridge, Mass.)* **30**, 449–458.

Dworetzky, S. I., and Feldherr, C. M. (1988). Translocation of RNA-coated gold particles through the nuclear pores of oocytes. *J. Cell Biol.* **106**, 575–584.

Featherstone, C. M., Darby, M. K., and Gerace, L. (1988). A monoclonal antibody against the nuclear pore complex inhibits nucleocytoplasmic transport of protein and RNA *in vivo. J. Cell Biol.* **107**, 1289–1297.

Feldherr, C. M., and Akin, D. (1990). The permeability of the nuclear envelope in dividing and nondividing cell cultures. *J. Cell Biol.* **111**, 1–8.

Feldherr, C., Kallenbach, E., and Schultz, N. (1984). Movement of a karyophilic protein through the nuclear pores of oocytes. *J. Cell Biol.* **99**, 2216–2222.

Finlay, D. R., and Forbes, D. J. (1990). Reconstitution of biochemically altered nuclear pores: Transport can be eliminated and restored. *Cell (Cambridge, Mass.)* **60**, 17–29.

Finlay, D. R., Newmeyer, D. D., Price, T. M., and Forbes, D. J. (1987). Inhibition of *in vitro* nuclear transport by a lectin that binds to nuclear pores. *J. Cell Biol.* **104**, 189–200.

Gerace, L., and Burke, B. (1988). Functional organization of the nuclear envelope. *Annu. Rev. Cell Biol.* **4**, 335–374.

Goldfarb, D. S. (1988). Karyophilic peptides: Applications to the study of nuclear transport. *Cell Biol. Int. Rep.* **12**, 809–832.

Goldfarb, D. S. (1989). Nuclear transport. *Curr. Op. Cell Biol.* **1**, 441–446.

Imamoto-Sonobe, N., Yoneda, Y., Iwamoto, R., Sugawa, H., and Uchida, T. (1988). ATP-dependent association of nuclear proteins with isolated rat liver nuclei. *Proc. Natl. Acad. Sci. U.S.A.* **85**, 3426–3430.

Kalderon, D., Roberts, B. L., Richardson, W. D., and Smith, A. E. (1984). A short amino acid sequence able to specify nuclear location. *Cell (Cambridge, Mass.)* **39**, 499–509.

Kalinich, J. F., and Douglas, M. G. (1989). *In vitro* translocation through the yeast nuclear envelope. *J. Biol. Chem.* **264,** 17979–17989.

Lanford, R. E., and Butel, J. S. (1984). Construction and characterization of an SV40 mutant defective in nuclear transport of T antigen. *Cell (Cambridge, Mass.)* **37,** 801–813.

Lanford, R. E., Kanda, P., and Kennedy, R. C. (1986). Induction of nuclear transport with a synthetic peptide homologous to the SV40 T antigen transport signal. *Mol. Cell Biol.* **8,** 2722–2729.

Markland, W., Smith, A. E., and Roberts, B. L. (1987). Signal dependent translocation of Simian Virus 40 large T antigen into rat liver nuclei in a cell-free system. *Mol. Cell. Biol.* **7,** 4255–4265.

Meier, U. T., and Blobel, G. (1990). A nuclear localization signal binding protein in the nucleolus. *J. Cell Biol.* **111,** 2235–2245.

Newmeyer, D. D., and Forbes, D. J. (1988). Nuclear import can be separated into distinct steps *in vitro*: Nuclear pore binding and translocation. *Cell (Cambridge, Mass.)* **52,** 641–653.

Newmeyer, D. D., and Forbes, D. J. (1990). An N-ethylmaleimide-sensitive cytosolic factor necessary for nuclear protein import: Requirement in signal-mediated binding to the nuclear pore. *J. Cell Biol.* **110,** 547–557.

Newmeyer, D. D., Finlay, D. R., and Forbes, D. J. (1986). *In vitro* transport of a fluorescent nuclear protein and exclusion of non-nuclear proteins. *J. Cell Biol.* **103,** 2091–2102.

Paine, P. L., Moore, L. C., and Horowitz, S. B. (1975). Nuclear envelope permeability. *Nature (London)* **254,** 109–114.

Parniak, V. K., and Kennady, P. K., (1990). Nuclear transport of proteins translated *in vitro* from SP6 plasmid-generated mRNAs. *Mol. Cell. Biol.* **10,** 1287–1292.

Reichelt, R., Holzenburg, A., Buhle, E. L., Jarnik, M., Engel, A., and Aebi, U. (1990). Correlation between structure and mass distribution of nuclear pore complex components. *J. Cell Biol.* **110,** 883–894.

Richardson, W. D., Mills, A. D., Dilworth, S. M., Laskey, R. A., and Dingwall, C. (1988). Nuclear protein migration involves two steps: Rapid binding at the nuclear envelope followed by slower translocation through the nuclear pores. *Cell (Cambridge, Mass.)* **52,** 655–664.

Roberts, B. L., Richardson, W. D., and Smith, A. E. (1987). The effect of protein context on nuclear location signal function. *Cell (Cambridge, Mass.)* **50,** 465–475.

Silver, P., Sadler, I., and Osborne, M. A. (1989). Yeast proteins that recognize nuclear localization sequences. *J. Cell Biol.* **109,** 983–989.

Yoneda, Y., Imamoto-Sonobe, N., Yamaizumi, M., and Uchida, T. (1987). Reversible inhibition of protein import into the nucleus by wheat germ agglutinin injected into cultured cells. *Exp. Cell Res.* **173,** 586–595.

Part V.   Genetic Approaches

Chapter 19

Mutations That Affect Chromosomal Proteins in Yeast

M. MITCHELL SMITH

Department of Microbiology
School of Medicine
University of Virginia
Charlottesville, Virginia 22908

METHODS IN CELL BIOLOGY, VOL. 35

I. Introduction

Chromosomal proteins include a large and diverse class of molecules, from histones, to high-mobility group proteins, from nonhistone chromosomal proteins, to specific transcription factors. Many of these have been extensively studied for their physical and structural biochemistry. However, a genetic understanding of chromosomal proteins is still in its infancy. For example, the roles of the histones and high-mobility group proteins in cell physiology are only now beginning to be understood. There are good reasons for this. In the first place, many of these proteins are highly conserved during evolution such that natural variants are not available for exploitation. The genes for the major core histones, at least, are also present in many copies per genome in metazoans which makes it difficult to conduct an experimental genetics. Finally, the chromosomal proteins undoubtedly participate in some of the most fundamental aspects of chromosome organization and function. Thus, mutations in these proteins are expected to be pleiotropic and their pheno- types difficult to investigate. Even in the case of specific transcription fac- tors, where molecular genetics has played a major role in elucidating their function, it is as yet unclear how these proteins interact with the architecture of the chromosome.

In this chapter we focus on genetic approaches to studying chromosomal proteins with the emphasis primarily on the model organism *Saccharomyces cerevisiae*. The yeasts *S. cerevisiae* and *Schizosaccharomyces pombe* are par- ticularly well suited to studies of chromosomal proteins because of their small genome sizes, the excellent molecular genetic tools that have been de- veloped over the past decade, and the ability to combine these molecular approaches with classical techniques of genetics, cell biology, and biochemis- try (Botstein and Fink, 1988). Many of the known chromosomal proteins, such as the histones and the high-mobility group proteins, are evolution- arily conserved between the yeast and higher eukaryotic cells. In addition, chromosomal proteins have recently been identified in yeast that function pleiotropically in such diverse activities as nucleosome positioning, tran- scriptional activation, gene silencing, and telomere formation (Buchman *et al.,* 1988a, b; Chasman *et al.,* 1990; Diffley and Stillman, 1989; Ju *et al.,* 1990; Longtine *et al.,* 1989; Shore and Nasmyth, 1987). These proteins will likely have at least functional counterparts in mammalian cells. Therefore, genetic studies in yeast are bound to complement biochemical studies in higher eukaryotic cells to the benefit of both disciplines.

In the following sections, we first consider briefly strategies for the muta- genesis of chromosomal protein genes. Next, we examine genetic tools for manipulating recessive mutations in essential genes in yeast to facilitate their study. Finally, we summarize a few fundamental assays of relevance for chromosomal protein mutations and consider strategies for further genetic manipulations. For detailed discussions of individual techniques and those not covered explicitly in this review, the reader is referred to the excellent laboratory manuals on recombinant DNA techniques (Ausubel *et al.,* 1989; Sambrook *et al.,* 1989) and yeast molecular genetics (Guthrie and Fink, 1990; Rose *et al.,* 1990).

II. Mutagenesis

A. Deletions and Domain Swapping

Frequently, DNA sequence information or biochemical studies of a chromosomal protein will suggest the presence of functional domains whose importance can be tested directly by deletion or by swapping the domain with another of known function. Examples of such domains include the nucleic acid binding and activation domains of transcription factors (Brent and Ptashne, 1985; Hope and Struhl, 1986; Ma and Ptashne, 1987; Pfeifer *et al.,* 1989), and the N-terminal domains of the histone proteins (Kayne *et al.,* 1988; Morgan *et al.,* 1991; Schuster *et al.,* 1986; Wallis *et al.,* 1983). Deletion of these domains, or their interchange between related proteins, can often provide important evidence defining the role of the protein *in vivo.*

It is sometimes possible to generate deletion derivatives or swap homologous domains between proteins using existing restriction sites. Such mutations are quick and easy to construct. Where natural restriction sites are lacking, there may be a sequence that can be converted to a suitable restriction site by oligonucleotide-directed mutagenesis. For example, a deletion of the N-terminus of histone H4 was constructed using this approach (Megee *et al.,* 1990). One naturally occurring restriction site was present near the second codon of the gene, an *Msp*I site, but a second site downstream was missing. However, a single oligonucleotide-directed base change within the codon for Gln27 created a *Taq*I restriction site that could then be used to create an in-frame deletion with the *Msp*I site. When a new restriction site results in an amino acid alteration, this mutation must itself be checked to ensure that it is phenotypically neutral. In the case of the histone H4 N-terminal deletion, the creation of the restriction site changed amino acid 27 from Gln to Glu. No phenotypes were detected for this mutation on its own.

In the past, when suitable restriction sites were not available, deletion alleles were usually constructed by creating a large number of nested deletions and then sequencing and sifting through them looking for appropriate deletion end points. Such progressive deletions can be generated by *Bal*31 exonuclease, or the combined actions of exonuclease III and a single-stranded nuclease such as S1 or mung bean nuclease (Henikoff, 1987). However, the advent of polymerase chain reaction (PCR) protocols has made the construction of deletion and domain swap alleles rapid and simple. Figure 1 shows an example of a PCR strategy for the construction of a histone H3 N-terminal domain deletion. PCR amplification procedures are presented in several standard laboratory manuals (Innis *et al.*, 1990; Sambrook *et al.*, 1989).

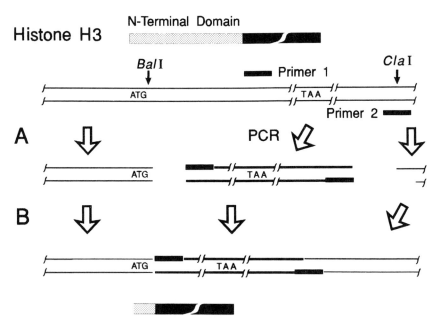

FIG. 1. Construction of a deletion of the N-terminal domain of histone H3 by PCR amplification. Relevant features of histone H3 and its gene are shown in the upper part of the diagram. Two restriction sites were used: a *Bal*I restriction site located in the codon for the N-terminal alanine, and a downstream *Cla*I site located past the end of the gene. The locations of the two primers used for PCR amplification are shown. "Primer 1" was designed to form an in-phase deletion of the N-terminal domain when ligated to the blunt-end *Bal*I restriction site. "Primer 2" was synthesized to span the downstream *Cla*I site. Step A consisted of two parts. In the first step, a plasmid containing the cloned H3 gene was cut with *Bal*I and *Cla*I and the backbone vector fragment was purified for subsequent cloning steps. In the second step, the two primers were used to amplify the H3 DNA fragment truncated for the N-terminal domain. In step B, the vector fragment and the amplified PCR fragment were then combined to create the deletion allele. The PCR fragment was cut with *Cla*I to provide the 3' junction with the *Cla*I site of the vector fragment, while the 5' end was joined to the *Bal*I site by blunt-end ligation. The final H3 allele was precisely deleted for the N-terminal domain from amino acids 1–28.

If restriction sites are incorporated into the PCR primers to facilitate subsequent cloning, it is important to remember that many restriction endonucleases cannot bind and/or cut the target site if it is located at the extreme end of the DNA sequence. Four or five additional nucleotides must be added to PCR primers on the 5' side of the restriction site sequence to facilitate efficient digestion.

In addition to its polymerase activity, *Taq* DNA polymerase also has a terminal transferase activity that can add a nontemplated nucleotide (probably an A) onto the 3' ends of at least a fraction of the products. This can interfere with blunt-end cloning of PCR-amplified fragments. While we have successfully cloned blunt-end PCR fragments without further treatment, it is now recommended that fragments be treated with the Klenow fragment of DNA polymerase I or T4 polymerase to remove the 3' overhang and improve the efficiency of cloning.

Finally, cloned PCR fragments may contain point mutations resulting from misincorporation by *Taq* DNA polymerase. Repeated cycles of amplification cause any mistakes in base incorporation to accumulate with each successive round of the reaction. These random errors are not detectable in the bulk amplified fragments, but an individual clone may carry an unexpected mutation. Obviously, the preferred way to address this problem is to sequence the cloned PCR fragment completely to confirm its fidelity. Where this is not practical, a less desirable solution is to examine multiple clones from separate PCR reactions to ensure that they present identical phenotypes.

B. Site-Specific Point Mutations

Increasingly, structure–function studies of nuclear and chromosomal proteins require the creation of specific amino acid substitutions. Such an approach is critical for understanding the roles and importance of specific protein motifs and of posttranslational modifications such as ubiquitin conjugation (Swerdlow *et al.*, 1990) or lysine acetylation of the histones (Megee *et al.*, 1990). The mutagenesis of specific amino acid residues is now easily accomplished using synthetic oligonucleotides and single-strand bacteriophage vector clones. Many institutions have service support facilities for making custom oligonucleotides, and there are now commercial services that provide oligonucleotides both rapidly and relatively inexpensively.

Mutagenic oligonucleotides should be 18 nucleotides or longer and be designed with a minimum of about 8 perfectly matching bases on the 3' and 5' sides of the mismatched nucleotide (Gillam and Smith, 1979). It is also useful to scan the vector and gene target sequences by computer for other potential homologous sequences and design the oligonucleotide to minimize the number of these and their match if possible. We try where possible to arrange the sequence at the ends of the oligonucleotide so that the hybrid has terminal G-C base pairs; the efficacy of this strategy has not been studied systematically, however. If difficulties are encountered recovering mutations, it may be helpful to increase the length and thus the specificity of the oligonucleotide. For example, the DNA sequence coding for the N-terminal 16 amino acids of histone H4 has partial internal redundancy, and we encountered problems altering individual lysine residues using 18 to 20-base oligonucleotides. However, multiple mutations in the region were successfully constructed using a 60-base oligonucleotide carrying four separate point mutations.

While there are several excellent strategies for oligo-directed mutagenesis, the protocol described by Kunkel (Kunkel, 1985; Kunkel et al., 1987; Sambrook et al., 1989), is relatively simple and efficient and has been the workhorse for site-specific mutagenesis in our laboratory. This protocol relies on the ability of ung^+ Escherichia coli to enrich for the mutation by selecting against a wild-type template containing deoxyuridine. In this genetic background the desired mutant derivatives can be isolated at frequencies as high as 50% of the progeny plaques. The correct clone is rapidly identified by direct sequence analysis. For screening purposes, usually only a single nucleotide sequencing reaction needs to be performed and 15–30 templates can be scored at once. Occasionally, for unknown reasons, we have failed to recover an oligo-directed mutation in the first set of templates screened by sequence analysis. In such cases, the desired mutation can usually be identified by screening a larger number of templates by hybridization and differential melting using the radiolabeled oligonucleotide (Zoller and Smith, 1982).

C.   Random Point Mutations

While specific oligo-directed point mutations are appropriate for addressing sharply defined problems, random mutagenesis has the advantage that fewer assumptions are involved; the interesting mutants are determined by the biology of the system rather than the guess of the investigator! The choice of protocol for mutagenesis depends on the purpose of the experiment. If the objective is to recover all possible mutations in a defined region of the protein, and the investigator is without a genetic screen for mutants, then mutagenesis using oligonucleotide pools of random mismatch bases provides the most efficient method of accumulating variant alleles and is definitely preferred (Horwitz and Dimaio, 1990; Sambrook et al., 1989). However, if a genetic screen can be used to identify mutations of biological interest, then other protocols are easier and cheaper to carry out over the larger length of the whole gene. Unlike systematic DNA sequencing screens for mutant alleles, in a genetic screen it is possible to examine tens of thousands of colonies and the fact that amino acid substitutions may be achieved in only 10% of the colonies is not a severe limitation.

Traditional methods of random mutagenesis involve treating whole cells with chemical mutagens, ultraviolet light, or radiation. There are numerous protocols for inducing mutations *in vivo* and the reader is referred to current laboratory manuals (Guthrie and Fink, 1990; Rose *et al.*, 1990) and previous volumes in this series (Kilbey, 1975; Pringle, 1975) for discussions and methods. These approaches are particularly useful for extending the functional hierarchy of related genes by suppressor, or pseudorevertant, analysis. However, if the gene of interest is already cloned, then there are more efficient and specific mutational strategies available.

1.   DOUBLE-STRAND PLASMID MUTAGENESIS

A major advantage of using double-stranded DNA plasmids for mutagenesis is simplicity. The protocol can be carried out on the same plasmid construct destined to be used *in vivo*.

Bacterial Mutator Strains.   Mutator strains of *E. coli* can be used to induce mutations in a plasmid construct simply by propagating it in the strain. The *mutD* gene of E. coli encodes the ε-subunit of the *Pol*III holoenzyme which carries out 3' to 5' proofreading during DNA replication (Echols *et al.*, 1983). The *mutD5* mutation is one of the strongest mutators in *E. coli*, increasing the frequency of mutation by several orders of magnitude (Cox and Horner, 1983). One of the advantages of *mutD5* is that it generates both transition and transversion base substitutions, although the ratio of these is affected by growth conditions (Schaaper, 1988; Wu *et al.*, 1990). Most of the mutations are base substitutions with a few percent of single-base deletions.

Since *mutD5* is a mutator, growth of the strains results in the frequent loss of the mutator character, and it is important to test single colonies for mutagenic activity before the start of the experiment. A simple test is to spread isolates on plates containing 20 μg/ml of nalidixic acid, checking for the frequent appearance of resistant colonies. Once a good mutator isolate is in hand, competent cells can be made and transformed with the recombinant plasmid as usual. Transformed cells are grown up in batch culture and a pooled mutagenized plasmid DNA preparation is isolated. These mutant plasmid pools can then be used to transform yeast and screen for phenotype. Note that independent mutant isolates are only guaranteed when they are derived from separate plasmid pools.

Hyroxylamine Treatment.   Hydroxylamine causes the deamination of cytosine bases resulting in C-to-T transition mutations following transformation and replication (Chu *et al.*, 1979; Volker and Showe, 1980). The following protocol can be scaled up to increase the number of independent mutant pools.

Approximately 10 μg of double-stranded plasmid DNA is added to 0.5 ml of 0.45 M sodium hydroxide, 1 M hydroxylamine, and 5 mM EDTA. The reaction is incubated at 37°C for 18–20 hours. After incubation, the reaction is stopped by adjusting the buffer to 10 mM sodium chloride, 100 μg/ml bovine serum albumin, and precipitating the DNA with 1 ml of ethanol. The mutagenized DNA is reprecipitated with ethanol twice from 0.3 M sodium acetate, 10 mM Tris-HCl (pH 7.9), 1 mM EDTA, and finally dissolved in 10 μl of 10 mM Tris-HCl (pH 7.9), 1 mM EDTA.

The mutagenized DNA can be used directly to transform competent yeast cells. Following transformation, cells should be outgrown in rich medium for 2–4 hours to allow homozygosis of the plasmids. Transformants are then spread for single colonies on selective plates. Again, independent mutant isolates are only guaranteed for separate tranformation pools.

2.   SINGLE-STRANDED TEMPLATE MUTAGENESIS

For saturation mutagenesis, *in vitro* chemical mutagenesis using single-stranded DNA templates may be more efficient than using mutator strains or hydroxylamine treatment of recombinant plasmids, since the coding sequences of the gene may be specifically targeted independent of the final plasmid vector.

Bisulfite Mutangenesis.   Sodium bisulfite is a single-strand-specific reagent that will deaminate cytosine residues to generate C-to-T transitions in DNA (Shortle and Botstein, 1983; Weiher and Schaller, 1982). Make a fresh "bisulfite reagent" by dissolving 156 mg of sodium bisulfite and 64 mg of sodium sulfite in 430 μl of water. For mutagenesis, mix together about 50 μg (20–25 pmol) of single-stranded DNA in 20μl with 320 μl of bisulfite reagent, and 12 μl of 50 mM hydroquinone and 48 μl of water. Wrap the reaction in aluminum foil to exclude light and incubate at 37°C for 30 minutes to 2 hours, depending on the extent of mutagenesis required. After the reaction, remove the bisulfite reagent by putting the sample over a 10-ml Sephadex G-50 column in 10 mM potassium phosphate buffer (pH 6.8). Collect 0.5-ml samples and read the optical density at 260 nm on the spectrophotometer, pooling the peak fractions. Adjust the sample to 0.2 M Tris-HCl (pH 9.2), 50 mM NaCl, and 2 mM EDTA and incubate for 8–16 hours at 37°C. Desalt the sample over a Sephadex G-50 column in 0.1 M ammonium acetate and lyophilize the mutagenized DNA.

The gene fragment can then be recovered for subcloning into the appropriate vector by synthesizing the complementary DNA strand *in vitro* using a synthetic oligonucleotide. If the insert is subcloned in one of the M13 mp series vectors, then the standard universal sequencing primer is perfectly adequate. The treated DNA template and a 4- to 5-fold excess of primer (70–100 pmol) are dissolved in approximately 100 μ of 50mM Tris-HCl (pH 7.5), 10 mM magnesium chloride, 50 mM sodium chloride, and 1 mM dithiothreitol. The primer is annealed by heating to 85°C for 5 minutes and then cooling slowly to room temperature. All four deoxynucleotides are then added to 100 μM and the reaction is initiated by adding 10 units of the Klenow fragment of DNA polymerase I or Sequenase. The polymerase reaction is incubated at 37°C for 30 minutes. Following phenol extraction and ethanol precipitation, the synthetic duplex is then digested with appropriate restriction enzymes, and the gene fragment is purified by gel electrophoresis and subcloned into a suitable yeast vector.

Other Base-Specific Mutagens.   Myers *et al.* (1985) have adapted the base-specific chemical reactions used for Maxam–Gilbert DNA sequencing to a mutagenesis protocol useful for single-stranded DNA templates. The reactions induce the damage of specific base residues in the single-stranded DNA template: nitrous acid deaminates C, A, and G residues; formic acid depurinates the DNA; while hydrazine breaks pyrimidines. After the base cleavage reactions, avian reverse transcriptase is used to synthesize the complementary strand incorporating random nucleotides opposite the damaged base.

For each reaction, 40 μg (approximately 18–20 pmol) of single-stranded DNA template containing the target sequence is incubated in a 100-μl reaction. The nitrous acid reaction contains 250 mM sodium acetate (pH 4.3) and 1.0 M sodium nitrite (prepared from a fresh 2.0 M stock) and is incubated for about 60 minutes at room temperature. The formic acid reaction is carried out in 12 M formic acid and is incubated for about 10 minutes at room temperature. The hydrazine reaction is carried out in 60% hydrazine and is also incubated for about 10 minutes at room temperature. Remove the cleavage reagents from the DNA immediately after the incubations by putting the samples over 0.5-ml Sephadex G-50 columns equilibrated with 10 mM Tris-HCl (pH 7.9), 10 mM EDTA. The samples are then adjusted to 1.0 M sodium acetate (pH 7), 20 μg of carrier tRNA is added, and the nucleic acids are ethanol precipitated. The DNA is precipitated twice more with ethanol to remove residual chemical reagents.

A synthetic oligonucleotide is used to prime the synthesis of the complementary DNA strand as described for the bisulfite reaction. The mutagenized DNA template and primer are dissolved in about 100 μl of 7 mM Tris-HCl (pH 7.5), 7 mM magnesium chloride, 50 mM sodium chloride, and 2 mM dithiothreitol. The primer is annealed by heating to 85°C for 5 minutes and then cooling slowly to room temperature. All four deoxynucleotides are added to 100 μM and the reactions is initiated with the addition of 20 U of avian reverse transcriptase (Life Sciences, 2900 72nd Street N., St. Petersburg, FL 33718). Reverse transcriptase is used in this reaction, even though the template is DNA, because of its ability to misincorporate nucleotides at the sites of damage in the template. After incubation at 37°C for 1 hour, the DNA is purified by phenol extraction and ethanol precipitated. The mutagenized gene fragment can now be excised by digestion with the approriate restriction enzymes and subcloned into a yeast vector for transformation.

III.   Manipulation of Mutant Alleles

There are several strategies available for introducing mutant alleles into a host strain for testing. If the gene of interest is not essential for viability, of course, the wild-type gene can be disrupted and specific alleles or libraries of mutated alleles can be introduced by DNA transformation. For dominant mutants it is not necessary to eliminate the normal wild-type gene. An excellent example of this approach for mammalian cells is provided by the expression of histone H5 in tissue culture cells (Bergman *et al.,* 1988; Sun *et al.,* 1989). In many cases, however, recessive mutations are sought in essential genes, thus requiring some additional trickery!

A.   Analysis of the Null Allele

The characterization of the null allele of a gene is a critical first step in understanding its functional role. The phenotypes of point mutations can be quite complicated, including partial loss, gain, or alteration of function. If the phenotype of the null allele is unknown, it can be difficult to acertain whether a gene is acting positively, negatively, directly, or indirectly.

Gene disruptions in yeast were originally accomplished using a two-step protocol (Hicks *et al.,* 1979; Scherer and Davis, 1979), and this method is still useful where it is undersirable to introduce foreign sequences at the site of the disruption (Smith and Stirling, 1988). The strategy is shown in Fig. 2A. A circular plasmid carrying the disrupted gene and a selectable marker such as *URA3* is integrated into the chromosome by homolgous recombination. Then, cells in which the plasmid has excised back out of the genome are recovered, as indicated by the loss of the plasmid marker (*URA3* in this example). Homologous excision can have two general outcomes: retention of the disrupted allele or retention of the wild-type allele. These can be distinguished by genomic Southern blot analysis. It is now more common to

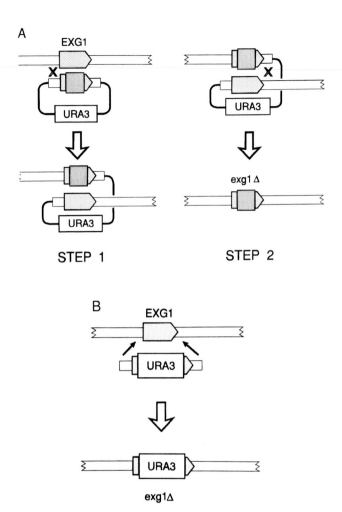

FIG. 2. Gene disruption strategies for *Example Gene 1* (*EXG1*). (A) Two-step gene disruption. Step 1 shows the integration of a disruption plasmid. The disruption allele of *EXG1* is carried on the plasmid and is represented by the black box interrupting the gene; the disruption may be a gene deletion, the insertion of a selectable marker, a transposon, or other suitable disruption. The disruption plasmid is integrated by recombination into a *ura3⁻* host by DNA transformation with selection for *URA3⁺* colonies. For this example, the cross-over is assumed to have occurred within homologous sequences to the left of the gene. Step 2 shows the excision of plasmid sequences. Cells that have undergone excision of the plasmid are selected by growth on 5-fluoroorotic acid (5FOA) medium. Crossing-over between homologous sequences to the left of the gene will simply restore the wild-type gene. In the example, the excision cross-over is shown between homologous sequences to the right of the gene. The resulting chromosome then carries the disruption allele, *exg1Δ*. (B) Single-step gene disruption. The disruption allele *exg1Δ* gene was constructed by the replacement of *EXG1* sequences with a selectable marker, the *URA3* in this case. Cells are transformed with a linear fragment spanning the disruption, selecting for Ura⁺ colonies. The ends of the linear DNA fragment are recombinagenic and mitotic gene conversion results in the direct substitution of the disrupted *exg1Δ* allele for the wild-type gene.

use a one-step gene disruption protocol (Rothstein, 1983) in which a linear fragment carrying the disrupted allele is integrated directly by gene conversion of the chromosomal locus (Fig. 2B). The free ends of the fragment are recombinagenic and target the replacement to the homolgous DNA sequences in the chromosome (Orr-Weaver et al., 1981).

If it is not known whether the gene is essential, disruptions are constructed in a diploid host and then tested by sporulation and tetrad analysis. If the gene is nonessential, then tetrads yield four viable spore colonies. However, if the gene is essential for viability, then the tetrads segregate two live and two dead spore colonies, with the dead segregants always linked to the selectable marker used in the disruption (Shortle et al., 1982).

Null alleles have been constructed in a number of essential chromosomal protein genes. There are two nonallelic sets of H2A and H2B histone gene pairs and two nonallelic sets of H3 and H4 histone gene pairs in the yeast haploid genome (Hereford et al., 1979; Smith and Murray, 1983). Cells are viable with the deletion of any single set of histone genes, but complete loss of either both H2A–H2B or both H3–H4 loci is lethal (Choe et al., 1982; Kolodrubetz et al., 1982; Smith and Stirling, 1988). The yeast ACP2 gene encodes a protein with homology to calf high-mobility group proteins HMG1 and HMG2 (Haggren and Kolodrubetz, 1988) and disruption of ACP2 is lethal. Other yeast chromosomal proteins that have recently been shown to be essential by gene disruption include ABF1 (Rhode et al., 1989) and RAP1 (Shore and Nasmyth, 1987). In favorable cases, disruption alleles can be recovered in higher eukaryotic cells. For example, the H2AvD gene codes for a highly conserved histone H2A.Z variant and is present as a single-copy gene at locus 97D1-5 on chromosome 3 in Drosophila (van Daal et al., 1988). An ethyl methanesulfonate (EMS)-induced mutation that deletes the second exon of the H2AvD gene has been isolated and the deficiency is lethal. The mutant can be rescued, however, by P-element transformation with a cloned fragment containing the H2AvD gene (Elgin and van Daal, 1990).

B. Integration of Mutated Gene Fragments

By mutating a fragment truncated at the 5' or 3' end of the gene it is possible to screen for mutations by integrative transformation (Fig. 3). This "hemizapper" approach was first used with the yeast actin gene (Shortle et al., 1984) and has been applied to several others (Holm et al., 1985; Huffaker et al., 1988). This is a convenient method for isolating recessive mutants in one step and has the added advantage that mutant alleles are integrated and are thus stable; artifacts due to variations in plasmid copy number or loss are eliminated. However, we have found that this approach can suffer from

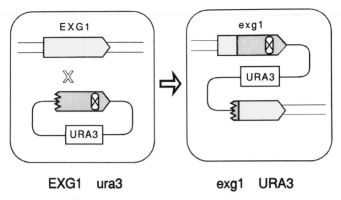

EXG1   ura3   **exg1   URA3**

FIG. 3.   Recovery of mutations by integration of a gene fragment. A strategy for recovering mutations in *Example Gene 1* (*EXG1*) is shown. The host yeast strain is wild type for *EXG1* and is auxotrophic for a selectable marker gene, in this case the *ura3* gene. Mutations are made in the truncated copy of the *EXG1* gene carried on an integrative plasmid. In the diagram, the plasmid copy of *EXG1* is deleted at the 5' end, and the site of a mutation is indicated by the X within the gene. The mutagenized plasmid DNA is then introduced into the host yeast by transformation, selecting for Ura$^+$ colonies. Integration of the plasmid into the chromosome by homologous recombination produces a nontandem duplication of one full-length functional gene and one truncated gene. If the location of the cross-over is to the left of the mutation in the plasmid, as indicated in the figure, then the functional copy of the gene carries the mutation. Thus, tranformed Ura$^+$ colonies can be screened directly for mutations in the *EXG1* gene.

recombinational repair. During the integration of the plasmid construct, mutations in the gene fragment can undergo mismatch correction to the wild-type sequence, reducing efficiency (Ford and Smith, 1985).

C.   Meiotic Segregation

An alternative approach for analyzing recessive mutations is to utilize autonomously replicating centromeric plasmids—circular minichromosomes. In the meiotic segregation protocol a diploid strain, hemizygous for the gene or genes of interest, is transformed with a plasmid or plasmid library carrying the mutated gene fragment (Megee *et al.,* 1990; Park and Szostak, 1990; Riles and Olson, 1988). The mutant allele can then be uncovered by sporulation of the diploid (Fig. 4). The chromosomal alleles will segregate 2:2 and half of the spores will carry the wild-type allele and half will carry the disrupted or deleted allele. For essential genes the progeny with the deletion allele will be inviable unless they also carry a functional copy of the gene on the transforming plasmid. These spore segregants can be identified because they are unable to produce colonies lacking the transforming plasmid. This

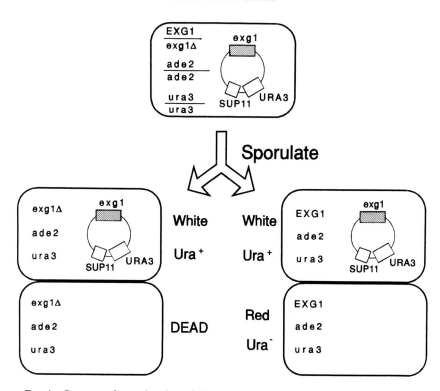

FIG. 4.   Recovery of mutations by meiotic segregation. The strategy for recovering mutations in *EXG1* is shown. Mutations are made in a copy of the gene carried on a *CEN/ARS* plasmid, that is, one containing a centromeric fragment (*CEN*) and an autonomously replicating sequence (*ARS*), a yeast origin of DNA replication. The plasmid also contains the marker genes *URA3* and *SUP11*. Plasmid DNA is used to transform a diploid yeast strain that is hemizygous for *EXG1*, that is, *EXG1/exg1Δ*, and homozygous for *ade2* and *ura3*, selecting for Ura$^+$ colonies. When this diploid is sporulated, the *EXG1/exg1Δ* chromatids segregate 2:2 to produce spore colonies that are either *exg1Δ* or *EXG1*. Colonies produced from the *EXG1* spores, shown in the right-hand pathway, do not depend on the plasmid to provide *EXG1* function. Loss of the plasmid give rise to sectors within the colony; cells that retain the plasmid are Ura$^+$ and white, cells that lose the plasmid are Ura$^-$ and red. Colonies produced from the *exg1Δ* spores, shown in the left-hand pathway, require the presence of the plasmid to provide *EXG1* function, in this case the mutant allele *exg1*. Loss of the plasmid is lethal and thus the colonies are non-sectored; only cells that retain the plasmid can grow, producing uniformly white Ura$^+$ colonies. Thus, nonsectoring white Ura$^+$ colonies can be screened for *exg1* mutations.

is particularly covenient if the rescue plasmid carries the marker *URA3*, the gene encoding ornithine monophosphate decarboxylase (Odc). The advantage of *URA3* is that it is subject to both forward and backward selection (Boeke *et al.*, 1984, 1987). In a strain that is auxotrophic for uracil because of a *ura3* mutation (Odc$^-$), a plasmid carrying the *URA3* gene confers the ability to grow on medium lacking uracil. On the other hand, cells that express the wild-type *URA3* gene (Odc⁺) are unable to produce colonies in the presence of the drug 5-fluoroorotic acid (5FOA) (PCR Inc., P.O. Box 1466, Gainesville, FL 32602). (Boeke *et al.,* 1984). Thus, in the meiotic segregation protocol, the desired spore colonies, those dependent exclusively on the plasmid copy of the gene of interest, fail to grow on 5FOA plates. They fail to grow since any cells *with* the plasmid are inviable on the 5FOA medium because of the *URA3* expression from the plasmid, and any cells *without* the plasmid are inviable because the plasmid has the only copy of the essential chromosomal protein gene.

In addition to *URA3*, we have found it useful to incorporate the ochre suppressor tRNA genes *SUP11* on the rescue plasmid vector and include the ochre mutation *ade2-101* in the strain background. Haploid *ade2-101* strains are red on limiting adenine because of the accumulation of pigmented metabolic intermediates derived from phosphoribosylaminoimidazole (Jones and Fink, 1982). However, in the presence of *SUP11*, the cells are white. A haploid segregant that is deleted for an essential gene remains entirely white since every viable cell must contain the plasmid and its linked *SUP11* gene. A haploid segregant that has retained the essential gene on the chromosome is viable in the absence of the plasmid and produces colored sectors in the colony under nonselective growth—white sectors for those cells containing the plasmid and red sectors for those cells lacking the plasmid. In practice, therefore, after the diploid transformants are sporulated, haploid segregants rescued by the plasmid copy of the gene can be rapidly identified as nonsectoring white colonies that are completely 5FOA sensitive. These can then be screened for relevant phenotypes.

An advantage of this sporulation approach is that it does not require propagating multiple plasmids in the same strain and thus avoids potential recombination between the plasmids. Recombination between the chromosomal wild-type gene and the plasmid-borne allele in the diploid is still possible, however. The method is also well suited to replicate plate screens of mutagenized libraries. A disadvantage of the meiotic exposure protocol is that it requires germination and outgrowth of the haploid segregants. This may limit the recovery of some mutations if they confer a defect in these functions.

D. Plasmid Shuffle

The substitution of mutant for wild-type alleles can also be accomplished in haploids by exchanging plasmids, a technique dubbed a "plasmid shuffle" (Boeke *et al.,* 1987). The strategy is shown in Fig. 5. The starting yeast strain is constructed by the meiotic segregation strategy outlined in the previous

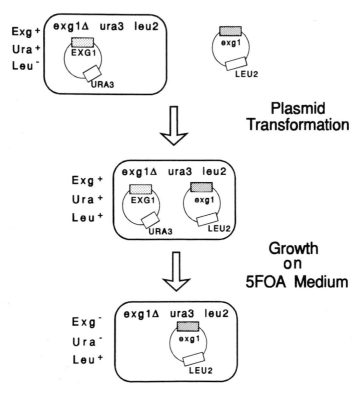

FIG. 5. Recovery of mutations by plasmid shuffle. A plasmid shuffle strategy for the mutagenesis of *EXG1* is shown. The starting host strain for the plasmid shuffle protocol has the genotype *exg1Δ ura3 lue2* and carries a *URA3/CEN/ARS* plasmid with the wild-type *EXG1* gene. The phenotype of the strain is Exg⁺, Ura⁺, and Leu⁻. Mutagenesis is carried out on a copy of the *EXG1* gene cloned in a second *LEU2/CEN/ARS* plasmid. This mutagenized plasmid library is used to transform the host yeast, selecting for Leu⁺ Ura⁺ colonies. The plasmid shuffle is completed by growing colonies on 5FOA medium to select for cells that have lost the original *EXG1/URA3* plasmid. The resulting colonies are Ura⁻ Leu⁺ and now depend on the gene derived from the mutagenized plasmid library, in this case, the mutant allele *exg1*. Thus, these plasmid shuffle colonies can be screened for expression of mutant *exg1* alleles.

section. The desired strain is a haploid deleted for the chromosomal copy of the gene of interest and rescued by an autonomously replicating centromeric plasmid containing the wild-type gene. The plasmid must contain a marker that can be selected *against* on appropriate medium. The *URA3* gene and 5FOA medium is again a useful combination and will be used as the example here (Boeke *et al.*, 1984, 1987). Potential alternatives include the *LYS2* gene, which can be selected against on α-aminoadipate medium (Chattoo *et al.*,

1979), and the *SUP11* gene, which can be selected against on high-osmolarity medium (Singh, 1977).

The mutant library or specific experimental allele is subcloned into a second plasmid carrying a different marker, such as *LEU2*, *HIS3*, or any other selectable marker suitable for the genetic background of the strain. This plasmid is introduced into the starting strain by DNA transformation. We usually select for maintenance of both plasmids at this step since it is often important to have the double plasmid transformant as a control for checking the dominance properties of any new mutant alleles. The colonies are then replicated onto plates permissive for mitotic loss of the plasmid with the wild-type allele. In the present example, the plates would be supplemented with uracil. The resulting colonies are then replicate plated a second time onto 5FOA medium to select actively against the first plasmid. At this point the plasmid shuffle is complete. The intermediate step of growth on nonselective medium enriches for spontaneous loss of the *URA3* plasmid and improves the signal obtained during the subsequent negative selection. Finally, the derived plasmid shuffle colonies are replicate plated and screened for appropriate phenotypes.

There are a number of advantages to the plasmid shuffle protocol. As mentioned above, it can be used with haploid strains under vegetative growth conditions. The plasmids used are simple to construct and do not require special gene fusions. The manipulations are straightforward and are suited to large-scale replicate plating screens.

The negative selection against the Odc$^+$ phenotype by 5FOA is not completely cytocidal and isolates picked from 5FOA or subsequent plates should be further purified from single colonies and confirmed to be *ura3$^-$*. The protocol also requires the propagation of two related plasmids in the cells at the same time. Plasmid–plasmid recombination frequencies, converting a mutant allele to wild type, can be as high as 10% of the 5FOA-resistant colonies. In practice, this means that there will usually be a background of wild-type colonies among the apparent plasmid shuffle isolates. For mutant screens this is not a problem, but for the test of a specific site-directed mutant false positive colonies will be present. Therefore, it is important to recover the plasmid from the putative plasmid shuffle isolate and confirm that it carries the correct allele, particularly if there is no phenotype associated with the plasmid exchange.

A subtle problem with the plasmid shuffle protocol is that the efficiency with which a mutation is recovered decreases if the allele confers an unconditional growth defect on the cell. The frequency of plasmid loss during mitotic segregation is several percent per generation, normally about 3–5% (Carbon, 1984; Hieter *et al.*, 1985; Koshland *et al.*, 1985; Murray and Szostak, 1983), and therefore plasmid shuffle segregants are generated at a relatively high rate in a colony or culture. However, these segregants are only recovered

once they are phenotypically Odc⁻, and there is a lag between the time the
URA3 plasmid is lost and the time the gene product becomes inactive in the
cell. For example, imagine that cells with a particular allele of a chromosomal
protein gene have a generation time twice as long as their wild-type sisters. If
the phenotypic lag for becoming Odc⁻ is four cell divisions (the exact figure
will vary with the particular plasmid construct), then during the lag period
the plasmid shuffle segregant will divide four times, while the wild-type cells
will double 8 times; a relative cell ratio of mutant to wild-type of approximately 1:16 by the time selection is possible. In this case, the frequency of
plasmid shuffle segregants in the population will be closer to 0.1–0.3% than
the 2–5% expected in the absence of a difference in growth rate. This complication can be limiting under conditions where the expression of *URA3* is
high and thus the phenotypic lag for Odc⁻ is long, where cells containing
just the mutant allele have a severe growth defect, where the phenotypic lag
for the chromosomal protein is very short, and where there is significant sequence homology for plasmid–plasmid recombination. Using the plasmid
shuffle protocol, we have successfully isolated mutants that have a generation time approximately four times longer than wild type.

E.   Conditional Expression

The phenotypes of recessive mutations can also be examined if expression
of the wild-type allele is conditional (Fig. 6). To accomplish conditional ex-

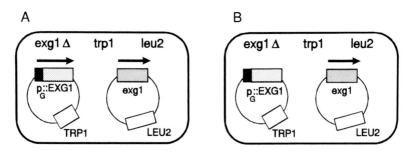

FIG. 6.   Recovery of mutations by conditional expression. An example of the conditional
expression strategy for *EXG1* is shown. The host yeast strain is deleted for the chromosomal copy
of the *EXG1* gene and has the genotype *exg1Δ trp1 leu2*. Wild-type *EXG1* function is provided
by a fusion of the *GAL1* promoter to the *EXG1* coding sequences; in this example, the fusion
is carried on a *TRP1/CEN/ARS* plasmid. Mutations in *EXG1* are constructed in a copy of
the gene carried on a second *LEU2/CEN/ARS* plasmid. The expression of *EXG1* is shown for
growth on (A) galactose and (B) glucose medium. On galactose, where the fusion gene is active,
the cells are functionally *EXG1/exg1* and have a wild-type phenotype because of the expression
of the fusion gene. However, on glucose, where the fusion gene is repressed, recessive *exg1*
mutations are unrecovered.

pression, a fusion is engineered that places the wild-type gene under control of a regulated promoter. The promoter commonly used for this purpose is *GAL1*, a tightly regulated promoter that is activated when cells are grown on galactose and is repressed when cells are grown on glucose (Johnston, 1987). A plasmid bearing this fusion construct is then used to rescue cells deleted for the chromosomal copies of the gene as described previously. The resulting strain is viable when grown on medium containing galactose but dead when grown on glucose. A second plasmid is then used to introduce a specific allele or mutagenized library of fragments into this strain. The function of the mutant allele can be tested by patching or replicate plating cells onto glucose medium to turn off expression from the wild-type gene.

The conditional expression system is useful for analyzing the terminal phenotype of null mutants. Figure 7 shows the results of a microcolony assay (Weinert and Hartwell, 1988) to test the phenotype of the null allele of histone H4, a modification of experiments first reported by Han *et al.* (1987) and Kim *et al.* (1988). The starting strain was deleted for all chromosomal copies of the histone H3 and H4 genes and carried a *LEU2* plasmid expressing histone H3 and a second *TRP1* plasmid with a P_{GAL}: *HHF1* fusion for the conditional expression of H4. These cells were viable when grown on galactose medium but died when shifted to glucose as a carbon source because the only copy of histone H4 was repressed. For the microcolony assay, cells from a culture grown in galactose were spread for single colonies on a agar slab containing glucose as the carbon source. Fields of cells were photographed immediately after spreading and then again after incubation for 36 hours. A comparison of the two panels shows that cells that were unbudded at the time of the shift arrested in the first cell division with a large bud. Cells that had started to bud at the time of the shift were able to complete the first division but arrested in the next round as two cells each with a large bud. Thus, the histone H4 null allele has a first-cycle-arrest phenotype with a transition point near the time of bud emergence. This is a reasonable result, since histone gene expression occurs late in G_1 and early S phases (Cross and Smith, 1988; Hereford *et al.*, 1981, 1982), a time that corresponds roughly with the beginning of bud emergence (Fangman and Zakian, 1981; Pringle and Hartwell, 1981). These results show that cells do not contain a pool of histone H4 capable of supporting a complete round of the cell division cycle. The cellular and nuclear morphology of the arrested cells, and an analysis of DNA content, as described below, showed that cells having the null allele of histone H4 completed or nearly completed DNA replication and blocked in the G_2 phase of the division cycle prior to nuclear division (Kim *et al.*, 1988; Morgan *et al.*, 1991). This is an excellent example in which a simple analysis of the null allele suggested a role of nucleosome density and chromatin structure in chromosome segregation. Indeed, Saunders *et al.* (1990) have since shown that nucleosome depletion directly or indirectly results in an alteration of centromere structure.

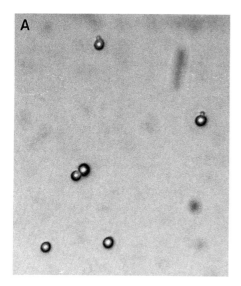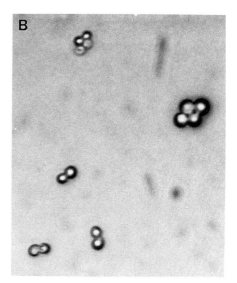

FIG. 7.   Microcolony assay of histone H4 depletion. The analysis of a conditional histone H4 null mutant is illustrated in a microcolony assay. The yeast strain used in this experiment carried deletions in the chromosomal copies of the H3 and H4 genes. Wild-type histone H4 function was provided by a fusion gene consisting of the *GAL1* promoter fused to the histone H4 gene (*HHF1*); the fusion was cloned on a *TRP/CEN/ARS* plasmid. Histone H3 function was provided by a wild-type histone H3 gene (*HHT1*) carried on a *LEU2/CEN/ARS* plasmid. Therefore, the cells were viable on galactose medium where both histone H3 and H4 were expressed, but inviable on glucose medium where histone H4 expression was repressed. In the experiment shown, cells were grown on galactose to early exponential growth and then plated for single cells on a glucose agar slab. In A, a field of cells at the time of plating on glucose is shown, and B shows the same field of cells approximately 36 hours later. A comparison of the two panels shows that cells that were unbudded at the time of the shift arrested in the first division cycle as large budded cells. Cells that had already budded at the time of the shift completed one division cycle and arrested in the next cycle as two large budded cells.

A major advantage of conditional expression is that all of the cells in the population are simultaneously deprived of wild-type gene expression and therefore mutants that have a growth defect are less likely to be overgrown by wild-type cells before they can be detected. Another advantage of the method is that, in principle, the wild-type allele can be turned back on by transferring the cells back to galactose medium, thus potentially permitting reciprocal shift experiments. In practice, there is a significant time lag in reexpression of the gene when going from glucose to galactose. Thus, the wild-type allele actually turns on some time *after* the shift to galactose medium. This makes it unreliable to link the timing of a defect to a particular point in the cell cycle using synchronous cultures, for example.

A minor disadvantage of the conditional expression system is that it requires additional recombinant DNA manipulations to put the wild-type gene under conditional control. However, the fusion only needs to be made once for the analysis of any mutant allele. Another disadvantage of the method, as often implemented, is that both the wild-type plasmid and the mutant plasmid are coresident in the host throughout the experiment. Unlike the plasmid shuffle protocol, the wild-type gene is not evicted from the cell but is simply repressed. Thus, plasmid–plasmid recombination can produce a significant background of false positive colonies. Where there is little sequence homology flanking the mutation, as in N-terminal or C-terminal deletions, this is not a major problem (Kayne *et al.*, 1988).

Finally, there is potential for a subtle problem in the case where mutant alleles affect transcription. The conditional expression screen depends on the complete repression of the wild-type gene for analysis of the mutant allele. However, if the mutant allele causes a general derepression of transcription, then the regulated promoter may become leaky enough to give low-level expression of the wild-type gene, complicating the analysis. It should be possible to overcome most of the limitations of the conditional expression screen by combining it with the principles of the plasmid shuffle. If the *GAL1* promoter fusion is carried on a *URA3* plasmid, then it can be selected against following the shift to glucose medium; only cells that happen to have lost the *URA3* plasmid during growth will be able to form colonies on 5FOA medium. Thus, isolates dependent on the mutant allele can be selected free of the original plasmid bearing the regulated wild-type gene.

F.   Overexpression of Mutant Alleles

Overexpression of a chromosomal protein gene provides another approach to understanding its function *in vivo*. The simplest way to attempt to overexpress a gene is to subclone the fragment onto a high-copy-number plasmid such as those containing segments of the *S. cerevisiae* 2-μm plasmid (Rose and Broach, 1990). Following transformation the 2-μm plasmid replication and copy number functions result in the establishment of from 10 to 30 copies of the plasmid per cell (Hsu and Kohlhaw, 1982; Jayaram *et al.*, 1983; Kikuchi, 1983), and up to 200–300 copies with the *leu2-d* selectable marker (Futcher and Cox, 1984). This increased gene dosage may be sufficient to produce a significant overexpression of the gene of interest.

Overexpression of the histone genes, for example, has been useful in understanding their role in chromosome transmission and transcription. Meeks-Wagner and Hartwell (1986), reasoning from the genetics of phage assembly, showed that the fidelity of mitotic chromosome transmission depended on the proper stoichiometry of histone gene expression. When either an H2A and H2B gene set or an H3 and H4 gene set was expressed on a high-copy plasmid, the frequency of chromosome loss increased as much as 10- to 50-fold. Overexpression of all four core histone genes together on the same high-copy plasmid, however, did not increase loss rates. Using this chromosome loss assay and a yeast genomic library in a high-copy plasmid vector, two new genes involved in chromosome transmission were subsequently isolated (Meeks-Wagner et al., 1986).

Histone gene stoichiometry is also important for proper transcription initiation. Clark-Adams et al. (1988) have shown that promoter selection at Ty1 transposition sites can also depend on the proper stoichiometry of histone gene expression. The unbalanced overexpression of the core histone genes is able to suppress the promoter interference effects of Ty1 insertions at HIS4 and LYS2.

High-copy plasmids may not achieve overexpression of a particular gene if there is strong transcriptional feedback control or if the mRNA or gene product is under posttranscriptional regulation. For example, histone gene mRNA is almost certainly subject to posttranscriptional turnover under cell cycle control (Lycan et al., 1987; Xu et al., 1990). Therefore, a negative result with a high-copy plasmid system is inconclusive in the absence of biochemical evidence for overexpression.

IV.   Analysis of Mutants

The initial characterization of a new gene will likely concentrate on the phenotype of the null allele and on assigning the functions of the gene to major nuclear pathways. Subsequent genetic analysis will then focus on the hierarchy of interacting gene functions, how the protein interacts with other components of the pathway, and detailed structure–function studies. The power of current molecular genetics derives from the ability to combine genetic manipulation of a gene in vivo and in vitro with detailed cytology and biochemical analysis.

A.   Growth Rates

The growth rate of a nuclear protein mutant is a rather global measure of phenotype, but is therefore also a sensitive measure. When combined with an assessment of cell cycle distribution, growth rate measurements can help identify phenotypic defects in specific cell cycle pathways. For conditional lethal mutations it is also important to know whether a particular allele arrests immediately upon shift to restrictive conditions or only after several divisions (Pringle and Hartwell, 1981).

Growth rates can be measured in very simple or very sophisticated ways depending on the purposes of the experiment. A simple measure of cell density with time, by light scatter in a spectrophotometer for example, can provide a rough estimate of generation time. However, changes in cell volume can confuse these measurements. Some conditional mutants will continue to grow in size without additional cell division when shifted to the nonpermissive temperature. The increase in cell volume produces an increase in light scatter that may be erroneously interpreted as continued cell division. More accurate measurements are obtained by direct cell counts, either manually using a hemocytometer or by an electronic particle counter.

The number of cell divisions that can be followed using liquid culture growth is relatively small, usually 3–4 generations, and may not be sensitive to small differences in growth rate. Colony size can provide a more sensitive measure of growth rates since a single cell lineage may go through 20 or more generations during the growth of the colony. For colony measurements, several dozen unbudded cells of isogenic mutant and wild-type strains are micromanipulated 5 mm apart side-by-side onto an agar slab and incubated for 2 to 3 days. A growth defect in the mutant is usually obvious by visual inspection. Semiquantitative results can be obtained by picking colonies, sonicating briefly to separate cells and counting.

It is worth recalling that culture growth rates may not provide a true measure of the cell cycle generation time. If the mutant produces a fraction of dead or noncycling cells each generation, then the doubling time of the culture will be longer than the cell cycle generation time. Therefore, it is important to assay the plating efficiency of any mutant under study so that growth rates can be interpreted cautiously if the strain generates inviable or noncycling cells. In such cases, it may be necessary to resort to time-lapse photomicrography to determine cell cycle generation times.

B. Cell and Nuclear Morphology

The cellular and nuclear morphology of a new mutant is another general but diagnostic assay useful in sorting out the function of the gene. For example, a mutant allele of histone H4, *hhf1-10*, in which the lysine residues at positions 5, 8, 12, and 16 were changed to glutamine residues to mimic the hyperacetylated state, produced cells with a dramatic and informative morphology (Megee *et al.*, 1990). A proportion of the cells in the culture exhibited very large and irregular cell and bud shapes, with multiple or segmented nuclei per cell. These results suggest a role of the N-terminal domain, and perhaps lysine deacetylation, in normal nuclear division and chromosome dynamics.

The *S. cerevisiae* cell cycle proceeds by budding, and there are landmark morphologies characteristic of particular arrest points in the division cycle (Pringle and Hartwell, 1981). Cells with defects in the division cycle tend to delay at these positions at least in part because of negative "checkpoint" regulatory controls (Hartwell and Weinert, 1989; Weinert and Hartwell, 1988). A typical arrest morphology, for example, is the cell with a large bud, single nucleus at the neck of the bud, and a 2*N* DNA content as described previously for the histone H4 null allele experiment.

Cellular morphology is easily examined microscopically, and nuclear morphology can be visualized using the fluorescent dye 4′, 6-diamidino-2-phenylindole (DAPI) (Williamson and Fennell, 1975). Approximately 1×10^6 to 1×10^7 cells are pelleted in a 1.5-ml Eppendorf test tube. The cells are washed in 1% sodium chloride (saline), the wash is decanted, and the pellet is resuspended in the residual saline solution. About 1.0 ml of methanol : acetic acid (3 : 1) is added and the cells are fixed at room temperature for 45 minutes. The fixed cells are centrifuged and then washed three time with saline solution. The cells may adhere to the side of the plastic centrifuge tube and care must be taken not to lose them while decanting the saline. The nuclear and mitochondrial DNA are stained by resuspending the cells in 0.5 ml of 1 μg/ml DAPI and incubating for 45 minutes at room temperature. Finally, the cells are washed three time with saline and resuspended in a final volume of 0.1 ml of saline. The cells are wet-mounted under a coverslip and examined using a fluorescence microscope with a 365-nm UV excitation filter and a 420-nm long-pass filter.

C.   Flow Cytometry

Detailed information on the cell age distribution of a mutant can be obtained by collecting the DNA histogram of an exponentially growing culture using flow cytometry. Figure 8 shows the histograms for a pair of plasmid shuffle strains isogenic except for the histone H4 allele carried on the plasmid. Figure 8A shows the DNA histogram for cells from an exponential culture carrying the wild-type histone H4 gene, while Fig. 8B shows the results for cells carrying the glutamine substitution allele *hhf1-10* described above. The mutant shows a significantly larger percentage of postreplication cells in the population as the result of a delay in the G_2 + M portions of the division cycle. These results strengthen the hypothesis that the N-terminal domain of histone H4, and perhaps lysine deacetylation, plays a role in chromosome segregation.

In the past, the DNA content of yeast, less than 1% relative to a mammalian cell nucleus, made it difficult to obtain histograms for haploid yeast strains (Slater *et al.*, 1977; Smith and Stirling, 1988), but improvements in instrumentation and protocol have improved sensitivity, and we now routinely

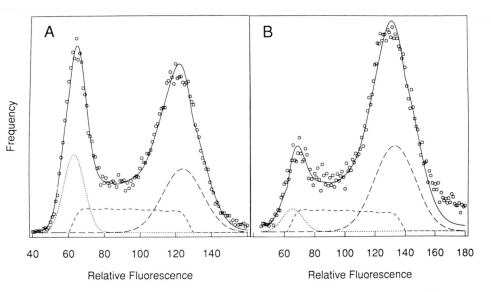

FIG. 8. DNA histograms of cells expressing wild-type and mutant histone H4. The distributions of cellular DNA content in exponential cultures are shown. The histograms were determined from fluorescence flow cytometry of cells stained with propidium iodide as described in the text. (A) The results obtained for cells expressing the wild-type histone H4 gene. (B) The results obtained for cells expressing the mutant H4 allele *hhf1-10* in which the four lysines in the N-terminal domain were changed to glutamine. The open circles show the relative number of cells (Frequency) plotted as a function of DNA content (Relative Fluorescence). The solid line shows the fitted curve for the overall distribution of cells. The broken lines show the component distributions calculated for G_1 (dotted), S (short dash), and $G_2 + M$ (long dash) plotted at one-half the actual y-axis scale. A comparison of the two panels shows that the cells expressing the mutant *hhf1-10* allele have a much larger proportion of cells in the $G_2 + M$ postreplication phases of the division cycle.

obtain good results even with haploid stains. Approximately $1-5 \times 10^7$ cells are pelleted and washed with 50 mM Tris-HCl (pH 7.5) buffer (Tris buffer). The cells are resuspended in 0.3 ml of water, fixed by the addition of 0.7 ml of absolute ethanol, and incubated at room temperature for 1 hour. The fixed cells are washed twice in Tris buffer and sonicated briefly to achieve a single-cell suspension. The cells are then pelleted and resuspended in 1 ml of 1 mg/ml RNase in Tris buffer. The cells are incubated for 1 hour at 37°C and then 18–24 hours at 4°C. The next day the cells are pelleted, resuspended in 1 ml of a fresh pepsin solution (5 mg/ml pepsin in 55 mM HCl), and incubated for 5 minutes at room temperature. The cells are pelleted and then stained in the dark in 1.5 ml of 50 μg/ml propidium iodide in 0.18 M Tris-HCl (pH 7.5), 0.18 M sodium chloride, 70 mM magnesium chloride. Next, the cells are incubated in the dark for 1 hour at room temperature and then at 4°C overnight. Just prior to flow cytometry, the stained cells are pelleted and resuspended at approximately 1×10^6 cells/ml in Tris buffer and kept in the dark on ice. If the cells are clumped when examined under the microscope, the samples should be sonicated for 5–10 seconds to give a single-cell suspension. If the RNase digestion was complete, the cells should show strong nuclear staining with little or no cytoplasmic fluorescence under the microscope. For flow cytometry, we currently use a Coulter EPICS model 753 fluorescence activated cell sorter. The laser is driven at 1 W with an excitation wave length of 488 nm and a 550-nm long-pass filter. A confocal lens block is used to focus the laser beam on the cell flow. Data are collected from approximately 30,000 to 100,000 cells for histograms.

D. Genetic Assays for Transcription

There are many well-characterized transcriptional systems in yeast that can be assayed for regulated expression (Struhl, 1989). Prominent examples include *HIS3* (Struhl, 1989), *HIS4* (Arndt et al., 1987; Donahue et al., 1983; Nagawa and Fink, 1985), *CYC1* (Guarente, 1987), *HO* (Herskowitz, 1989), and *GAL1–10* (Johnston, 1987). In several cases, the roles of nucleosomes and chromatin structure have been studied. The effect of histone stoichiometry on promoter selection in Ty1 elements has already been cited (Clark-Adams et al., 1988). The *PHO5* gene is another example where chromatin structure has been shown to play a role in transcription and has been exploited as an assay system (Almer et al., 1986; Fascher et al., 1990; Han et al., 1988).

The yeast silent mating type loci *HML* and *HMR* provide an attractive system for analyzing permanent gene repression (Herskowitz, 1988). The two silent mating type loci contain duplicate copies of the DNA encoding the **a**-mating (*HMR*) and α-mating (*HML*) information. They are normally never transcribed and mating and ability is controlled by the expressed information at the single *MAT* locus; *MATa* cells mate with α cells, *MATα* cells mate with **a** cells, and diploids heterozygous at *MAT* do not mate and are capable of entering meiosis. There is good biochemical evidence that the nucleosomes at the silent loci are in a compact conformation different from that of bulk chromatin (Chen et al., 1991) and that the chromatin structure at the silent loci is regulated by the genes that maintain silencing (Nasmyth, 1982).

The permanent repression of the silent loci is under the control of several genes, including *SIR1–SIR4* (Rine and Herskowitz, 1987), the N-terminal domain of histone H4 (Kayne et al., 1988; Megee et al., 1990; Park and Szostak, 1990), the *NAT1/ARD1* protein acetylation system perhaps via the N-terminus of histone H2B (Mullen et al., 1989), the sequence-specific binding proteins RAP1 and and ABF1 (Diffley and Stillman, 1989; Kimmerly et al., 1988; Shore and Nasmyth, 1987), and cis-acting sequences having autonomous replication activity (Brand *et al.,* 1985, 1987; Mahoney and Broach, 1989). Loss of any one of these participating factors can result in derepression of the mating type genes at the silent loci. This transcription in turn results in codominant expression of both **a**-mating and α-mating information in the same cell, which therefore becomes phenotypically sterile. Thus, haploid mating ability provides a rapid first-line screen for potential transcriptional derepression at the silent loci. Because there are many other defects that affect mating ability (Herskowitz, 1989), true derepression of the silent loci must be confirmed by other experiments, such as Northern blot analysis (Kayne *et al.,* 1988; Mullen *et al.,* 1989).

E.   Assays for Recombination

There are many types of recombination: mitotic, meiotic, reciprocal recombination, gene conversion, equal and unequal sister chromatid exchange, intrachromosomol and interchromosomal exchange, nonallelic recombination, etc. Detailed studies can become highly specialized, requiring considerable skill in setting up and evaluating the experiment (Orr-Weaver and Szostak, 1985). However, a number of simple genetic assays have been developed that provide satisfactory initial screens for mutations affecting recombination.

There are several schemes for assaying intrachromosomal mitotic recombination. Homologous integration of a circular plasmid carrying *URA3* as a selectable marker results in the nontandem duplication of chromosomal sequences interrupted by plasmid vector sequences and the *URA3* gene. Recombination between the flanking chromosomal sequences results in the excision of the *URA3* gene, which is readily scored by the ability to form colonies on 5FOA medium. While simple to set up, this assay sometimes overloads the use of *URA3*, if combined with a plasmid shuffle protocol, for example.

The disruption of a scorable nonessential gene provides another approach that is flexible with respect to selectable markers. Integration of a circular plasmid carrying a fragment internal to the coding sequences of a gene results in the nontandem duplication of two defective gene sequences. The selectable marker on the disrupting plasmid can be any gene convenient for the strain genotype and is merely used to select the initial disrupted transformants. Recombination between the shared homologous sequences restores the target gene and its function. For example, if *LEU2* is disrupted in an appropriate strain by an integrative plasmid carrying an internal fragment of the *LEU2* gene, with *HIS3* as the selectable marker, the resulting tranformant is Leu⁻ and His⁺. Recombination frequencies can then be measured by scoring colonies that are Leu⁺ and His⁻.

Interchromosomal mitotic recombination is conveniently assayed using a recessive drug resistance gene (Hartwell and Smith, 1985). The *CAN1* gene codes for an arginine permease and the recessive *can1* allele confers resistance to the amino acid analog canavinine. The assay is conducted with a diploid strain carrying heterozyous markers on the two arms of chromosome V. One homolog is *CAN1 HOM3* and the other is *can1 hom3*. The *HOM3* gene encodes aspartate kinase and is required for the biosynthesis of methionine and threonine (Jones and Fink, 1982). The starting heterozygous diploid is then canavinine sensitive (Can^S) and methionine prototrophic (Met$^+$). Recombination between *CAN1* and its centromere can give rise to colonies that are resistant to canavinine. These can be scored on canavivine plates as colonies that are Can^R but still methionine independent (Met$^+$).

Why the test for methionine auxotrophy? The other frequent event that can give rise to canavinine-resistant colonies is chromosome loss. If the *CAN1 HOM3* homolog of chromosome V is lost, the daughter cell will be canavinine resistant without a recombination event. However, in this case, the recessive *hom3* allele will be uncovered on the right arm and the colony will also be auxotrophic for methionine. This assay, then, provides a convenient measure of both recombination and chromosome loss (Hartwell and Smith, 1985).

In practice, the test diploid is inoculated at low density in rich medium and grown for approximately 10 generations. Dilutions of the culture are then plated for single colonies on canavinine plates and resistant colonies are allowed to grow up, the products of both loss and recombination. The resistant colonies are then replicated to medium lacking methionine to distinguish between the two classes of events.

F.   Chromosome Loss

We have already considered one assay for chromosome loss in the previous section. A number of other convenient assays have been developed recently. The color colony phenotypes imparted by mutations in the adenine metabolic pathway provide particularly useful tools for following mitotic chromosome segregation (Koshland and Hieter, 1987). Hieter *et al.* (1985) have described a color colony assay for plasmid loss. The suppression phenotype obtained in a diploid homozygous for *ade2-101* depends on the copy number of the *SUP11* gene: in the absence of *SUP11* the cells are red, with one copy of *SUP11* the cells are pink, and with two or more copies of *SUP11* the cells are white. Koshland *et al.* (1985) have described a similar scheme using a leaky allele of *ADE3* called *ade3-2p*. The *ADE3* gene functions prior to *ADE2*, and a mutation in *ADE3* prevents accumulation of the red pigment seen in an *ade2* single mutant. In *ade2 ade3* strains, the leaky *ade3-2p* allele allows some flux of metabolites through to the *ade2* block, and the copy number of a plasmid carrying the *ade3-2p* allele can be determined by colony color morphology.

In the *SUP11* assay, for example, diploids are transformed with a plasmid carrying *SUP11* and a pink colony is isolated carrying one copy of the plasmid per cell. This isolate is then cultured to early log phase and spread for single colonies on plates with limiting adenine (Koshland and Hieter, 1987). Events taking place in the first cell division on the plates are identified as exact half-sectored colonies. If the plasmid either fails to replicate or one daughter chromosome is lost in the first division, the resulting colony will be half pink and half red, a "1:0" event. If the plasmid replicates but daughter chromosomes fail to disjoin, the colony will be half white and half red, a "2:0" event.

The exact number and nature of the events that produce 1:0 segregants are not known. Replication failure is probably one. Similarly, the origins of 2:0 events can be ambiguous, since the plasmid copy number in white sectors can be two or more per cell. Recombination between duplicate or replicated plasmid copies can also complicate the interpretation (Koshland and Hieter, 1987). Nevertheless, if starting strains are constructed with appropriate adenine gene markers in their background, the color colony assay is a rapid and sensitive indicator of defects in major nuclear pathways.

G.   Pathways, Complexes, and Interacting Genes

Genetic analysis of a chromosomal protein gene can yield important information on how the protein functions in the cell and how it interacts with other components of the chromosome. Interactions are frequently detected by combining mutations in the gene of interest with those in other genes known to be involved in the same functional pathway to create double mutants, or by seeking extragenic suppressors of the phenotype of the mutated gene. Such genetic evidence can confirm interactions and complexes suspected on biochemical grounds, and suggest the existence of new pathways, functions, and physical interactions. There are several important considerations in developing an experimental strategy for further genetic analysis.

1. The phenotypes of mutants in the primary gene should be thoroughly characterized prior to embarking on a search for secondary genetic interactions. The greater the understanding of the behavior of mutants in the primary gene, the more likely that significant modifications of these properties will be detected.

2. When comparing phenotypes among different strains, it is important to keep the genetic backgrounds isogenic except for the alleles being studied. Strains of *S. cerevisiae* differ considerably between laboratories, even among

those derived from the same "wild-type" standard. One of the advantages of the plasmid shuffle and conditional expression schemes is that the resulting strains are isogenic, differing only in the allele of interest on the plasmid. Frequently, other gene mutations needed in a strain background can be obtained as cloned alleles and incorporated by one-step or two-step gene replacement. In the absence of cloned alleles, test strains constructed by genetic crosses should be extensively backcrossed to a standard parent.

3. It is important to have multiple alleles of the primary gene for investigation. This requirement is obvious if the protein has multiple functions; a whole repertoire of activity may be overlooked if only a single allele is examined. In addition, the case for a physical interaction between two gene products is strengthened if their phenotypes are specific for certain allelic combinations. That is, the function of interacting suppressors or enhancers of a mutated gene may depend on the specific structural defect in the gene products.

4. New extragenic suppressors, synthetic lethals, and other mutations in interacting gene products are most useful if they confer new phenotypes of their own. It is from these phenotypes that the specificity and mechanism of the interaction can be deduced. For example, starting with a temperature-sensitive mutation in a chromosomal protein, it is useful to screen for extragenic suppressors of temperature sensitivity that themselves confer a cold-sensitive phenotype. The properties of the cold-sensitive strain can then be compared with those of the temperature-sensitive strain. Similar characteristics would strengthen the hypothesis that the two genes code for proteins that interact on the same pathway.

1.   DOUBLE MUTANTS

If preliminary characterization of a new mutant succeeds in associating the gene with a major nuclear function, such as recombination or chromosome segregation, more detailed information can be obtained by combining it with other mutations affecting the same pathway. For example, given the effect of nucleosome depletion on centromere structure (Saunders et al., 1990), it will be interesting to examine mutations of the histone genes together with mutations in centromere-binding proteins as they become available.

The properties of a double mutant are only easily interpreted if the phenotypes of the two individual mutants can be distinguished. Synergy can often be detected, however, even if mutations in the two genes have similar effects, provided the assay is quantitative. If the component genes participate separately in ordered pathways, then the interpretation of a double mutant is relatively straightforward. If the double mutant has a phenotype identical to one but not the other of the single mutants, then the gene function of the former is epistatic or upstream in the same functional pathway. If the phenotype of the double mutant is more extreme than either of the single mutants alone, then it is likely that the two mutations impair two separate pathways. The interpretation is more complicated if the two proteins are part of a physical complex. One can imagine a protein complex with sufficient function to support growth if only one of its subunits is defective, but failing if it has defects in two different subunits.

2. SYNTHETIC LETHAL MUTATIONS

Mutations in two separate genes are termed synthetic lethals if a cell containing both mutations is inviable under conditions where cells with either single mutation are alive. Synthetic lethality for chromosomal protein mutations is illustrated by the the N-terminal domain mutations of the histone genes. Cells with single N-terminal deletion mutations are viable, but double deletions of the H2A–H2B domains (Schuster *et al.*, 1986) or the H3–H4 domains (Morgan *et al.*, 1991) are lethal. Currently, it is not known if this synthetic lethality extends to N-terminal deletions in one dimer and one tetramer histone, an H2B gene and an H4 gene, for example. Examples of other synthetic lethal mutations in yeast include the α-tubulin and β-tubulin genes (Huffaker *et al.*, 1987), secretory pathway genes (Kaiser and Schekman, 1990), and CDC genes (Bender and Pringle, 1991). Synthetic phenotypes must be interpreted with caution. If the individual mutations are functionally unrelated but both confer severe growth defects on the cell, lethality may result simply from the compound effect of poor growth.

A screen for synthetic lethal mutations is easily set up using the plasmid shuffle or conditional expression systems. The starting mutant allele is carried on the chromosome or a plasmid, while the wild type is carried on a second plasmid with the URA3 selectable marker or under the control of the GAL1 promoter. The strain is mutagenized (Guthrie and Fink, 1990; Kilbe, 1975; Pringle, 1975; Rose *et al.*, 1990) and screened for colonies that fail to grow on 5FOA medium for the plasmid shuffle system, or glucose medium for the conditional expression system. Among such colonies are those that require the presence of the plasmid with the wild-type gene and its expression. Further genetic tests are necessary to rule out false mutations, such as plasmid–plasmid recombination, and to show that the new synthetic lethal mutation is unlinked to the original mutation. For example, if the strain is cured of the plasmid with the primary mutation and retransformed with a fresh unmutagenized copy, then the wild-type gene should still be required for viability if the synthetic lethality is unlinked to the primary gene.

3.   EXTRAGENEIC SUPPRESSORS

Suppressor, or "pseudorevertant," analysis is a powerful approach to identifying interacting gene products. It is also a complicated approach because of the multiple levels at which suppression can occur (Hartman and Roth, 1973). Direct or "informational" suppression can occur via suppressor tRNAs, resulting in the production of the wild-type protein. Extragenic suppressors may alter the level or stability of the mutated gene product, either up or down, to restore more normal function. Bypass suppressors can activate alternate pathways or break down normal compartmentalization in the cell, allowing existing pathways to take over the lost function.

It is not trivial to identify protein–protein interaction suppressors among a collection of revertants. Bypass suppressors are insensitive to the particular allele of the primary gene and may even suppress a deletion of the gene, something beyond the ability of an interaction suppressor. Suppression by interacting gene products, on the other hand, should show allele specificity. In addition, suppressors of loss-of-function mutations should be dominant, in the simplest interpretation, since the altered suppressor protein may interact even in the presence of its wild-type gene product (Adams et al., 1989; Adams and Botstein, 1989). However, recessive allele-specific extragenic suppressors have also been isolated (Novick et al., 1989).

Despite its difficulties, suppression analysis can provide important insights into gene interactions. For example, Johnson et al. (1990) recently selected for suppressors of mating sterility caused by mutations in the N-terminal domain of histone H4. Of several complementation groups, one class of extragenic suppressor was found to be in the SIR3 gene. These sir3 suppressors were unable to suppress the N-terminal deletion allele of histone H4 consistent with an interaction between the two proteins. However, several different H4 alleles with point mutations in residues 16–19 were all suppressed in the sir3 mutants. These results suggest an interaction between the product of the SIR3 gene and histone H4, although the nature of that interaction, direct or indirect, is not yet clear.

4.   NONCOMPLEMENTING MUTATIONS

One way to search directly for mutations in interacting proteins is by screening for new unlinked noncomplementing mutations (Stearns and Botstein, 1988). Normally, when two strains with recessive mutations in different genes mate, the wild-type alleles on the opposing homologs will complement each other in the diploid. Two strains with unlinked noncomplementing mutations in opposition fail to produce viable diploids when mated with each other, although either alone mates with a wild type to give viable diploids.

A screen for noncomplementing mutations is relatively simple to implement. First, a marked strain wild type for the primary gene of interest is mutagenized. Colonies are then mated to two different strains: one that is also wild type for the primary gene, and a second that carries the mutated experimental allele. Colonies are screened for those that mate with the wild-type strain but not the mutant. Additional tests are then necessary to ensure that the noncomplementing mutation is unlinked to the primary gene. It is useful to conduct the search with conditional alleles of the primary gene and screen for conditional noncomplementation. The ability to produce viable diploids with the wild-type tester under nonpermissive conditions eliminates colonies that have picked up conditional dominant lethal mutations, mutations in mating type genes, or mutations in nutritional markers.

The simplest interpretation of unlinked noncomplementation is that the two gene products are components of a multiprotein complex and that mixed complexes are formed in the mutants. These heteromeric complexes either reduce the levels of active complex below that required for viability or are directly toxic to the cell (Stearns and Botstein, 1988). There are several potential advantages to the noncomplementation strategy: The new noncomplementing mutation is obtained on its own in a haploid background, facilitating further analysis of its properties. In a suppressor analysis, the suppressor is isolated in the same genome as the original mutation and must be crossed out for analysis. In addition, the noncomplementation screen is conducted with diploids, so that recessive nonspecific secondary mutations that might affect growth in the haploid are covered and do not influence the screen. For these reasons, the unlinked noncomplementation strategy is a promising approach for the analysis of interacting protein genes.

ACKNOWLEDGMENTS

I thank Ru Chih Huang for first introducing me to chromosomal proteins and Noreen and Ken Murray for adding the discipline of molecular genetics. I thank my colleagues in the field for many helpful discussions and for generously sharing technical methods, particularly Dan Burke, Gerry Fink, Kim Arndt, Mike Christman, Elaine Elion, Clarence Chang, and Alison Adams. Work conducted in our laboratory has been supported by U. S. Public Health Service grant GM28290 and grant MV176 from the American Cancer Society.

REFERENCES

Adams, A. E., and Botstein, D. (1989). Dominant suppressors of yeast actin mutations that are reciprocally suppressed. *Genetics* **121**, 675–683.

Adams, A. E., Botstein, D., and Drubin, D. G. (1989). A yeast actin-binding protein is encoded by *SAC6*, a gene found by suppression of an actin mutation. *Science* **243**, 231–233.

Almer, A., Rudolph, H., Hinnen, A., and Horz, W. (1986). Removal of positioned nucleosomes from the yeast PHO5 promoter upon PHO5 induction releases additional upstream activating DNA elements. *EMBO J.* **5**, 2689–2696.

Arndt, K. T., Styles, C., and Fink, G. R. (1987). Multiple global regulators control HIS4 transcription in yeast. *Science* **237**, 874–880.

Ausubel, F. M., Brent, R., Kingston, R. E., Moore, D. D., Seidman, J. G., Smith, J. A., and Struhl, K. (1989). "Current Protocols in Molecular Biology." Wiley, New York.

Bender, A., and Pringle, J. R. (1991). Use of a screen for synthetic lethal and multicopy suppressee mutants to identify two new genes involved in morphogenesis in *Saccharomyces cerevisiae*. *Mol. Cell. Biol.* **11**, 1295–1305.

Bergman, M. G., Wawra, E., and Winge, M. (1988). Chicken histone H5 inhibits transcription and replication when introduced into proliferating cells by microinjection. *J. Cell Sci.* **91**, 201–209.

Boeke, J. D., LaCroute, F., and Fink, G. R. (1984). A positive selection for mutants lacking orotidine-5'-phosphate decarboxylase activity in yeast: 5-Fluoro-orotic acid resistance. *Mol. Gen. Genet.* **197**, 345–346.

Boeke, J. D., Trueheart, J., Natsoulis, B., and Fink, G. R. (1987). 5-Fluoroorotic acid as a selective agent in yeast molecular genetics. *In* "Methods in Enzymology" (R. Wu and L. Grossman, eds.), Vol. 154, pp. 164–175. Academic Press, Orlando, Florida.

Botstein, D., and Fink. G. R. (1988). Yeast: An experimental organism for modern biology. *Science* **240**, 1439–1443.

Brand, A. H., Breeden, L., Abraham, J., Sternglanz, R., and Nasmyth, K. (1985). Characterization of a "silencer" in yeast: A DNA sequence with properties opposite to those of a transcriptional enhancer. *Cell (Cambridge, Mass.)* **41**, 41–48.

Brand, A. H., Micklem, G., and Nasmyth, K. (1987). A yeast silencer contains sequences that can promote autonomous plasmid replication and transcriptional activation. *Cell (Cambridge, Mass.)* **51**, 709–719.

Brent, R., and Ptashne, M. (1985). A eukaryotic transcriptional activator bearing the DNA specificity of a prokaryotic repressor. *Cell (Cambridge, Mass.)* **43**, 729–736.

Buchman, A. R., Lue, N. F., and Kornberg, R. D. (1988a). Connections between transcriptional activators, silencers, and telomeres as revealed by functional analysis of a yeast DNA-binding protein. *Mol. Cell. Biol.* **8**, 5986–5099.

Buchman, A. R., Kimmerly, W. J., Rine, J., and Kornberg, R. D. (1988b). Two DNA-binding factors recognize specific sequences at silencers, upstream activating sequences, autonomously replicating sequences, and telomeres in *Saccharomyces cerevisiae*. *Mol. Cell. Biol.* **8**, 210–225.

Carbon, J. (1984). Yeast centromeres: Structure and function. *Cell (Cambridge, Mass.)* **37**, 351–353.

Chasman, D. I., Lue, N. F., Buchman, A. R., LaPointe, J. W., Lorch, Y., and Kornberg, R. D. (1990). A yeast protein that influences the chromatin structure of UASG and functions as a powerful auxiliary gene activator. *Genes Dev.* **4**, 503–514.

Chattoo, B. B., Sherman, F., Azubalis, D. A., Fjellstedt, T. A., Mehnert, D., and Ogur, M. (1979). Selection of lys2 mutants of the yeast *Saccharomyces cerevisiae by the utilization of α-mainoadipate*. *Genetics* **93**, 51–65.

Chen, T. A., Smith, M. M., Le, S., Sternglanz, R., and Allfrey, V. G. (1991). Nucleosome fractionation by mercury-affinity chromatography. Contrasting distribution of transcriptionally active DNA sequences and acetylated histones in nucleosome fractions of wild-type yeast cells and cells expressing a histone H3 gene altered to encode a cysteine-110 residue. *J. Biol. Chem.* **266**, 6489–6498.

Choe, J., Kolodrubetz, D., and Grunstein, M. (1982). The two yeast histone *H2A* genes encode similar protein subtypes. *Proc. Natl. Acad. Sci. U.S.A.* **79**, 1484–1487.

Chu, C., Parris, D. S., Dixon, R. A. F., Farber, F. E., and Schaffer, P. A. (1979). Hydroxylamine mutagenesis of HSV DNA and DNA fragments: Introduction of mutations into selected regions of the viral genome. *Virology* **98**, 168–181.

Clark-Adams, C. D., Norris, D., Osley, M. A., Fassler, J. S., and Winston, F. (1988). Changes in histone gene dosage alter transcription in yeast. *Genes Dev.* **2**, 150–159.

Cox, E. C., and Horner, D. L. (1983). Structure and coding properties of a dominant *Escherichia coli* mutator gene, mutD. *Proc. Natl. Acad. Sci. U.S.A.* **80**, 2295–2299.

Cross, S. L., and Smith, M. M. (1988). Comparison of the structure and cell cycle expression of mRNAs encoded by two histone H3-H4 loci in *Saccharomyces cerevisiae*. *Mol. Cell. Biol.* **8**, 945–954.

Diffley, J. F., and Stillman, B. (1989). Similarity between the transcriptional silencer binding proteins ABF1 and RAP1. *Science* **246**, 1034–1038.

Donahue, T. F., Daves, R. S., Lucchini, G., and Fink, G. R. (1983). A short nucleotide sequence required for regulation of HIS4 by the general control system of yeast. *Cell (Cambridge, Mass.)* **32**, 89–98.

Echols, H., Lu, C., and Burgers, P. M. (1983). Mutator strains of *Escherichia coli*, mutD and dnaQ, with defective exonucleolytic editing by DNA polymerase III holoenzyme. *Proc. Natl. Acad. Sci. U.S.A.* **80**, 2189–2192.

Elgin, S. C. R., and van Daal, A. (1990). The highly conserved histone H2A variant H2A.Z is essential in *Drosophila*. *J. Cell Biol.* **111**, 250a.

Fangman, W. L., and Zakian, V. A. (1981). Genome structure and replication. *In* "The Molecular Biology of the Yeast *Saccharomyces*" (J. N. Strathern, E. W. Jones, and J. R. Broach, eds.), pp. 27–58. Cold Spring Harbor Laboratory, Cold Spring Harbor, New York.

Fascher, K. D., Schmitz, J., and Horz, W. (1990). Role of transactivating proteins in the generation of active chromatin at the PHO5 promoter in *S. cerevisiae*. *EMBO J.* **9**, 2523–2528.

Ford, C. F., and Smith, M. M. (1985). Use of an oligonucleotide probe to detect transplacement of an amber mutation into a yeast histone H3 gene. *Gene* **37**, 45–52.

Futcher, A. B., and Cox, B. S. (1984). Copy number and the stability of 2-micron circle-based artificial plasmids of *Saccharomyces cerevisiae*. *J. Bacteriol.* **157**, 283–290.

Gillam, S., and Smith, M. (1979). Site-specific mutagenesis using synthetic oligodeoxyribonucleotide primers: I. Optimum conditions and minimum oligodeoxyribonucleotide length. *Gene* **8**, 81–97.

Guarente, L. (1987). Regulatory proteins in yeast. *Ann. Rev. Genet.* **21**, 425–452.

Guthrie, C., and Fink, G. R. (1990). "Guide to Yeast Genetics and Molecular Biology." Academic Press, San Diego, California.

Haggren, W., and Kolodrubetz, D. (1988). The *Saccharomyces cerevisiae ACP2* gene encodes an essential HMG1-like protein. *Mol. Cell. Biol.* **8**, 1282–1289.

Han, M., Chang, M., Kim, U., and Grunstein, M. (1987). Histone H2B repression causes cell-cycle-specific arrest in yeast: Effects on chromosomal segregation, replication, and transcription. *Cell (Cambridge, Mass.)* **48**, 589–597.

Han, M., Kim, U. J., Kayne, P., and Grunstein, M. (1988). Depletion of histone H4 and nucleosomes activates the *PHO5* gene in *Saccharomyces cerevisiae*. *EMBO J.* **7**, 2221–2228.

Hartman, P. E., and Roth, J. R. (1973). Mechanisms of suppression. *Adv. Genet.* **17**, 1–105.

Hartwell, L. H., and Smith, D. (1985). Altered fidelity of mitotic chromosome transmission in cell cycle mutants of *S. cerevisiae*. *Genetics* **110**, 381–395.

Hartwell, L. H., and Weinert, T. A. (1989). Checkpoints: Controls that ensure the order of cell cycle events. *Science* **246**, 629–634.

Henikoff, S. (1987). Unidirectional digestion with exonuclease III in DNA sequence analysis. *In* "Methods in Enzymology" (R. Wu, ed.), Vol. 155, pp. 156–165. Academic Press, Orlando, Florida.

Hereford, L., Fahrner, K., Woolford, Jr., J., Rosbash, M., and Kaback, D. (1979). Isolation of yeast histone genes *H2A* and *H2B*. *Cell (Cambridge, Mass.)* **18**, 1261–1271.

Hereford, L. M., Osley, M. A., Ludwig II, J. R., and McLaughlin, C. S. (1981). Cell-cycle regulation of yeast histone mRNA. *Cell (Cambridge, Mass.)* **24**, 367–375.

Hereford, L., Bromley, S., and Osley, M. A. (1982). Periodic transcription of yeast histone genes. *Cell (Cambridge, Mass.)* **30**, 305–310.

Herskowtiz, I. (1988). Life cycle of the budding yeast *Saccharomyces cerevisiae*. *Microbiol. Rev.* **52**, 536–553.

Herskowitz, I. (1989). A regulatory hierarchy for cell specialization in yeast. *Nature (London)* **342**, 749–757.

Hicks, J. B., Hinnen, A., and Fink, G. R. (1979). Properties of yeast transformation. *Cold Spring Harbor Symp. Quant. Biol.* **43**, 1305–1313.

Hieter, P., Mann, C., Snyder, M., and Davis, R. W. (1985). Mitotic stability of yeast chromosomes: A colony color assay that measures nondisjunction and chromosome loss. *Cell (Cambridge, Mass.)* **40**, 381–392.

Holm, C., Goto, T., Wang, J. C., and Botstein, D. (1985). DNA topoisomerase II is required at the time of mitosis in yeast. *Cell (Cambridge, Mass.)* **41**, 553–563.

Hope, I. A., and Struhl, K. (1986). Functional dissection of a eukaryotic transcriptional activator protein, GCN4 of yeast. *Cell (Cambridge, Mass.)* **46**, 885–894.

Horwitz, B. H., and Dimaio, D. (1990). Saturation mutagenesis using mixed oligonucleotides and M13 templates containing uracil. *In* "Methods in Enzymology" (D. V. Goeddel, ed.), Vol. 185, pp. 599–611. Academic Press, San Diego, California.

Hsu, Y. P., and Kohlhaw, G. B. (1982). Overproduction and control of the *LEU2* gene product, beta-isopropylmalate dehydrogenase, in transformed yeast strain. *J. Biol. Chem.* **257**, 39–41.

Huffaker, T. C., Hoyt, M. A., and Botstein, D. (1987). Genetic analysis of the yeast cytoskeleton. *Annu. Rev. Genet.* **21**, 259–284.

Huffaker, T. C., Thomas, J. H., and Botstein, D. (1988). Diverse effects of beta-tubulin mutations on microtubule formation and function. *J. Cell Biol.* **106**, 1997–2010.

Innis, M. A., Gelfand, D. H., Sninsky, J. J., and White, T. J. (1990). "PCR Protocols: A Guide to Methods and Applications." Academic Press, San Diego, California.

Jayaram, M., Li, Y. Y., and Broach, J. R. (1983). The yeast plasmid 2mu circle encodes components required for its high copy propagation. *Cell (Cambridge, Mass.)* **34**, 95–104.

Johnson, L. M., Kayne, P. S., Kahn, E. S., and Grunstein, M. (1990). Genetic evidence for an interaction between SIR3 and histone H4 in the repression of the silent mating loci in *Saccharomyces cerevisiae*. *Proc. Natl. Acad. Sci. U.S.A.* **87**, 6286–6290.

Johnston, M. (1987). A model fungal gene regulatory mechanism: The *GAL* genes of *Saccharomyces cerevisiae*. *Microbiol. Rev.* **51**, 458–476.

Jones, E. W., and Fink, G. R. (1982). Regulation of amino acid and nucleotide biosynthesis in yeast. *In* "The Molecular Biology of the Yeast *Saccharomyces*: Metabolism and Gene Expression" (J. N. Strathern E. W. Jones, and J. R. Broach, eds.), pp. 181–299. Cold Spring Harbor Laboratory, Cold Spring Harbor, New York.

Ju, Q. D., Morrow, B. E., and Warner, J. R. (1990). REB1, a yeast DNA-binding protein with many targets, is essential for growth and bears some resemblance to the oncogene *myb*. *Mol. Cell. Biol.* **10**, 5226–5234.

Kaiser, C. A., and Schekman, R. (1990). Distinct sets of *SEC* genes govern transport vesicle formation and fusion early in the secretory pathway. *Cell (Cambridge, Mass.)* **61**, 723–733.

Kayne, P. S., Kim, U. J., Han, M., Mullen, J. R., Yoshizaki, F., and Grunstein, M. (1988). Extremely conserved histone H4 N terminus is dispensable for growth but essential for repressing the silent mating loci in yeast. *Cell (Cambridge, Mass.)* **55**, 27–39.

Kikuchi, Y. (1983). Yeast plasmid requires a cis-acting locus and two plasmid proteins for its stable maintenance. *Cell (Cambridge, Mass.)* **35,** 487–493.

Kilbey, B. J. (1975). Mutagenesis in yeast. *Methods Cell Biol.* **12,** 209–231.

Kim, U. J., Han, M., Kayne, P., and Grunstein, M. (1988). Effects of histone H4 depletion on the cell cycle and transcription of *Saccharomyces cerevisiae. EMBO J.* **7,** 2211–2219.

Kimmerly, W., Buchman, A., Kornberg, R., and Rine, J. (1988). Roles of two DNA-binding factors in replication, segregation and transcriptional repression mediated by a yeast silencer. *EMBO J.* **7,** 2241–2253.

Kolodrubetz, D., Rykowski, M. C., and Grunstein, M. (1982). Histone H2A subtypes associate interchangeably *in vivo* with histone H2B subtypes. *Proc. Natl. Acad. Sci. U.S.A.* **79,** 7814–7818.

Koshland, D., and Hieter, P. (1987). Visual assay for chromosome ploidy. *In* "Methods in Enzymology (R. Wu, ed.), Vol. 155, pp. 351–372. Academic Press, Orlando, Florida.

Koshland, D., Kent, J. C., and Hartwell, L. H. (1985). Genetic analysis of the mitotic transmission of minichromosomes. *Cell (Cambridge, Mass.)* **40,** 393–403.

Kunkel, T. A. (1985). Rapid and efficient site-specific mutagenesis without phenotypic selection. *Proc. Natl. Acad. Sci. U.S.A.* **82,** 488–492.

Kunkel, T. A., Roberts, J. D., and Zakour, R. A. (1987). Rapid and efficient site-specific mutagenesis without phenotypic selection. *In* "Methods in Enzymology" (R. Wu and L. Grossman, eds.), Vol. 154, pp. 367–382. Academic Press, Orlando, Florida.

Longtine, M. S., Wilson, N. M., Petracek, M. E., and Berman, J. (1989). A yeast telomere binding activity binds to two related telomere sequence motifs and is indistinguishable from RAP1. *Curr. Genet.* **16,** 225–239.

Lycan, D. E., Osley, M. A., and Hereford, L. M. (1987). Role of transcriptional and post-transcriptional regulation in expression of histone genes in *Saccharomyces cerevisiae. Mol. Cell. Biol.* **7,** 614–621.

Ma, J., and Ptashne, M. (1987). Deletion analysis of GAL4 defines two transcriptional activating segments. *Cell (Cambridge, Mass.)* **48,** 847–853.

Mahoney, D. J., and Broach, J. R. (1989). The HML mating-type cassette of *Saccharomyces cerevisiae* is regulated by two separate but functionally equivalent silencers. *Mol. Cell. Biol.* **9,** 4621–4630.

Meeks-Wagner, D., and Hartwell, L. H. (1986). Normal stoichiometry of histone dimer sets is necessary for high fidelity of mitotic chromosome transmission. *Cell (Cambridge, Mass.)* **44,** 43–52.

Meeks-Wagner, D., Wood, J. S., Garvik, B., and Hartwell, L. H. (1986). Isolation of two genes that affect mitotic chromosome transmission in *S. cerevisiae. Cell (Cambridge, Mass.)* **44,** 53–63.

Megee, P. C., Morgan, B. A., Mittman, B. A., and Smith, M. M. (1990). Genetic analysis of histone H4: Essential role of lysines subject to reversible acetylation. *Science* **247,** 841–845.

Morgan, B. A., Mittman, B. A., and Smith, M. M. (1991). The highly conserved N-termini of histones H3 and H4 are required for normal cell cycle progression. *Mol. Cell. Biol.* **11,** 4111–4120.

Mullen, J. R., Kayne, P. S., Moerschell, R. P., Tsunasawa, S., Gribskov, M., Colavito-Shepanski, M., Grunstein, M., Sherman, F., and Sternglanz, R. (1989). Identification and characterization of genes and mutants for an N-terminal acetyltransferase from yeast. *EMBO J.* **8,** 2067–2075.

Murray, A. W., and Szostak, J. W. (1983). Pedigree analysis of plasmid segregation in yeast. *Cell (Cambridge, Mass.)* **34,** 961–970.

Myers, R. M., Lerman, L. S., and Maniatis, T. (1985). A general method for saturation mutagenesis of cloned DNA fragments. *Science* **229,** 242–247.

Nagawa, F., and Fink, G. R. (1985). The relationship between the "TATA" sequence and transcription initiation sites at the HIS4 gene of *Saccharomyces cerevisiae. Proc. Natl. Acad. Sci. U.S.A.* **82,** 8557–8561.

Nasmyth, K. A. (1982). The regulation of yeast mating-type chromatin structure by SIR: An action at a distance affecting both transcription and transposition. *Cell (Cambridge, Mass.)* **30**, 567–578.

Novick, P., Osmond, B. C., and Botstein, D. (1989). Suppressors of yeast actin mutations. *Genetics* **121**, 659–674.

Orr-Weaver, T. L., and Szostak, J. W. (1985). Fungal recombination, *Microbiol. Rev.* **49**, 33–58.

Orr-Weaver, T. L., Szostak, J. W., and Rothstein, R. J. (1981). Yeast transformation: A model system for the study of recombination. *Proc. Natl. Acad. Sci. U.S.A.* **78**, 6354–6358.

Park, E., and Szostak, J. W. (1990). Point mutations in the yeast histone H4 gene prevent silencing of the silent mating type locus *HML*. *Mol. Cell. Biol.* **10**, 4932–4934.

Pfeifer, K., Kim, K. S., Kogan, S., and Guarente, L. (1989). Functional dissection and sequence of yeast HAP1 activator. *Cell (Cambridge, Mass.)* **56**, 291–301.

Pringle, J. R. (1975). Induction, selection, and experimental uses of temperature-sensitive and other conditional mutants of yeast. *Methods Cell Biol.* **12**, 233–272.

Pringle, J. R., and Hartwell, L. H. (1981). The *Saccharomyces cerevisiae* cell cycle. *In* "The Molecular Biology of the Yeast *Saccharomyces*" (J. N. Strathern, E. W. Jones, and J. R. Broach, eds), pp. 97–142. Cold Spring Harbor Laboratory, Cold Spring Harbor, New York.

Rhode, P. R., Sweder, K. S., Oegema, K. F., and Campbell, J. L. (1989). The gene encoding ARS-binding factor I is essential for the viability of yeast. *Genes Dev.* **3**, 1926–1939.

Riles, L., and Olson, M. V. (1988). Nonsense mutations is essential genes of *Saccharomyces cerevisiae*. *Genetics* **118**, 601–607.

Rine, J., and Herskowitz, I. (1987). Four genes responsible for a position effect on expression from HML and HMR in *Saccharomyces cerevisiae*. *Genetics* **116**, 9–22.

Rose, A. B., and Broach, J. R. (1990). Propagation and expression of cloned genes in yeast: 2-μm circle-based vectors. *In* "Methods in Enzymology" (D. V. Goeddel, ed), Vol. 185, pp. 234–279. Academic Press, San Diego, California.

Rose, M. D., Winston, F., and Hieter, P. (1990). "Methods in Yeast Genetics: A Laboratory Course Manual." Cold Spring Harbor Laboratory, Cold Spring Harbor, New York.

Rothstein, R. J. (1983). One-step gene disruption in yeast. *In* "Methods in Enzymology" (R. Wu, L. Grossman, and K. Moldave, eds.), Vol. 101, pp. 202–211. Academic Press, New York.

Sambrook, J., Fritsch, E. F., and Maniatis, T. (1989). "Molecular Cloning: A Laboratory Manual." Cold Spring Harbor Laboratory, Cold Spring Harbor, New York.

Saunders, M. J., Yeh, E., Grunstein, M., and Bloom, K. (1990). Nucleosome depletion alters the chromatin structure of *Saccharomyces cerevisiae* centromeres. *Mol. Cell. Biol.* **10**, 5721–5727.

Schaaper, R. M. (1988). Mechanisms of mutagenesis in the *Escherichia coli* mutator mutD5: Role of DNA mismatch repair. *Proc. Natl. Acad. Sci. U.S.A.* **85**, 8126–8130.

Scherer, S., and Davis, R. W. (1979). Replacement of chromosome segments with altered DNA sequences constructed *in vitro*. *Proc. Natl. Acad. Sci. U.S.A.* **76**, 4951–4955.

Schuster, T., Han, M., and Grunstein, M. (1986). Yeast histone H2A and H2B amino termini have interchangeable functions: *Cell (Cambridge, Mass.)* **45**, 445–451.

Shore, D., and Nasmyth, K. (1987). Purification and cloning of a DNA binding protein from yeast that binds to both silencer and activator elements. *Cell (Cambridge, Mass.)* **51**, 721–732.

Shortle, D., and Botstein, D. (1983). Directed mutagenesis with sodium bisulfite. *In* "Methods in Enzymology" (R. Wu, L. Grossman, and K. Moldave, eds.), Vol. 100, pp. 457–468. Academic Press, New York.

Shortle, D., Haber, J. E., and Botstein, D. (1982). Lethal disruption of the yeast actin gene by integrative DNA transformation. *Science* **217**, 371–373.

Shortle, D., Novick, P., and Botstein, D. (1984). Construction and genetic characterization of temperature-sensitive mutant alleles of the yeast actin gene. *Proc. Natl. Acad. Sci. U.S.A.* **81**, 4889–4893.

Singh, A. (1977). Nonsense suppressors of yeast cause osmotic-sensitive growth. *Proc. Natl. Acad. Sci. U.S.A.* **74,** 305–309.

Slater, M. L., Sharrow, S. O., and Gart, J. J. (1977). Cell cycle of *Saccharomyces cerevisiae* in populations growing at different rates. *Proc. Natl. Acad. Sci. U.S.A.* **74,** 3850–3854.

Smith, M. M., and Murray, K. (1983). Yeast H3 and H4 histone messenger RNAs are transcribed from two non-allelic gene sets. *J. Mol. Biol.* **169,** 641–661.

Smith, M. M., and Stirling, V. B. (1988). Histone H3 and H4 gene deletions in *Saccharomyces ceresivisiae. J. Cell Biol.* **106,** 557–566.

Stearns, T., and Botstein, D. (1988). Unlinked noncomplementation: Isolation of new conditional-lethal mutations in each of the tubulin genes of *Saccharomyces cerevisiae. Genetics* **119,** 249–260.

Struhl, K. (1989). Molecular mechanisms of transcriptional regulation in yeast. *Annu. Rev. Biochem.* **58,** 1051–1077.

Sun, J. M., Wiaderkiewicz, R., and Ruiz-Carrillo, A. (1989). Histone H5 in the control of DNA synthesis and cell proliferation. *Science* **245,** 68–71.

Swerdlow, P. S., Schuster, T., and Finley, D. (1990). A conserved sequence in histone H2A which is a ubiquitination site in higher eucaryotes is not required for growth in *Saccharomyces cerevisiae. Mol. Cell. Biol.* **10,** 4905–4911.

van Daal, A., White, E. M., Gorovsky, M. A., and Elgin, S. C. R. (1988). *Drosophila* has a single copy of the gene encoding a highly conserved histone H2A variant of the H2A.F/Z type. *Nucleic Acids Res.* **16,** 7487–7497.

Volker, T. A., and Showe, M. K. (1980). Induction of mutations in specific genes of bacteriophage T4 using cloned restriction fragments and marker rescue. *Mol. Genet.* **177,** 447–452.

Wallis, J. W., Rykowski, M., and Grunstein, M. (1983). Yeast histone H2B containing large amino terminus deletions can function *in vivo. Cell (Cambridge, Mass.)* **35,** 711–719.

Weiher, H., and Schaller, H. (1982). Segment-specific mutagenesis: Extensive mutagenesis of a lac promoter/operator element. *Proc. Nat. Acad. Sci. U.S.A.* **79,** 1408–1412.

Weinert, T. A., and Hartwell, L. H. (1988). The *RAD9* gene controls the cell cycle response to DNA damage is *Saccharomyces cerevisiae. Science* **241,** 317–322.

Williamson, D., and Fennell, D. (1975). The use of fluorescent DNA-binding agent for detecting and separating yeast mitochondrial DNA. *Methods Cell Biol.* **20,** 335–351.

Wu, T. H., Clarke, C. H., and Marinus, M. G. (1990). Specificity of *Escherichia coli* mutD and mutL mutator strains. *Gene* **87,** 1–5.

Xu, H. H., Johnson, L., and Grunstein, M. (1990). Coding and noncoding sequences at the 3' end of yeast histone H2B messenger RNA confer cell cycle regulation. *Mol. Cell. Biol.* **10,** 2687–2694.

Zoller, M. J., and Smith, M. (1982). Oligonucleotide-directed mutagenesis using M13-derived vectors: An efficient and general procedure for the production of point mutations in any fragment of DNA. *Nucleic Acids Res.* **10,** 6487–6500.

Chapter 20

Mutations That Affect Nuclear Organization in Yeast

ANN O. SPERRY, BARBARA R. FISHEL, AND W. T. GARRARD

University of Texas Southwestern Medical Center at Dallas
Dallas, Texas 75235

I. Introduction

Eukaryotic chromatin is highly organized within both the interphase nucleus and the metaphase chromosome. At the first level of organization, the DNA duplex is packaged into nucleosomes to form a 100 Å fiber which is in turn compacted into a 300 Å fiber. At the highest level of organization, the 300 Å fiber forms topologically constrained loops averaging 50 to 100 kb (Benyajati and Worcel, 1976; Cook and Brazell, 1976; Paulson and Laemmli, 1977; Vogelstein *et al.,* 1980; Hancock, 1982). The first indication of a looped DNA structure within the nucleus came from light microscopic studies of lampbrush chromosomes more than 100 years ago (reviewed by Callan, 1986). More recently, the loop-domain model of DNA organization was supported

METHODS IN CELL BIOLOGY, VOL. 35

by the demonstration that DNA within the interphase nucleus is also organized into looped domains (Cook and Brazell, 1976) and by electron microscopic studies of histone-depleted metaphase chromosomes where unconstrained loops of DNA radiate from the mitotic chromosomes (Paulson and Laemmli, 1977).

The loop-domain model of DNA organization is particularly attractive, as it allows for subdivision of the genome into discrete, functional domains. Indeed, work in a number of laboratories using different experimental approaches indicates that nuclear functions such as replication (Berezney and Coffey, 1974; Vogelstein et al., 1980; Vaughn et al., 1990), transcription (Jackson and Cook, 1985; reviewed in Nelson et al., 1986), and RNA processing and transport (Fu and Maniatis, 1990; Lawrence et al., 1989; Spector, 1990) occur in functional compartments of the nucleus. It has been proposed that the nuclear substructure, or matrix, provides the framework for these activities (reviewed in Berezney, 1984; Nelson et al., 1986), as well as providing the attachment sites for DNA loops.

Different experimental approaches used to identify DNA sequences at the bases of the DNA loops have revealed the same class of evolutionarily conserved DNA sequences. We developed an in vitro assay for matrix association region (MARs) that relies on the ability of such sequences to bind specifically to isolated nuclear matrices (Cockerill and Garrard, 1986a). This approach can be complemented by the nuclear "halo" assay of Laemmli and co-workers, where endogenous scaffold-attached regions (SARs) fractionate with the residual nuclear matrix after extraction of histones and restriction endonuclease digestion of DNA loops (Mirkovitch et al., 1984). Significantly, both methods identified the same class of sequences. Typically, MARs (or SARs) are at least 200 bp long, contain topoisomerase II (topoII) consensus sequences, are very AT-rich, and are evoluntarily conserved (Cockerill and Garrard, 1986a,b; Cockerill et al., 1987; Gasser and Laemmli, 1986a,b; Amati and Gasser, 1988; Izaurralde et al., 1988; Mirkovitch et al., 1986, 1988). In addition, MARs often reside next to transcriptional regulatory sequences (Cockerill and Garrard 1986a; Cockerill et al., 1987; Gasser and Laemmli, 1986a). In yeast, MARs have been found associated with other cis-acting sequences such as centromeres (CENs) and autonomously replicating sequences (ARSs) (Amati and Gasser, 1988). MARs have been identified in specific genetic loci in cellular DNA derived from human, mouse, hamster, chicken, rabbit, Drosophila, and yeast (reviewed in Garrard, 1990).

While the DNA elements of loop attachment complexes have been fairly well characterized, the proteins that interact with these sequences are less well understood. MARs contain topoII consensus sequences, and this enzyme has been shown to bind specifically to MARs in vitro (Sperry et al., 1989; Adachi et al., 1989). However, topoII binding alone does not specify a MAR since

DNA sequences that are not MARs can also bind topoII (Sperry *et al.,* 1989). Certainly, other proteins must participate in loop organization. Other proteins that bind MAR sequences *in vitro* include histone H1 (Izaurralde *et al.,* 1989) and the yeast protein RAP-1 (Hofmann *et al.,* 1989). Since most of the information gained concerning protein interaction with MARs has come from *in vitro* experiments, it seems clear that the development of a system to probe loop organization *in vivo* would provide vital information regarding the function of these elements.

This chapter describes a genetic approach to investigate nuclear order in yeast. We have developed a screen for mutations that disrupt nuclear organization. This approach relies on the observation that loop attachment complexes are evolutionarily conserved (Cockerill and Garrard, 1986a,b; Cockerill *et al.,* 1987; Amati and Gasser, 1988). The MAR we initally characterized in the mouse κ-immunoglobulin gene binds specifically to isolated yeast nuclear matrices (Cockerill and Garrard, 1986b). Furthermore, when inserted into a yeast promoter (see below), the mouse κ-MAR interferes with transcription *in vivo* by interaction with yeast proteins. We have used these observations to develop a genetic screen based on alleviation of the inhibitory effect of the MAR. This method is also applicable to the identification and/or characterization of other DNA/protein complexes (discussed in Section IV).

II. Rationale

Our experimental approach exploits an observation of Brent and Ptashne (1984). They found that insertion of a protein recognition sequence between the UAS and the TATA box of a yeast *GAL1-lacZ* fusion gene inhibits *lacZ* expression if the DNA-binding protein is present in yeast. Similarly, when the κ-MAR is inserted within this promoter, we observe inhibition of β-galactosidase activity (see Fig. 1), indicating that proteins in yeast recognize the mouse sequence. This inhibition of *lacZ* can also be observed when cells are plated on X-gal indicator media; cells without the MAR insert are blue, while those with the insert are white. Inhibition is specific for the κ-MAR insertion because a similarly sized fragment from pBR322 has no effect. Also, if the MAR is placed upstream of the UAS, no inhibition is seen. We have taken advantage of the inhibition conferred by protein recognition of the MAR *in vivo* to isolate conditional mutants that express *lacZ* at the nonpermissive condition, since these can be easily scored as blue colonies on X-gal indicator plates. We reasoned that some of these mutants may be defective in proteins that interact directly with the MAR, while others may confer lesions in the

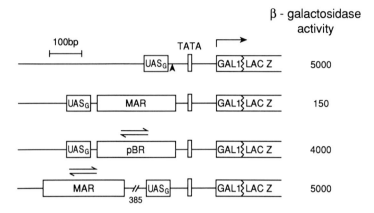

FIG. 1.   The mouse κ gene MAR inhibits the *GAL1* promoter. The first construct shows the initial plasmid into which the MAR fragment was inserted (at the arrowhead) between the upstream activating sequence (UAS$_G$) and the TATA box (TATA). Not shown are the yeast sequences responsible for replication (the 2μ origin of replication) and selection (the *URA3* gene) present on each plasmid. In the right-hand column are the corresponding β-galactosidase activities determined in yeast cultures after permeabilizing the cells with chloroform and sodium dodecyl sulfate as described (Guanrente, 1983).

formation or stability of matrix attachment sites. Once conditional mutants are obtained, it is straightforward to clone the genes that complement the mutations.

III.   Methods

A flowchart of the experimental scheme for mutagenesis and screening is shown in Fig. 2. Yeast cells transformed with the recombinant assay plasmid containing the MAR between the UAS and TATA box (Fig. 1) are mutagenized with ethyl methanesulfonate (EMS) according to a standard protocol (Reed, 1980) given below. Conditional lethal mutants affecting protein interaction with the MAR are then detected by screening for appearance of blue color at the nonpermissive temperature (Section III,B,2).

A.   Ethyl Methanesulfonate Mutagenesis

1.   MEDIA

Minimal medium.   Numbers given before the ingredient indicate the final concentration of that ingredient; numbers on the right indicate the amount required.

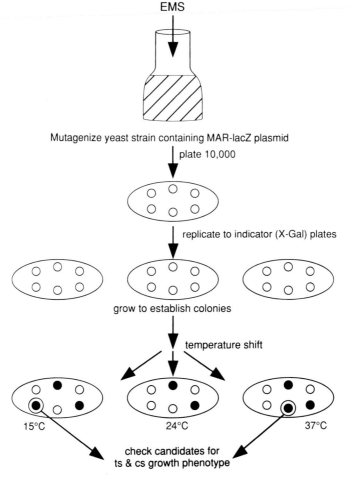

FIG. 2. Flowchart of experimental approach for mutagenesis and screening. Wild-type yeast colonies containing the assay plasmid are mutagenized with EMS as described in the text. Ten thousand colonies are screened for conditional β-galactosidase induction and conditional growth by replica plating.

0.67% Bacto-yeast nitrogen base without amino acids (Difco)	6.7 g
2% Glucose	20 g
Distilled water	1000 ml

Mix together in a 2-liter flask and autoclave. For solid minimal medium, add 2% Bacto-agar before autoclaving and pour about 30 ml per petri dish. For supplemented minimal medium add required nutrient additives.

Yeast peptone dextrose (YPD):

1% Bacto-yeast extract	10 g
2% Bacto-peptone	20 g
2% Glucose	20 g
Distilled water	1000 ml

Mix together in a 2-liter flask and autoclave. For solid medium add 2% Bacto-agar before autoclaving.

2.   YEAST TRANSFORMATION PROTOCOL

The protocol given here is derived from that of Ito *et al.* (1983).

Step 1.   Yeast cells are grown to mid-log phase (A_{600} = 0.8 to 1.0) in YPD. Ten milliliters of culture is used for each transformation. The cells are collected by centrifugation at room temperature and washed with 1 ml of sterile 0.1 M lithium acetate. The cells are then resuspended in 1 ml of 0.1 M lithium acetate and incubated at 30°C for 2 hours with agitation. The cells for each transformation are pelleted, resuspended in 0.1 ml of 0.1 M lithium acetate, and transferred to a 1.5-ml tube.

Step 2.   Two to 5 µg of plasmid DNA and 50 µg of sheared, deproteinized, salmon sperm DNA are added to each sample and the tube is incubated at 30°C for 30 minutes. Then 0.5 ml of 40% polyethylene glycol 4000 in 0.1 M lithium acetate is added, the transformation mixture is mixed well, and the tube is incubated at 30°C for 1 hour.

Step 3.   The cells are then heat-shocked by incubation at 42°C for 8 minutes. Next, 0.5 ml of sterile water is added, and, after mixing, the cells are collected by centrifugation in a microcentrifuge for 1 minute. The cells are washed twice in 0.5 ml of sterile water, resuspended in 0.2 ml of sterile water, and spread on selective plates. Colonies usually appear in 2–3 days at 30°C.

3.   MUTAGENESIS PROTOCOLS

Step 1.   Wild-type yeast are grown to stationary phase in 50 ml of medium selective for maintenance of the recombinant assay plasmid. Selective medium consists of minimal medium plus nutrient additions to supplement the auxotrophies present in the specific yeast strain employed. For example, we use the yeast strain W3031 (*MATa, ade2*-1, *ura3*-1, *his3*-11, *trp1*-11, *leu2*-3, 112, *can1*-100) containing the recombinant assay plasmid bearing a *URA3* gene; therefore, the selective medium we use is minimal medium supplemented with 0.0025% adenine, 0.005% histidine, 0.005% tryptophan, and 0.005% leucine but lacking uracil. All operations are conducted under sterile conditions.

Step 2.   An aliquot of the stationary culture is diluted and the exact cell density determined by counting in a counting chamber. The culture is split into four 10-ml aliquots, pelleted, and rinsed in 10 ml of water. Three of the aliquots are incubated with 4% EMS in 10 ml of 0.05 M potassium phosphate buffer (pH 7.0) for time points varying from 30 to 60 minutes at 30°C. The remaining tube is the control lacking EMS.

Step 3.   The EMS-treated cells and the control cells are collected by centrifugation, resuspended in 10 ml of 10% sodium thiosulfate, and incubated at room temperature for 10 minutes to inactivate the residual EMS. The cells are then pelleted, resuspended in selective medium, and grown at room temperature for 4 hours with agitation.

Step 4.   To determine the amount of killing, an aliquot of each culture (including the no-EMS control) is diluted to 10^3 cells/ml and 0.1 ml is spread on YPD plates (about 100 colonies per plate). This is done in duplicate. The plates are incubated at 30°C for 2 days until colonies appear. The experimental plates from cultures treated with EMS are then compared to the controls to obtain the killing estimate. The EMS incubation period that results in 40–50% killing is selected for the genetic screen described below.

B.   Primary Genetic Screen

1.   MEDIA

Galactose X-gal (gal X-gal) indicator plates.   We have used two recipes, both given below. While the first is more time-consuming to prepare, it is more sensitive in our hands. This recipe is adapted from the laboratory manual of the Cold Spring Harbor Yeast Genetics Course (Rose *et al.,* 1988).

Recipe 1

Make the following solutions:

Solution A

10 × Phosphate butter (recipe below)	100 ml
1000 × Mineral stock (recipe below)	1 ml
Distilled water	324 ml

10 × Phosphate buffer

1 M KH$_2$PO$_4$	136.1 g
0.15 M (NH$_4$)$_2$SO$_4$	19.8 g
0.75 M KOH	42.1 g
Distilled water	1000 ml

Adjust to pH 7.0 with 1 M KOH and autoclave.

1000 × Mineral stock

2 mM FeCl$_3$	32 mg
0.8M MgSO$_4$·7H$_2$O	19.72 mg
Distilled water	100 ml

Autoclave. After autoclaving, a yellow-orange precipitate will appear which should be resuspended before use.

Solution B (in a 2-liter flask)
Bacto-agar	20 g
Distilled water	500 ml

Autoclave each solution separately and let cool to 60°C. Add the following to solution A: 40 ml of 50% galactose (autoclaved), 5 ml of amino acid supplement mix (1% each necessary amino acid), 10 ml of 100 × vitamin stock (recipe below), 2 ml of 20 mg/ml 5-bromo-4-chloro-3-indolyl-β-D-galactopyranoside (X-gal; Boerhringer Mannheim, cat. no. 100-08; dissolved in dimethyl formamide). Pour solution A into solution B with gentle mixing, then pour 30 ml per plate.

100 × Vitamin stock
0.04 mg/ml thiamine	4 mg
2 μg/ml biotin	0.2 mg
0.04 mg/ml pyridoxine	4 mg
0.2 mg/ml inositol	20 mg
0.04 mg/ml pantothenic acid	4 mg
Distilled water	100 ml

Filter sterilize.

Recipe 2
0.67% Bacto-yeast nitrogen base without amino acids	6.7 g
2% Galactose	20 g
2% Bacto-agar	20 g
Distilled water	1000 ml

Autoclave and let cool to 60°C. Add nutrient supplements, 1 M potassium phosphate (pH 7) to 70 mM final concentration, and 20 mg/ml X-gal in dimethyl formamide to 40 μg/ml final concentration. Pour approximately 30 ml per plate.

2.   GENETIC SCREENING PROTOCOL

Step 1.   The EMS-treated culture (0.1 ml) resulting in 50% killing is inoculated into 10 ml of selective medium (as described in Section III,A,2) and grown overnight at 24°C to allow recovery of the cells.

Step 2.   After determining the cell density by counting in a counting chamber, the cells are diluted to about 3000 cells/ml and 0.1-ml portions are spread on each of 30 selective plates. This should give about 10,000 colonies to screen. The plates are then incubated at the permissive temperature (24°C) for 2 to 3 days or until colonies appear.

Step 3.   The colonies are tested for conditional growth by transfer to three YPD plates by replica plating. To do this, an impression of the master plate is made by pressing the plate onto a velveteen pad and then pressing the velveteen onto three YPD plates in succession. The YPD plates are immediately incubated at 15, 37 and 24°C.

Step 4.   At the same time, the colonies are tested for conditional induction of β-galactosidase by transfer to three gal X-gal indicator plates. Galactose is used as the sole energy source in the medium and therefore serves to induce *lacZ* expression from *GAL1* promoter in the MAR assay plasmid. The gal X-gal plates are incubated at 24°C for 2 days to establish colonies and then two plates are shifted to 15 and 37°C to assess *cs*- and *ts*-sensitive β-galactosidase induction, respectively. An 8- to 16-hour incubation at 37°C is usually sufficient to detect *ts* β-galactosidase induction; however, incubation at 15°C requires several days to detect *cs* induction.

Step 5.   The plates are then compared and those colonies that show both a growth defect at high or low temperature along with β-galactosidase induction are selected for further analysis. In a typical experiment, we find that 3% of the mutagenized cells are *ts* lethal, with 15% of these also exhibiting *ts* induction of β-galactosidase. The mutant candidates are streaked out to isolate single colonies and three colonies of each mutant are retested for *ts* or *cs* growth and β-galactosidase induction by replica plating. In our experiments, only about 30% of the candidates proved to be conditional for growth and β-galactosidase induction upon further testing.

Figure 3 shows a comparison of duplicate gal X-gal plates, one incubated at 24°C (Fig. 3A) while the other was shifted to 37°C (Fig. 3B). Several phenotypes are apparent. Most of the colonies remained white at the higher temperature. One colony (colony a) turned blue when the indicator plates were incubated at 37°C while others were constitutively blue (colony b is an example).

3.   DESIGN OF THE RECOMBINANT ASSAY PLASMID

Several aspects of the recombinant assay plasmid shown in Fig. 1 should be discussed. Since the screen is based on blocking transcription, a variety of different promoter elements could be used. It is important that a relatively strong promoter be employed to facilitate detection of β-galactosidase at the nonpermissive temperature. We have also used a similar construct containing the *CYC1* promoter fused to *lacZ* (Guarente and Mason, 1983). This plasmid gives an overall lower level of β-galctosidase transcription than the *GAL1* promoter, but in theory may be useful in eliminating mutants affecting the galactose induction machinery rather than the MAR sequence specifically.

There are limits to the length of the DNA sequence to be inserted into the assay plasmid. We have found that sequences longer than 300 bp inhibit transcription from the particular plasmid we employed simply due to the

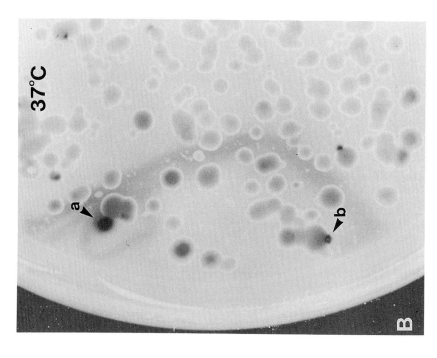

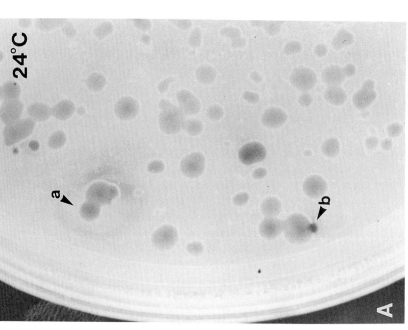

FIG. 3. Gal X-gal plate assay of mutagenized yeast cells. A master plate was replicated to two indicator plates as described in the text. A shows a plate after incubation at 24°C, while B shows the duplicate plate after incubation at 37°C. Colony a shows temperature-sensitive induction of β-galactosidase, while colony b is one of several smaller colonies that shows β-galactosidase activity at both temperatures.

increased distance between the UAS and the TATA box. In addition, an important control in this experiment, if available, is a mutation in the DNA sequence of interest that disrupts its specific protein-binding site. If the β-galactosidase inhibition seen with the wild-type sequence is the result of specific protein interactions, then the mutated sequence will not inhibit transcription when inserted in place of the wild-type sequence.

C. Further Genetic Characterization of Mutants

1. MUTANT PURIFICATION

Treatment of cells with a strong dose of mutagen (as described in Section III,A,2) can result in primary mutants that also carry secondary mutations that might interfere with subsequent characterization. Therefore, it is necessary to segregate the desired mutation away from other contaminating mutations. This is accomplished by a series of genetic crosses. The first cross is to a wild-type strain containing different complementable auxotrophies (an outcross), followed by two or three crosses of the progeny of each cross to the original parent strain (backcrosses). For example, when we mutagenized strain W3031a (*MATa*, ade2-1, ura3-1, his3-11, leu2-3, 112, can1-100), we chose F809 (*MATα*, ade6, ura3-52, his4-519) for the outcross.

The genetic crosses are conducted as follows. Two strains of opposite mating type are mixed together on YPD solid medium by transferring a colony of each to a YPD plate and smearing them together with a wooden applicator. The plate is then incubated at room temperature from 4–5 hours to overnight. Zygotes should be visible under the phase microscope after 4–5 hours of mating. Diploid cells are selected by complementation of the auxotrophies of the parental strains by streaking the mating mixture on selective plates. In our case, diploids of the W3031a and F809 cross will grow on medium lacking histidine and adenine due to complementation of the respective markers. The cells are induced to form spores by starvation as follows. Diploids are grown to an A_{600} of 0.3 in YPD. The cells are then washed three times in sterile water and resuspended in 10 ml of 0.3% potassium acetate so that the A_{600} is 0.05. Asci are visible after 2 to 3 days of incubation at room temperature with agitation. The pelleted, sporulated culture is resuspended in a small volume of sterile water (about 0.3 ml). Spore progeny of the outcross and the first backcross are isolated by a "random spore" protocol as follows. A small aliquot (40 μl) of the sporulated culture is incubated with 1 μl of glusulase (Dupont, cat. no. NEE-154) for 1 hour at room temperature to remove the ascus wall. The culture is diluted into 5 ml of sterile water and the spores dispersed by brief sonication. The spores are then serially diluted 10^{-2} and 10^{-3} in sterile water and 0.1 ml of each dilution is plated on YPD. The plates are incubated at the permissive temperature (24°C) until colonies appear. The colonies are replica-plated to duplicate YPD plates followed by incubation of one plate at the permissive temperature and one at the non-permissive temperature. Mutants are identified by comparison of the two plates. The genotype of each colony arising from a single spore is determined by testing its growth on a series of plates, each lacking just one nutrient requirement of the parental strain. We select an **a** and α of each mutant that have complementing auxotrophies (*his3* and *his4*, for example). The haploid progeny of one of the crosses should be isolated by micromanipulation to ascertain that the mutation segregates 2 : 2, which is indicative of a single nuclear mutation. Furthermore, the *ts* growth phenotype and β-galactosidase induction should still cosegregate.

2.   ASSIGNMENT OF COMPLEMENTATION GROUPS

To determine if any of the mutants are allelic, it is necessary to place them into complementation groups. This is made easier if **a** and α cells of complementing auxotrophies (such as *his3* and *his4*) are isolated during the mutant segregations described above. Mutants that are *MATa* are streaked in parallel lines across the surface of a YPD plate using a sterile tootpick. *MATα* cells are streaked in a similar pattern on a separate YPD plate. The plates are incubated at 24°C for 2 to 3 days. An impression of one of the plates is made by pressing it onto a velveteen pad. The second plate is then pressed onto the same velveteen such that the stripes are perpendicular to one another. The checkerboard pattern is then transferred to a new YPD plate by pressing the velveteen to the plate. The plate is then incubated at 24°C for 1 day. Mating will occur at the intersection of the lines. Diploids are then selected by replica-plating the grid pattern to a plate containing all nutrient additives except histidine. Complementation of the *ts* phenotype in the diploids is determined by replicating the diploids to two YPD plates and incubating one at 37°C and one at 24°C; for *cs* phenotypes, incubate one at 15°C and one at 24°C.

It is often useful to determine if the mutants isolated in this screen interact genetically with already characterized mutants. This is made easier if the mutant shares a phenotype with a group of already characterized mutants. For example, several of the mutants in our collection display a cell division cycle (*cdc*) phenotype (Hartwell *et al.*, 1970). Therefore, we mated these mutants with several *cdc* mutants and tested for complementation (as described above).

D.   Secondary Screens with Other Assay Plasmids

The genetic screen described here is by nature very general. Therefore, it is absolutely necessary to conduct further genetic screens and biochemical tests to characterize the mutant collection further. Detailed information about the function of the DNA sequence under investigation will facilitate the design of secondary screens. Because the presumed function of MAR sequences is in nuclear organization, it is difficult to predict the phenotypes of mutations in the proteins that recognize these sequences. Mutations that alter nuclear structure might be expected to have pleiotropic effects. Therefore, as one approach, we recommend transforming the mutants with constructs containing other protein-binding sites inserted into the *GAL1* promoter under the rationale that β-galactosidase transcription from these plasmids might not be affected by mutations in genes specifying proteins that specifically interact with the MAR.

We have utilized two alternative plasmids for our studies. One contains the centromere from chromosome 3 (CEN3) in place of the MAR (Ng *et al.,* 1986). Because CEN sequences are matrix associated (Amati and Gasser, 1988), we anticipated that a subset of mutants that induce β-galactosidase at the nonpermissive temperature from the MAR plasmid would also induce from the CEN3 alternative plasmid. The second construct contains the α2-repressor binding site from the yeast *STE6* gene (Johnson and Herskowitz, 1985) and therefore represents a different type of protein recognition site. Each mutant is transformed with the test constructs using a standard lithium acetate protocol (Section III,A,3). The transformants are then tested for β-galactosidase induction at the nonpermissive temperature on X-gal indicator plates as described above (Section III,B,2). We have found that the mutants give different patterns of β-galactosidase induction from the different plasmids and thus we were able to subdivide our mutant collection further.

E. Additional Characterization Techniques

1. GENETIC

The most straightforward way to characterize a mutant is to isolate the gene responsible for the mutant phenotype. This is easily done using the conditional mutants obtained in Section III,B,2. The yeast mutant is transformed with a yeast genomic CEN library (Rose and Broach, 1991) followed by growth of the transformants at the nonpermissive temperature. Colonies that grow at the nonpermissive temperature have received a plasmid containing a gene that complements the mutation. The complementing plasmid can then be isolated from the yeast and further characterized by physical mapping and sequence analysis. This technique usually yields the gene allelic to that containing the original mutation; however, second-site suppression can occur, especially if the genomic library is cloned in a high-copy-number vector. It is therefore necessary to show, by site-directed integration followed by tetrad analysis, that the cloned gene is allelic to the original mutation. Once the gene is cloned, other alleles of the original mutation can be obtained using the plasmid shuffling technique (Budd and Campbell, 1987).

A standard approach to identify other loci that interact genetically with the mutant is to isolate second-site suppressors of the original mutation. This is done by asking for pseudoreversion of the mutation (either spontaneously, by plating large numbers of cells at the nonpermissive temperature, or by treatment with a mutagen) and ascertaining by genetic linkage studies that the suppressor is extragenic to the original mutation and not simply a revertant. The suppressor is more easily characterized if it has a mutant phenotype of its own when genetically isolated from the original mutant.

2.   Cell Biological and Biochemical

A variety of techniques can be used to assess the integrity of the nucleus in the mutants. These include electron microscopy and fluorescent microscopy. For example, the nucleus can be labeled with a fluorescent tag by targeting β-galactosidase to the nucleus with an H2B nuclear targeting signal (Moreland *et al.,* 1987) and using an anti-β-galactosidase antibody in conjunction with indirect immunofluorescence. It is also possible to examine alterations in structural aspects of the nucleus in the mutants using, for example, antitubulin antibodies to label the mitotic spindle or antinucleoporin antibodies to label the nuclear pores (Davis and Fink, 1990). We have also used matrix binding assays (Cockerill and Garrard, 1986a,b) to investigate alterations in DNA binding of matrices isolated from yeast mutants incubated at the nonpermissive temperature. The nuclear halo assay can also be employed (Amati and Gasser, 1988).

IV.   Further Applications of the Technique

The method described in this chapter can be adapted to investigate the interaction between any DNA-binding protein and its recognition sequence, provided the protein is found in or can be expressed in yeast cells. For example, the sequence requirements of binding can be investigated by saturation mutagenesis of the recognition site followed by transformation of yeast with the pool of mutagenized plasmid and assay for disruption of *in vivo* association with the trans-acting factor. Clones that show derepression of *lacZ* expression can be detected on X-gal indicator plates, and the plasmid

DNA isolated and sequenced to determine the nature of the mutation. Conversely, the DNA-binding domain of a protein can be mapped by deletion of portions of the protein followed by assay of its ability to bind DNA *in vivo* as measured by β-galactosidase activity.

The technique can also be used to clone DNA-binding proteins from mammalian cells. This can be done if yeast does not contain proteins which recognize the sequence of interest. Yeast containing an assay plasmid with the relevant sequence inserted into the *GAL1* promoter test site are transformed with the mammalian cDNA expression library and β-galactosidase activity of the transformants assayed on X-gal indicator plates. Cells that have received an appropriate DNA-binding activity will appear white on the indicator plates due to formation of a DNA–protein complex that blocks transcription.

Acknowledgments

This work was supported in part by grants from the National Institutes of Health (GM22201, GM29935, and GM31689) and the Robert A. Welch Foundation (I-823) to WTG, and by a National Institutes of Health Postdoctoral Fellowship Award (GM12547) to AOS. The authors would like to thank Patrick Pfaffle, Lita Freeman, and Myeong Sok-Lee for critical review of the manuscript and Mong-Duyen Tran for excellent technical assistance.

References

Adachi, Y., Kas, E., and Laemmli, U. K. (1989). Preferential cooperative binding of DNA topoisomerase II to scaffold-associated regions. *EMBO J.* **8,** 3997–4006.

Amati, B. V., and Gasser, S. M. (1988). Chromosomal ARS and CEN elements bind specifically to the yeast nuclear scaffold. *Cell (Cambridge, Mass.)* **54,** 967–978.

Benyajati C., and Worcel, A. (1976). Isolation, characterization, and structure of folded interphase genome of *Drosophila melanogaster. Cell (Cambridge, Mass.)* **9,** 393–407.

Berezney, R. (1984). Organization and functions of the nuclear matrix. *In* "Chromosomal Nonhistone Proteins-Structural Associations" (L. S. Hnilica, ed.), Vol. 4, pp. 119–180. CRC Press, Boca Raton, Florida.

Berezney R., and Coffey D. (1974). Identification of a nuclear protein matrix. *Biochem. Biophys. Res. Commun.* **60,** 1410–1417.

Brent, R., and Ptashne, M. (1984). A bacterial repressor protein or a yeast transcriptional terminator can block upstream activation of a yeast gene. *Nature (London)* **312,** 612–615.

Budd, M., and Campbell, J. L. (1987). Temperature-sensitive mutations in the yeast DNA polymerase I gene. *Proc. Natl. Acad. Sci. U.S.A.* **84,** 2838–2842.

Callan, H. G. (1986). Historical introduction. *In* "Lampbrush Chromosomes" (M. Solioz, ed.), pp. 1–24. Springer-Verlag, Berlin.

Cockerill, P. N., and Garrard, W. T. (1986a). Chromosomal loop anchorage of the kappa immunoglobulin gene occurs next to the enhancer in a region containing topoisomerase II sites. *Cell (Cambridge, Mass.)* **44,** 273–282.

Cockerill, P. N., and Garrard, W. T. (1986b). Chromosomal loop anchorage sites appear to be evolutionarily conserved. *FEBS Lett.* **204,** 5–7.

Cockerill, P. N., Yuen, M.-H., and Garrard, W. T. (1987). The enhancer of the immunoglobulin heavy chain is flanked by presumptive chromosomal loop anchorage elements. *J. Biol. Chem.* **262,** 5394–5397.

Cook, P. R., and Brazell, I. A. (1976). Conformational contraints in nuclear DNA. *J. Cell Sci.* **22,** 287–302.

Davis, L. I., and Fink, G. R. (1990). The NUP1 gene encodes an essential component of the yeast nuclear pore complex. *Cell (Cambridge, Mass.)* **61,** 965–978.

Fu, X.-D., and Maniatis, T. (1990). Factor required for mammalian spliceosome assembly is localized to discrete regions in the nucleus. *Nature (London)* **343,** 437–441.

Garrard, W. T. (1990). Chromosomal loop organization in eukaryotic genomes. *In* "Nucleic Acids and Molecular Biology" (F. Eckstein and D. M. J. Lilly, eds.), Vol. 4, pp. 163–175. Springer-Verlag, Berlin.

Gasser, S. M., and Laemmli, U. K. (1986a). Cohabitation of scaffold binding regions with upstream/enhancer elements of three developmentally regulated genes of *D. melanogaster. Cell (Cambridge, Mass.)* **46,** 521–530.

Gasser, S. M., and Laemmli, U. K. (1986b). The organization of chromatin loops: Characterization of a scaffold attachment site. *EMBO J.* **5,** 511–518.

Guarente, L. (1983). Yeast promoters and lacZ fusions designed to study expression of cloned genes in yeast. *In* "Methods in Enzymology" (R. Wu, L. Grossman, and K. Moldave, eds.), Vol. 101, pp. 181–191. Academic Press, New York.

Guarente, L., and Mason T. (1983). Heme regulates transcription of the *CYC1* gene of *Saccharomyces cerevisiae* via an upstream activation site. *Cell (Cambridge, Mass.)* **32,** 1279–1286.

Hancock, R. (1982). Topological organization of interphase DNA: The nuclear matrix and other skeletal structures. *Biol. Cell* **46,** 105–122.

Hartwell, L. H., Culloti, J., and Reid, B. (1970). Genetic control of the cell division cycle in yeast I. Detection of mutants. *Proc. Natl. Acad Sci. U.S.A.* **66,** 352–359.

Hofmann, J. F.-X., Laroch, T., Brand, A. H., and Gasser, S. M. (1989). RAP-1 factor is necessary for DNA loop formation *in vitro* at the silent mating type locus HML. *Cell (Cambridge, Mass.)* **57,** 725–737.

Ito, H., Funkuda, Y., Murata, K., and Kimura, A. (1983). Transformation of intact cells treated with alkali cations. *J. Bacteriol.* **153,** 163–168.

Izaurralde, E., Mirkovitch, J., and Laemmli, U. K. (1988). Interaction of DNA with nuclear scaffolds *in vitro. J. Mol. Biol.* **200,** 111–125.

Izaurralde, E., Kas, E., and Laemmli, U. K. (1989). Highly preferential nucleation of histone H1 assembly on scaffold-associated regions. *J. Mol. Biol.* **210,** 573–585.

Jackson, D. A., and Cook, P. R. (1985). Transcription occurs at a nucleoskeleton. *EMBO J.* **4,** 919–925.

Johnson, A. J., and Herskowitz, I. (1985). A repressor (MATα2) product and its operator control expression of a set of cell type specific genes in yeast. *Cell (Cambridge, Mass.)* **42,** 237–247.

Lawrence, J. B., Singer, R. H., and Marsille, L. M. (1989). Highly localized tracks of specific transcripts within interphase nuclei visualized by *in situ* hybridization. *Cell (Cambridge, Mass.)* **57,** 493–502.

Mirkovitch, J., Mirault, M.-E., and Laemmli, U. K. (1984). Organization of the higher-order chromatin loop: Specific attachment sites on nuclear scaffold. *Cell (Cambridge, Mass.)* **39,** 223–232.

Mirkovich, J., Spierer, P., and Laemmli, U. K. (1986). Genes and loops in 320,000 base-pairs of the *Drosophila melanogaster* chromosome. *J. Mol. Biol.* **190,** 255–258.

Mirkovich, J., Gasser, S. M., and Laemmli, U. K. (1988). Scaffold attachment of DNA loops in metaphase chromosomes. *J. Mol. Biol.* **200,** 101–109.

Moreland, R. B., Langevin, G. L., Singer, R. H., Garcea, R. L., and Hereford, L. M. (1987). Amino acid sequences that determine the nuclear localization of yeast histone H2B. *Mol. Cell. Biol.* **7,** 4048–4057.

Nelson, W. G., Pienta, K. J., Barrack, E. R., and Coffey, D.S. (1986). The role of the nuclear matrix in the organization and function of DNA. *Annu. Rev. Biophys. Biophys. Chem.* **15,** 457–475.

Ng, R., Ness, J., and Carbon, J. (1986). Structural studies on centromeres in the yeast, *Saccharomyces cerevisiae. In* "Extrachromosomal Elements in Lower Eukaryotes" (R. Wickner, *et al,* eds.), pp. 479–492. Plenum, New York.

Pardoll, D. M., Vogelstein, B., and Cooffey, D. S. (1980). A fixed site of DNA replication in eucaryotic cells. *Cell (Cambridge, Mass.)* **19,** 527–536.

Paulson, J. R., and Laemmli, U. K. (1977). The structure of histone-depleted metaphase chromosomes. *Cell (Cambridge, Mass.)* **12,** 817–828.

Reed, S. I. (1980). The selection of *S. cerevisiae* mutants defective in the START event of cell division. *Genetics* **95,** 561–577.

Rose, M. D., and Broach, J. R. (1991). Cloning genes by complementation in yeast. *In* "Methods in Enzymology" (R. A. Johnson and J. D. Corbin, eds.), Vol. 194, pp. 195–230. Academic Press, San Diego, California.

Rose, M. D., Winston, F., and Heiter, P. (1988). "Laboratory Course Manual for Methods in Yeast Genetics." Cold Spring Harbor Laboratory, Cold Spring Harbor, New York.

Spector, D. L. (1990). Higher order nuclear organization: Three-dimensional distribution of small nuclear ribonucleoprotein particles. *Proc. Natl. Acad. Sci. U.S.A.* **87,** 147–151.

Sperry, A. O., Blasquez, V. C., and Garrard, W. T. (1989). Dysfunction of chromosomal loop attachment sites: Illegitimate recombination linked to matrix association regions and topoisomerase II. *Proc. Natl. Acad. Sci. U.S.A.* **86,** 5497–5501.

Vaughn, J. P., Dijkwel, P. A., Mullenders, L. H. F., and Hamlin, J. L. (1990). Replication forks are associated with the nuclear matrix. *Nucleic Acids Res.* **18,** 1965–1969.

Vogelstein, B., Pardoll, D. M., and Coffey, D. S. (1980). Supercoiled loops and eukaryotic DNA replication. *Cell (Cambridge, Mass.)* **22,** 79–85.

Chapter 21

Mutations Affecting Cell Division in Drosophila

MAURIZIO GATTI

Dipartimento di Genetica e Biologia Molecolare
Università di Roma "La Sapienza"
00185 Rome, Italy

MICHAEL L. GOLDBERG

Section of Genetics and Development
Cornell University
Ithaca, New York 14853

I. Introduction
II. Cell Division in *Drosophila*
III. Cytological Analysis of Mitosis and Meiosis
    A. Embryos
    B. Larval Neuroblasts
    C. Female Germ Cells
    D. Male Germ Cells
IV. Strategies for Isolation of Mitotic Mutants
    A. Meiotic Mutants
    B. Mutagen-Sensitive Mutants
    C. Temperature-Sensitive Lethals
    D. Methods of Analysis
    E. Late Lethals

I. Introduction

The basic features of mitotic cell division are remarkably similar among eukaryotic organisms. A complex series of exquisitely precise events is required to ensure that the chromosome complement of a dividing cell is equally partitioned between the two daughter cells. These events involve elaborate morphological transformations of both the chromosomes and the cytoplasmic organelles that comprise the mitotic apparatus. Although substantial progress has been made in the description and definition of mitotic organelles, we still know very little about their structure and interactions at the molecular level.

One approach to obtain insight into the mechanisms of mitosis is the isolation and genetic analysis of mutations affecting this process, coupled with the cloning and molecular characterization of the genes defined by these mutations. The rewards of this approach are evident from the studies investigating the regulation of cell-cycle events in *Saccharomyces cerevisiae* (Hartwell, 1974; Pringle and Hartwell, 1981; Moir *et al.,* 1982; Moir and Botstein, 1982) and more recently, in *Schizosaccharomyces pombe* (Fantes, 1984; Nurse, 1985; Hirano and Yanagida, 1988). In these systems, many genes involved in cell division have been isolated and ordered in functional pathways. Moreover, the nature of the products of some of these genes has been elucidated, permitting the construction of molecular models for the execution and regulation of certain aspects of the cell cycle.

In this chapter, we suggest that a combined genetic and molecular analysis of cell division can easily be carried out in *Drosophila melanogaster*. As relatively high-resolution cytological observations can be made on a variety of mitotically and meiotically active *Drosophila* tissues, this approach has the potential to yield considerable novel information. We first review available strategies for the isolation and characterization of *Drosophila* mitotic mutants. We then describe the phenotypes of the mitotic mutants thus far isolated, showing that lesions in many different aspects of cell division can be detected. Finally, we outline a rapid and efficient protocol for the isolation of mitotic mutants that facilitates subsequent molecular cloning of the genes of interest.

II. Cell Division in *Drosophila*

Knowledge of when and where cell division occurs in *Drosophila* is necessary to understand the strategies used for the isolation of mitotic mutants. After fertilization, the first 13 rounds of embryonic mitoses take place very rapidly in a syncytial manner. Each of these cell cycles consists of an S and an M phase, together lasting about 10 minutes; there are no detectable gap periods. These divisions occur in the almost complete absence of transcription of the zygotic genome. Extensive transcription is first detected during cycle 14, when an extended interphase first occurs and the nuclei become cellularized (Fig. 1; Foe and Alberts, 1983). After cycle 14, there are 3–4 additional rounds of replication which occur asynchronously in characteristic mitotic domains (Foe, 1989). At the completion of these latter cycles, all of the cells needed to form a larva have been generated. Subsequent larval growth occurs by an increase in cell size, which may be accompanied by polyploidization or polytenization in several tissues.

The only mitotically active tissues in larvae are the imaginal disks, the abdominal histoblasts, and the neuroblasts (Fig. 2). The imaginal disks are dispensable for larval viability, since individuals devoid of these tissues survive till the larval/pupal transition (Shearn *et al.,* 1971)—the stage during which these cell populations differentiate to produce adult structures. The central nervous system, which contains different types of mitotically active neuroblasts, is one of the few tissues common to the larva and to the adult. Mitosis in these neuroblasts can be easily examined, providing an important tool for the

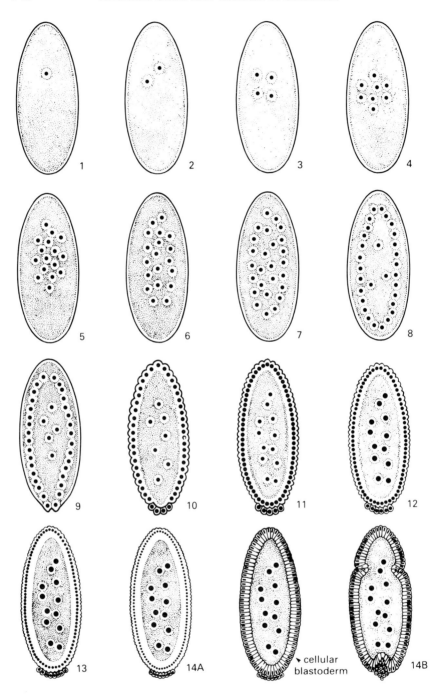

cellular blastoderm

identification and characterization of mitotic mutants (see below). Finally, in adult flies, cell division takes place almost exclusively in the germline, where both male and female gonial cells undergo several rounds of mitotic division before entering meiosis.

III.   Cytological Analysis of Mitosis and Meiosis

A large repertoire of cytological procedures is currently available for the analysis of mitotic cell division in *Drosophila*. Techniques applicable to a variety of tissue types are discussed below; detailed descriptions of specialized procedures mentioned only briefly here can be found in the recently published Ashburner manual (1989). We also summarize methodology for the investigation of meiotic cell divisions, as mitotic mutants may also disrupt meiotic processes (see below).

A.   Embryos

Because embryonic nuclear divisions prior to cell cycle 14 occur nearly synchronously, examination of a single embryo allows comparison of many nuclei in the same phase of the cell cycle. The mitotic domains characteristic of later rounds of replication allow visualization of defined groups of cells at different stages of the cell cycle, though analysis is more complex because of the greater number of cells. Embryonic cell division can be analyzed either in fixed material or *in vivo*. Any of several good fixation techniques for

---

FIG. 1.   Early embryonic stages of *Drosophila melanogaster*. Numbers indicate division cycles, which are pictured at the start of interphase. All nuclei (black circles) are shown for stages 1–5; only a subset is shown thereafter. During cycles 1–7, nuclei divide exponentially in the central region of the egg. During cycles 8 and 9, nuclei begin migrating to the egg periphery, leaving behind future yolk nuclei in the central region of the egg. At stage 10, migrating nuclei have reached the embryonic periphery. In addition, pole cells (germline progenitors) are pinched off at the posterior end; further division of the pole cells is asynchronous with respect to the syncytial nuclei. During mitotic cycles 10–13, syncytial nuclei divide almost synchronously. Entrance into each of these cell cycles occurs as a wavefront ("mitotic wave") through the nuclei, with those at the anterior and posterior poles first starting division and nuclei near the equator being the last to enter mitosis. At the start of cycle 14, plasma membranes begin to separate the nuclei (stage 14A), forming a cellular blastoderm. Gastrulation now begins (stage 14B); about 15 min later, mitosis of cycle 14 starts asynchronously and in a region-specific manner. Reproduced, with permission, from Foe and Alberts (1983).

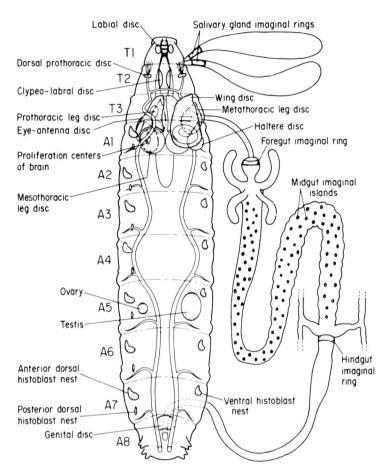

FIG. 2. A schematic diagram of a *Drosophila* third instar larva with particular emphasis on the imaginal primordia (i.e., imaginal discs and histoblast nests). Some of these structures are displayed on only one side of the embryo, but all are paired in the animal with the exception of the genital disc and the gut primordia. T1–T3 and A1–A8 designate thoracic and abdominal segments, respectively. The labial discs in T1 are attached to dark, chitinous mouth parts that extend into T2. The two brain hemispheres ("proliferation centers of the brain") are partially obscured by various imaginal discs in A1, and are connected to the ventral ganglion extending down the midline of the larva from A1 to A2. [In actuality, the brain hemispheres are found in T3, and the ventral ganglion extends to the end of A1 (Bodenstein, 1950).] Reproduced from Bryant and Levinson (1985) with permission.

Drosophila embryos yield preparations in which the spindle can be visualized by indirect immunofluorescence with anti-tubulin antibodies, while the chromosomes can be followed by DAPI (or Hoechst 33258) staining (see [10], by Rykowski, this volume; see also Karr and Alberts, 1986 for a specific example). Antibody reagents directed against centrosome-associated *Drosophila* antigens are also currently available (Frasch *et al.*, 1986; Whitfield *et al.*, 1988; Kellogg *et al.*, 1989). In order to understand the three-dimensional relationship of mitotic organelles better, several laboratories are routinely employing confocal computer-interfaced light microscopy for optical sectioning of embryonic preparations (Miller *et al.*, 1989; Whitfield *et al.*, 1990, Gonzalez et al., 1990).

Remarkably, these mitotic structures can also be visualized in living embryos, where their morphological transformations can be followed for several cell cycles. This is achieved by injecting syncytial embryos with either rhodamine-labeled histones or rhodamine-labeled tubulin. These molecules are quickly incorporated into chromatin or into microtubules, fluorescently marking these structures upon excitation with the proper wavelength (Kellogg *et al.*, 1988; Minden *et al.*, 1989). These *in vivo* techniques have been developed only recently, and in only one case to date have they been used to define the primary defect of a mitotic mutation (Sullivan *et al.*, 1991). However, we believe that examination of living embryos will soon gain wide application, because it offers the possibility of following aberrant phenomena in real time and through several cell cycles.

B. Larval Neuroblasts

The tissue providing the best material for analyzing mitotic divisions by light microscopy is the larval brain. To characterize fully various aspects of mitosis, larval brain squashes should be prepared by three different experimental regimes. First, dissected brains can be immediately squashed and stained with acetoorcein. This procedure (protocol I, presented in detail below) allows the observation of all phases of mitosis, and permits evaluation of the mitotic index and the frequency of anaphase cells (Gatti and Baker, 1989). However, chromosomal morphology is only poorly defined in these squashes. In the second regime (protocol II), pretreatment of brains with hypotonic solution prior to squashing and acetoorcein staining improves the spreading of metaphase chromosomes and separates sister chromatids, allowing an evaluation of chromosome condensation and integrity. However, it should be noted that clear anaphase figures are absent in hypotonic-treated preparations because hypotonic shock disrupts these mitotic figures (Brinkley *et al.*, 1980). Finally, a third method (*in vitro* incubation of brains with colchicine, followed by treatment with hypotonic solution; protocol III below) provides a large number of well-spread metaphase figures that can be analyzed for defects in chromosome morphology or number, and for the presence of chromosome aberrations.

Although the squashing procedures outlined above leave the morphology of most components of the mitotic apparatus intact, the three-dimensional organization of cells undergoing mitosis is obviously disrupted. These spatial arrangements in brain neuroblasts can be determined by examination of serial sections prepared for electron microscopy [A. T. C. Carpenter (University of California, San Diego), personal communication]. Recently, a whole-mount fixation technique followed by immunological staining has been developed for the analysis of cell division in larval brains by confocal microscopy. Using this procedure, Gonzalez and co-workers were able to visualize the spatial arrangement of the chromosomes, the spindle, and the centrosomes in dividing neuroblasts of wild-type brains as well as neuroblasts homozygous for the mitotic mutation *abnormal spindle* (Gonzalez et al., 1990).

1. Squash Techniques for Mitotic Chromosome Preparations

Because of their wide applicability, we outline here the squashing procedures routinely used in our laboratories (Gatti et al., 1974; Gatti and Baker, 1989), but note that alternative techniques have been developed by other laboratories (see Ashburner, 1989). For all the regimens we employ, brains of third instar larvae are first dissected in saline (0.7% NaCl in distilled water) as described in the following section. For protocol I, the brains are then immediately fixed at room temperature in a freshly prepared mixture of acetic acid/methanol/distilled water in the ratio 11:11:2 until they become transparent (usually about 20 seconds). In protocol II, brains are treated for 10 minutes at room temperature with hypotonic solution (0.5 g of sodium citrate·2H$_2$O added to 100 ml of distilled water) prior to fixation. For protocol III, brains are first incubated for 1.5 hours at 25°C in a covered, small plastic petri plate (35 × 10 mm) containing ∼2 ml of saline supplemented with one drop (about 100 μl) of 10^{-3} M colchicine. Brains are then transferred to hypotonic solution for 10 minutes at room temperature, and finally fixed as above.

After fixation, between one and four brains are transferred to a corresponding number of small drops (1–2 μl) of acetoorcein placed on a very clean, 20 × 20 or 22 × 22 mm coverslip. [Acetoorcein is prepared by boiling commercial orcein powder (Gurr) in 45% acetic acid for 30–45 minutes in a reflux condenser; we usually make 5% orcein, which is subsequently diluted to 2% with 45% acetic acid. Residual particulate matter is then removed by filtration through blotting paper.] After the brains are transferred to the acetoorcein drops, a very clean slide is lowered onto the coverslip, inverted, and squashed very hard under 3–4 sheets of blotting paper. The coverslip is then sealed with depilatory wax, and may be stored for 1–2 months at 4°C.

Certain variations of these basic protocols expand their utility. For example, such chromosome preparations can be successfully stained by a variety of banding techniques providing high-resolution longitudinal differentiation of heterochromatin. Slides containing fixed brains, squashed under a siliconized coverslip in drops of 45% acetic acid without the orcein, are frozen in liquid nitrogen or on dry ice. After removal of the coverslip with a razorblade and air drying, the slides can be immediately stained with Giemsa, Hoechst 33258, quinacrine, or using an N-banding technique (Gatti et al., 1976; Pimpinelli et al., 1976; Gatti and Pimpinelli, 1983).

2. DISSECTION OF LARVAL BRAINS

An actively crawling third instar larva is transferred to an ~ 50-μl drop of saline placed on a siliconized slide. Using two forceps (preferably Dumond number 5 Biologie), the larva is grabbed just posterior to the black mouth parts in the head (near the T1–T2 boundary in Fig. 2), and at a position approximately two-thirds of the body length posterior to the head (circa A5 in Fig. 2). The larval body is then pulled apart. The translucent brain is usually attached to the head portion and is recognized by its characteristic morphology, consisting of two brain hemispheres connected at their point of fusion with a long ventral ganglion (Figs. 2 and 3A). At this point, the brain is often associated with one or more imaginal disks, which are usually about one-third the volume of a brain hemisphere. The brain is removed with the forceps from the remainder of the head and transferred into a fresh drop of saline to be washed, and is then immediately fixed or otherwise treated as described above. A small amount of residual matter such as imaginal disks will not interfere with squashing, but care must be taken to remove completely the more rigid mouth parts.

C. Female Germ Cells

Female gonial mitoses can be analyzed in squash preparations made with the same acetoorcein staining procedures described above for larval brains (M. Gatti and co-workers, unpublished results). In particular, we have found that oogonia from early pupae constitute a rich source of good mitotic figures. It is worth noting that beautiful camera-lucida drawings of gonial metaphases were published by Bridges (1916) in his classic study on chromosomal nondisjunction. However, despite the good quality of oogonial preparations, this tissue has not been routinely used for the analysis of cell division in mitotic mutants.

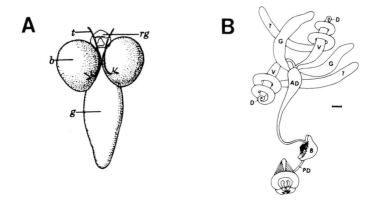

FIG. 3.   (A) The larval brain, showing the two brain hemispheres (b) and the ventral ganglion (g), reproduced from Bodenstein (1950). Similarity of shape with that of certain aspects of human male anatomy has often been mentioned as a guide for the recognition of this organ. rg, ring gland; t, trachea. (B) The adult male reproductive tract [reproduced from Lindsley and Tokuyasu (1980)]. In the adult fly, the large coiled testes (T) are easily found, and are connected to the testicular ducts which lead to the seminal vesicles and other structures derived from the genital disc. (D, testicular duct; V, seminal vesicle; G, accessory gland; AD, anterior ejaculatory duct; B, ejaculatory bulb; PD, posterior ejaculatory duct.)

Cytological studies of meiotic chromosome behavior in females have historically been difficult to perform. However, thanks to the efforts of Davring, Puro, and their co-workers, procedures that permit visualization of female meiotic chromosomes in the light microscope have recently been developed (for complete protocols, see Ashburner, 1989). In addition, the exacting procedure of assembling serial sections viewed by electron micro-copy has been employed by Carpenter to examine the structure and behavior of the synaptonemal complex and recombination nodules in wild-type females and in females homozygous for the recombination-defective mutants *mei-9*, *mei-41*, and *mei-218* (Carpenter, 1975, 1979a,b). These elegant studies showed that mutations at the *mei-9* and *mei-218* loci exhibit clear lesions in recom-bination nodules, suggesting that these structures mediate several aspects of meiotic recombination.

D.   Male Germ Cells

Acetoorcein squash preparations made in a manner identical to that presented above for larval neuroblasts and female gonial cells have also proved to be useful for the analysis of male meiotic chromosomes. Both the reductional and the equational division in male spermatocytes can be readily followed. The same squashing procedure also provides good preparations of spermatogonial cells to examine premeiotic gonial mitoses (M. Gatti and co-workers, unpublished results; see also Ashburner, 1989).

Male meiosis can also be analyzed in squashed preparations of living testes viewed by phase-contrast microscopy. This technique is suitable for the visualization of meiotic spindles in primary and secondary spermatocytes, and of postmeiotic onion stage spermatids. Onion stage spermatids observed in phase contrast present a very regular appearance: each cell contains a round, light-colored nucleus paired with a round, dark mitochondrial derivative (nebenkern) of approximately the same size. This peculiar cellular morphology depends on the correct segregation of chromosomes and the regular, equal partitioning of mitochondria during meiosis. Meiotic failures may result in spermatid nuclei and nebenkern of irregular size (for review, see Fuller, 1986).

Fixed testis preparations stained with Hoechst 33258 and anti-tubulin antibodies provide a third way of examining male meiosis. Because spermatocyte nuclei of *D. melanogaster* have spindles much larger than somatic diploid (i.e., brain or imaginal disc) nuclei, this system is particularly amenable to detailed observations of the meiotic spindle by immunofluorescence. There are currently two effective protocols for fixation and immunostaining of *Drosophila* testes. The first, recently developed by Casal *et al.* (1990a,b), employs formaldehyde and pretreatment with taxol. The second procedure, described in detail below, was originally developed for the analysis of Y chromosome loops. This protocol involves fixation with methanol/acetone, allowing excellent preservation of the cytoskeleton and spindle even in the absence of taxol (Pisano and Gatti, in Ashburner, 1989). Both of these techniques allow the simultaneous visualization of the chromosomes and spindle fibers, and, in several cases, have permitted the identification of the primary defect present in mutants with abnormal meiotic divisions (see below). It should be noted here that the most commonly used immunological reagents for centrosomes [the monoclonal antibody Bx63 and the polyclonal Rb188, both of which recognize the same centrosome-associated antigen (Frasch *et al.*, 1986; Whitfield *et al.*, 1988)] do not mark the centrosomes participating in male meiosis. Although these antibodies immunostain the centrosomes of embryonic cells, larval neuroblasts, and spermatogonia, the corresponding antigen is scattered in the cytoplasm of primary and secondary spermatocytes (Casal *et al.*, 1990a).

Finally, it should be noted that reconstruction of the three-dimensional structure of male meiosis has been achieved by detailed examination of serial sections with the electron microscope (Church and Lin, 1982, 1985). However, this procedure has not yet been used to define the defects present in mutants affecting meiotic cell division.

1. A FIXATION TECHNIQUE ALLOWING IMMUNOSTAINING OF MALE MEIOSIS

We present here details of the fixation procedure mentioned above, developed by Pisano and Gatti. Larval, pupal, or adult testes are dissected in testis buffer as described below, and transferred to a small drop of the same buffer (2 μl) on a coverslip. A clean slide is then gently placed over the coverslip without pressing, and the sandwich is inverted. Very mild squashing is obtained by removing excessive buffer from the edges of the coverslip with a piece of blotting paper. Slides are then frozen in liquid nitrogen. After removal of the coverslip with a razorblade, the slides are fixed by immersion for 5 minutes in methanol precooled to $-20°C$, followed by 1 minute immersion in acetone precooled to the same temperature. Slides are then incubated for 10 minutes in PBS containing 0.5% acetic acid and 1% Triton X-100. After three subsequent 10-minute washes in PBS (Dulbecco's formula modified: 1 mM CaCl$_2$, 2.6 mM KCl, 1.5 mM KH$_2$PO$_4$, 0.5 mM MgCl$_2$, 137 mM NaCl, 8.1 mM Na$_2$HPO$_4$), slides can be incubated with the desired primary antibody for indirect immunofluorescence.

2. DISSECTION OF LARVAL AND ADULT TESTES

Male larvae are recognized because of the presence of the testes, which can be seen from the outside of the larva as two transparent spots located laterally at two-thirds of the body length (Fig. 2). The testes are much larger than the presumptive ovaries located in the corresponding position of the female. A male third instar larva is transferred to an $\sim 50\mu$l drop of testis buffer (0.183 M KCl, 47 mM NaCl, 10 mM Tris-HCl, pH 6.8, 1 mM DMSF, 1mM EDTA) on a siliconized slide. The larva is grabbed with forceps at a position at one-half body length (around the A3–A4 border in Fig. 2) and at the tail (A7–A8). The larva is now pulled apart, and the testes are isolated. These appear as transparent, disk-like bodies surrounded by a ring of fatty material (Fig. 2). As much of the fatty material as possible should be removed, and the cleaned testes are now transferred to a 2-μl drop of testis buffer placed on a coverslip for squashing as described above.

Dissection of adult testes is simplified because these organs are quite large and occupy a considerable portion of the male abdominal cavity. Etherized, 1- to 2-day-old males are transferred to a drop of testis buffer, and the abdomens are opened with forceps. Testes appear as coiled bodies (Fig. 3B) with a pale yellow pigmentation. The testes are then transferred to a 2-μl drop of testis buffer for squashing; it is often helpful to cut through the testes with the forceps to favor the spreading of the various component cell types.

IV. Strategies for Isolation of Mitotic Mutants

Mutants affecting cell division do not have unambiguous external characteristics which permit them to be distinguished from nonmutant individuals or from strains carrying lesions in other important cellular processes. Consequently, strategies for effective mitotic mutant isolation demand selection of mutant classes potentially enriched in mitotic defects by other criteria; collections of such mutants must then be screened by a quick and reliable method for mitotic lesions. The categories of mutants screened for mitotic abnormalities are reported below.

A. Meiotic Mutants

The first class of mutants screened for effects on mitotic cell division were meiotic mutants (for reviews see Baker *et al.,* 1976a, 1987; Lindsley and Sandler, 1977). This choice was based on the expectation that many meiotic functions are also needed for normal mitotic chromosome behavior. Meiotic mutants, orginally identified because of abnormal chromosome segregation exhibited in appropriate genetic crosses, were initially screened for the occurrence of mitotic errors in imaginal disc and histoblast cells. The consequences of such errors (chromosome breakage, nondisjunction, or loss) were detected as somatic spots on the cuticle of adult flies homozygous for the meiotic mutation under study and heterozygous for suitable somatic cell markers (for details, see Baker *et al.,* 1978; Baker and Smith, 1979; Smith *et al.,* 1985). This analysis showed that the frequency of somatic spots was significantly increased in 10 out of 13 meiotic mutants, fulfilling the prediction that many such loci are also involved in mitotic cell division.

Some of these meiotic mutants were subsequently examined for cytologically observable effects on brain neuroblast chromosomes. Two mutants, *mei-9* and *mei-41,* which are severely defective in meiotic recombination in females, also showed a high spontaneous frequency of chromosome aberrations (Gatti, 1979).

B. Mutagen-Sensitive Mutants

Since *mei-9* and *mei-41* are hypersensitive to killing with mutagens because of strong defects in DNA repair (Baker *et al.,* 1976b), additional mutagen-sensitive mutants (Boyd *et al.,* 1976, 1981; Snyder and Smith, 1982) were examined for effects on mitotic chromosome stability. Analysis of somatic spots and examination of neuroblast chromosomes showed that many mutagen-sensitive mutants control mitotic chromosome integrity in non-mutagenized cells [Baker and Smith, 1979; Gatti, 1979; Gatti et al., 1983a,b; C. Bove and M. Gatti (University of Rome), unpublished results]. Alleles of at least 11 different mutagen-sensitive loci show substantial increases in chromosome breakage, presumably because of a failure to repair spontaneous chromosome damage.

C. Temperature-Sensitive Lethals

Mutations in most mitotic functions would be expected to identify genes required for *Drosophila* development and viability. Smith et al. (1985) reasoned that conditional alleles of such essential loci could be utilized to generate adult individuals with decreased gene activity. If grown at semirestrictive temperatures, surviving escapers carrying temperature-sensitive alleles of essential mitotic genes might show defects in cell division. A collection of 168 temperature-sensitive lethal and semilethal mutations raised under such conditions was therefore examined for the appearance of cuticular clones. This analysis yielded 15 mutants with elevated frequencies of somatic spots, as described above; these were subsequently examined for neuroblast chromosome integrity and behavior after growth at restrictive temperature. Most of these mutants showed low levels of chromosome breakage, suggesting that they identify genes with peripheral roles in the cell cycle. However, the three strongest mutants in this collection appeared to be defective in essential mitotic functions needed for chromosome condensation [*mus-101^{tsl}* and *mit (1)4*] or for chromosome segregation [*1(1) zw10*].

D. Methods of Analysis

Several mitotic mutants discovered by the screening methods described above have been examined both genetically for the presence of somatic spots and cytologically for the occurrence of chromosomal abnormalities. The relative advantages of these two analytical methods may therefore be compared. In general, there is reasonable concordance between the relative increases in the frequencies of somatic spots and of chromosome aberrations caused by the various mutants. However, the absolute frequency of chromosome aberrations tends to be approximately two orders of magnitude higher than the appearance of somatic spots, suggesting that many cytologically detectable chromosome aberrations do not give rise to marked clones in the cuticle (Smith et al., 1985). Because a wing contains about 30,000 cells, genetic analysis can nonetheless be very effective at detecting low levels of chromosome instability. However, the somatic spot technique has two major limitations, as it allows neither precise definition of the event causing chromosome instability, nor the screening of mutant individuals that do not survive to adulthood. This latter limitation is particularly serious because many mitotic genes indeed have organismal-lethal alleles. The majority of recent work on mitotic mutants has thus most heavily utilized cytological screening and characterization.

E. Late Lethals

Lethal alleles at the mutagen-sensitive loci *mus-105* and *mus-109* exhibit very high frequencies of chromosome aberrations and produce larvae with very small imaginal discs which die around the time of pupariation (Baker *et al.,* 1982). Baker and co-workers anticipated that many other mitotic mutants would also display small imaginal discs and would survive only until the larval/pupal transition. These authors hypothesized that most, if not all, divisions needed to form a larva are accomplished using maternal products packaged into the egg. Divisions of the imaginal discs, the abdominal histoblasts, and the neuroblasts, which occur later in development, would be under the control of the zygotic genome. As a result, zygotes homozygous for a recessive mutation in an essential mitotic function would be able to develop into larvae due to material supplied by their heterozygous mothers. Such individuals would die only at the end of the larval period, when imaginal cells begin to replace larval cells. Depending on the severity of the mitotic defect, mutant larvae should also exhibit more-or-less defective imaginal discs.

Based on this hypothesis, 59 lethals dying at the larval/pupal transition, including 41 with defective imaginal discs, were screened for effects on mitotic cell division (Gatti and Baker, 1989). Of this collection, 30 showed aberrant mitotic chromosome behavior in brain neuroblast squashes. A variety of mitotic defects were observed, including chromosome breakage, abnormal chromosome condensation, endoreduplication, metaphase arrest, and the formation of giant polypoid cells. With a single exception, all of the mitotic mutants identified also displayed missing, degenerated, or small imaginal discs. It therefore appears that the phenotypes of late lethality and poorly developed imaginal discs are together almost diagnostic of mutations in essential mitotic genes.

In spite of the correlation between mitotic lesions and defective discs, our recent efforts to identify new mitotic mutants have involved the screening of late lethal collections without preselection for abnormal imaginal discs. The cytological phenotypes of even relatively weak mitotic mutants can be extremely clear, even if imaginal discs do not exhibit a correspondingly obvious defect (M. Gatti and M. Goldberg, unpublished observations). Moreover, the preparation and analysis of brain squashes are not significantly more time consuming than examination of imaginal discs in mutant larvae.

Several groups interested in cell division have screened for mitotic defects by the cytological analysis of late lethal mutations, resulting in the identification to date of some 60 mitotic genes affecting different aspects of chromosome morphology or chromosome movement [Gatti *et al.*, 1983a; Ripoll *et al.*, 1987; Gatti and Baker, 1989; D. Glover (The University, Dundee, U.K.) and R. Karess (NYU Medical School), personal communication; M. Fuller (Stanford University), personal communication; M. Gatti and M. Goldberg, unpublished results]. Owing to differences in mutagens employed, criteria for selection of late lethals, cytological techniques, and distribution of good fortune, experiences with this approach have been quite varied. However, as a very rough rule of thumb, late larval/pupal lethals amenable to cytological analysis comprise about a third of all lethal mutations. Between 2 and 10% of these late lethals exhibited detectable mitotic phenotypes.

F.   Female Sterile Mutants

A corollary to the hypothesis that early embryonic divisions rely on maternal products deposited in the egg is that many cell cycle genes might be anticipated to have female sterile alleles. Two different classes of such female sterile mutations would be expected: those affecting mitotic genes needed throughout development, and those identifying genes specifically required for the specialized syncytial divisions of early embryogenesis (see Baker *et al.*, 1987). Zygotes homozygous for weak mutations in genes essential throughout development might have sufficient gene activity to undergo adult morphogenesis, but resultant homozygous females would deposit insufficient product into the egg to support embryonic mitosis.

A collection of 70 female sterile mutations provided by D. L. Lindsley (University of California, San Diego) and C. Nusslein (Max Planck Institut für Entwicklungsbiologie, Tubingen, Germany) was therefore screened for potential effects on neuroblast chromosome behavior. Examination of the homozygous mutant progeny of heterozygous mothers showed that 10 of these female sterile mutations significantly increase the frequency of chromosome abnormalities in dividing neuroblasts [Gatti *et al.*, 1983a; M. Gatti, S. Ciafrè, and G. Belloni (University of Rome), unpublished]. Interestingly, D. Glover and co-workers (The University, Dundee, U.K.; personal communication) have examined embryos produced by females homozygous for many of these same female sterile mutations; they identified five mutants with observable defects in early embryonic mitoses. Of these, three affect cell division both in embryos and in neuroblasts, while only the embryonic cell cycle is disrupted by the remaining two mutations. One of the two mutants in this latter class, called *giant nuclei* (*gnu*), has been characterized in considerable detail, and is thought to identify a function specifically needed during embryogenesis (Freeman *et al.*, 1986; Freeman and Glover, 1987; see also below).

G. Male Sterile Mutants

Only a fraction of the male sterile mutants analyzed display defects specific to spermatogenesis. The remaining male sterile mutations appear to be leaky alleles at essential loci (Lifschytz, 1987). It is thus reasonable to assume that male sterile and semisterile mutants can identify two kinds of functions affecting cell division: those which are specifically involved in meiosis, and those that are required for cell division in both the germline and soma.

Cytological analysis of testes from male sterile mutants has identified disruptions in several different aspects of meiosis (Lifschytz and Meyer, 1977). Meiotic defects resulting in morphologically abnormal spermatids have also been observed in mutants of the testis-specific β_2-tubulin genes as well as in genes encoding products that interact with β_2-tubulin (for review, see Fuller, 1986; Regan and Fuller, 1988). In a more recent study, M. Fuller and co-workers (Stanford University) have screened a collection of male sterile mutations for those that produce aberrant spermatids. They have found mutations simultaneously affecting both chromosome segregation and mitochondrial distribution during meiosis, as well as a second class of mutants, which only disrupt mitochondrial partitioning because of a probable defect in meiotic cytokinesis (personal communication).

It is also of interest to note several well-documented examples in which late lethal mutations severely defective in mitotic chromosome behavior also show effects on male meiosis. Meiosis can be viewed in testes preparations from late larvae, allowing observation of meiosis in animals homozygous for mutations in mitotic genes. Using this technique, several late lethals showing aberrant mitotic neuroblast cytology [*asp, mgr, polo, 1(1) d. deg-4,* and *1(3) 7m62*] have also been shown to disrupt male meiosis, producing spermatid nuclei and nebenkern of irregular size and morphology (Ripoll *et al.*, 1985; Gonzalez *et al.*, 1988; Sunkel and Glover, 1988; M. Gatti and co-workers, unpublished results).

H. Different Types of Mutants Can Identify the Same Gene

Although the screening procedures described above were designed to detect different abnormalities in cell division, they can in fact identify overlapping sets of genes. For example, mutant alleles of *mei-9* and *mei-41* were detected in screens for meiotic mutants, for mutagen-sensitive mutants, and for female sterile mutants (for review, see Baker *et al.*, 1987). Similarly, alleles at the *mus-105* and *mus-109* loci were found among mutagen-sensitive, female sterile, and larval/pupal boundary lethal mutations (Baker *et al.,* 1982; Gatti and Baker, 1989). Finally, different mutants at the *abnormal spindle* (*asp*) locus exhibit late lethality, female sterility, and severe defects in male meiosis (Ripoll *et al.,* 1985; Gonzalez *et al.,* 1990). These observations suggest that both viable and lethal alleles of many cell division genes may be detected by several different kinds of screens. However, genes required solely for particular types of cell division, such as meiosis or syncytial embryonic cleavages, would be identified only by the appropriate procedure (s). As an example, it would be unexpected to obtain lethal alleles of genes specifically involved in meiosis because the gene products should not be required for viability.

V.   Cytological Phenotypes Displayed by Mitotic Mutants

The cytological analysis of cell division in brain neuroblasts has shown that mitotic mutants exhibit a surprisingly wide range of defects. Moreover, some of the abnormalities observed in these mutants are sufficiently distinctive to suggest possible roles of the wild-type gene products. Here, we group the best-characterized *Drosophila* mitotic mutants into seven classes based on differences in neuroblast cytological phenotype we believe to be of functional significance. Mitotic mutants have previously been categorized in alternative ways by Ripoll *et al.* (1987) and by Glover (1989); their classification schemes also include female sterile mutations yielding embryonic phenotypes (such as *gnu*), which are not discussed below.

A.   Mutants Affecting Progression through Interphase

Several mutants have been described which display very few or no dividing cells in larval brains. The rare divisions observed in these mutants exhibit different abnormalities, indicating that they specify a variety of functions needed for progression of the cell cycle (Gatti and Baker, 1989). One of the most interesting mutants of this class is *1(3)13m-281*, in which all mitotic figures are endoreduplicated. Since endoreduplication is believed to result from the occurrence of successive S phases in the absence of intervening mitotic division (Nagl, 1978), the wild-type allele of *1(3)13m-281* may specify a function required either for progression through G_2 or for the onset of mitosis.

B.   Mutants Affecting Chromosomal Integrity

A large class of genes, currently including some 18 loci, appears to be involved in the maintenance of chromosomal integrity (for review, see Gatti *et al.,* 1983a; Baker *et al.,* 1987; Ripoll *et al.,* 1987; Glover, 1989). Mutants at these loci exhibit elevated frequencies of chromosomal aberrations (Fig. 4) which occur spontaneously in the absence of mutagen treatments. These mutants may differ significantly in the type of breaks and exchanges produced. For example, mutations at the *mei-9* locus produce almost exclusively chromatid breaks, in contrast with mutations at other loci which generate both chromatid and isochromatid breaks (Fig. 4). Moreover, the type of exchanges and the proportion of breaks relative to interchanges induced by different mutations in this class may also vary significantly (for detailed discussion, see Gatti, 1979; Gatti *et al.,* 1980; Baker *et al.,* 1982, 1987). These findings suggest that the various mutants produce different chromosomal lesions which can lead to the formation of different types of chromosome aberrations.

Another interesting difference between mutants in this class concerns the regional distribution of breaks and exchanges along the chromosomes. There

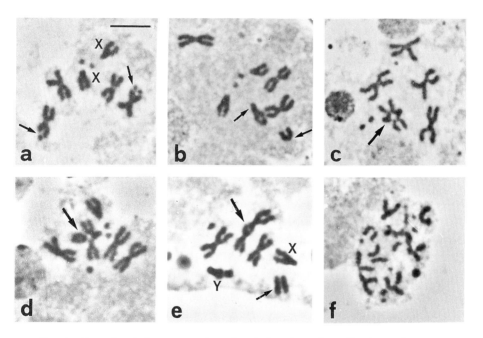

FIG. 4.   Examples of chromosome aberrations in larval ganglion cells of *D. melanogaster.* Brains have been incubated with colchicine for 1.5 hours and hypotonically treated before fixation (see text). (a) Chromatid break (left) and isochromatid break (right), both involving autosomal euchromatin; (b) isochromatid break involving the euchromatin–heterochromatin junction of a major autosome; (c) symmetrical exchange between X chromosomes; (d) asymmetrical exchange between an autosome and X chromosome; (e) dicentric between autosomes with associated fragment; (f) Extensive chromosome breakage in *mus-109[1S].* Bar, 5 μm.

are mutants that preferentially produce euchromatic breaks (*mus-105*), mutants in which aberrations are clustered in the heterochromatin [*mus-109* and *fs(3)820*], and mutants displaying no regional specificity (see Fig. 4). The mutations affecting heterochromatin can even be further subdivided. In *mus-109*, most of the breakpoints appear to involve the junctions between euchromatin and heterochromatin. In contrast, most of the breaks caused by *fs(3)820* are located at or near the nucleolus organizer, suggesting specific involvement of this latter locus in ribosomal DNA metabolism (Gatti, 1979; Baker *et al.*, 1982; Gatti *et al.*, 1983a). Together, these results indicate that the integrity of different chromosomal regions is under separate genetic control.

C. Mutants Affecting Chromosome Condensation

Mutants at more than 20 loci have substantial effects on neuroblast chromosome condensation (Gatti *et al.*, 1983b; Smith *et al.*, 1985; Ripoll *et al.*, 1987; Gatti and Baker, 1989; Glover, 1989). The best-characterized gene in this class is *mus-101*, originally identified by viable mutagen-sensitive alleles defective in postreplication repair (Boyd *et al.*, 1976). Cytological analysis of a temperature-sensitive allele of *mus-101* (*mus-101^{tsl}*) revealed that this locus is also involved in the control of chromatin condensation. When individuals hemizygous or homozygous for *mus-101^{tsl}* are exposed to restrictive temperature, there is a striking undercondensation of all heterochromatic regions, while the euchromatin appears normally condensed, as shown in Fig. 5a–c (Gatti *et al.*, 1983b). It has been shown that this defect is not related to gross defects in the replication of heterochromatic DNA (Gatti *et al.*, 1983b). These results suggested the hypothesis that mutagen sensitivity and defective DNA repair are secondary effects of a primary defect in heterochromatin condensation. However, the finding that an allele of the *mus-101* locus (*mus101^{K451}*) interferes with the amplification of the euchromatically located chorion genes (Orr *et al.*, 1984; Baker *et al.*, 1987) indicates that the function specified by *mus101^{+}* is not solely involved in the control of heterochromatic regions.

The other mutants in this class affect both euchromatin and heterochromatin, producing swollen and unevenly condensed chromosomes (see Fig. 5d–f) (Smith *et al.*, 1985; Gatti and Baker, 1989). In many of these strains, irregular chromosome condensation is accompanied by additional mitotic defects, including low levels of chromosome breakage and polyploidy. The relationship between abnormal chromosome condensation and other mitotic aberrations in these mutants remains unclear.

D. Mutants Affecting Chromosome Structure

Alleles of the locus *parallel sister chromatids* [*pasc*; also known as *1(1)C204*] cause a unique phenotype that cannot easily be included in the other categories discussed here. In wild type, sister chromatids of metaphase chromosomes are well separated in euchromatin but closely apposed in heterochromatin. This determines the characteristic morphology of the metacentric autosomes and the telocentric X chromosome, which appear X-shaped and V-shaped, respectively. In contrast, sister chromatids found within neuroblasts of larvae hemizygous or homozygous for *pasc* mutations are widely separated in the heterochromatic region, being connected only by a tiny thread of chromatin at the centromeres. As a result, the sister chromatids appear parallel to each other in squash preparations made by a variety of protocols (Perrimon *et al.*, 1985; M. Gatti, F. Verni, R. Gandhi, and M. Goldberg, unpublished results). Interestingly, the cytological phenotype observed in *pasc* is very similar to that associated with Roberts syndrome, a condition causing severe multiple phocomelia in humans (German, 1979).

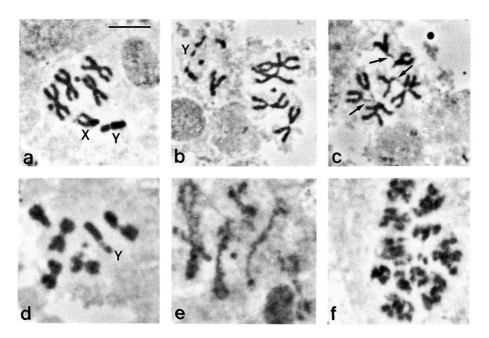

Fig. 5. Colchicine-treated metaphases of mutants affecting chromosome condensation. (a) Normal male metaphase; (b and c) undercondensation of heterochromatin in *mus-101^tsl*. In (b) the entirely heterochromatic Y chromosome is elongated and exhibits several stretched areas; in (c) centric heterochromatin of both pairs of autosomes and X heterochromatin are clearly affected. (d) Fuzzy and swollen chromosomes in *(1)d.deg-9*; (e) poorly condensed metaphase chromosomes in *l(3)e20*; (f) irregularly condensed metaphase chromosomes in *l(3)2612*. Bar, 5 μm.

E. Metaphase Arrest Mutants

Mutants that block cell division at metaphase are characterized by a high mitotic index; most mitotic figures are metaphases with overcontracted chromosomes, while anaphase figures are rare or absent. All of these mutants also generate polyploid cells, indicating that some cells blocked in metaphase can revert to interphase and undergo subsequent chromosome replication. Metaphase arrest mutants also exhibit additional cytological traits which we have used to group them into three subclasses.

The first subclass consists of two mutants, *asp* and *1(1)d. deg-4*. Most metaphase figures observed in these mutant brains, untreated with colchicine, have contracted and scattered chromosomes, closely resembling the appearance of colchicine-induced metaphase figures in wild-type neuroblasts (Fig. 6). These observations have suggested that *asp* and *1(1)d. deg-4* identify functions necessary for spindle formation or function (Ripoll *et al.*, 1985; Gatti and Baker, 1989). A recent analysis by confocal microscopy has shown that arrested *asp* neuroblasts frequently exhibit hemispindles associated with a single centrosome (Gonzalez *et al.*, 1990).

The mutants of the second class, *merry-go-round* (*mgr*) and *polo*, display a peculiar metaphase morphology in which the chromosomes are arranged in a circle. It has been suggested that these circular figures are caused by monopolar spindles, and that both *mgr* and *polo* identify functions required for correct centrosome behavior (Gonzalez *et al.*, 1988; Sunkel and Glover, 1988). These circular metaphase figures are destroyed either by colchicine treatment or by the presence of *asp* in the double mutant combination *asp mgr*. Based on these results, it has been hypothesized that the hemispindles observed in *asp* may be static and incapable of organizing chromosomal movement, whereas the putative monopolar spindles of *mgr* and *polo* may actually be pulling the chromosomes (Gonzalez *et al.*, 1990).

The third subclass of metaphase arrest mutants includes the loci *1(1)d. deg-3* and *1(1)d. deg-10*. Metaphase figures from brains of these mutants that have not been treated with colchicine exhibit widely scattered and extremely contracted chromosomes with separated sister chromatids (Fig. 6g). The degree of chromosomal contraction is higher than that produced by prolonged colchicine treatments in wild type. This distinctive terminal phenotype does not suggest a simple hypothesis for the primary defect present in these mutants.

F. Mutants Affecting the Fidelity of Chromosome Segregation

The mutants included in this class can be subdivided into two categories: those which exhibit frequent failures in the segregation of individual chromo-

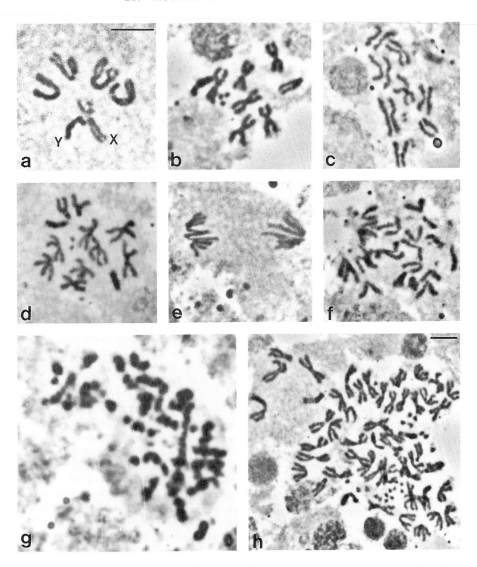

FIG. 6. Cytological phenotypes of mutants affecting chromosome segregation. (a) Control male metaphase from a brain without colchicine and hypotonic pretreatments; (b and c) colchicine-treated aneuploid metaphases in *l(1)zw10*. The metaphase figure in (c) shows precocious sister chromatid separation. (d) Noncolchicine, nonhypotonic-treated tetraploid metaphase from *l(1)d.deg-4*; (e) normal control anaphase; (f) abnormal anaphase in *l(1)d.deg-4*; (g) polyploid/hyperploid metaphase from *l(1)d.deg-10* without colchicine or hypotonic pretreatment; (h) polyploid/hyperploid metaphase treated with hypotonic solution but not with colchicine from *l(3)7m-62*. Magnification in (a–g) is the same; the polyploid metaphase in (h) has a lower magnification. Bars, 5 μm.

somes [$1(1)zw10$ and *rough deal* (*rod*)], and those which fail to segregate entire chromosome complements [$1(1)d. deg-11$, $1(3)7m-62$, and *spaghetti-squash* (*sqh*)].

Larval brains within individuals hemizygous or homozygous for mutations at both the $1(1)zw10$ and *rod* loci exhibit very high frequencies of aneuploid cells resulting from nondisjunction, as seen in Fig. 6b,c (Smith *et al.*, 1985; Karess and Glover, 1989). Both also show defects at anaphase, primarily the appearance of lagging chromosomes which remain near the center of the spindle apparatus even until telophase. Despite these similarities, it is possible that the problems caused by these two mutations are quite different. Many neuroblasts in hypotonic-treated preparations of $1(1)zw10$ brains display premature separation of sister chromatids at prophase/metaphase (Fig. 6c) (Smith *et al.*, 1985; M. Goldberg, unpublished results). This suggests that $1(1)zw10$ may encode a centromere function required to maintain sister chromatid attachment until the onset of anaphase. Precocious sister chromatid separation is not seen in mitotic figures from *rod* mutants; instead, some anaphase aberrations (anaphase bridges) and the appearance of broken chromosomes imply that the separation of sister chromatids in *rod* cells is actually delayed rather than accelerated.

Mutants at the $1(1)d. deg-11$, $1(3)7m-62$, and *sqh* loci exhibit high frequencies of polyploid cells (Fig. 6h). However, these mutants have mitotic indices and frequencies of anaphase figures comparable to controls, indicating that they do not cause a mitotic block [Gatti and Baker, 1989; Karess *et al.*, 1991; R. Karess (NYU Medical School), personal communication]. The degree of polyploidy seen in these mutants is much higher than that present in the metaphase arrest mutants described in the previous section. For example, giant polyploid cells containing 500–1000 chromosomes are relatively common in $1(1)d. deg-11$ and $1(3)7m-62$. These observations suggest that this group of mutants identifies genes involved in cytokinesis or the completion of chromosome segregation.

G. Mutants with Anaphase Defects Leading to Chromosome Breakage

Both nod^{DTW} and $fs(3)1755$ affect chromosome behavior during anaphase, leading to increased chromosomal breakage or fragmentation. nod^{DTW}, formerly designated $1(1)TW-6^{CS}$, is a cold-sensitive lethal dominant allele of the *nod* locus [Wright, 1974; Zhang *et al.*, 1990; R. S. Hawley (A. Einstein Medical School), personal communication]. Upon exposure to the restrictive temperature, homologous chromatids in nod^{DTW} mutants frequently form anaphase bridges. The subsequent breakage of these bridges produces either daughter cells containing broken chromosomes without the corresponding acentric fragments, or daughter cells with a normal chromosome complement plus extra fragments (M. Gatti and B. Baker, unpublished data). The stickiness of homologous chromatids observed in nod^{DTW} appears to result from the presence of a poisonous product encoded by this neomorphic mutation [R. S. Hawley (Albert Einstein College of Medicine), personal communication; see below]. The other mutation in this class, $fs(3)2755$, is less well characterized, but clearly exhibits chromosome fragmentation specifically occurring during anaphase, producing segmentally aneuploid metaphases analogous to those seen in nod^{DTW} (Gatti et al., 1983a).

VI.   Molecular Cloning of *Drosophila* Mitotic Genes

As is evident from the discussion above, cytological analysis of mutant phenotypes can be quite informative, but knowledge of the associated bio-chemical defects may nonetheless remain elusive. Further progress thus clearly requires molecular characterization of mitotic genes and the proteins they encode. This has often been hampered by the difficulty of cloning loci associated with mitotic lesions. The worst case scenario is presented by a gene defined by a single EMS point mutation that does not lie within a genomic region well-characterized in either genetic or molecular terms. The gene of interest may reside some distance from the nearest available piece of cloned DNA, necessitating a time-consuming chromosomal walk. This particular aspect of the problem should be less onerous in the future, as more of the *Drosophila* genome becomes available in clone banks such as those produced by yeast artificial chromosome (YAC) vectors (Garza et al., 1989) or micro-dissection cloning (Siden-Kiamos et al., 1990). Nevertheless, considerable effort must still be expended to generate chromosomal rearrangements re-quired to delimit the gene within a cloned region.

The recent introduction of powerful methodology for single-hop P element transposon mutagenesis in *Drosophila* potentially offers a much faster, direct route from the identification of a mitotic gene to its isolation in recombinant DNA. The basic idea concerns the mobilization of a genetically marked P element transposon from a preexisting site of integration in the genome to a new location. The function of a mitotic gene may be disrupted if the trans-poson lands in its vicinity; genes defined by such mutations can then be cloned, taking advantage of the P element as a molecular tag.

This kind of insertional mutagenesis is generally initiated by generating flies containing two types of P elements. The first is a marked P element which has the two P element ends required for movement, but which is defective in its ability to encode the transposase function needed for mobilization. The second is a P element (the driver) unable itself to transpose but able to potentiate movement of the marked P element by supplying transposase activity. Genomes in which the marked P element has hopped to a new location can be identified because the P element marker is no longer linked to the chromosome on which it was originally found. Slightly different genetic schemes are thus required to isolate novel mutations on the X chromosome or on the autosomes.

We have successfully identified mitotic lesions among collections of P element-induced lethal mutations generated in the laboratories of both Dr. W. Gehring (O'Kane and Gehring, 1987; Bellen et al., 1989; Wilson et al., 1989) and Dr. J. Merriam (University of California, Los Angeles, personal communication). The mutagenesis schemes developed by these investigators and by others (Robertson et al., 1988; Cooley et al., 1988), which differ in the characteristics and original location of the marked P element, the source of transposase, and the precise genetic methodology for detecting new transpositions, are compared and reviewed in detail by Ashburner (1989).

A.   A Procedure for Selection of P Element Hops

Here, we take the opportunity to propose a system for insertional mutagenesis of the autosomes, based on recent work by Drs. D. Lindsley, E. Bier, and colleagues (University of California, San Diego, personal communication; see also Hamilton et al., 1991), that we believe to be of potential advantage for the isolation of mitotic mutants.

This strategy employs the marked P element $P[lacW]$ (Bier et al., 1989) and the driver element $P[ry^+(\Delta 2\text{-}3)](99B)$ (Robertson et al., 1988), the latter providing a high level of transposase function both in somatic cells and in the germ line. $P[lacW]$, which is marked with the eye color marker $white^+$ (w^+), incorporates two useful refinements (Fig. 7A). First, it contains an Escherichia coli lacZ gene under the control of a weak promoter near a P element end. When such a construct becomes inserted near an enhancer sequence regulating a gene, transcription of lacZ will be activated so as to be detectable by histochemical staining (an "enhancer trap"). This theoretically allows analysis of the tissue and developmental specificity by which a nearby gene is controlled. Second, $P[lacW]$ includes a bacterial plasmid within the marked P element, permitting rapid cloning of Drosophila sequences near novel insertion points by plasmid rescue (see Fig. 7B and discussion below).

In the protocol diagrammed in Fig. 8, females are generated which have a third chromosome bearing the driver element $P[ry^+(\Delta 2\text{-}3)](99B)$ and an attached X chromosome, each homolog of which bears four $P[lacW]$ insertions

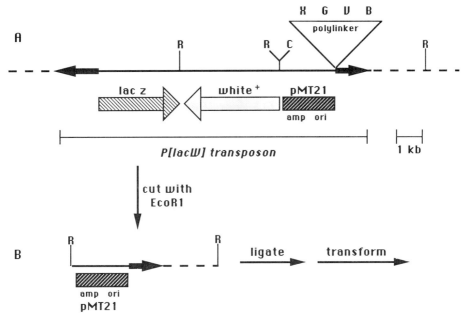

FIG. 7. (A) Structure of the marked P element transposon P[lacW], inserted into Dro-sophila genomic DNA (dotted line) (redrawn from Bier et al., 1989). The 5' and 3' ends of the P element (black arrows) denote the unit transposed upon mobilization as shown. The white⁺ gene (actually a "mini"-white⁺ gene reduced in size by removal of part of a large intron from the wild-type locus) identifies flies carrying this construct. A P–lacZ fusion gene under the control of the weak P element transposase promoter is located near the 5' end of the P construct, allowing detection of tissue-specific enhancers found near new insertion sites. P[lacW] also contains the plasmid vector pMT21, as well as the following restriction enzyme sites useful for plasmid rescue: R-EcoRI, C- SacII, X-XbaI, G-BgIII, V-PstI, B-BamHI. An EcoRI site in adjacent genomic DNA is also shown. (B) Plasmid rescue of Drosophila DNA next to a P[lacW] insertion site. Genomic DNA from the appropriate insert strain is digested to completion with the enzyme of choice from the list above; use of EcoRI is illustrated here. A resultant EcoRI fragment contains the entire plasmid vector pMT21, one P element end (black arrow), and Drosophila DNA adjacent to the insertion site (dotted line). After phenol/chloroform extraction and ethanol precipitation, the genomic DNA digest is ligated at a final concentration of approximately 1 μg/ 200 μl with T4 DNA ligase. Under these dilute conditions, intramolecular reactions leading to fragment circularization (see figure) are favored. After ligation, the DNA is again purified by phenol/chloroform extraction, and concentrated by ethanol precipitation. DNA is then used to transform highly competent E. coli cells [preferably of the XL1-Blue strain (Stratagene); see also Hanahan, 1985], selecting for the ampicillin resistance marker on pMT21. The only transformed colonies appearing should contain the circularized form of the fragment shown.

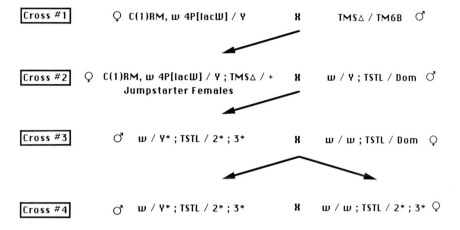

Fig. 8. Generation and identification of autosomal P element insertions, based on a protocol developed by D. Lindsley, E. Bier, and co-workers (University of California, San Diego, personal communication; also used by Hamilton *et al.,* 1991). In cross #1 females with *C(1)RM, w 4P[lacW]* (an attached X chromosome, each homolog of which contains four copies of the *P[lacW]* transposon), are mated with males heterozygous for (1) the third chromosome balancer *TM6B* and (2) *TMSΔ,* a derivative of the third chromosome balancer *MKRS* containing the dominant marker *Sb* (stubble bristles), the driver element *P[ry⁺ (Δ2-3)](99B),* and an additional inversion expanding the utility of this balancer [D. Lindsley and Simpson (University of California, San Diego), personal communication]. Alternatively, the driver element can also be supplied on the balancer *TM2* marked with *Ubx* and *ry* [J. Merriam (University of California, Los Angeles), personal communication; Hamilton *et al.,* 1991]. Female progeny of cross #1 with stubble bristles and pigmented eyes are designated "Jumpstarter" females. These females contain both the marked P element *P[lacW]* and the driver element in the same genome. In order to detect transpositions that have occurred in the germline of Jumpstarter females, individual Jumpstarter females are mated in cross #2 with males hemizygous for an X chromosome containing a *white* mutation (*w*), and heterozygous for two different chromosomes: (1) *TSTL,* the translocation *TSTL14* described in the text, which is marked with *Cy* (curly wings) and *Tb* (Tubby body), and (2) *Dom,* a second or third chromosome containing any easily recognized dominant marker. Most male progeny of this mating will display the mutant *w* eye color because they inherited an X chromosome from their father, but those males containing new autosomal (or Y chromosomal) transpositions of *P[lacW]* will have pigmented eyes. (An asterisk is used to designate chromosomes which could harbor a new *P[lacW]* transposition). Transposition-containing males phenotypically non-*Sb* (indicating absence of *TMSΔ* and thus the driver element) and *Cy* (showing the presence of *TSTL*) are selected. Although not shown, males with pigmented eyes and showing the *Dom* phenotype can be used as well. To ensure that all of the strains to be analyzed contain independent hops, a single transposition-containing son of each Jumpstarter female is now mated with females of the same genotype as their fathers (cross #3). Male and female progeny of cross #3 who have pigmented eyes and *Cy* wings are now mated with each other. If all the progeny of this mating have *Cy* wings and pigmented eyes, then the *P[lacW]* insertion has potentially generated an autosomal lethal mutation. Such a stock can be maintained in subsequent generations without further selection. The developmental stage of lethality can then be determined as explained in the text. If only the male progeny of cross #4 have pigmented eyes, it is probable that *P[lacW]* has inserted into the Y chromosome. Further information concerning the characteristics of the mutations and balancer chromosomes used in this and subsequent figures can be found in Lindsley and Grell (1968) and Lindsley and Zimm (1985, 1986, 1987, 1990). Most of the stocks required can be obtained from the *Drosophila* Stock Center at Indiana University (Department of Biology, Indiana University, Bloomington, Indiana 47401) or from the authors.

[this chromosome was recently contructed by D. L. Lindsley (University of California, San Diego)]. Within the germline of these "Jumpstarter" females, the P[lacW] element is mobilized. Jumpstarter females are next mated with males carrying a w mutation on their X chromosome. The male progeny produced by such a cross will inherit their X chromosome from their father; as a result, most of these males will have white mutant eye color. However, if one or more of the P[lacW] transposons moved to an autosomal location in the eggs produced by the Jumpstarter female, males with pigmented eyes (w^+) are generated. The frequency of transposition in this system appears to be quite high; moreover, a high proportion of such mutagenized genomes contain only a single P element, greatly simplifying subsequent molecular analysis (Hamilton et al., 1991).

Recall that homozygosity for mutations in mitotic genes often causes lethality during larval or pupal stages. For screening purposes based on this criterion, it is thus necessary to determine whether a P element insertion has caused lethality, and if so, at what time during development. For this purpose, we have chosen to place P element-mutagenized autosomes over the balancer TSTL14 (J. Casal and P. Ripoll, (Universidad Autonoma de Madrid, Spain, personal communication), which is a translocation between the second chromosome balancer SM5 [marked with the dominant mutation Curly wings (Cy)] and the third chromosome balancer TM6B. The TM6B moiety of TSTL14 contains the dominant marker Tubby (Tb); larvae or pupae possessing this chromosome appear short and squat relative to wild type. An autosomal P element-induced late lethal mutation is thus characterized by the absence of non-Cy adults among the progeny of cross #4 in Fig. 8 (indicating lethality), but the presence of non-Tb third instar larvae homozygous for the mutation. The brains of such non-Tb larvae can be analyzed cytologically for mitotic defects. We have previously successfully employed TSTL14 to balance autosomal mitotic mutants obtained from other single-hop P element-induced lethal collections (M. Gatti and M. Goldberg, unpublished data).

Transpositions of a marked P element onto the X chromosome can be obtained among the progeny of a Jumpstarter fly initially carrying the marked P element on an autosome (see Bellen et al., 1989). Selection is made first for individuals in which the marker no longer segregates with the original autosome (indicating that a jump has occurred), and secondarily for X linkage. In Fig. 9, Jumpstarter males heterozygous for a second chromosome containing P[lacW] and the second chromosome balancer CyO are mated with females carrying the X chromosome balancer FM7 [which has a white allele (w^a) producing a very light yellow eye color]. In the next generation, females with pigmented eyes and curly wings (i.e., CyO) are selected; a transposition of P[lacW] is required to generate this combination of markers. Finally, stocks carrying a lethal mutation on the X chromosome are identified by the

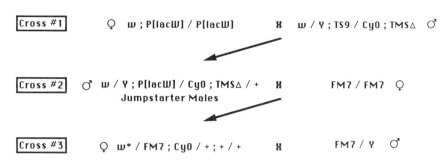

FIG. 9.   Generation and identification of X chromosome P element insertions, loosely based on Bellen *et al.* (1989). All stocks carry a *white* mutation (w^a in *FM7*), so that red eye pigmentation denotes presence of *P[lacW]*. In cross #1, females homozygous for a *P[lacW]* insertion on the second chromosome are mated with males heterozygous for (1) *CyO* and *TMSΔ*, carrying the transposase-encoding driver element (see Fig. 8) and (2) *TS9*, a translocation between autosomes 2 and 3 [*T (2;3) S9, brown ebony*]. The *CyO* and *TMSΔ* chromosomes will segregate together into the progeny of these males. Jumpstarter males containing both the marked P element and the driver are now mated in cross #2 with *FM7* females (the *FM7* X chromosome balancer is recognized by the markers *y* (yellow body), w^a (white apricot eye color), and *B* [Bar (kidney-shaped) eyes]. Although not shown, Jumpstarter females (*w/w*; *P[lacW]/CyO*; *+/TMSΔ*) could also be utilized, and mated in cross #2 with FM7/Y males. A transposition of *P[lacW]* in the germline of Jumpstarter males (or females) will be indicated in the form of progeny simultaneously carrying *CyO* and the transposon (curly wings, pigmented Bar eyes). Such females are selected and mated with *FM7/Y* males. Flies carrying a newly induced X-linked lethal mutation (indicated with an asterisk), as opposed to a jump onto *CyO* or a third chromosome, are identified by stocks in which there are no *w*/Y* males (i.e., no non-*FM7* males) among the progeny of cross #3.

inability of males hemizygous for the mutagenized X chromosome to survive. For subsequent cytological screening, mutations causing late lethality are identified by the presence of male larvae with pigmented Malpighian tubules. *FM7* males have essentially colorless Malpighian tubules, but the w^+ marker on *P[lacW]* confers pigmentation on these organs.

B.   Screening for Mitotic Mutants

In our experience to date, two collections totaling 500 single-hop P lethal insertions contained approximately 150 stocks exhibiting late lethality. Using brain cytology as a screen, we have identified six clear-cut mitotic mutants from this collection. Assuming 5,000–10,000 lethal complementation groups in the *Drosophila* genome, this suggests very roughly that some 100 mitotic genes could be detected by the above protocol.

It should not be overlooked that even nonlethal P element transpositions can also be screened for potential effects on mitosis in a variety of alternative ways. Insertions causing female sterility by maternal effects may represent

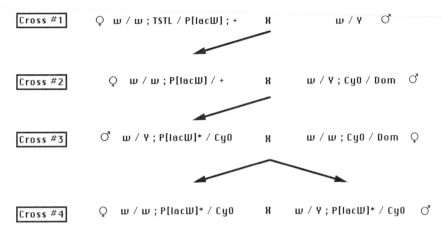

FIG. 10.   Testing cosegregation of P[lacW] and the lethal phenotype. Stocks carrying an autosomal lethal mutation potentially associated with a P[lacW] transposition (see Fig. 8) are mated with a w mutation strain. An example with P[lacW] on the second chromosome is shown. In female progeny of cross #1, free meiotic recombination will occur between the P[lacW]-bearing second chromosome and an unmarked second chromosome of normal gene order. Recombinant chromosomes carrying P[lacW] (marked by an asterisk), heterozygous with the second chromosome balancer CyO, are recovered in the progeny of cross #2; these will appear as flies with Cy wings and pigmented eyes. One hundred flies with recombinant chromosomes are individually mated with a CyO-bearing stock in cross #3, so as to generate both males and females heterozygous for the P[lacW]-containing autosome and CyO. These sons and daughters are now mated together (cross #4). If, in each of the 100 matings tested, all of the progeny of cross #4 have Cy wings (indicating that transposon-bearing autosomes cannot be made homozygous), then the lethal mutation must reside very close to the P[lacW] insertion (within 1 cM). If the transposon is inserted into the third chromosome, rather than the second chromosome as in this example, the treatment would be exactly the same except that a third chromosome balancer such as TM3, Sb would replace CyO.

weak mutant alleles of mitotic functions, incapable of supplying sufficient product to the developing embryo (see above). Transpositions may also be associated with defects in chromosome segregation during meiosis. Moreover, severe meiotic difficulties might lead to male sterility, and could be identified by examining testis squashes of mutant male larvae or adults. Finally, the lacZ enchancer trap present on the P[lacW] element could possibly be exploited to find transpositions in the vicinity of genes whose expression patterns, as revealed by β-galactosidase staining, might be characteristic of genes involved in mitosis (i.e., imaginal discs or proliferating germinal tissue).

C.   Does the Marked P Element Cause the Mitotic Lesion?

Several investigators have found that not all lethal mutations induced by P element mobilization in the above manner are actually caused by insertion of the P element. Some lethal mutations might preexist in the population of chromosomes to be mutagenized. Alternatively, subsequent to a P element hop, the transposon might be imprecisely excised from its new location, disrupting a required genetic function in the process. To ensure that mitotic genes identified in the genetic screens described above can be rapidly cloned, it is thus necessary to determine if a marked P element is indeed associated with the lethal lesion.

One method of evaluation is to check whether, after free recombination is allowed, chromosomes containing the P element cannot be made homozygous (Fig. 10; see also Bellen *et al.*, 1989). If the P element marker does not always cosegregate with a lethal complementation group, then the transposition cannot be responsible for the mutation. The reverse result does not definitively prove that the P element insertion causes the mitotic defect, as the mutation causing the abnormality may not correspond to the P element-induced lethal. It is thus advisable to examine whether larvae from several stocks homozygous for different *P[lacW]*-containing recombinant chromosomes still show the mitotic phenotype.

A second, more direct approach to this question is the generation of precise P element excisions by a second round of mobilization with the *P[ry⁺ (Δ2-3)]*

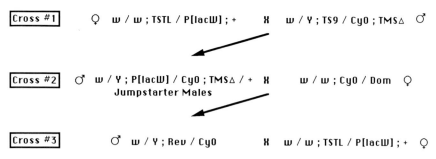

FIG. 11.   Remobilization of an autosomal *P[lacW]* insert; an example with *P[lacW]* on the second chromosome is shown. Stocks carrying a *P[lacW]* transposition presumed to be associated with a lethal mutation (see Fig. 8) are mated with a strain simultaneously supplying the chromosome *TMSΔ* [and thus the driver element *P[ry⁺ (Δ2-3)](99B)*] and the second chromosome balancer *CyO* (see also cross #1 in Fig. 9). The marked P element is mobilized in the genome of the resultant Jumpstarter males. (Though not shown, Jumpstarter females of the same phenotype may also be used because the autosomes are balanced.) Occasionally, *P[lacW]* may be lost from the genome (*Rev*); such instances are revealed as progeny of cross #2 that have curly wings, nonpigmented eyes, and fail to show the *Dom* phenotype (as before, any appropriate second chromosome dominant marker). Such flies are now mated with a strain carrying the original lethal mutation. If the excision is precise, and if there are no other marked P element transposons in the genome, the appearance of non-*Cy* individuals among the progeny of cross #3 indicates that the original insertion is responsible for the lethality. It is possible that, among several excisions, a few may show only *Cy* flies in the progeny of cross #3, due to the fact that imprecise excision has taken place. A similar strategy for remobilization of a third chromosome insertion may be inferred from the above and from the description in Bellen *et al.* (1989).

(99B) driver. If the insertion is in fact responsible for lethality, and if no other P elements are present in the genome, precise excision should repair the lethality simultaneously with loss of the P element marker (Fig. 11; again, see Bellen *et al.,* 1989). An additional advantage of this remobilization tactic is the possibility that imprecise excision of the P element may also occur, providing novel alleles of the mitotic gene in question. Imprecise excision would be recognized as chromosomes which lose the P element marker but remain homozygous lethal.

D. Molecular Cloning of a P Element-Induced Mitotic Mutant

The chromosomal region in the vicinity of a single-hop P element transposition may be readily cloned by a variety of techniques. For example, a recombinant library can be constructed using DNA from the mutagenized genome, and this can be screened with P element sequence probes. In addition, DNA adjacent to insertion sites can rapidly be obtained via "inverse" polymerase chain reaction (PCR) [Ochman *et al.,* 1988; B. Matthews, (University of California, Los Angeles), personal communication]. Genomic DNA containing the P element insert is cut with a restriction enzyme that cleaves as shown in Fig. 12, generating a fragment containing sequences within the insert, the P element ends, and the adjacent genomic region. This DNA is then circularized and used as the template for amplification by PCR, utilizing primer oligonucleotides from the P element ends. The amplified product can now be used to screen a genomic library in order to obtain genomic sequences on both sides of the marked P element insertion site.

However, if the marked P element contains bacterial plasmid sequences, as does *P[lacW]*, plasmid rescue provides the most efficient route to clone the gene of interest. The mutagenized genomic DNA is cleaved with a restriction enzyme cutting at one end of the plasmid sequences but not elsewhere in the plasmid vector. The DNA fragment generated by such cleavage can be circularized by ligation at low concentration, yielding a plasmid now containing *Drosophila* DNA next to the transposon site (see Fig. 7B). Recovery of the plasmid is simply effected by transformation of *E. coli* cells with the ligation mix, selecting for the plasmid drug resistance marker.

Subsequent molecular analysis of the mitotic gene of interest may unfortunately not prove to be trivial, even after the region containing the locus has been cloned. If the P element has jumped into a chromosomal region encoding several mRNAs, it is not necessarily immediately obvious which transcriptional unit corresponds to the mitotic locus.

The relative abundance of each transcript at different periods of development may provide some suggestive information. For example, the mRNA of

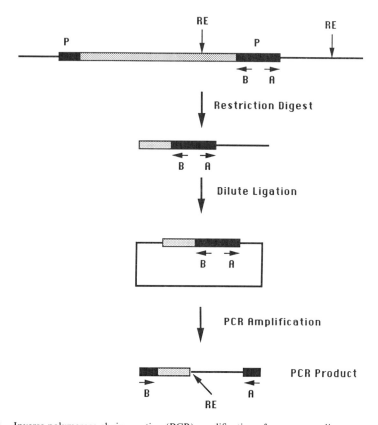

Fig. 12. Inverse polymerase chain reaction (PCR) amplification of sequences adjacent to a P element insert (Ochman *et al.,* 1988; B. Matthews, University of California, Los Angeles, personal communication). A marked P element transposon is pictured as a thick line and *Drosophila* genomic DNA as a thin line. Black regions correspond to P element ends and stippled regions depict sequences internal to the transposon, including marker genes such as *w* or *ry*, and the *E. coli lacZ* gene for enhancer trap mutagenesis, depending on the particular construct. Genomic DNA from an insert-containing strain is digested with a restriction enzyme (RE) that does not cut within the P element ends but does cleave other sequences within the transposon and in adjacent *Drosophila* DNA. The choice of restriction enzyme is dictated by the possibility of obtaining a fragment sufficiently small to amplify efficiently (usually less than 1 kb), and the utility of the restriction site for future subcloning; good candidates include *Sau*3A, *Hha*I, *Taq*I, *Hpa*II, and *Ssp*I. About 0.5 μg of the digested DNA is ligated in a volume of 400 μl in order to obtain circularized fragments. This is then used as a template for PCR reactions (40 cycles) using two oligonucleotide primers (A and B) which contain sequences from the P element ends. For P element ends derived from the Carnegie series of vectors (Rubin and Spradling, 1983), A is 5′ CGCTCTAGAATTCACTCGCACTTATTGCA and B is 5′ CCAG-AATTCTAACCCTTAGCATGTCCGTG [T_M of both sequences is approximately 61°C; underlined nucleotides are not homologous to P elements, but contain restriction sites (*Xba*I and *Eco*RI for A, *Eco*RI for B) useful for subsequent steps]. The amplified fragments may be used directly for screening genomic libraries for DNA surrounding the insertion site; alternatively, the fragments may be subcloned using these additional restriction sites. Cleavage with the original restriction enzyme (RE) may be used to remove unwanted sequences from the fragment. Because the probe contains P element material, genomic clones obtained on the basis of hybridization with these probes must be counterscreened with P element DNA. The inverse PCR strategy described above has been successfully employed by B. Matthew (University of California, Los Angeles, personal communication) for isolating DNA adjacent to hops of the marked P element in the plasmid pLacA92 (O'Kane and Gehring, 1987), and should be useful for any transposon

the mitotic gene *l(1)zw10* appears at high levels in adult females and embryos, at intermediate levels in third instar larvae and early pupae, and in lower quantities in adult males, the first two larval instars, and late pupae [B. Williams and M. L. Goldberg (Cornell University), unpublished results]. This type of pattern is expected for a mitotic gene adhering to the hypothesis described above: mitotic proteins are primarily required as maternal products to support nuclear proliferation in embryonic stages, and subsequently for growth of imaginal structures during late larval/early pupal periods. However, evidence is presently too fragmentary to conclude that this expression pattern is truly a paradigm for a variety of mitotic genes.

Identification of the proper transcriptional unit can also be aided by the study of additional alleles of the mitotic gene, some of which may already have been obtained by other investigators. In many cases, it should be possible to generate novel mutations associated with detectable chromosomal rearrangements by remobilizing the P element so as to obtain imprecise excisions (see above). If available, chromosomal rearrangements in the vicinity might also help delimit the mitotic locus on the cloned DNA.

Conclusive identification of the transcriptional unit for a mitotic gene requires the reintroduction of a DNA fragment back into the *Drosophila* germline by P element-mediated transformation. Genomic DNA may be cloned into a variety of transformation vectors and then injected into early cleavage stage embryos. Transposase may be supplied in any of several different ways (e.g., coinjection of a helper plasmid, or the presence of $P[ry^+ (\Delta 2\text{-}3)](99B)$ in the genome of injected embryos), allowing incorporation of the cloned DNA fragment into random chromosomal locations. Descriptions of various transformation technologies are given by Karess (1985), Pirrotta (1986), Spradling (1986), and Robertson *et al.* (1988); these methods are also enumerated and explained by Ashburner (1989). Once such a stock has been obtained, the ability of the transformed DNA to rescue both lethality and the cytological mitotic phenotype can be tested by appropriate crosses.

VII. Investigations of Mitotic and Meiotic Genes in *Drosophila*

Although the study of mitotic and meiotic genes in *Drosophila* is still in its infancy, enough information has already been obtained to verify its potential

---

containing P element ends originally derived from Carnegie 20 (Rubin and Spradling, 1983). However, it should be stressed that plasmid rescue (Fig. 7B) is likely to be a more efficient method when dealing with inserts of *P[lacW]* or other constructs (Bellen *et al.*, 1989; Wilson *et al.*, 1989) containing bacterial plasmid sequences internal to the transposon.

as a source of important insights into cell division. We present here a necessarily incomplete summary of highlights from cytological and biochemical investigations on *Drosophila* cell division genes. Progress has been achieved in a variety of ways. For example, even in the absence of molecular cloning, careful observation of the phenotype associated with mitotic and meiotic mutations has generated some remarkable findings, one of which is discussed in the section on *gnu* below. On the other hand, molecular characterization of only a few *Drosophila* cell division genes has been accomplished to date. The most rapid progress has been obtained in the analysis of *Drosophila* genes strongly homologous to known cell-cycle genes from other organisms; even here, quite intriguing novel information has been garnered. In addition, biochemical analysis of two genes essential for the fidelity of chromosome transmission during meiosis has uncovered a role for kinesin-related proteins in chromosome movement along the spindle. Finally, potential biochemical functions can now be ascribed to a few genes defined by mutations causing late lethality.

A.   Phenotypic Analysis of *gnu*

The mutation *giant nuclei* (*gnu*) is a maternal effect mutation affecting nuclear division during early embryogenesis. Females homozygous for *gnu* lay eggs which develop giant nuclei as a result of continued DNA replication in the absence of nuclear division. It is striking that centrosomes nonetheless continue to replicate in mutant embryos. These centrosomes migrate to the cortex of the mutant embryo, just as they would in wild type (Freeman *et al.,* 1986; Freeman and Glover, 1987). The *gnu* phenotype indicates that there is a centrosomal component of the cell cycle that can run independently of the nuclear division cycle.

B.   Molecular Analysis of *Drosophila* Genes Homologous to Yeast Cell Cycle Genes

Intriguing information has been garnered even in the study of *Drosophila* homologs of genes well-characterized in other organisms. For example, work in several organisms has suggested that time of entry into mitosis is regulated by the accumulation of cyclin proteins. O'Farrell and co-workers have recently obtained evidence that the accumulation of *Drosophila* cyclins does not appear to be rate limiting for entry into the fourteenth cell cycle of embryogenesis. Instead, transcription of the gene *string* seems to be the trigger controlling the timing of this mitosis (Edgar and O'Farrell, 1989; O'Farrell *et al.,* 1989). *string* encodes the *Drosophila* homolog of the *Schizosaccharomyces pombe* gene *cdc25*, whose product is known to activate mitosis promoting factor (MPF).

Although cyclins in other systems are periodically degraded during metaphase, *Drosophila* cyclins A and B are continually present in early cleavage embryos (prior to cycle 10). It may be that spatial redistribution of cyclins in the vicinity of syncytial nuclei may result in localized oscillations of cyclin levels (Lehner and O'Farrell, 1989, 1990; Whitfield *et al.*, 1990). However, in cellularized embryos and larval brains, both cyclin A and B exhibit clear patterns of cycling. Cyclin A reaches a peak level in prophase and is degraded during metaphase, whereas cyclin B is degraded at the metaphase/anaphase boundary. These observations indicate that the two *Drosophila* cyclins control different transition points during the cell cycle (Whitfield *et al.*, 1990). It is also curious that transcripts of *Drosophila* cyclin B, but not of cyclin A, are highly enriched at the posterior end of developing embryos, where they are incorporated into germline pole cells (Whitfield *et al.*, 1989; Lehner and O'Farrell, 1989, 1990). Cyclin B protein is not correspondingly enriched in these cells, suggesting the existence of translational controls (Lehner and O'Farrell, 1990).

C. *Drosophila* Genes Encoding Kinesin-Related Microtubule Motors

Molecular analysis of two *Drosophila* genes identified because of a meiotic phenotype has been rewarded by the discovery that their protein products are similar to molecules with known biochemical properties. Work on the mutations *no distributive disjunction* (*nod*) and *claret non-disjunctional* (*ca^{nd}*) has shown that members of the kinesin superfamily of proteins are required for accurate chromosome segregation during meiosis (Endow *et al.*, 1990; McDonald and Goldstein, 1990; Zhang *et al.*, 1990). Mutants at the *ca^{nd}* and *nod* loci are not substantially defective in meiotic recombination but exhibit high levels of nondisjunction at the first meiotic division. *nod* is defective in the disjunction of nonexchange chromosomes but does not affect the meiotic behavior of recombinant chromosomes. *nod* function therefore appears to be limited to the distributive segregation system. In contrast, in mutants at the *ca^{nd}* locus, both recombinant and nonrecombinant chromosomes segregate abnormally. The product of *ca^{nd}* thus serves as a component of both exchange pairing and distributive pairing disjunction systems in *Drosophila*.

Both of these molecules appear to play a role in mitosis as well. The product of *ca^{nd}* also functions during the first few mitotic divisions in the zygote, where the loss of (almost exclusively) maternally inherited chromosomes is elevated. This suggests that the *ca^{nd}* protein remains associated with maternal chromosomes during these early cell cycles. On the other hand, although null mutations at the *nod* locus do not exhibit any mitotic effects, *nod^+* appears to be expressed in mitotically active tissues [Zhang *et al.*, 1990; R. S. Hawley, (A. Einstein School of Medicine), personal communication]. Moreover, an antimorphic mutation in *nod* (*nod^{DTW}*) leads to anaphase bridges, resulting in chromosome breakage and loss during mitosis in brain neuroblasts [M. Gatti (University of Rome) and B. Baker (Stanford University), unpublished data; R. S. Hawley (Albert Einstein College of Medicine), personal communication]. These data indicate that, although the *nod* product by itself is not necessary for mitotic cell division, it may nontheless be part of a redundant system playing some unknown mitotic role.

Finally, it should be noted that recent investigations on *ca^{nd}* have been instrumental in altering views of kinesin function. The *ca^{nd}* protein was unexpectedly discovered to move toward microtubule minus ends, in contrast to previously studied kinesins (McDonald *et al.,* 1990; Walker *et al.,* 1990).

D.   Molecular Analysis of Late Lethal Mutations

The criterion of late larval/pupal lethality has been used to identify a few mitotic genes whose protein products are cognates of genes with known biochemical functions.

1.   THE LOCUS *abnormal wing disks* ENCODES AN NDP KINASE

Homozygosity for mutations of the gene *abnormal wing disks* (*awd*) causes the appearance of abnormal mitotic arrest phenotypes in brain neuroblasts. The *awd* gene codes for a microtubule-associated nucleoside diphosphate (NDP) kinase; this enzyme catalyzes the *trans*-phosphorylation of tubulin-bound GDP that is presumably required for polymerization of spindle microtubules (Rosengard *et al.,* 1989). However, a variety of additional phenotypic defects in *awd* mutant larvae have been described, suggesting that this NDP kinase is required by other cellular processes in addition to its role in mitosis. Interestingly, *awd* is homologous to the mammalian *nm23* gene, whose expression is correlated with the metastatic potential of tumor cells. It has been proposed on this basis that the *nm23/awd* gene may be required for normal growth regulation (Rosengard *et al.,* 1989).

2.   MUTATIONS IN A GENE ENCODING A PROTEIN PHOSPHATASE CAUSE MITOTIC ABNORMALITIES

Larvae homozygous for null mutations in the late lethal complementation group *1(3)ck19* show several defects characteristic of some degree of mitotic arrest, including abnormal sister chromatid segregation, excessive chromosome condensation, and defective spindle organization (Axton *et al.,* 1990). The *ck19* gene is one of four loci in *Drosophila* encoding a type 1 serine/threonine phosphatase; one can thus hypothesize that this enzyme is required either for the inactivation of MPF after metaphase, or for the dephosphorylation of proteins that had been phosphorylated either directly or indirectly by MPF kinase activity. Interestingly, examination of *ck19* mutant brains by indirect immunofluorescence microscopy reveals an abnormal mitotic spindle organization, with unusually densely aggregated spindle microtubules (Axton *et al.*, 1990).

3. A MYOSIN REGULATORY LIGHT CHAIN IS REQUIRED FOR THE COMPLETION OF CYTOKINESIS

Highly polyploid cells accumulate in normally diploid tissues of larvae homozygous for the mutation *spaghetti-squash* (*sqh*). It has recently been found that this gene encodes the regulatory light chain (RLC) of nonsarcomeric myosin, which modulates both myosin filament assembly and the actin-activated myosin ATPase in a phosphorylation-dependent manner [Karess *et al.*, 1991; R. Karess (NYU Medical School) and D. Kiehart (Harvard University), personal communication]. These results suggest that the RLC molecule may regulate assembly of the contractile ring essential for the completion of cytokinesis.

VIII.   Conclusions and Perspectives

The wide variety of mitotic mutants that can be identified in the screens described above give confidence that investigations on cell division in *Drosophila* will continue to reveal many new insights. Potentially the richest source of cell division mutants are those isolated through cytological analysis of collections of strains dying at the larval/pupal boundary, but this source as yet remains virtually untapped for molecular analysis. Given the efficiency by which genes identified by P element-induced mitotic mutations can be cloned, we anticipate that the rate of progress will further accelerate. In the future, immunocytochemical localization using antibodies directed against mitotic gene products will increasingly supplement information obtained from DNA sequencing; the variety of tissue types amenable to cytological analysis provides great flexibility in this regard. Finally, given the clear indications that many components of the mitotic apparatus are conserved in evolution, we believe that *Drosophila* genes and immunological reagents will be of value for the study of cell division in other organisms. Many of the phenotypes displayed by *Drosophila* mitotic mutants are reminiscent of those seen in mammalian cell cycle mutants (Gatti and Baker, 1989). Common genetic threads are thus likely to weave together events of the cell cycle in organisms as disparate as *Drosophila* and man.

REFERENCES

Ashburner, M. (1989). "*Drosophila*: A Laboratory Handbook." Cold Spring Harbor Laboratory, Cold Spring Harbor, New York.

Axton, J. M., Dombradi, V., Cohen, P. T. W., and Glover, D. M. (1990). One of the protein phosphatase 1 isoenzymes in *Drosophila* is essential for mitosis. *Cell (Cambridge, Mass.)* **63**, 33–46.

Baker, B. S., and D. A. Smith (1979). The effects of mutagen sensitive mutants of *Drosophila melanogaster* in nonmutagenized cells. *Genetics* **92**, 833–847.

Baker, B. S., Carpenter, A. T. C., Esposito, M. S., Esposito, R. E., and Sandler, L. (1976a). The genetic control of meiosis. *Ann. Rev. Genet.* **10**, 53–134.

Baker, B. S., Boyd, J. B., Carpenter, A. T. C., Green, M. M., NguYen, T. D., Ripoll, P., and Smith, P. D. (1976b). Genetic control of meiotic recombination and somatic DNA metabolism in *Drosophila melanogaster*. *Proc. Natl. Acad. Sci. U.S.A.* **73**, 4140–4144.

Baker, B. S., Carpenter, A. T. C., and Ripoll, P. (1978). The utilization during mitotic cell division of loci controlling meiotic recombination and disjunction in *Drosophila melanogaster*. *Genetics* **90**, 531–578.

Baker, B. S., Smith, D. A., and Gatti, M. (1982). Region-specific effects on chromosome integrity of mutations at essential loci in *Drosophila melanogaster*. *Proc. Natl. Acad. Sci. U.S.A.* **79**, 1205–1209.

Baker, B. S., Carpenter, A. T. C., and Gatti, M. (1987). On the biological effects of mutants producing aneuploidy in *Drosophila*. *In* "Aneuploidy — Incidence and Etiology" (A. Sandberg and B. Vig, eds.), pp. 273–296. Alan R. Liss. New York.

Bellen, H. J., O'Kane, C. J., Wilson, C., Grossniklaus, U., Pearson, R. K., and Gehring, W. J. (1989). P element-mediated enhancer detection: A versatile method to study development in *Drosophila*. *Genes Dev.* **3**, 1288–1300.

Bier, E., Vaessin, H., Shepherd, S., Lee, K., McCall, K., Barbel, S., Ackerman, L., Carretto, R., Uemura, T., Grell, E., Jan, L. Y., and Jan, Y. N. (1989). Searching for pattern and mutation in the *Drosophila* genome with a *P-lacZ* vector. *Genes Dev.* **3**, 1273–1287.

Bodenstein, D. (1950). The postembryonic development of *Drosophila*. *In* "Biology of *Drosophila*" (M. Demerec, ed.), pp. 275–367. Wiley, New York.

Boyd, J. B., Golino, M. D., Nguyen, T. G., and Green, M. M. (1976). Isolation and characterization of X-linked mutants of *Drosophila melanogaster* which are sensitive to mutagens. *Genetics* **84**, 485–508.

Boyd, J. B., Golino, M. D., Shaw, K. E. S., Osgood, C. L., and Green, M. M. (1981). Third chromosome mutagen-sensitive mutants of *Drosophila melanogaster*. *Genetics* **97**, 607–623.

Bridges, C. B. (1916). Non-disjunction as a proof of the chromosome theory of heredity. *Genetics* **1**, 1–52 and 107–163.

Brinkley, B. R., Cox, S. M., and Pepper, D. A. (1980). Structure of the mitotic apparatus and chromosomes after hypotonic treatment of mammalian cells *in vitro*. *Cytogenet. Cell Genet.* **26**, 165–176.

Bryant, P. J., and Levinson, P. (1985). Intrinsic growth control in the imaginal primordia of *Drosophila*, and the autonomous action of a lethal mutation causing overgrowth. *Dev. Biol.* **107**, 355–363.

Carpenter, A. T. C. (1975). Electron microscopy of meiosis in *Drosophila melanogaster* females. I. Structure, arrangement, and temporal change of the synaptonemal complex in wild type. *Chromosoma* **51**, 157–182.

Carpenter, A. T. C. (1979a). Synaptonemal complex and recombination nodules in wild type *Drosophila melanogaster* females. *Genetics* **92**, 511–541.

Carpenter, A. T. C. (1979b). Recombination nodules and synaptonemal complex in recombination-defective females of *Drosophila melanogaster*. *Chromosoma* **75**, 259–292.

Casal, J., Gonzalez, C., and Ripoll, P. (1990a). Spindles and centrosomes during male meiosis in *Drosophila melanogaster*. *Eur. J. Cell Biol.* **51**, 38–44.

Casal, J., Gonzalez, C., Wandosell, F., Avila, J., and Ripoll, P. (1990b). Abnormal meiotic spindles cause a cascade of defects during spermatogenesis in *asp* males of *Drosophila*. *Development* **108**, 251–260.

Church, K., and Lin, H. P. P. (1982). Meiosis in *Drosophila melanogaster*. The prometaphase-I kinetochore microtubule bundle and kinetochore orientation in males. *J. Cell Biol.* **93**, 365–373.

Church, K., and Lin, H. P. P. (1985). Kinetochore microtubules and chromosome movement during prometaphase in *Drosophila melanogaster* spermatocytes studied in life and with the electron microscope. *Chromosoma* **92**, 273–282.

Cooley, L., Kelley, R., and Spradling, A. C. (1988). Insertional mutagenesis of the *Drosophila* genome with single P elements. *Science* **239**, 1121–1128.

Edgar, B. A., and O'Farrell, P. H. (1989). Genetic control of cell division patterns in the *Drosophila* embryo. *Cell (Cambridge, Mass.)* **57**, 177–187.

Endow, S. A., Henikoff, S., and Soler-Niedziela, L. (1990). Mediation of meiotic and early mitotic chromosome segregation in *Drosophila* by a protein related to kinesin. *Nature (London)* **345**, 81–83.

Fantes, P. (1984). Cell-cycle control in *Schizosaccharomyces pombe*. *In* "The Microbial Cell Cycle" (P. Nurse and E. Streiblova eds.), pp. 109–125. CRC Press, Boca Raton, Florida.

Foe, V. E. (1989). Mitotic domains reveal early commitment of cells in *Drosophila* embryos. *Development* **107**, 1–22.

Foe, V. E., and Alberts, B. M. (1983). Studies of nuclear and cytoplasmic behavior during the five mitotic cycles that precede gastrulation in *Drosophila* embryogenesis. *J. Cell Sci.* **61**, 31–70.

Frasch, M., Glover, D. M., and Saumweber, H. (1986). Nuclear antigens follow different pathways into daughter nuclei during mitosis in early *Drosophila* embryos. *J. Cell Sci.* **82**, 155–172.

Freeman, M., and Glover, D. M. (1987). The *gnu* mutation of *Drosophila* causes inappropriate DNA synthesis in unfertilized and fertilized eggs. *Genes Dev.* **1**, 924–930.

Freeman, M., Nusslein-Volhard, C., and Glover, D. M. (1986). The dissociation of nuclear and centrosomal division in *gnu*, a mutation causing giant nuclei in *Drosophila*. *Cell (Cambridge, Mass.)* **46**, 457–468.

Fuller, M. T. (1986). Genetic analysis of spermatogenesis in *Drosophila*: The role of testis-specific β-tubulin and interacting genes in cellular morphogenesis. *In* "Gametogenesis and the Early Embryo" (J. G. Gall, ed.), pp. 19–41. Alan R. Liss, New York.

Garza, D., Ajioka, J. W., Burke, D. T., and Hartl, D. L. (1989). Mapping the *Drosophila* genome with yeast artificial chromosomes. *Science* **246**, 641–646.

Gatti, M. (1979). Genetic control of chromosome breakage and rejoining in *Drosophila melanogaster*. I. Spontaneous chromosome aberrations in X-linked mutants defective in DNA metabolism. *Proc. Natl. Acad. Sci. U.S.A.* **76**, 1377–1381.

Gatti, M., and Baker, B. S. (1989). Genes controlling essential cell-cycle functions in *Drosophila melanogaster*. *Genes Dev.* **3**, 438–453.

Gatti, M., and Pimpinelli, S. (1983). Cytological and genetic analysis of the Y chromosome of *Drosophila melanogaster*. I. Organization of the fertility factors. *Chromosoma* **88**, 349–373.

Gatti, M., Tanzarella, C., and Olivieri, G. (1974). Analysis of the chromosome aberrations induced by X-rays in somatic cells of *Drosophila melanogaster*. *Genetics* **77**, 701–719.

Gatti, M., Pimpinelli, S., and Santini, G. (1976). Characterization of *Drosophila* heterochromatin. I. Staining and decondensation with Hoechst 33258 and quinacrine. *Chromosoma* **57**, 351–375.

Gatti, M., Pimpinelli, S., and Baker, B. S. (1980). Relationships among chromatid interchanges, sister chromatid exchanges, and meiotic recombination in *Drosophila melanogaster. Proc. Natl. Acad. Sci. U.S.A.* **77,** 1575–1579.

Gatti, M., Pimpinelli, S., Bove, C., Baker, B. S., Smith, D. A., Carpenter, A. T. C., and Ripoll, P. (1983a). Genetic control of mitotic cell division in *Drosophila melanogaster. In* "Genetics – New Frontiers. Proceedings of the XV International Congress of Genetics" (V. L. Chopra, B. C. Joshi, R. P. Sharma, and H. C. Bansal, eds.). Vol. III, pp. 193–204. Oxford and IBH Publishing Co., New Delhi.

Gatti, M., Smith, D. A., and Baker, B. S. (1983b). A gene controlling condensation of heterochromatin in *Drosophila melanogaster. Science* **221,** 83–85.

German, J. (1979). Roberts syndrome. I. Cytological evidence for a disturbance in chromatid pairing. *Clin. Genet.* **16,** 441–447.

Glover, D. M. (1989). Mitosis in *Drosophila. J. Cell Sci.* **92,** 137–146.

Gonzalez, C., Casal, J., and Ripoll, P. (1988). Functional monopolar spindles caused by the mutation in *mgr,* a cell division gene of *Drosophila melanogaster. J. Cell Sci.* **89,** 34–47.

Gonzalez, C., Saunders, R. D. C., Casal, J., Molina, I., Carmena, M., Ripoll, P., and Glover, D. M. (1990). Mutations at the *asp* locus of *Drosophila* lead to multiple free centrosomes in syncytial embryos but restrict centrosome duplication in larval neuroblasts. *J. Cell Sci.* **96,** 605–616.

Hamilton, B. A., Palazzolo, M. J., Chang, J. H., VuayRhaghavan, K., Mayeda, C. A., Whitney, M. A., and Meyerowitz, E. M. (1991). Large scale detection of transposon insertions into cloned loci by plasmid rescue of flanking DNA. *Proc. Natl. Acad. Sci. U.S.A.* **88,** 2731–2735.

Hanahan, D. (1985). Techniques for transformation of *E. coli. In* "DNA Cloning: A Practical Approach" (D. M. Glover, ed.), Vol. 1, pp. 109–135. IRL Press, Oxford.

Hartwell, L. H. (1974). *Saccharomyces cerevisiae* cell cycle. *Bacteriol. Rev.* **38,** 164–198.

Hirano, T., and Yanagida, M. (1988). Controlling elements in the cell division cycle of *Schizosaccharomyces pombe. In* "Molecular and Cell Biology of Yeasts" (E. F. Walton, ed.), pp. 223–245. Blakie Press, Glasgow, Scotland.

Karess, R. E. (1985). P element mediated germ line transformation of *Drosophila. In* "DNA Cloning: A Practical Approach" (D. M. Glover, ed.), pp. 121–141. IRL Press, Oxford.

Karess, R. E., and Glover, D. M. (1989). *rough deal*: A gene required for proper mitotic segregation in *Drosophila. J. Cell Biol.* **109,** 2951–2961.

Karess, R. E., Chang, X.-j., Edwards, K. A., Kulkarni, S., Aguilera, I., and Kiehart, D. P. (1991). The regulatory light chain of nonmuscle myosin is encoded by *spaghetti-squash,* a gene required for cytokinesis in *Drosophila. Cell (Cambridge, Mass.)* **65,** 1177–1189.

Karr, T. L., and Alberts, B. M. (1986). Organization of the cytoskeleton in early *Drosophila* embryos. *J. Cell Biol.* **102,** 1494–1509.

Kellogg, D. R., Mitchison, T. J., and Alberts, B. M. (1988). Behavior of microtubules and actin filaments in living *Drosophila* embryos. *Development* **103,** 675–686.

Kellogg, D. R., Field, C. M., and Alberts, B. M. (1989). Identification of microtubule-associated proteins in the centrosome, spindle, and kinetochore of the early *Drosophila* embryo. *J. Cell Biol.* **109,** 2977–2991.

Lehner, C. F., and O'Farrell, P. H. (1989). Expression and function of *Drosophila* cyclin A during embryonic cell cycle progression. *Cell (Cambridge, Mass.)* **56,** 957–968.

Lehner, C. F., and O'Farrell, P. H. (1990). The roles of *Drosophila* cyclins A and B in mitotic control. *Cell (Cambridge, Mass.)* **61,** 535–547.

Lifschytz, E. (1987). The developmental program of spermiogenesis in *Drosophila*: A genetic analysis. *Int. Rev. Cytol.* **109,** 211–254.

Lifschytz, E., and Meyer, G. F. (1977). Characterization of male meiotic-sterile mutations in *Drosophila melanogaster. Chromosoma* **64,** 371–392.

Lindsley, D. L., and Grell, E. H., (1968). Genetic variations of *Drosophila melanogaster. Carnegie Inst. Washington Publ.* 627.

Lindsley, D. L., and Sandler, L. (1977). The genetic analysis of meiosis in female *Drosophila melanogaster. Philos. Trans. R. Soc. London* **277**, 295–312.

Lindsley, D. L., and Tokuyasu, K. T. (1980). Spermatogenesis. *In* "The genetics and biology of *Drosophila*" (M. Ashburner and T. R. F. Wright, ed.), pp. 225–294. Academic Press, New York.

Lindsley, D. L., and Zimm, G. (1985). The genome of *Drosophila melanogaster*. Part 1: Genes A–K. *Drosophila Inf. Serv.* **62**, 1–227.

Lindsley, D. L., and Zimm, G. (1986). The genome of *Drosophila melanogaster*. Part 2: Lethals, maps. *Drosophila Inf. Serv.* **64**, 1–158.

Lindsley, D. L., and Zimm, G. (1987). The genome of *Drosophila melanogaster*. Part 3: Rearrangements. *Drosophila Inf. Serv.* **65**, 1–224.

Lindsley, D. L., and Zimm, G. (1990). The genome of *Drosophila melanogaster*. Part 4: Genes L–Z, balancers, transposable elements. *Drosophila Inf. Serv.* **68**, 1–382.

McDonald, H. B., and Goldstein, L. S. B. (1990). Identification and characterization of a gene encoding a kinesin-like protein in *Drosophila. Cell (Cambridge, Mass.)* **61**, 991–1000.

McDonald, H. B., Stewart, R. J., and Goldstein, L. S. B. (1990). The kinesin-like *ncd* protein of *Drosophila* is a minus end-directed microtubule motor. *Cell (Cambridge, Mass.)* **63**, 1159–1165.

Miller, K. G., Field, C. M., and Alberts, B. M. (1989). Actin-binding proteins from *Drosophila* embryos: A complex network of interacting proteins detected by F-actin affinity chromatography. *J. Cell Biol.* **109**, 2963–2975.

Minden, J. S., Agard, D. A., Sedat, J. W., and Alberts, B. M. (1989). Direct cell lineage analysis in *Drosophila melangoster* by time lapse three-dimensional optical microscopy of living embryos. *J. Cell Biol.* **109**, 505–516.

Moir, D., and Botstein, D. (1982). Determination of the order of gene function in the yeast nuclear division pathway using *cs* and *ts* mutants. *Genetics* **100**, 565–577.

Moir, D., Stewart, S. E., Osmond, B. C., and Botstein, D. (1982). Cold-sensitive cell-division cycle mutants of yeast: Isolation, properties, and pseudoreversion studies. *Genetics* **100**, 547–563.

Nagl. W. (1978). "Endopolyploidy and polyteny in differentiation and evolution." North-Holland, Amsterdam.

Nurse, P. (1985). Cell-cycle control in yeast. *Trends Genet.* **1**, 51–55.

Ochman, H., Gerber, A. S., and Hartl, D. L. (1988). Genetic applications of an inverse polymerase chain reaction. *Genetics* **120**, 621–623.

O'Farrell, P. H., Edgar, B. A., Lakich, D., and Lehner, C. F. (1989). Directing cell division during development. *Science* **246**, 635–640.

O'Kane, C. J., and Gehring, W. J. (1987). Detection *in situ* of genomic regulatory elements in *Drosophila. Proc. Natl. Acad. Sci. U.S.A.* **84**, 9123–9127.

Orr, W., Komitopoulu, K., and Kafatos, F. C. (1984). Mutants suppressing *in trans* chorion gene amplification in *Drosophila. Proc. Natl. Acad. Sci. (U.S.A.)* **81**, 3773–3777.

Perrimon, N., Engstrom, L., and Mahowald, A. P. (1985). Developmental genetics of the 2C-D region of the *Drosophila* X chromosome. *Genetics* **111**, 23–41.

Pimpinelli, S., Santini, G., and Gatti, M. (1976). Characterization of *Drosophila* heterochromatin. II. C- and N-banding. *Chromosoma* **57**, 377–386.

Pirrotta, V. (1986). Cloning *Drosophila* genes. *In* "*Drosophila*: A Practical Approach" (D. B. Roberts ed.), pp. 83–110. IRL Press, Oxford.

Pringle. J. R., and Hartwell, L. H. (1981). The *Saccharomyces cerevisiae* cell cycle. In "Molecular Biology of the Yeast *Saccharomyces*: Life Cycle and Inheritance" (J. N. Strathern, E. W. Jones, and J. R. Broach, eds.), pp. 97–142. Cold Spring Harbor Laboratory, Cold Spring Harbor, New York.

Regan, C. L., and Fuller, M. T. (1988). Interacting genes that affect microtubule function: The *nc2* allele of the *haywire* locus fails to complement mutations in the testes-specific β-tubulin gene of *Drosophila. Genes Dev.* **2**, 82–92.

Ripoll, P., Pimpinelli, S., Valdivia, M. M., and Avila, J. (1985). A cell division mutant of *Drosophila* with a functionally abnormal spindle. *Cell (Cambridge, Mass.)* **41,** 907–912.

Ripoll, P., Casal, J., and Gonzalez, C. (1987). Towards the genetic dissection of mitotis in *Drosophila. BioEssays* **7,** 204–210.

Robertson, H. M., Preston, C. R., Phillis, R. W., Johnson-Schlitz, D., Benz, W. K., and Engles, W. R. (1988). A stable genomic source of P element transposase in *Drosophila melanogaster. Genetics* **118,** 461–470.

Rosengard, A. M., Krutzsch, H. C., Shearn, A., Biggs, J. R., Barker, E., Margulies, I. M. K., King, C. R., Liotta, L. A., and Steeg, P. S. (1989). Reduced NM23/awd protein in tumour metastasis and aberrant *Drosophila* development. *Nature (London)* **342,** 177–180.

Rubin, G. M., and Spradling, A. C. (1983). Vectors for P element-mediated gene transfer in *Drosophila. Nucleic Acids Res.* **11,** 6341–6351.

Shearn, A., Rice, T., Garen, A., and Gehring, W. (1971). Imaginal disk abnormalities in lethal mutants of *Drosophila. Proc. Natl. Acad. Sci. U.S.A.* **68,** 2594–2598.

Siden-Kiamos, I., Saunders, R. D. C., Spanos, L., Majerus, T., Trenear, J., Savakis, C., Louis, C., Glover, D. M., Ashburner, M., and Kafatos, F. C. (1990). Towards a physical map of the *Drosophila melanogaster* genome: Mapping of cosmid clones within defined genomic divisions. *Nucleic Acids Res.* **18,** 6261–6270.

Smith, D. A., Baker, B. S., and Gatti, M. (1985). Mutations in genes controlling essential mitotic functions in *Drosophila melanogaster. Genetics* **110,** 647–670.

Snyder, R. D., and Smith, P. D. (1982). Mutagen sensitivity of *Drosophila melanogaster.* V. Identification of second chromosome mutagen-sensitive strains. *Mol Gen. Genet.* **188,** 249–255.

Spradling, A. C. (1986). P element-mediated transformation. In "*Drosophila*: A Practical Approach" (D. B. Roberts ed.), pp. 175–197. IRL Press, Oxford.

Sullivan, W., Minden, J. S., and Alberts, B. M. (1990). *daughterless-abo like,* a *Drosophila* maternal effect mutation that exhibits abnormal centrosome separation during the late blastoderm divisions. *Development* **110,** 311–323.

Sunkel, C., and Glover, D. M. (1988). *polo,* a mitotic mutant of *Drosophila* displaying abnormal mitotic spindles. *J. Cell Sci* **89,** 25–38.

Walker, R. A., Salmon, E. D., and Endow, S. A. (1990). The *Drosophila* claret segregation protein is a minus end-directed motor molecule. *Nature (London)* **347,** 780–782.

Whitfield, W. G. F., Millar, S. E., Saumweber, H., Frasch, M., and Glover, D. M. (1988). Cloning of a gene encoding an antigen associated with the centrosome in *Drosophila. J. Cell Sci.* **89,** 467–480.

Whitfield, W. G. F., Gonzalez, C., Sanchez-Herrero, E., and Glover, D. M. (1989). Transcripts of one of two *Drosophila* cyclin genes become localized in pole cells during embryogenesis. *Nature (London)* **338,** 337–340.

Whitfield, W. G. F., Gonzalez, C., Maldonado-Codina, G., and Glover, D. M. (1990). The A- and B-type cyclins of *Drosophila* are accumulated and destroyed in temporally distinct events that define separable phases of the G2-M transition. *EMBO J.* **9,** 2563–2572.

Wilson, C., Pearson, R. K., Bellen, H. J., O'Kane, C. J., Grossniklaus, U., and Gehring, W. J. (1989). P element-mediated enhancer detection: An efficient method for isolating and characterizing developmentally regulated genes in *Drosophila. Genes Dev.* **3,** 1301–1313.

Wright, T. R. F. (1974). A cold-sensitive zygotic lethal causing high frequencies of nondisjunction during meiosis I in *Drosophila melanogaster* females. *Genetics* **76,** 511–536.

Zhang, P., Knowles, B. A., Goldstein, L. S. B., and Hawley, R. S. (1990). A kinesin-like protein required for distributive chromosome segregation in *Drosophila. Cell (Cambridge, Mass.)* **62,** 1053–1062.

Chapter 22

Position-Effect Variegation—An Assay for Nonhistone Chromosomal Proteins and Chromatin Assembly and Modifying Factors

T. GRIGLIATTI

Department of Zoology
University of British Columbia
Vancouver, British Columbia
V6T 1Z4, Canada

METHODS IN CELL BIOLOGY, VOL. 35

I. Introduction

The extensive sizes of the genomes of most eukaryotic organisms necessitate the partitioning of the genetic material into smaller linkage groups both to facilitate replication and to maintain the fidelity of the system. The DNA in these linkage units is compacted into the typical 30-nm fibers seen in interphase nuclei by packaging it as chromatin. The proper segregation and equal distribution of the genome to daughter cells requires, or at least is greatly facilitate by, the further compaction of chromatin into chromosomes. This elaborate and multifarious infrastructure of chromatin and chromosomes makes both the processes of DNA replication and transcriptional activation of specific genes more complex. It naturally leads to the question of whether this infrastructure demands an additional stratum of gene regulation, or whether this infrastructure has been adapted, at least in some cases, to regulate gene expression in eukaryotes. Thus, for several decades, the role that chromatin structure played in the regulation of gene expression was often debated.

In the past decade or so, numerous studies have shown that chromatin decondensation is closely correlated with activation of transcription. Many studies have demonstrated that actively transcribed genes are more sensitive than nontranscribed genes to digestion by endonucleases such as DNase I. The nucleases sensitivity of a particular gene is limited to the tissue in which that gene is expressed. Finally, the increased sensitivity to nuclease can be detected just prior to and at transcription. The advent of transformation in a number of higher organisms has demonstrated that the genomic environment in which a gene is placed can strikingly influence its expression. More recently, there has been good evidence for the existence of chromatin domains within the euchromatin. In some cases, the limits of these domains are defined by either specific nonhistone chromosomal proteins (Hoffmann *et al.,* 1989) or DNA sequences (Udvardy *et al.,* 1985; Kellum and Schedl, 1991). Finally, extensive analyses of the homeotic genes in *Drosophila* have suggested that chromatin structure plays a fundamental role in the control of segment determination and differentiation (Peifer *et al.,* 1987). Hence, chromatin structure is now thought to play an important role in the regulation of gene expression. It has become clear that the formation of transcriptionally competent domains is an essential first step in proper gene expression. Yet, very little is known about the structural components of chromatin, other than histones and a few HMG proteins, and even less is known about the control and mechanics of chromatin assembly at the time of cell division and how chromatin is modulated to allow gene expression during other portions of the cell cycle.

To understand how modifications of chromatin domains are brought about, it is clear that we must identify: (1) the nonhistone proteins that are structural components of chromatin, (2) those proteins that play a role in chromatin assembly at the time of DNA replication, and (3) those proteins that modify chromatin structure to facilitate proper gene expression. There are two approaches which can be taken to identify the structural components of chromatin. One is the biochemical fractionation of chromatin. Although this approach has met with limited success over the past two decades, the advent of new techniques such as those presented in Section III suggest that this approach will now be more rewarding. The second approach is the identification of genes that encode nonhistone chromosomal proteins (NHCPs) and chromatin assembly or modification factors. The genetic approach is limited by the availability of a good assay system for the detection of mutations in NHCPs or chromatin assembly or modification factors.

An appropriate assay system for the recovery of mutants in genes that encode NHCPs or chromatin modification factors must have the following characteristics. First, there must be a reporter gene whose activity is easily assayed. Second, alterations in the expression of the reporter gene must result from a distinct alteration in chromatin structure. Third, one must be able to identify strains in which the altered reporter gene function results from an alteration in chromatin structure, not simply a mutation in the reporter gene itself. The second and third criteria are not easily met. For example, the heat-shock response in *Drosophila* is associated with massive puffing of the bands encoding the heat-shock genes. While this might presumably meet the second requirement, mutants in the heat-shock response could easily involve a number of processes other than alterations in chromatin domains. The formation of lampbrush loops on the Y chromosome of *Drosophila hydeii* might be a useful model system, but attempts to isolate point mutations that disrupt loop formation has been fraught with frustration. The phenomenon of X-inactivation in mammals might be the paragon assay system for the isolation of mutations in genes that encode NHCPs. In the somatic tissue of mammals, all but one X chromosome is heterochromatinized and thus functionally inactivated during the first trimester of embryogenesis. In eutherian mammals, the choice of which X chromosome remains transcriptionally competent (euchromatic) appears to be made randomly, and, more intriguingly, once an X chromosome is inactivated that homolog remains inactive in all daughter cells in the lineage. Hence, the initial decision is propagated clonally and the somatic memory or chromosomal imprinting of that initial decision may be examined. However, the isolation of mutations that alter X-inactivation in mammals would prove to be an Herculean task. In contrast, given an appropriate set of chromosomal rearrangements, the identification and isolation of the DNA site from which X-inactivation spreads are quite plausible in mammals (Brown *et al.*, 1991).

II.   Position-Effect Variegation—The Phenomenon

The phenomenon of position-effect variegation (PEV) resembles X-inactivation in mammals. Hence, it may serve as suitable assay system for the recovery of mutations in NHCPs. PEV is the mosaic expression of a gene, associated with a chromosomal or genomic rearrangement. In most instances, PEV occurs when a gene, normally located in euchromatin, is placed adjacent to a broken segment of heterochromatin. The variegating gene is inactivated in some cells, but remains transcriptionally competent in other cells, giving a mosaic phenotype (Fig. 1). PEV has been most extensively studied in *Drosophila*, but it has been described in vertebrates (Cattanach, 1974), lower eukaryotes (Clutterbuck and Spathas, 1984), and plants (Catcheside, 1938, 1947). Hence, by inference the phenomenon appears to be widespread among eukaryotes.

There have been several extensive reviews of the literature on PEV (Lewis, 1950; Baker, 1986; Spofford, 1976) and several shorter reviews (Eissenberg, 1989; Henikoff, 1990; Spradling and Karpen, 1990). Therefore, what follows is simply a summary of the general properties of PEV. In *Drosophila*, virtually any gene appears to be susceptible to PEV. The transposition event which

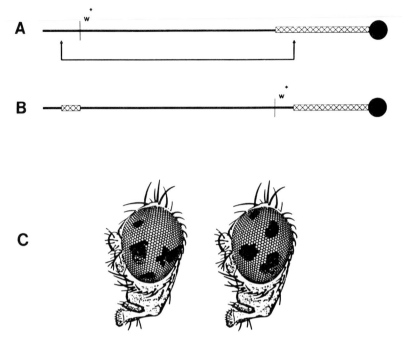

FIG. 1. An example of position-effect variegation. The *white* (w^+) gene, which is required for deposition of pigments into the eye pigments cells, is shown in its normal configuration (A) . A chromosomal rearrangement such as an inversion (shown in this example), translocation, or transposition which juxtaposes the w^+ allele to a broken piece of heterochromatin may result in a mosaic expression of the w^+ gene, an example of which is shown in B. The solid line represents euchromatin; the cross-hatched region represents heterochromatin; the filled circle represents the centromere. Two examples of variegated eyes are shown in C. The dark areas represent pigmented areas (w^+ gene expressed); the light areas represent unpigmented areas (w^+ gene inactive).

juxtaposes a broken piece of heterochromatin to the normally euchromatic segment of DNA can be any chromosomal rearrangement, such as an inversion or translocation. More recently, novel insertions into heterochromatic regions, resulting from DNA transfection, also give variegated phenotypes which resemble PEV. Furthermore, most segments of heterochromatin seem capable of causing PEV. Hence, the phenomenon appears to be associated with the juxtaposition of a broken segment of heterochromatin with a normally euchromatic locus. The juxtaposition of euchromatin and a disrupted segment of heterochromatin appear to be one characteristic that distinguishes PEV from stable types of position effects that result from the placement of a gene into a new location within euchromatin, which often occurs as a consequence of transfection experiments.

The developmental interval during which PEV, or the inactivation of the reporter gene, occurs is not well defined. Many of the visible PEV phenotypes comprise large spot mosaics, suggesting that the transcriptional fate of the variegating gene (the gene which reports on the variegation phenomenon) is determined early during development and that, like X-chromosome inactivation, it is propagated clonally. However, some of the PEV phenotypes are fine grain mosaics, suggesting that, in some cases, transcriptional competency of the variegating gene may be determined later in development, or that the initial inactivation event occurs early but its propagation is considerably less stable. There has been some discussion as to whether the gene inactivation resulting from PEV is an all-or-none phenomenon (complete inactivation) or simply reflects reduced expression (repression) at the level of the individual cell. While this is an intriguing question, it is not important to the use of PEV as an assay system for the identification of NHCPs or chromatin assembly and modifying factors.

The two most extensively characterized affectors of PEV are temperature and the number of Y chromosomes in the genome. Generally, allowing development to proceed at a high temperature (25–30°C) suppresses PEV, that is, the proportion of cells in which the variegating gene is repressed is decreased. Allowing the flies to develop at a low temperature (17–20°C) generally enhances PEV, that is, the variegating gene is repressed in a greater proportion of cells. The effects of temperature on PEV are often consistent, but are not always very dramatic. For example, in the strain $Inversion(1)white^{motted\ 4}$ (w^{m4}), which variegates for the deposition of pigments in the eye, the amount of eye pigment is about 5, 15, and 25–30% of wild-type in flies which proceed through development at 17, 22, and 29°C, respectively.

Alterations in the amount of heterochromatin in the genome also affect PEV. This is generally measured by altering the number of Y chromosomes in the genome, since they are easy to manipulate, but similar effects are seen by altering the amount of autosomal heterochromatin. Addition of Y chromosome (or large segment of heterochromatin) to the genome of either males or females generally suppresses PEV, while removal of a Y chromosome (heterochromatin) enhances PEV. For example, the amount of eye pigment in X0, XY, and XYY w^{m4} males is 1, 15, and 30% respectively, at 22°C.

The molecular basis for the gene inactivation associated with PEV has not been established. However, it is well documented that the allele that is inactivated is generally cis to the broken segment of heterochromatin (Judd, 1955) [an exception may be *Brown* dominant variegation; see Henikoff and Dreesen (1989)], and that inactivation is not caused by somatic loss of the variegating allele (Henikoff, 1981). Furthermore, there is a good correlation between the frequency with which a variegating gene is inactivated in one tissue and the frequency with which that region is packaged as heterochromatin in polytene salivary gland chromosomes (Hartmann-Goldstein, 1967). Thus, it appears that PEV involves some type of cis-inactivation of genes juxtaposed to heterochromatin and that this inactivation is mediated by an alteration in chromatin packaging.

PEV varies considerably in the distance over which the gene inactivation spreads. In some variegating rearrangements, only the gene adjacent to the heterochromatic breakpoint appears to be inactivated, an intrusion of heterochromatin for a distance of only about 25 kb. At the other extreme, genes approximately 80 chromosome bands, or nearly 2000 kb, away from the variegation-inducing breakpoint are inactivated.

III.   A Model for Position-Effect Variegation

The following is a simple model describing the assembly of chromatin in standard chromosomes and how PEV occurs in strains with the appropriate genomic rearrangements. It is meant to be a general model with, undoubtedly, many exceptions. This type of model provided the impetus to use PEV as an assay system to identify mutations in genes that encode NHCPs and chromatin assembly and modifying factors. It also serves as one focus for the interpretation of the phenotypes of the mutants that were recovered.

It is well established that PEV is a consequence of placing a gene next to a broken piece of heterochromatin. This has led many people to speculate that the gene inactivation that is symptomatic of PEV occurs by the illicit spread of heterochromatin, or heterochromatic elements, beyond the newly created, and artifical, euchromatin–heterochromatin junction. The normally euchromatic segment of DNA adjacent to the novel junction is packaged as heterochromatin (compaction) in some cells, and thus the transcriptional competency of the genes in this DNA segment is repressed. In neighboring cells, this region might remain euchromatic, and consequently the genes in this region of the DNA would remain transcriptionally competent. The mosaicism is easily explained by postulating a slight variation, from one cell to another, in the distance over which the intruding heterochromatin spreads. Indeed, cytogenetic evidence, based on polytene chromosome morphology in variegating strains, suggests that the distance over which heterochromatinization spreads varies from one cell to another [for examples, see Hartmann-Goldstein (1967) and Hayashi et al. (1990)]. Since alterations in the ploidy of the Y chromosome produce striking effects on PEV (extra copies suppress and loss enhances PEV), Zuckerkandl (1974) and others have suggested that at least some of the structural components of heterochromatin are limited. The Y chromosome simply acts as a "sink" for these heterochromatic elements. In other words, it competes for the elements and can influence the likelihood or distance of spread.

A. Nucleation Sites for the Formation of Chromatin

What initiates the formation of heterochromatin and euchromatin in their respective regions of the genome? First, one might postulate that DNA sequences exist that act as initiation or nucleation sites for the formation of heterochromatin and for the formation of euchromatin. The sequences which initiate the formation of heterochromatin must differ from those that initiate the formation of euchromatin. Once initiated at a particular site, the formation of heterochromatin and euchromatin might continue as a self-assembly process. Like PEV, such self-assembly processes are often temperature sensitive (cold sensitive). Tartof and colleagues have produced a very elegant model for the self-assembly of chromatin based on the mass action model of chemistry and the assumption that the NHCPs are produced in limited amounts. The reader is directed to their manuscript for a very lucid discussion of the self-assembly process and its implications for PEV (Locke *et al.*, 1988).

The assembly of chromatin could start at a variety of initiation sites which are distributed throughout heterochromatin and euchromatin. These initiation or nucleation sites would serve as the start sites for the formation of the appropriate type of chromatin and, once initiated, the process may continue largely, or at least in part, as a self-assembly process. Like DNA replication, the assembly of chromatin could continue in a bidirectional fashion until two domains of euchromatin (or heterochromatin), assembling in opposite direction, meet and fuse. This model would not require stop signals within the normal boundaries of heterochromatin or euchromatin.

B. The Junction between Euchromatin and Heterochromatin

How is the junction between heterochromatin and euchromatin established in a wild-type cell? The normal junction between heterochromatin and euchromatin might simply be a buffer zone in which there are either no genes or only a few genes which are relatively insensitive to their chromatin environment. In normal chromosomes this butter zone might allow for plasticity in the actual junction between heterochromatin and euchromatin. Hence, the actual site of interface between heterochromatin and euchromatin may be slightly fickle, even under normal circumstances, and the juction could "breathe," like a telomere segment. Indeed, this region of plasticity might be the region that has been termed β-heterochromatin. [Heitz (1934) suggested that there are two distinct types of heterochromatin in polytene chromosomes: α-heterochromatin, which is the highly compacted, unbanded, and underreplicated central heterochomatin, and β-heterochromatin, which is loosely compacted, poorly or variably banded, peripheral heterochromatin which resides between euchromatin and α-heterochromatin.] Such a model of plasticity at chromatin junctions allows for the accumulation of transposable elements in these locals. Mobile elements might occasionally insert in these regions when they are packaged as euchromatin. Subsequently, their movement would be restricted by the occasional formation of heterochromatin and they would eventually become entrapped by mutation to loss of function. The entrapment of middle repeats would only add to the buffer zone between true euchromatin and heterochromatin. Furthermore, with this model, one need not postulate the existence of heterochromatin and euchromatin stop signals at the euchromatin–heterochromatin junction; however, weak heterochromatin and euchromatin stop signals at the interface of euchromatin and heterochromatin would be salutary.

C. Genomic Rearrangements, Position-Effect Variegation, and Chromatin Invasion

What prevents the spread, or invasion, of one form of packaging into the territory of the other at these junctions? In PEV, one of the chromosome breaks would occur between two of the many nucleation sites in heterochromatin (or between the "buffer zone" and the first heterochromatin nucleation site). There would be no "buffer zone" at the newly established junction between euchromatin and heterochromatin, and this would allow heterochromatic packaging elements (a subset of NHCPs) to invade, and thereby repress,the genes in the normally euchromatic segment of DNA in some cells.

The initiation sites within the heterochromatic segments of the genome (as well as those for euchromatin formation) need not be identical. Indeed, heterochromatin (like euchromatin) is likely a mosaic of chromatin domains (see Section IX) and thus there may be distinctly different sets of chromatin nucleation sites distributed throughout the heterochromatic regions of the genome. The distance of spread, or degree of invasiveness, of heterochromatin in a particular variegation strain might depend on the distance of the rearrangement breakpoint from the flanking heterochromatic and euchromatic nucleation sites and/or the strength of either the heterochromatic or euchromatic initiation site. The strength of a particular initiation or nucleation site might depend on the sequence of the initiator, the number of tandem copies of that sequence at a particular site, or the density of initiators in a specific segment of DNA.

An alternative model might have the initiation or nucleation sites being directional, so that packaging occurs in one direction only and stop sites would be associated with the adjacent initiation signals. However, in this case, inversions wholly within euchromatin would inevitably fall between two initiation sites and create new junctions with the initiation sites oriented in opposite direction. At one junction, the initiation sites would be oriented toward each other. At the other junction they would face away from one another and the segment delimited by them would not be packaged properly. Hence, we do not favor such a model.

D. Stop Signals for Chromatin Assembly

A number of people have referred to chromatin assembly beginning at some initiating signal and terminating at a stop signal. Within the boundaries of euchromatin and heterochromatin, are there DNA sequences that signal the cessation or termination of chromatin assembly?

There is little doubt that specific chromatin environments are crucial for the proper expression of some genes and disturbance of these domains alters gene expression (Gyurkovics *et al.*, 1990). DNA sequences that act as boundaries for specific chromatin domains have been described in several systems from yeast cassettes to hemoglobin genes (for a recent example, see Kellum and Schedl, 1991). Accordingly, it appears that sequences which delimit the boundaries of specialized types of chromatin domains do exist within euchromatin. One might also suggest that such sequences and specialized domains also exist within heterochromatin, for example, for the nucleolus organizer region. However, stop signals for the general formation of heterochromatin and euchromatin may not be efficacious and, in fact, postulating the existence of such chromatin termination signals within the boundaries of euchromatin and heterochromatin only serves to complicate chromatin assembly.

If chromatin assembly is initiated at a specific nucleation site and proceeds to a termination site, then these initiation and termination sites must be interspersed at regular intervals throughout both euchromatin and heterochromatin. We might then ask what consequence an inversion, or any chromosome rearrangement, would have on chromatin structure. This is easily seen by examining the diagram shown in Fig. 2. An inversion whose breaks occur between an initiation site and its two flanking stop sites would have a minimum impact on chromatin packaging. The same is true for breaks that flank a stop site (Fig. 2-I). An inversion whose distal (left) break point is to the left of one start site and whose proximal (right) break point is located to the right of another start site would also have a minimal impact of chromosome structure (Fig. 2-II). The same is true for an inversion whose break points flank adjacent or distantly located stop sites within the euchromatin

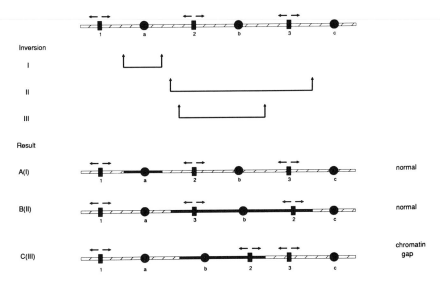

FIG. 2. A model in which chromatin assembly proceeds from a chromatin nucleation or initiation site and progresses to an assembly stop signal and its consequences for chromatin structure. The top line depicts a small segment of a chromosome which may be either from euchromatin or heterochromatin (not both). The filled boxes represent nucleation sites for the initiation of chromatin assembly. They are assigned numbers to indicate relative order on the chromosome segment and may indeed represent nonidentical sequences. The filled circles symbolized stop sequences which function to terminate chromatin assembly; they are assigned letters to indicate relative order along the segment of the chromosome. The hatched line represents DNA. The nucleation and stop sites have been placed equidistant from each other only for the sake of simplicity. Varying the distance alters the frequency with which certain events would occur. I, II, and III represent inversions with different break points (flanking stop sites only, I; flanking nucleation sites only, II; flanking an initiation and stop sites, III). The size of the inversion is not relevant. Consequences of these representative chromosome rearrangements are depicted below, as A, B, and C, respectively. The solid line repressents the inverted segment of the chromosome.

(or heterochromatin). In contrast, if the break point of an inversion flank adjacent start and stop signals, or multiples thereof, a serious problem would result (Fig. 2-III). At one terminus of the inverted segment there would exist a DNA segment delimited by two stop sites. If stop sites actually terminate chromatin assembly, a "chromatin gap" would occur in this DNA interval. Of course, this does not mean that such stop signals do not exist. They may exist, and "chromatin gaps" that are formed as a consequence of certain rearrangements or transpositions may indeed be deleterious. Consequently, rearrangements leading to "chromatin gaps" might occur and simply be selected against. This would only limit the types of rearrangements that survive.

On the other hand, the DNA interval immediately adjacent to the other break point depicted in situation III would be delimited by two initiations sites. Chromatin assembly could commence at this initiation site and spread until it encounters chromatin which is being assembled from the adjacent nucleation site. If this does not create a problem, then there is no *a priori* reason for the existence of stop signals.

Hence, the existence of chromatin assembly stop signals within the boundaries of euchromatin and heterochromatin would be unnecessary in theory and postulating their existence appears to add a complication to chromosome replication, chromatin assembly, and gene regulation. Again, this does not undermine the notion that specialized chromatin domains are defined by a specific subset of unidirectional chromatin assembly start sites or stop sites.

The existence of chromatin assembly start and stop signals throughout the chromosome also has implications for PEV. For example, simple inversions with one break point in euchromatin and the other in heterochromatin should variegate at one of the heterochromatic breaks only. The euchromatin placed adjacent to the other broken segment of heterochromatin would be protected by a stop signal. Complicated chromosomal rearrangements may display other patterns.

While it is unnecessary to postulate the existence of chromatin assembly stop signals within the boundaries of euchromatin and heterochromatin, the existence of such signals at the interface of euchromatin and heterochromatin would be salutary. However, they may also be nonessential for the formation of a serviceable interface (see above).

IV.   Alternative Models for Position-Effect Variegation

A.   Underreplication of the Variegating Genes

Several researchers have suggested that PEV results from underreplication of the euchromatic DNA segment that is juxtaposed to heterochromatin. In polytene nuclei, the α-heterochromatic region of the chromosome, in addition to being late replicating, is underreplicated by about 10-fold relative to the euchromatic sequences. Variegating genes would be underreplicated as a consequence of their position adjacent to heterochromatin, and this large reduction in template number would be responsible for the apparent repression. In a very elegant study, Karpen and Spradling (1990) examined the DNA organization at the interface of heterochromatin and euchromatin in a *Drosophila* minichromosome, and X chromosome in which over 90% of the euchromatic material had been removed by a deletion. The minichromosome, which consisted of the distal tip of the X juxtaposed to a broken segment of the centromeric X heterochromatin, variegates for the bristle genes *acheate* and *yellow*. There is a gradation of underreplicated DNA which encompasses the normally euchromatic material adjoining the heterochromatin in this mini-chromosome. This is a striking contrast from the situation in the standard X chromosome in which this segment of DNA is normally euchromatic and fully replicated. Hence, in this system, there is a good correlation between PEV in the bristle cells and underreplication of the region encoding these variegating loci in the polytene chromosomes of the salivary gland. Since bristle cells are polytenized, it is quite possible that underreplication (reduced template number) accounts for the repression of the *acheate* locus.

The underreplication and chromatin packaging models for PEV are not necessarily mutually exclusive. Clearly, the compaction of DNA into heterochromatin might cause the variegating DNA segment to become underreplicated in polytene chromosomes simply because heterochromatic regions are late replicating. If the determinative decision for PEV is made early enough, then the illicitly heterochromatinized regions may become underreplicated in some tissues.

B.   Anti-Sense RNA

A third model suggests that the variegating gene is placed adjacent to a transcription enhancer in an inverted configuration and that the noncoding strand is now transcribed, making anti-sense RNA. For a detailed discussion of this model and its impact on PEV, see Frankham (1988).

C.   Adjacent Mobile Elements Interfere with Transcription

Finally, one might also suggest that PEV occurs as a result of juxtaposition of the variegating gene to a tranposable element. Transcription from the adjacent transposable element might interfere with the normal expression of the variegating gene (Spradling and Karpen, 1990). Indeed, repeated elements are known to exist at the euchromatin–heterochromatin interface of several *white* variegating strains (Tartof *et al.,* 1984).

With the abundance of variegating genes that exist in *Drosophila* and other systems, it is possible, and clearly quite likely, that examples of several different mechanisms of variegation exist. We simply argue that PEV often involves an alteration in chromatin structure. Thus, we believe that PEV can be used as an assay system for detecting and analyzing genes that encode NHCPs and/or chromatin assembly and modifying factors. While not all mutations that influence PEV will necessarily identify NHCPs, it is likely that a large subset would. However, mutations that alter the attachement of chromosomes to the nuclear envelope, or that alter the spatial arrangement of chromosomes within the nucleus (Foe and Alberts, 1983; Foe, 1989), or that control the level of polyteny would also be fascinating.

V. Position-Effect Variegation as an Assay System for Chromosomal Proteins and Chromatin Assembly and Modification Factors

Over a decade ago, we speculated that PEV might be used as an assay system to identify genes that encode nonhistone chromosomal proteins and chromatin assembly or modifying factors. The hypothesis that PEV results from alterations in chromatin packaging makes a simple prediction. PEV should be sensitive to alterations in the availability of histones, or other chromatin structural proteins, at the time of DNA replication and cell division. Indeed, a 50% reduction in the number of histone genes suppresses PEV (Moore et al., 1979, 1983). The interpretation of this observation is that limiting the availability of histones reduces the spread of heterochromatin into euchromatin. Since the heterochromatic regions of the genome are late-replicating, they would be particularly sensitive to loss of these building blocks. Furthermore, the addition of butyrate, which interferes with histone deacetylation, also suppresses PEV (Mottus et al., 1980). Again, the reduction in the availability of histones for DNA binding at the time of chromosome replication would limit the rate at which chromatin could assemble, and hence limit the spread of the late-replicating heterochromatin.

These observations strongly suggested that mutations in genes encoding NHCPs, or alterations in the dosage of such loci, would also lead to an alterations in the gene inactivation associated with PEV. For example, a completely amorphic mutation (complete loss of function) in a gene encoding a NHCP that is a component of heterochromatin should act like a functional deletion and suppress PEV. An antimorphic mutation in such a nonhistone chromosomal protein should also suppress PEV (an antimorphic mutation makes a product that opposes, or interferes with, the action of the wild-type product of that gene). If an abnormal chromosomal protein is incorporated into chromatin at the time of DNA replication, it would disrupt the normal chromatin assembly process and hence reduce the rate of assembly and spread of heterochromatin. Antimorphic mutations should also identify genes that are reiterated or genes that are part of a multigene family; such genes might not be sensitive to loss of function mutations (amorphs or hypomorphs).

Likewise, amorphic or antimorphic mutations in genes that encode chromosomal proteins necessary for the formation or assembly of euchromatin might result in an enhancement of PEV.

Using this line of reasoning, two laboratories independently commenced to select dominant mutations that either suppress [Su(var)] or hence [E(var)] the gene inactivation associated with PEV. Our laboratory (University of British Columbia) selected for mutations that suppressed or enhanced w^{m4}. A simple protocol for selecting Su(var)s and E(var)s is shown in Fig. 3. Dr. Reuter's laboratory, at Martin Luther University in Germany, selected for mutations that suppressed a modified w^{m4} strain that carried an enhancer

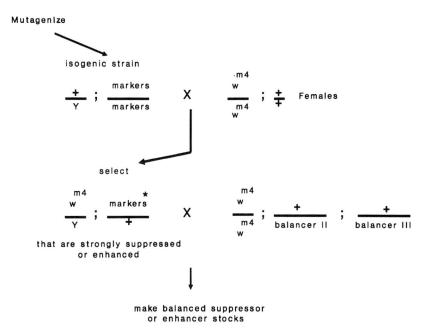

FIG. 3. A simple protocol for the isolation of Su(var) or E(var) mutations. The males carrying appropriate genetic markers for the second or third chromosome are treated with a mutagen, such as EMS or X rays, and mated with females carrying an easily scored variegation reporter gene, such as w^m (white-mottled). Males in which variegation is suppressed or enhanced are recovered among the F1 individuals. They are mated to females carrying appropriate inversions for either the second (balancer II) or the third (balancer III) chromosome. Balancer refers to an inversion to eliminate recombinant products (II = second, III = third chromosome, respectively). The suppressed or enhanced progeny are recovered and tested for segregation on chromosome II or on chromosome III, and balanced stocks are established from the appropriate lines.

A

Mutangenize

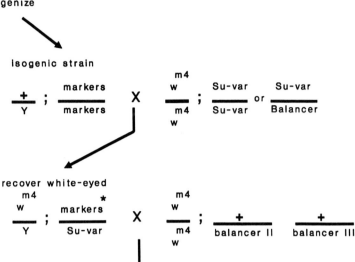

B

Mutagenize

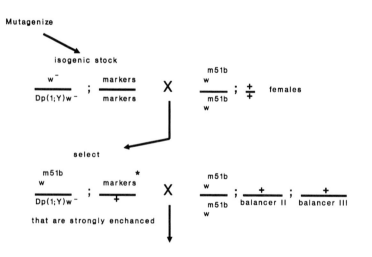

Note: the enhancer mutation may cause lethality or sterility in individuals lacking the duplication. This problem can be avoided by crossing the males to attached XX / Dp(1;Y)w⁻ females with appropriate balancers to maintain the stock.

of variegation (Reuter and Wolff, 1981). Using this approach, in several different screens, each group has recovered over 100 mutations that suppress PEV. Both laboratories have also recovered mutations that enhance PEV [E(var)]. E(var) mutations are recovered at a lower frequency than Su(var) mutations, possibly because they cause other loci to variegate, including loci essential for viability or fertility. However, appropriately designed screens allow the recovery of this class of mutation (for example, see Fig. 4).

A. Properties of Su(var) and E(var) Mutations

The dominant suppressors of variegation have the following general properties. First, they strongly suppress PEV. They usually restore over 50%, and often restore nearly full, activity of the variegating allele. Second, they act on the phenomenon of PEV, not simply on the expression of one variegating allele or gene. We have tested over 50 Su(var) mutations on several different variegating strains, including four different *white* variegating strains (w^{m4}, w^{mj}, w^{mMc}, and w^{m51b}) and a *brown* variegating strain (bw^{VDe2}), both of which affect eye pigment deposition; *Stubble* (Sb^V), which is a mutant allele causing abnormal bristle morphology; and *Bar* of Stone (B^{SV}), a mutant allele of the *Bar* locus (tandem duplication) that causes abnormal eye morphology. The Su(var) mutations suppress the variegation of all of these genes, although they do not necessarily act identically on each of them. For example, a particular Su(var) allele may be very strong in w^{m4} but only moderately strong in bw^V. Nonetheless, the Su(var) mutations appear to act on a variety of variegating genes, regardless of whether the gene is a wild-type or mutant allele (for example, *white*+ in w^{m4} and *Sb*− in Sb^V), and regardless of the type of heterochromatin with which it is associated (X heterochromatin for w^{m4}, heterochromatin of the right arm of chromosome 2 for Sb^V, or the Y chromosome for B^{Sv}) (Sinclair *et al.*, 1983, 1991; Hayashi *et al.*, 1990).

FIG. 4. A simple protocol for the isolation of E(var) mutations using (A) selection against a Su(var) or (B) selection against a variegating strain carrying a duplication for genes distal to the variegating reporter locus. Protocol B protects against inactivation of a neighboring gene that is required for normal vitality or fertility. Su-var refers to either second or third chromosome dominant suppressor of PEV. The presence of the Su-var mutation, in protocol A, gives the w^{m4} strain a normal red eye phenotype. Dp(1, Y) refers to a Y chromosome with a small segment of the X chromosome translocated to it. In this example, it carries a mutant allele of *white* (w^-) to allow monitoring of the variegating w^{m51b} allele and it carries the wild-type alleles of those genes that are distal to the w^+ locus on the variegating rearrangement (this allows inactivation to spread distally without loss of function of these neighboring genes). w^{m51b} is a variegating strain in which the w^+ locus is active in most eye pigment cells, giving a red eye phenotype; enhancement of variegation is easily detected as white eyes.

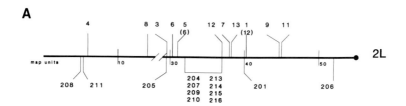

A

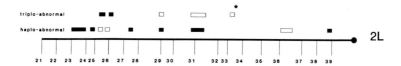

B

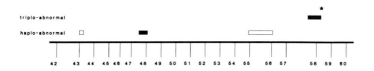

C

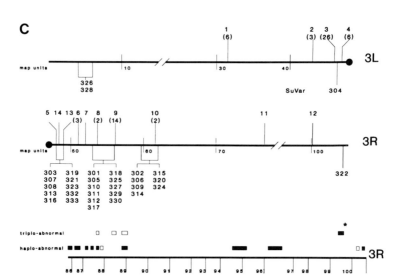

B. Distribution of Su(var) Loci

Many of the Su(var) mutations have been mapped. Using 91 dominant Su(var) mutations, Reuter's group has identified 23 loci, 12 on chromosome 2 and 11 on chromosome 3. Using 54 Su(var) mutations we have mapped 16 genes to chromosome 2 and 33 to chromosome 3. A composite of the results from the two laboratories is shown in Fig. 5. The second chromosome Su(var) mutations are distributed throughout the chromosome, with one cluster at 32–35 cM on the genetic map (bands 31A–32A on the polytene chromosome). Many of the third chromosome Su(var) mutations fall into one of three clusters in the proximal region of the right arm, with seven or eight genes located elsewhere on chromosome 3.

The clustered mutations may represent several discrete loci, in which case these genes may be evolutionarily conserved and represent related functions. Alternatively, these clustered mutations may represent alleles of a single locus, in which case these loci may be hypermutable. This issue can be addressed since many of the dominant Su(var) mutations have recessive phenotypes; some are homozygous lethal while others are homozygous male and/or female sterile. These recessive phenotypes have been used in complementation studies to determine allelism. We have identified one Su(var) with 10 mutant alleles. Approximately 12 of the 23 Su(var) loci identified by Dr. Reuter's collection are represented by 2 or more alleles, with one gene represented by 26 alleles (Wustman et al., 1989). Clearly, some Su(var) genes represent highly mutable loci.

No attempt has been made to consolidate the genetic data between Reuter's laboratory and the University of British Columbia laboratory. However, such analyses are planned for the near future. At that time, we should have a better

---

FIG. 5.   Distribution of Su(var) point mutations (genetic map) and genomic regions for which PEV is dosage sensitive (cytogenetic map). Maps depict mutations on the left arm of chromosome 2 (A); the right arm of chromosome 2 (B); and the left (3L) and right (3R) arms of chromosome three (C). Map units are given on the genetic map (upper chromosome in A and B; the upper two chromosomes in C) and bands are given on the cytogenetic map (bottom chromosome). The numbers above the chromosome refer to Su(var) alleles isolated in Reuter's laboratory, the numbers in parentheses refer to the number of alleles; the numbers below the chromosome refer to Su(var) isolated at University of British Columbia, with the exception of SuVar, which refers to the first suppressor of variegation, isolated by Dr. J. Spofford. Shaded boxes refer to genomic segments that suppress; open boxes refer to genomic segments that enhance when aneuploid. The data on dosage-sensitive regions are compiled from data published by Tartof, Reuter, and Henikoff (Locke et al., 1988; Wustmann et al., 1989; Henikoff, 1981) and data from our laboratory (UBC). The asterisk indicates segments of the genome for which the reciprocal condition (duplication or deletion) has not been tested.

estimate of the number of Su(var) loci in the genome and their cytogenetic as well as genetic map positions. In addition, both laboratories are collaborating on an analysis of E(var) loci.

C. Hypermutable Loci and the Nature of Su(var) Proteins

The hypermutable phenotype may reflect the conserved nature of these genes. Nonhistone chromosomal proteins or chromatin assembly/modification factors might have a very specific three-dimensional structure. These proteins may tolerate very little change and still be able to function properly. Consequently, loci that encode NHCPs may have little latitude for mutational alterations without causing a mutant phenotype. As an extension of this argument, one might hypothesize that the some Su(var) and E(var) loci identified in *Drosophila* will be highly conserved and thus may be used as probes to recover homologous loci in vertebrates and perhaps even plants.

Many of the Su(var) genes in the 2L cluster appear to complement each other on the basis of their recessive lethality (Sinclair *et al.,* 1983; 1991). While one must be careful not to over interpret cis–trans tests when using dominant mutations, based on this criterion alone, these mutations appear to identify several loci which map close together. The recessive lethality of these Su(var) mutations was used to position the mutations physically on the polytene chromosome map. Five mutations have been positioned in the 31A–C region, two mutations in the 31C–D region, and two mutations in the 31E–32A region (Sinclair *et al.,* 1991). The five mutations in the 31A–C region show three complementation groups based on recessive lethality, but they also show strong interaction phenotypes among the survivors. A conservative interpretation is that these mutations are alleles of a single locus and some *trans*-heterozygotes are viable but show recessive phenotypes. Alternatively, these mutations may identify reiterated genes that are functionally related. *trans*-Heterozygotes between two dominant mutations in a multigene family may survive, but have an aberrant phenotype. The interaction phenotype may be caused by either the reduction in functional product or the accumulation of mutant product in a macromolecular complex.

VI.   Su(var) Mutations and Chromatin Structure

A.   Su(var) and E(var) Mutations Can Alter Chromosome Structure

The Su(var) mutations have been tested for their effect on chromosome and chromatin structure. The *white*⁺ gene is located at band 3C2, which is near the tip of the X chromosome. In the w^{m4} strain, an inversion with a euchromatic break at the interface of the 3B–C regions has relocated w^+ to a position approximately 25 kb distal to the broken piece of X heterochromatin (Tartof *et al.*, 1984). In this strain [in the absence of Su(var) mutations], the 26 bands in the 3C–E region are packaged as euchromatin in only about 17% of the nuclei (the remainder appear to be heterochromatic—unbanded and amorphous in structure). The effect of eight different dominant Su(var) mutations on the morphology of this 26-band region has been examined, and in all cases the proportion of nuclei in which this variegating region was packaged as euchromatin increased up to 50 to 100% (Fig. 6), depending on the Su(var)

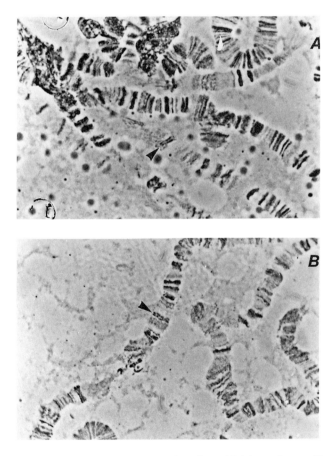

FIG. 6. Salivary gland chromosome preparations from third instar larvae. Chromosomes from (A) a w^{m4} strain and (B) a w^{m4} strain heterozygous for a dominant suppressor of variegation. The w^+ gene is located at band 3C2, which is not visible in A but is visible in B. In each case, the arrowhead indicates bands in the 3E region; bands in the 3C and D regions are to the left. They are amorphous (heterochromatic) in A and clearly banded in B.

mutation used (Hayashi *et al.*, 1990). This alteration in chromosome morphology, from heterochromatin-like to euchromatin-like, might reflect either modification in DNA packaging or an alteration in the level of polytenization in this segment of the DNA.

In situ hybridization of the w^+ gene (two probes from the coding region were used) to the polytene chromosomes resulted in a much stronger hybridization signal (40–85% of the wild-type strain) in the variegation-suppressed strains [w^{m4} heterozygous for one of the Su(var) mutations] than in the variegating strain (w^{m4}). This change in hybridization signal, like the change in chromosome morphology, could result either from an alteration in chromatin structure or a change in the number of w^+ gene templates (DNA replication).

Southern blot analyses of DNA isolated from the salivary glands of wild-type, variegating, and variegation-suppressed strains demonstrated that the *white* gene is not underreplicated in the variegating strain (relative to wild type). Furthermore, the Su(var) mutations do not cause overreplication of the *white* gene DNA (Hayashi *et al.*, 1990). Hence, it appears that the Su(var) mutations have altered the accessibility of the probe to its template in the *in situ* hybridization studies. This implies that the Su(var) mutations actually alter chromosome structure in the area of variegating genes. There was no evidence for any effect of the Su(var) mutations on the morphology or replication of euchromatic loci. (For those interested in cytological analyses of, or *in situ* hybridization to, polytene chromosomes, the general methods are described in Ashburner, 1989.)

B.   Su(var) Mutations and Chromatin Structure

The sensitivity of the *white*$^+$ gene to digestion by DNase I was used to assess, more directly, the effect of Su(var) mutations on chromatin structure [procedures for assays of this type can be found in "Methods in Enzymology" (Ballard *et al.*, 1989)]. The *white* gene is less sensitive to DNase I digestion in the variegating strain w^{m4} than it is in the wild-type strain (Oregon R). The addition of a dominant Su(var) mutation restored a more wild-type DNase I digestion pattern (Hayashi *et al.*, 1990). While the differences in the DNase I sensitivity of the *white* gene between wild-type, variegating, and variegation-suppressed strains were consistent in a number of different isolates and lines tested, the differences in the phenotype were slight. In contrast, no differences in DNase I sensitivity were detected between the wild-type, *white* variegating, and variegation-suppressed strains for three nonvariegating loci: (1) the histone genes (euchromatic location, actively transcribed); (2) the larval serum protein gene 1 gamma (euchromatic location, transcribed in third instar larvae); and (3) type I insert of the rDNA genes (heterochromatic location, not transcribed). Nuclei isolated from the embryonic and third instar larval stages produced identical results (A. Ruddell, 1986, University of British Columbia, unpublished observations). One might expect that the differences in chromosome morphology detected in the polytene chromosomes reflect differences in chromatin structure. However, the changes in chromatin structure, as measured by DNase I sensitivity, are slight when compared to the rather dramatic changes in chromosome morphology. This may suggest that the differences in chromatin structure, in diploid interphase nuclei, between variegating and variegation-suppressed strains are subtle at best, or alternatively, that nuclease digestion studies, while capable of distinguishing between actively transcribed genes and inactive genes from a given tissue, are not sensitive enough to detect differences in chromatin packaging at a single locus in nuclei isolated from an organism which itself is mosaic for chromatin packaging of the reporter locus.

C. Su(var) Loci and Nucleosome Spacing

The nucleosome spacing in the *white* gene was also examined in wild-type, variegating, and variegation-suppressed strains [for methods see Chapters 12 and 13, this volume, and (Lutter, 1989)]. The data are shown in Table I. While there is a detectable difference in the internucleosome distance of the *white* gene between wild-type and w^{m4} strains, the effect of two Su(var) mutations on this parameter of chromatin structure is somewhat equivocal. Su(var)215, which is one of the weaker Su(var) mutations, had little effect on nucleosome spacing. In contrast, a single copy of the Su(var)306 mutation, which restores nearly complete *white*[+] gene activity in the eye pigment cells and has a striking effect on chromosome morphology (Hayashi *et al.*, 1990), restores a

TABLE I

White Gene and Type 1 Insert Sequence Nucleosome Spacing in Wild-Type, Variegating, and Variegation-Suppressed Embryos

Strain	*White* gene[a]	Type 1 insert[b]
Oregon R	160 ± 1.4	177 ± 1.8
In(1)w^{m4}	171 ± 1.1	178 ± 1.4
In(1)w^{m4}; Su(var)215/CyO	170 ± 1.1	177 ± 1.2
In(1)w^{m4}; Su(var)306/Su(var)306	164 ± 1.4	177 ± 1.8

[a] The *white* gene nucleosome spacing, in bp ± standard error, was an average of measurements determined from 11 separate nuclear preparations, each run digested at three different nuclease concentrations ($n = 33$). Each of the 33 measurements was corrected to a histone gene internal standard.

[b] rDNA type 1 insert nucleosome spacing was determined as in *a*.

somewhat more wild-type internucleosome distance to the *white* locus. Based on these two results alone, it is difficult to assess the effect of the Su(var) mutations on nucleosome spacing.

VII. Dosage Effects of Su(var)⁺ and E(var)⁺ Loci

A. Position-Effect Variegation Is Sensitive to Alterations in the Dose of Su(var)⁺ Loci

Deletions that remove the Su(var)⁺ locus, or loci, in 31A/E [that is, one copy of Su(var)⁺ remains] suppress variegation and duplications for this region [three copies of Su(var)⁺] enhance variegation. This result suggests that at least some Su(var) loci are dosage sensitive and implies that their products are produced in a limited quantity. Furthermore, the addition of three copies of the Su(var)⁺ locus not only enhances the likelihood of inactivation of the *white* locus, but also increases the likelihood that neighboring loci are repressed. In the w^{m51b} strain, the w^+ gene is repressed in only about 5% of the eye pigment cells; the gene *roughest* (rst^+), which is about four chromosome bands distal to the w^+ gene (about 100 kb), does not variegate in this strain. In the w^{m51b} strain carrying three copies of the Su(var)⁺ locus at 31A/E (duplication), the w^+ gene is inactivated in approximately 75% of the eye pigment cells, and the rst^+ gene is inactivated in 30% of the eye cells. Furthermore, the clones of *rst* cells are always located within a subset of the *white* clones. The rst^+ gene is not inactivated in a cell in which the w^+ gene is active (Fig. 7). A similar pattern of gene inactivation is observed with the w^{mMc} strain (T. Grigliatti, unpublished observations). This suggests that the Su(var) products act on a segment of the chromosome and not on a particular gene. Second, and more importantly, overproduction of the Su(var) product is associated with an increase in the distance over which PEV (heterochromatinization) can spread. This supports the notion that some Su(var)⁺ products may be produced in limited amounts.

B. Aneuploidy of Euchromatic Segments of the Genome and Position-Effect Variegation

Using a set of deletions and duplications, Tartof and colleagues were able to identify about a dozen segments of the genome for which PEV is dosage sensitive (Locke *et al.*, 1988). These were divided into two classes based on their effect on PEV. Class I regions enhance PEV when duplicated and

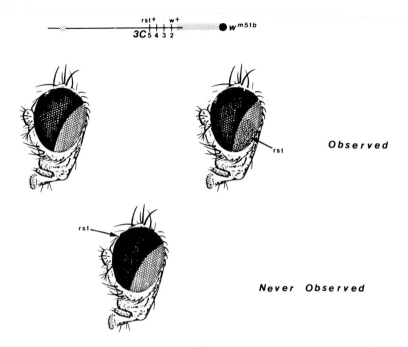

FIG. 7.   Position-effect variegation for w^{m5lb} with a duplication of cytogenetic region 31 of chromosome 2. The w^{m5lb} rearrangement is shown in the schematic diagram at the top; w^+, the wild-type allele of *white*; rst^+ the wild-type allele of *roughest*. Dark and light shading represents pigmented and nonpigmented areas, respectively, and the stippling represents the mutant roughest phenotype. Note *rst* is not inactivated in clones of cells in which *w* is active.

suppress variegation when deleted. Class II regions behave oppositely; they suppress variegation when duplicated and enhance variegation when deleted. Reuter's group has conducted a similar set of a segmental aneuploid studies. In addition to class I and class II phenotypes, they found segments of the genome which are either haplo-abnormal suppressors or enhancers of PEV, but have no effect when duplicated. Thus, using aneuploidy for regions of the euchromatin, one can identify at least four distinct classes of effects on PEV (Table II). It is tempting to speculate on the nature of these dosage-sensitive regions, or the genes located therein. However, the duplications and deletions that were used to define these regions vary in size considerably. In some cases, not all of the deletions (or duplications) in an overlapping series elicit the effect. It has been argued that this inconsistency reflects the existence of two loci with opposite effects, that is, a Su(var)$^+$ and an E(var)$^+$ locus, that map in close proximity to each other. This may indeed be true, but it will require the isolation of much smaller duplications and deletions, or point mutations in these presumptive genes, to identify such loci unequivocally.

TABLE II

THE EFFECT OF DOSAGE OF VARIOUS
EUCHROMATIC SEGMENTS OF THE
D. melanogaster GENOME ON
POSTION-EFFECT VARIEGATION

Class	Three doses	One dose
I	Enhance	Suppress
II	Suppress	Enhance
III	No phene	Suppress
IV	No phene	Enhance
V	Suppress	Not tested

There is good correspondence between the haplo/triplo-abnormal regions and the EMS and γ-ray-induced dominant Su(var) mutations isolated in our laboratory and Reuter's laboratory for both the left arm of chromosome 2 and the right arm of chromosome 3 (see Fig. 5). Class I and class II regions may comprise Su(var) and E(var) genes whose products are among the structural components of chromatin. The products of these loci may be made in stoichiometric amounts and hence be sensitive to either addition or loss of template number. Classes III and IV may represent Su(var) and E(var) loci whose products are part of a larger, heteropolymeric complex that plays a role in chromatin assembly or modification. Overproduction of one component of such a multimeric unit may have little or no effect on chromatin structure and PEV, while loss of function would have an effect. In contrast, loss of function would alter the rate of assembly of heterochromatin or euchromatin, or alter the ability of chromatin structure to be modulated at the time of transcription. Alternatively, they could represent reiterated loci; again, reduction of template number below some threshold could elicit an effect whereas duplication would not.

VIII. Analyses of Cloned Su(var) Loci

To date, three Su(var) genes have been cloned: Su(var)205, Su-var(3)7, and Su(var)216.

A. Su(var)205

Using monoclonal antibodies to fractionated chromatin, Elgin and co-workers identified a heterochromatin-specific protein called HP-1 which bound to a monoclonal antibody C1A9. The cDNA clones for this protein were recovered by screening an expression library. *In situ* hybridization with the genomic clones indicated that the gene which encodes this protein maps to the polytene bands 29A (James and Elgin, 1986), which is where Su(var)205 maps (Sinclair *et al.*, 1983). Northern blot analyses show that the size of the HP-1 mRNA is altered in the Su(var)205/+ strain, suggesting that Su(var)205 encodes the heterochromatin-specific protein HP-1. This mutant allele has now been cloned and sequenced. Su(var)205 has a G to A transition of the first nucleotide of the last intron, causing missplicing of the HP-1 mRNA (Eissenberg *et al.*, 1990). A second allele, Su-var(2)5, contains a nonsense mutation (T. Harnett, G. Reuter, and J. C. Eissenberg, personal communication). Hence, these two Su(var) alleles appear to be loss-of-function mutations. Transformation studies with Su(var)205[+] have not been done, since neither a full-length genomic clone nor an appropriate cDNA is available at this time. The inability to recover an intact genomic clone of Su(var)205 and the variablity at the 5' end of cDNA clones have lead to an investigation of the possibility of trans-splicing of the 205 mRNA or genomic rearrangement of the Su(var)205 gene (V. Ngan and T. C. James, personal communication).

Antibody binding to polytene chromosomes indicates that HP-1 is associated with β-heterochromatin, a banded portion of the fourth chromosome, and a few specific regions in the euchromatin (See Figure 7, [8], this volume). The predicted amino acid sequence of the Su(var)205 gene does not suggest any DNA-binding domain. Hence, this protein may not bind to DNA directly, but may be a chromatin protein which binds to other NHCPs. The HP-1 protein contains a 37-amino acid region in the amino terminus that is homologous to a similar domain in the Polycomb protein of *Drosophila* (Paro and Hogness, 1991). The *Polycomb* gene is also a nuclear protein (see Section X,C).

Curiously, one of the euchromatic regions to which HP-1 antibody binds is the left arm of chromosome 2 at bands 31–32, where several other Su(var) genes are located. Whether the binding of HP-1 to a region in which several Su(var) genes are located has any functional significance is unknown at this time.

B.   Su-var(3)7

Su-var(3)7 from Reuter's group was localized to the region between the *rosy* (*ry*) and *acetyl cholinesterase* (*Ace*) genes which had been cloned previously by a molecular walk. Spierer and colleagues used transformation studies to identify a segment of DNA from this region that, when present in three or more copies, enhanced the gene inactivation associated with PEV. The addition of an extra copy of this region also ameliorates (but does not eliminate) the Su-var(3)7 mutant phenotype (Reuter *et al.*, 1990). The Su-var(3)7 protein has five zinc-fingers which are separated by 40 to over 100

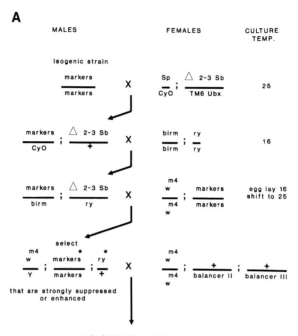

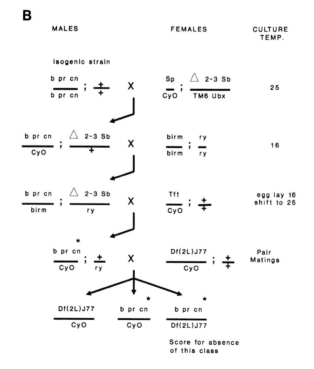

residues. This is a rather novel arrangement for zinc-finger motifs, since the zinc-fingers of the transcription factor class of proteins are usually contiguous (Struhl, 1989). Spierer and colleagues have postulated that the unusual organization of these motifs might allow the protein to interact with widely spaced or folded regions of DNA. Therefore, this protein might function to stabilize chromatin conformations. In addition, the Su-var(3)7 gene product has a number of putative phosphorylation sites which might allow its binding to DNA or chromatin to be regulated.

Since the Su-var(3)7 gene as well as other chromatin proteins have putative phosphorylation sights, we expect that some Su(var) genes within the class that controls chromatin assembly or modifies chromatin domains might encode kinases, phosphatases, or other potential PEV effectors such as deacetylases.

C. Su(var)216

Nine dominant Su(var) and two recessive su(var) mutations map in the left arm of chromosome 2 to bands 31A to 32A. These mutations define a

---

FIG. 8.   A scheme for the isolation of P transposable element-induced mutations. (A) A general method for the recovery of P-induced Su(var) or E(var) mutations. The culture temperature is given in °C; the transposase of the Δ2-3 strain is very active, and high developmental temperatures cause excessive somatic movement, resulting in low fertility and viability. The use of the isogenic strain makes future molecular analyses considerably simpler. "Markers" refers to any set of convenient genetic markers; birm, a stock called Birmingham which contains 17 defective P elements which are able to move in response to transposase (in this case provided by Δ2-3), but which do not encode transposase; ry, rosy, a third chromosome recessive mutation; Sp, Sterno-pleural, a second chromosome dominant mutation; w^{m4}, Inversion (1) white-mottled 4; Δ2-3 Sb, a third chromosome containing a P element in which the intron between exons 2 and 3 has been removed artificially; this chromosome is identified by the presence of the dominant mutation Stubble (Sb); CyO, Curly of Oster, a multiply inverted second chromosome balancer which carries the dominant mutation Curly wings (Cy); balancers II and III refer to any multiply inverted second and third chromosome, respectively; TM6 Ubx, a multiply inverted third chromosome balancer which carries the dominant marker Ultrabithorax. (B) A protocol for the recovery of P-induced lethal mutations for any segment of the genome delimited by a deletion. For example, the protocol shown here can be used to select lethal mutations that are allelic to Su(var) or E(var) mutations. At generation 4, the b, pr, cn*/CyO males are mated singly (one male per vial) with females that are genotypically deletion/CyO; +/+. (In this example, each male in this generation represents a different P "exposed" second chromosome; a similar scheme can be established for the third chromosome). The progeny from this cross are scored for the absence of b, pr, cn*/deletion class. The mutation is recovered from the siblings with the genotype b, pr, cn*/CyO, and a stock is established from males and females of this genotype. b, black body; pr, purple eyes; cn, cinnabar eyes which are recessive mutations on chromosome 2 and are representative of what is labeled "markers" in other protocols; Df(2)J77 is a deletion for bands 31C–E; Tft, Tuft a second chromosome dominant bristle mutation.

minimum of five genes. A combination of a molecular walk and P transposable element mutagenesis was used to clone the genes this region. P elements were placed in the banded region 31 and 32 by selecting for P element-induced lethal mutations over a deletion. [Figure 8 shows a protocol for the isolation of Su(var) and E(var) mutations. A modified form of this protocol can be used to isolate lethal mutations in a specific region delimited by a deletion.] To date, we have obtained P-tagged alleles of 216, a recessive su(var) gene which maps at 31E, and Su(var)214 (the latter was provided by Dr. Reuter as part of a collaborative effort). In addition, P-tagged alleles of several other nonsuppressor loci were recovered in these screens. The P-induced lethal alleles of the Su(var) loci are revertible by remobilization of the P element. Su(var)216 was the first suppressor for which a P allele was recovered, and it has been cloned and sequenced. The recessive su(var) mutation in 31E, which is a rather novel class of mutants, has also been cloned by P-tagging, but it has not yet been sequenced.

Su(var)216 is a fairly strong dominant suppressor of PEV; in addition, it has a recessive lethal phenotype. Several pieces of evidence suggest that the original dominant allele of Su(var)216 (216A) is an antimorph. First, the single mutant allele [Su(var)/+] is a stronger suppressor of PEV than is the deletion of this gene (deletion/+). Second, the Su(var)216 homozygote is completely lethal, whereas the hemizygote [Su(var)⁻/deletion] is male lethal but allows some females, which are sterile, to survive. Third, a duplication [Su(var)216/ +/+] ameliorates the dominant suppression of PEV but does not completely eliminate it. Based on the antimorphic nature of the mutant, we hypothesized that the 216 gene product is only one component of a multimeric complex. The 216 antimorph produces a protein which is still capable of interacting with other components to form the multimeric complex, but it disrupts the function of the complex. If this hypothesis is true, one might find the genes that encode the other components of this complex by screening for mutations that interact with, or suppress, Su(var)216; in subsequent screens we have found such loci (for protocols, see Fig. 3).

The 216-P allele fails to complement Su(var)216 for lethality, but as a dominant mutation (216P/+) does not suppress PEV. The P insert is located between two coding elements that are transcribed in opposite orientation. Both genes were cloned and sequenced. The P element is inserted in the 5' untranslated leader sequence of a gene that is homologous to the S. cerevisiae protein kinase encoding gene cdc2 (about 70% conservation). This protein kinase plays an essential role in cell division (see Norburg and Nurse, 1989, for a review) and has histone H1 as one of its substrates. Hence, it is an obvious candidate for a protein that plays a role in chromatin assembly and also PEV. The wild-type allele of this gene rescues the lethality associated with Su(var)216 homozygotes; however, it diminishes the dominant Su(var) phenotype only slightly. The lack of a stronger response on the suppression phenotype is unsettling and may imply that the Su(var)216 and *cdc-2* genes are separate loci located close together. Larger constructs of *cdc*+, and inserts in other areas of the genome, are currently being tested to determine their effects on the Su(var) phenotype.

Ten alleles of Su(var)216 were recovered in a screen for recessive lethal mutations in region. Two of these mutations are weak dominant suppressors of PEV; the other eight do not suppress PEV. Two of the weaker *cdc2-*Su(var)216 alleles have been cloned and sequenced and both show only a single nucleotide change from the wild-type allele (the mutant alleles were induced in a strain isogenic for the second chromosome to minimize polymorphisms and facilitate sequence comparison). Both alterations are in the 3′ coding element; they are not alterations in the ATP binding, PSTAIR, or enzymatic domains of this protein kinase. The original antimorphic allele of 216 as well as the two strongest alleles of 216 are currently being cloned and analyzed. They will be transformed back into *Drosophila* to determine if one dose of the mutant allele suppresses PEV.

In summary, some of the Su(var) loci encode proteins that appear to be structural components of chromatin, such as HP-1. Others appear to be involved in chromatin assembly or modification. Clearly, a large set of Su(var) mutations needs to be cloned and sequenced. We are currently in the process of cloning and sequencing the P-tagged Su(var)214 and the recessive suppressor of variegation in the 31E/F region. Two other Su(var) genes are located in 31B–D; however, these have not been P-tagged yet.

D. Inducing Su(var) and E(var) Mutations by P Element Insertion

Three laboratories have separately engaged in the isolation of P element-induced Su(var) and E(var) mutations. The mutations isolated in Tartof's laboratory turned out to be either deletions or duplications, with no P elements associated with the suppressor or enhancer phenotype (Locke *et al.*, 1988). This may suggest that Su(var) mutations are not recoverble by P mutagenesis. [P elements often insert in regions 5′ to the coding elements or in introns and perhaps the P alleles rarely have a Su(var) phenotype.] However, since their attempt at P tagging Su(var) loci was done prior to the new technology of P element mutagenesis, it may be that Tartof's group induced their P mutants on the chromosomes of the Harwich (P-containing) strain, and that deletions were created by the simultaneous excision of neighboring elements. In a recent set of P element screens, we have recovered over 30 dominant mutations that visibly suppress PEV (for protocols, see Fig. 8).

The Su(var) phenotype of at least one of these P-induced mutations can be reverted by remobilization of the P element (the remainder are currently being tested for reversion), suggesting that the Su(var) phenotype is associated with a P insertion. In addition, Reuter's laboratory has isolated a set of P-induced enhancer loci (G. Reuter, personal communication).

IX. Competition for Structural Components of Heterochromatin

Since the heterochromatic portion of the genome mostly comprises satellite and middle repetitive DNA sequences, one might ask whether the components of chromatin that complex with DNA to form heterochromatin are distributed uniformly in this segment of the geneome, or whether different regions of heterochromatin are formed from different protein components or at least different molar ratios of these components. Two series of experiments suggest that segments of heterochromatin are not packaged uniformly.

A. Segmental Aneuploidy for the Y Chromosome and Position-Effect Variegation

One of the most striking features of PEV is that it is sensitive to alterations in the amount of heterochromatin in the genome. Addition of more heterochromatin, such as an extra Y chromosome, generally suppresses PEV, while removal of the Y chromosome (or heterochromatin) generally enhances the gene inactivation associated with PEV. Several researchers have suggested that the Y chromosome effect on PEV is additive and simply results from alterations of the total amount of heterochromatin in the cell (Dimitri and Pisano, 1989).

A series of adjacent or overlapping deletions of the Y chromosome were used to ask whether the Y chromosome effect on PEV results from aneuploidy of a specific region (or regions) of the Y chromosome or whether it reflects simply the total amount of Y heterochromatin present in the cell. We have been able to show that the entire Y chromosome effect, in the variegating strain w^{m4}, is associated with aneuploidy of one specific region of the Y chromosome (Brock *et al.,* manuscript in preparation). This segment is not small; it represents about 15–20% of the Y chromosome. Nonetheless, aneuploidy for other regions, which represent up to 40% of the entire Y, have no effect on variegation in either males or females of the w^{m4} strain. Second, different variegating rearrangements of *white* (that is, different heterochromatic breakpoints) were sensitive to aneuploidy of different regions of the Y chromosome. Likewise, different variegating strains, for example, w^{m4} (associated with X heterochromatin) versus b^{VDe2} (associated with 2R heterochromatin), were sensitive to aneuploidy of different regions of the Y chromosome. We were never able to find very small deletions of the Y chromosome that could account for the full effect observed with the larger deletions. Hence, we believe that the Y chromosome effect of PEV does not result from altering the dose of a specific gene or even a specific DNA-binding site. Rather, it correlates with altering the dose of a specific segment, or subset of segments, of the Y chromosome. If the NHCPs are distributed nonrandomly along the Y chromosome, then certain segments of the Y may be rich in those NHCPs that are also found at high concentration in other heterochromatic regions of the genome. Hence, the effect that aneuploidy of a given segment of the Y chromosome has on a particular variegating strain may depend solely on the degree to which they share similar NHCPs. It will be interesting to correlate distribution of the satellite DNA sequences with those segments of the Y chromosome that influence variegation.

B. Competition for Structural Components of Heterochromatin between Variegating Rearrangements

In another series of experiments, strains that carried two different variegating rearrangements, such as w^{m4} and Sb^V, or Sb^V, and bw^{VDe2}, were constructed. These strains were used to address the question of whether two variegating rearrangements would compete with each other for a limited set of heterochromatic components. One of three simple results might be expected: (1) the two variegating breakpoints may compete with each other and one reporter gene would be suppressed while the other was enhanced; (2) the variegating response at both breakpoints might be diluted without bias; or (3) the two variegating rearrangements might act autonomously. Several "double-variegator" combinations were found in which the variegation at one breakpoint was suppressed while the variegation at another breakpoint was maintained or enhanced (Lloyd and Grigliatti, manuscript in preparation). We interpret this as competition between the variegation-causing DNA segments for a limited set of protein components. Other combinations showed little or no difference from their respective parental strains, which contained only one of the variegating rearrangements. The latter results suggest that some variegating breakpoints do not compete for the same components (or that the shared components are not present in limited amounts) while others do.

X. Chromatin Structure, Determination, and Somatic Memory

There are many parallels between the phenomenon of PEV, as it exists in *Drosophila*, and X chromosome inactivation in mammals. The principle differences between these two phenomena may be the limitation of inactivation-inducing sequences to the X chromosome in eutherian mammals and the distance over which the gene inactivation spreads. Alterations in chromatin structure have been invoked in several developmental abnormalities in humans. In fact, PEV has been used as a model for the Fragile-X syndrome in humans (Laird, 1987). PEV also resembles many other determinative events that occur during development. Many of the groups who work on the homeotic genes of *Drosophila*, and their homologs in other organisms, believe that the determination of segment identity is accomplished through alterations in chromatin packaging. It is postulated that the adjacent DNA domains of the Bithorax complex are sequentially activated, through alterations in chromatin structure, in successive posterior segments of the developing fly, and that this segment-specific activation controls segment identity. Once made, these determinative developmental decisions, like PEV, are stably maintained through many cell divisions. Can the phenomenon of PEV provide us with a general model for studying determination in eukaryotes?

If PEV is to be used as a model system for some of the determinative events that occur during the development of higher eukaryotes, three questions need to be addressed. First, when is the transcriptional fate of a variegating gene initially determined? Second, when do the Su(var) gene products act? Third, is there any commonalty between Su(var) or E(var) loci and those loci that regulate homeotic gene expression? These questions will be addressed in order.

A. Maternal Effect Mutations and Determination of Position-Effect Variegation

The developmental stage during which the transcriptional fate of a variegating gene is initially determined has been debated for some years. The consensus is that, for variegating genes whose phenotype is detected in the adult, the determinative decision occurs by the beginning of the second larval instar stage. Once made, this decision is then propagated with good fidelity in many cases (see Spofford, 1976, for a review). If the gene inactivation associated with PEV involves alterations in chromatin packaging, then, like X chromosome inactivation, it might be expected to occur rather early in development. In *Drosophila*, the first visible partitioning of the genome into

TABLE III

MATERNAL EFFECT: Su(var)205

	w^+ gene expression (%)	
	$\dfrac{w^{m4}}{\cdot Y}; \dfrac{Su(var)}{CyO} \times \dfrac{w^{m4}}{w^{m4}}; \dfrac{+}{+}$	$\dfrac{w^{m4}}{Y}; \dfrac{+}{+} \times \dfrac{w^{m4}}{w^{m4}}; \dfrac{Su(var)}{CyO}$
Su(var)/+		
Males	53 ± 10	100 ± 12
Females	66 ± 11	100 ± 14
Cyo/+		
Males	6 ± 2	79 ± 9
Females	4 ± 3	68 ± 7

heterochromatin and euchromatin occurs just prior to blastoderm formation. This is probably the earliest time at which the transcriptional fate of a variegating gene might be determined. We have screened many of our dominant Su(var) mutations for maternal effects. To date, 18 of these mutants have detectable maternal effects. In such cases, the expression of a variegating gene is suppressed in the offspring that do not carry a Su(var) mutation but were derived from a cross between a female that was heterozygous for a single dominant Su(var) mutation and a male that lacked a Su(var) mutation. An example of such a result is shown in Table III. Clearly, the maternal parent deposits material in the egg cytoplasm which can influence the initial determination event. Thus, the initial determinative event, for at least some variegating genes (whether the fate of all variegating genes in determined at the same time during development is not know), probably occurs during embryogenesis, although it is possible the Su(var) gene product perdures until later stages of development.

B. Altering the Somatic Memory of the Initial Determinative Decision.

If the transcriptional fate of a variegating gene can be influenced by products of the maternal genome, one must ask when the Su(var) genes in the zygote act. The Su(var) gene products synthesized from the zygote nucleus might act at any one, or more, of three times: (1) at the time of the initial determinative event (either along with maternally stored product or alone in those cases where no maternal product is stored or perdures), (2) on the maintenance of that decision (somatic memory) during subsequent cell divisions, and/or (3) at the time of transcription of the variegating gene. Several of our Su(var) mutations are temperature sensitive. The TSP has been

genes normally located in heterochromatin and imply that these Su(var)$^+$ products influence either the formation or modulation of chromatin.

B. Su(var) Mutations May Influence the Expression of Nonvariegating Genes

As mentioned previously, several of the Su(var) mutations have a strong interaction phenotype in trans-heterozygotes [Su(var) gene A^-, Su(var) gene B^+/Su(var) gene A^+, Su(var) gene B^-]. One of the strongest of these is a light eye phenotype that is reminiscent of It^v. However, this phenotype is observed in strains that are It^+/It^+. This interaction phenotype is exacerbated in a It^-/It^+ strain (a strain which does not variegate for It but contains only one functioning It^+ allele). Lethal alleles of It usually survive well as heterozygotes (It^l/It^+). However, strains that are It^l/It^+ and trans-heterozygotes for two interacting Su(var) mutations have reduced viability (N. Clegg and I. Whitehead, University of British Columbia, personal communication). One possibility is that the Su(var) mutations have additive effects and can disrupt the euchromatic–heterochromatic junction in a conventional chromosome (no variegating rearrangement). We are currently testing the effect of the Su(var) mutations on the production of the It^+ gene product.

XII. Summary

The past decade has brought us much information about gene regulation in eukaryotes and the importance of chromatin in gene expression. Clearly, we need to advance our understanding of chromatin assembly and modification. Hopefully, many Su(var) and E(var) loci will be cloned and their analysis will demonstrate that they are involved in this process. It will be interesting to see what classes of products they encode. Clearly, antibodies must be made against these proteins to determine: (1) the chromosomal distribution of those that encode chromatin structural proteins, (2) the tissue and temporal distribution of these proteins during development, and (3) the time of action during specific portions of the cell cycle for those genes encoding factors that regulate chromatin assembly or modifying proteins.

In addition, it would be intriguing to identify and isolate those sequences or segments of DNA that serve as nucleation sites for the assembly of euchromatin and heterochromatin. The necessary materials for identifying these sequences are in place with the large number of variegating strains and revertants of those rearrangements and Su(var) and E(var) mutations that exist.

From our current understanding of PEV, it appears that PEV provides a useful model for the study of determinative events that occur during development. This includes both the establishment of different lineages and the somatic memory or imprinting of the initial determinative decision. Clearly, PEV is more than a simple assay system for the identification of NHCPs, and ultimately the understanding of the mechanisms underlying PEV should lend insights into how such determinative events occur and are propagated.

ACKNOWLEDGMENTS

I would like to thank all members of the UBC *Drosophila* group, which currently includes, in alphabetical order: Nigel Clegg, Bob Lansman, Vett Llyod, Gerry Meister, Randy Mottus, Mike O'Grady, Greg Stromotich, Minto Vig, Ian Whitehead, and Keith Wollenberg. I would also like to express my appreciation to previous members of the group for their many contributions, and especially J. Brock, N. Hardin, S. Hayashi, A. Hedrick, R. Kroitzshi, N. Mawji, V. Ngan, A. Ruddell, D. Sinclair, R. Tejani, and J. Williams. I must thank my many colleagues for discussions and interesting insights, especially S. Elgin, M. Gatti, S. Henikoff, C. Laird, J. Locke, S. Pimpinelli, L. Sandler (now deceased), J. Spofford, K. Tartof, and B. Wakimoto. I would also like to thank K. Gorkoff and D. Kerfoot for typing the manuscript. This work was supported by an MRC Grant #10024 and a NSERC Operating Grant #3005.

REFERENCES

Ashburner, M. (1989). "*Drosophila*—A Laboratory Manual." Cold Spring Harbor Laboratory, Cold Spring Harbor, New York.

Baker, W. K. (1968). Position-effect variegation. *Adv. Genet.* **14,** 133–169.

Ballard, M., Dretzen. G., Giangrande, A., and Ramain, P. (1989). Nuclease digestion of transcriptionally active chromatin. *In* "Methods in Enzymology" (P. M. Wassarman and R. D. Kornberg, eds.), Vol. 170, pp. 317–346. Academic Press, San Diego, California.

Brock, J., Sinclair, D., and Grigliatti, T. A. The effect of Y chromosome heterochromation or position-effect variegation in *Drosophila melanogaster*. Manuscript in preparation.

Brown, C. J., Lafreniere, R. G., Powers, V. E., Sebastio, G., Ballabio, A., Pehigrew, A. L., Ledbetter, D. H., Levy, E., Craig, J. W., and Willard, H. F. (1991). Localization of the X inactivation centre on the human X chromosome in Xg13. *Nature (London)* **349,** 82–84.

Catcheside, D. G. (1938). A position-effect in *Oenothero*. *J. Genet.* **38,** 345–352.

Catcheside, D. G. (1947). The P-locus position effect in *Oerothera*. *J. Genet.* **48,** 31–42.

Cattanach, B. M. (1974). Position effect variegation in the mouse. *Genet. Res.* **23,** 291–306.

Clutterbuck, A. J., and Spathas, D. H. (1984). Genetic and environmental modification of gene expression in the *brlA12* variegated position effect mutant of *Aspargillus nidulans Genet. Res.* **43,** 123–138.

Dimitri, P., and Pisano, C. (1989). Position effect variegation in *Drosophila melanogaster*. Relationship between suppression effect and the amount of Y chromosome. *Genetics* **122,** 793–800.

Duncan, I., and Lewis, E. B. (1982). Genetic control of body segment differentiation in *Drosophila*. *In* "Developmental Order: Its Origin and Regulation" (S. Subtelny, ed.), pp. 533–554. Alan R. Liss, New York.

Eissenberg, J. C. (1989). Position effect variegation in *Drosophila*: Towards a genetics of chromatin assembly. *BioEssays* **11,** 14–17.

Eissenberg, J. C., James, T. C., Foster-Harnett, D. M., Harnett, T., Ngan, V., and Elgin, S. C. R. (1990). Mutation in a heterochromatin-specific chromosomal protein is associated with suppression of position-effect variegation is *Drosophila melanogaster. Proc. Natl. Acad. Sci. U.S.A.* **87**, 9923–9927.

Foe, V. E. (1989). Mitotic domains reveal early commitment of cells in *Drosophila* embryos. *Development* **107**, 1–22.

Foe, V., and Alberts, B. M. (1983). Studies of nuclear and cytoplasmic behavior during the five mitolic cycles that precede gastrulation in *Drosophila* embryogenesis. *J. Cell Sci.* **61**, 31–70.

Frankham, R. (1988). Molecular hypotheses for position-effect variegation: Anti-sense transcription and promotor occlusion. *J. Theor. Biol.* **135**, 85–107.

Gyurkovics, H., Gausz, J., Kummer, J., and Karch, F. (1990). A new homeotic mutation in the *Drosophila* bithorax complex removes a boundary separating two domains of regulation. *EMBO J.* **9**, 2579–2586.

Hartmann-Goldstein, I. J. (1967). On the relationship between heterochromatinization and variegation in *Drosophila* with special reference to temperature sensitive periods. *Genet. Res.* **10**, 143–159.

Hayashi, S., Ruddell, A., Sinclair, D., and Grigliatti, T. (1990). Chromosomal structure is altered by mutations that suppress or enhance position effect variegation. *Chromosoma* **99**, 391–400.

Hearn, M. G., Hedrick, A., Grigliatti, T. A., and Wakimoto, B. T. (1991). The effect of modifiers of position effect variegation on the variegation of heterochromatic genes of *Drosophila melanogaster. Genetics* **128**, 785–797.

Heitz, E. (1934). Uber alpha-heterchromatin sowie konstanz und bau der chromomeren bei *Drosophila. Biol. Zentralbl.* **45**, 588–609.

Henikoff, S. (1981). Position-effect variegation and chromosome structure of a heat shock puff in *Drosophila. Chromosoma* **83**, 381–393.

Henikoff, S. (1990). Position-effect variegation after 60 years. *Trends Genet* **6**, 422–426.

Henikoff, S., and Dreesen, T. D. (1989). Trans-inactivation of the *Drosophila* brown gene: Evidence for transcriptional repression and somatic pairing dependence. *Proc. Natl. Acad. Sci. U.S.A.* **86**, 6704–6708.

Hoffmann, X. Jr., Laroch, T., Brand, A. H., and Gasser, S. M. (1989). RAP-1 factor is necessary for DNA loop formation *in vitro* at the silent mating type locus HML. *Cell (Cambridge, Mass.)* **57**, 725–737.

James, T. C., and Elgin, S. C. R. (1986). Identification of a non-histone chromosomal protein associated with heterochromatin in *Drosophila melanogaster* and its gene. *Mol. Cell. Biol.* **6**, 3862–3872.

Judd, B. H., (1955). Direct proof of variegated-type position effect at the white locus in *Drosophila melanogaster Genetics* **40**, 739–744.

Karpen, G. H., and Spradling, A. C. (1990). Reduced DNA polytenization of a minichromosome region undergoing position-effect variegation in *Drosophila. Cell (Cambridge, Mass.)* **63**, 97–107.

Kellum, R., and Schedl, P. (1991). A position-effect assay for boundaries of higher order chromosomal domains. *Cell (Cambridge, Mass.)* **64**, 941–950.

Laird, C. D. (1987). Proposed mechanism of inheritance and expression of the human fragile-X syndrome of mental retardation. *Genetics* **117**, 587–599.

Lewis, E. B. (1950). The phenomenon of position effect. *Adv. Genet.* **3**, 73–115.

Lloyd, V., and Grigliatti, T. A. Different *Drosophila* genomic rearrangements compete for a limited factor in position-effect variegation. Manuscript in preparation.

Locke, J., Kotarski, M. A., and Tartof, K. D. (1988). Dosage-dependent modifiers of position effect variegation in *Drosophila* and a mass action model that explains their effect. *Genetics* **120**, 181–198.

Lutter, L. C., (1989). Digestion of nucleosomes with deoxyribonucleases I and II. *In* "Methods in

Enzymology" (P. M. Wassarman and R. D. Korberg, eds.), Vol. 170, pp. 264–269. Academic Press, San Diego, California.

Moore, G. D., Procunier, J. D., Cross, D. P., and Grigliatti, T. A. (1979). Histone gene deficiencies and position-effect variegation in *Drosophila*. *Nature (London)* **282,** 312–314.

Moore, G. D., Sinclair, D. A. R., and Grigliatti, T. A. (1983). Histone gene multiplication and position effect variegation in *Drosophila melanogaster*. *Genetics* **105,** 327–344.

Mottus, R., Reeves, R., and Grigliati, T. A. (1980). Butyrate suppression of position-effect variegation in *Drosophila melanogaster*. *Mol. Gen. Genet.* **178,** 465–469.

Norburg, C. J., and Nurse, P. (1989). Control of the higher eukaryote cell cycle by $p34^{cde2}$ homologues. *Biochim. Biophys. Acta* **989,** 85–95.

Paro, R., and Hogness, D. S. (1991). The polycomb protein shares a homologous domain with a heterochromatin-associated protein of *Drosophila*. *Proc. Natl. Acad. Sci. U.S.A.* **88,** 263–267.

Peifer, M., Karch, F., and Bender, W. (1987). The bithorax complex: Control of segmental identity. *Genes Dev.* **1,** 891–898.

Reuter, G., and Wolff, I. (1981). Isolation of dominant suppressor mutations for position-effect variegation in *Drosophila melanogaster*. *Mol. Genet.* **182,** 516–519.

Reuter, G., Giarre, M., Farah, J., Gausz, J., Spierer, A., and Spierer, P. (1990). Dependence of position-effect variegation in *Drosophila* on dose of a gene encoding an unusual zinc-finger protein. *Nature (London)* **344,** 219–223.

Sinclair, D. A. R., Mottus, R. C., and Grigliatti, T. A. (1983). Genes which suppress position-effect variegation in *Drosophila melanogaster* are clustered. *Mol. Gen. Genet.* **191,** 326–333.

Sinclair, D. A., Ruddell, A. A., Brock, J. K., Clegg, N. J., Lloyd, V. K., and Grigliatti, T. A. (1991). A cytogenetic and genetic characterization of a group of closely linked second chromosome mutations that suppress position effect variegation in *Drosophila melanogaster*. *Genetics* manuscript in preparation.

Sinclair, D., Clegg, N., Brock, H., and Grigliatti, T., manuscript in preparation.

Spofford, J. B. (1976). Position-effect variegation in *Drosophila*. In "The Genetics and Biology of *Drosophila*" (M. Ashburner and E. Novitski, eds.), Vol. 1c, pp. 955–1011. Academic Press, New York.

Spradling, A. C., and Karpen, G. H. (1990). Sixty years of mystery. *Genetics* **126,** 779–784.

Struhl, K. (1989). Helix-turn-helix, zinc-finger, and leucine zipper motifs for eukaryotic transcriptional regulatory proteins. *Trends Biochem. Sci.* **14,** 137–140.

Tartof, K. D., Hobbs, C., and Jones, M. (1984). A structural basis for variegating position effects. *Cell (Cambridge, Mass.)* **37,** 869–878.

Udvardy, A., Maine, E., and Schedl, P. (1985). The 87A7 chromomere: Identification of novel chromatin structures flanking the heat shock locus that may define the boundaries of higher order domains. *J. Mol. Biol.* **185,** 341–358.

Wakimoto, B. T., and Hearn, M. G. (1990). The effects of chromosome rearrangements on the expression of heterochromatic genes in chromosome 2L of *Drosophila melanogaster*. *Genetics* **125,** 141–154.

Wustmann, G., Szidonya, J., Taubert, H., and Reuter, G. (1989). The genetics of position-effect variegation loci in *Drosophila melanogaster*. *Mol. Gen. Genet.* **217,** 520–527.

Zink, B., and Paro, R. (1989). *In vivo* binding pattern of a trans-regulator of homeotic genes in *Drosophila melanogaster*. *Nature (London)* **337,** 468–471.

Zuckerkandl, E. (1974). Recherches sur les properties et l'activite biologique de la chromatine. *Biochemie* **56,** 937–954.

INDEX

A

Acetic acid, polytene chromosomes of *Drosophila* and, 205, 207–209, 215, 217

Acetoorcein, mutations affecting cell division and, 549–552

N-Acetoxy-2–aminoacetyfluorene (*N*-A-AAF), DNA sequence localization and, 10–11

Acetylation
 mRNA localization and, 48
 template-active nucleosomes and, 328, 330

Acid extraction of histones, nucleosome transcription and, 429

Actin, mRNA localization and, 38

β-Actin, fluorescent detection of RNA and DNA and, 75

Actinomycin, DNA sequence localization and, 25

Actinomycin D, fluorescent detection of RNA and DNA and, 89

ADE3, mutations affecting chromosomal proteins and, 512

Affinity chromatography
 nucleosome transcription and, 423
 template-active nucleosomes and, *see* Template-active nucleosomes, isolation of
 in vitro nuclear protein import and, 476–477

Affinity purification
 autoantibodies and, 162–164, 169
 polytene chromosomes of *Drosophila* and, 215
 yeast minichromosomes and, 301–304

Agar filtration, meiotic chromosome preparation and, 189–190, 193, 197, 199

Aldehyde
 autoantibodies and, 147, 151–152
 mRNA localization and, 41, 45–46

Aliquoting, autoantibodies and, 137, 141

Alkaline phosphatase
 autoantibodies and, 153–154
 fluorescent detection of RNA and DNA and, 85, 90–91
 mRNA localization and, 64

Alleles
 fluorescent detection of RNA and DNA and, 94
 monoclonal antibody libraries and, 245
 mutations affecting cell division and
  cloning of *Drosophila* genes, 573, 575, 577
  cytological phenotypes, 560, 562, 566–567
  strategies for isolation, 557–560
 mutations affecting chromosomal proteins and
  analysis, 506, 508, 512–514, 516–517
  manipulation, 494–506
  mutagenesis, 487, 490
 mutations affecting nuclear organization and, 536, 538
 position-effect variegation and, 592, 622, 624
  assay system, 603, 605–606
  Su(var) loci analyses, 616–617

Allophycocyanin, *in vitro* nuclear protein import and, 476–477, 480

α-Amanitin
 nucleosome transcription and, 435–436
 template-active nucleosomes and, 319, 331

AMCA, DNA sequence localization and, 5

Aminoacetyfluorene (AAF)
 DNA sequence localization and, 10–12, 21, 27
 electron microscopic *in situ* hybridization and, 119–122, 124, 128

Amino acids
 autoantibodies and, 165, 167
 fluorescent detection of RNA and DNA and, 96

CONTENTS OF RECENT VOLUMES

Volume 33

Flow Cytometry

Volume 34

Vectorial Transport of Proteins into and Across Membranes